# Techniques and Methods in Molecular Biology

# Techniques and Methods in Molecular Biology

Edited by Erik Pierre

## SYRAWOOD
PUBLISHING HOUSE

New York

Published by Syrawood Publishing House,
750 Third Avenue, 9th Floor,
New York, NY 10017, USA
www.syrawoodpublishinghouse.com

**Techniques and Methods in Molecular Biology**
Edited by Erik Pierre

International Standard Book Number: 978-1-68286-512-5 (Hardback)

**Cataloging-in-Publication Data**

Techniques and methods in molecular biology / edited by Erik Pierre.
    p. cm.
Includes bibliographical references and index.
ISBN 978-1-68286-512-5
1. Molecular biology. 2. Biology. I. Pierre, Erik.
QH506 .T43 2018
572.8--dc23

# TABLE OF CONTENTS

Preface ........................................................................................................................... VII

Chapter 1 **Analysis of Nidogen-1/Laminin γ1 Interaction by Cross- Linking, Mass Spectrometry, and Computational Modeling Reveals Multiple Binding Modes** .......................... 1
Philip Lössl, Knut Kölbel, Dirk Tänzler, David Nannemann, Christian H. Ihling, Manuel V. Keller, Marian Schneider, Frank Zaucke, Jens Meiler, Andrea Sinz

Chapter 2 **Identification of Functionally Important Residues of the Rat P2X4 Receptor by Alanine Scanning Mutagenesis of the Dorsal Fin and Left Flipper Domains** ............... 16
Vendula Tvrdonova, Milos B. Rokic, Stanko S. Stojilkovic, Hana Zemkova

Chapter 3 **Transcriptome-Based Identification of ABC Transporters in the Western Tarnished Plant Bug *Lygus hesperus*** ..................................................................... 28
J. Joe Hull, Kendrick Chaney, Scott M. Geib, Jeffrey A. Fabrick, Colin S. Brent, Douglas Walsh, Laura Corley Lavine

Chapter 4 **Pilon: An Integrated Tool for Comprehensive Microbial Variant Detection and Genome Assembly Improvement** ............................................................................. 48
Bruce J. Walker, Thomas Abeel, Terrance Shea, Margaret Priest, Amr Abouelliel, Sharadha Sakthikumar, Christina A. Cuomo, Qiandong Zeng, Jennifer Wortman, Sarah K. Young, Ashlee M. Earl

Chapter 5 **detectIR: A Novel Program for Detecting Perfect and Imperfect Inverted Repeats using Complex Numbers and Vector Calculation** .................................... 62
Congting Ye, Guoli Ji, Lei Li, Chun Liang

Chapter 6 **Fine Mapping and Candidate Gene Search of Quantitative Trait Loci for Growth and Obesity Using Mouse Intersubspecific Subcongenic Intercrosses and Exome Sequencing** ...................................................................................... 73
Akira Ishikawa, Sin-ichiro Okuno

Chapter 7 **Comparative Transcriptomic Analysis of the Response to Cold Acclimation in *Eucalyptus dunnii*** ................................................................................................. 86
Yiqing Liu, Yusong Jiang, Jianbin Lan, Yong Zou, Junping Gao

Chapter 8 **Complete Chloroplast Genome Sequence of Omani Lime (*Citrus aurantiifolia*) and Comparative Analysis within the Rosids** ...................................................... 97
Huei-Jiun Su, Saskia A. Hogenhout, Abdullah M. Al-Sadi, Chih-Horng Kuo

Chapter 9 **Characterization of the *Runx* Gene Family in a Jawless Vertebrate, the Japanese Lamprey (*Lethenteron japonicum*)** ..................................................................... 108
Giselle Sek Suan Nah, Boon-Hui Tay, Sydney Brenner, Motomi Osato, Byrappa Venkatesh

Chapter 10    **Transcriptome and Allele Specificity Associated with a 3BL Locus for Fusarium Crown Rot Resistance in Bread Wheat**.................................................................. 121
Jian Ma, Jiri Stiller, Qiang Zhao, Qi Feng, Colin Cavanagh, Penghao Wang, Donald Gardiner, Frédéric Choulet, Catherine Feuillet, You-Liang Zheng, Yuming Wei, Guijun Yan, Bin Han, John M. Manners, Chunji Liu

Chapter 11    **Variability among the Most Rapidly Evolving Plastid Genomic Regions is Lineage-Specific: Implications of Pairwise Genome Comparisons in *Pyrus* (Rosaceae) and other Angiosperms for Marker Choice**......................................... 132
Nadja Korotkova, Lars Nauheimer, Hasmik Ter-Voskanyan, Martin Allgaier, Thomas Borsch

Chapter 12    **Unusual Ratio between Free Thyroxine and Free Triiodothyronine in a Long-Lived Mole-Rat Species with Bimodal Ageing**......................................... 148
Yoshiyuki Henning, Christiane Vole, Sabine Begall, Martin Bens, Martina Broecker-Preuss, Arne Sahm, Karol Szafranski, Hynek Burda, Philip Dammann

Chapter 13    **The GCKIII Kinase Sps1 and the 14-3-3 Isoforms, Bmh1 and Bmh2, Cooperate to Ensure Proper Sporulation in *Saccharomyces cerevisiae***............................... 159
Christian J. Slubowski, Scott M. Paulissen, Linda S. Huang

Chapter 14    **Discrimination of *Escherichia* coli O157, O26 and O111 from other Serovars by MALDI-TOF MS based on the *S10*-GERMS Method**.................................. 173
Teruyo Ojima-Kato, Naomi Yamamoto, Mayumi Suzuki, Tomohiro Fukunaga, Hiroto Tamura

Chapter 15    **Is There Still Room for Novel Viral Pathogens in Pediatric Respiratory Tract Infections?**.................................................................................................... 184
Blanca Taboada, Marco A. Espinoza, Pavel Isa, Fernando E. Aponte, María A. Arias-Ortiz, Jesús Monge-Martínez, Rubén Rodríguez-Vázquez, Fidel Díaz-Hernández, Fernando Zárate-Vidal, Rosa María Wong-Chew, Verónica Firo-Reyes, Carlos N. del Río-Almendárez, Jesús Gaitán-Meza, Alberto Villaseñor-Sierra  Gerardo Martínez-Aguilar, Ma. del Carmen Salas-Mier, Daniel E. Noyola, Luis F. Pérez-Gónzalez, Susana López, José I. Santos-Preciado, Carlos F. Arias

Chapter 16    **Pdsg1 and Pdsg2, Novel Proteins Involved in Developmental Genome Remodelling in *Paramecium***................................................................................ 198
Miroslav Arambasic, Pamela Y. Sandoval, Cristina Hoehener, Aditi Singh, Estienne C. Swart, Mariusz Nowacki

Chapter 17    **Growth Hormone-Regulated mRNAs and miRNAs in Chicken Hepatocytes**.............................. 209
Xingguo Wang, Lei Yang, Huijuan Wang, Fang Shao, JianFeng Yu, Honglin Jiang, Yaoping Han, Daoqing Gong, Zhiliang Gu

**Permissions**

**List of Contributors**

**Index**

# PREFACE

This book has been an outcome of determined endeavour from a group of educationists in the field. The primary objective was to involve a broad spectrum of professionals from diverse cultural background involved in the field for developing new researches. The book not only targets students but also scholars pursuing higher research for further enhancement of the theoretical and practical applications of the subject.

Molecular biology encompasses all the processes that occur between biomolecules present in a cell. It is an important field as it contributes to the research as well as practice of allied fields like biotechnology, forensic biology, virology, medicine and biological systematics. The various studies that are constantly contributing towards advancing technologies and evolution of this field are examined in detail. This book on molecular biology includes fundamental insight into the technology and processes that are involved in gene sequencing. It brings forth some of the most innovative concepts and elucidates the unexplored aspects of molecular biology. Coherent flow of topics, student-friendly language and extensive use of examples make this book an invaluable source of knowledge.

It was an honour to edit such a profound book and also a challenging task to compile and examine all the relevant data for accuracy and originality. I wish to acknowledge the efforts of the contributors for submitting such brilliant and diverse chapters in the field and for endlessly working for the completion of the book. Last, but not the least; I thank my family for being a constant source of support in all my research endeavours.

**Editor**

# Analysis of Nidogen-1/Laminin γ1 Interaction by Cross-Linking, Mass Spectrometry, and Computational Modeling Reveals Multiple Binding Modes

**Philip Lössl**[1¤], **Knut Kölbel**[1], **Dirk Tänzler**[1], **David Nannemann**[2], **Christian H. Ihling**[1], **Manuel V. Keller**[3], **Marian Schneider**[4], **Frank Zaucke**[3], **Jens Meiler**[2], **Andrea Sinz**[1]*

1 Department of Pharmaceutical Chemistry and Bioanalytics, Institute of Pharmacy, Martin Luther University Halle-Wittenberg, Halle (Saale), Germany, 2 Department of Chemistry and Center for Structural Biology, Vanderbilt University, Nashville, TN, United States of America, 3 Center for Biochemistry, Medical Faculty, University of Cologne, Cologne, Germany, 4 Research Group Artificial Binding Proteins, Institute of Biochemistry and Biotechnology, Martin Luther University Halle-Wittenberg, Halle (Saale), Germany

## Abstract

We describe the detailed structural investigation of nidogen-1/laminin γ1 complexes using full-length nidogen-1 and a number of laminin γ1 variants. The interactions of nidogen-1 with laminin variants γ1 LEb2–4, γ1 LEb2–4 N836D, γ1 short arm, and γ1 short arm N836D were investigated by applying a combination of (photo-)chemical cross-linking, high-resolution mass spectrometry, and computational modeling. In addition, surface plasmon resonance and ELISA studies were used to determine kinetic constants of the nidogen-1/laminin γ1 interaction. Two complementary cross-linking strategies were pursued to analyze solution structures of laminin γ1 variants and nidogen-1. The majority of distance information was obtained with the homobifunctional amine-reactive cross-linker *bis*(sulfosuccinimidyl)glutarate. In a second approach, UV-induced cross-linking was performed after incorporation of the diazirine-containing unnatural amino acids photo-leucine and photo-methionine into laminin γ1 LEb2–4, laminin γ1 short arm, and nidogen-1. Our results indicate that Asn-836 within laminin γ1 LEb3 domain is not essential for complex formation. Cross-links between laminin γ1 short arm and nidogen-1 were found in all protein regions, evidencing several additional contact regions apart from the known interaction site. Computational modeling based on the cross-linking constraints indicates the existence of a conformational ensemble of both the individual proteins and the nidogen-1/laminin γ1 complex. This finding implies different modes of interaction resulting in several distinct protein-protein interfaces.

**Editor:** Bostjan Kobe, University of Queensland, Australia

**Funding:** KK is funded by the BMBF project. AS acknowledges financial support from the DFG (projects Si 867/13-1, 15-1, and 16-1), the BMBF (Pro-Net-T3), and the region of Saxony-Anhalt. JM is funded through NIH (R01 GM080403, R01 MH090192, R01 GM099842, and R01 DK097376) and NSF (CHE 1305874). The funders had no role in study design, data collection and analysis, decision to publish, or preparation of the manuscript.

**Competing Interests:** The authors have declared that no competing interests exist.

\* Email: andrea.sinz@pharmazie.uni-halle.de

¤ Current address: Biomolecular Mass Spectrometry and Proteomics, Bijvoet Center for Biomolecular Research and Utrecht Institute for Pharmaceutical Sciences, Utrecht University, Padualaan 8, CH Utrecht, The Netherlands

## Introduction

Laminins are the major non-collagenous proteins of basement membranes that are known to form networks through crucial non-covalent self-interactions [1,2]. Each member of the laminin protein family consists of three polypeptide chains with one copy of the α, β, and γ chain. Since the discovery of laminin [3], several nomenclatures have been developed, which are, however, not always completely systematic [4]. In this work, we apply the laminin nomenclature introduced by Aumailley *et al.* [5]. Electron microscopic studies of laminin-111 (α1-, β1-, and γ1-subunits) reveal a cross-shaped protein structure [6] with three subunits being connected within the central part according to a coiled-coil arrangement ('long arm'). The *N*-terminal regions of the laminin subunits are free ('short arms') (Figure 1). The globular domains at the *N*-termini of all three chains (LN domains) are required for

efficient polymerization as deletion mutants with two or fewer LN domains fail to form networks [2]. Furthermore, laminin-111 harbors more centrally located globular domains (α1 L4a, α1 L4b, β1 LF, and γ1 L4) as well as several 'laminin-type epidermal growth factor-like' (LE) modules. Three-dimensional structures of certain laminin domains are available, such as X-ray structures of the *C*-terminal LG domains (PDB entries 2JD4, 2WJS, 1QU0, 1DYK), the nidogen-binding region γ1 LEb2–4 (PDB entries 1KLO, 1NPE), and the α5, β1, and γ1 LN domains (PDB entries 2Y38, 4AQT, 4AQS), the latter of which were found to be in good agreement with previously reported computational models [7].

Nidogens (entactins) are sulfated monomeric glycoproteins that are ubiquitously present in basement membranes of higher organisms. While invertebrates possess only one nidogen, two nidogen isoforms, namely nidogen-1 (ca. 135 kDa) and nidogen-2

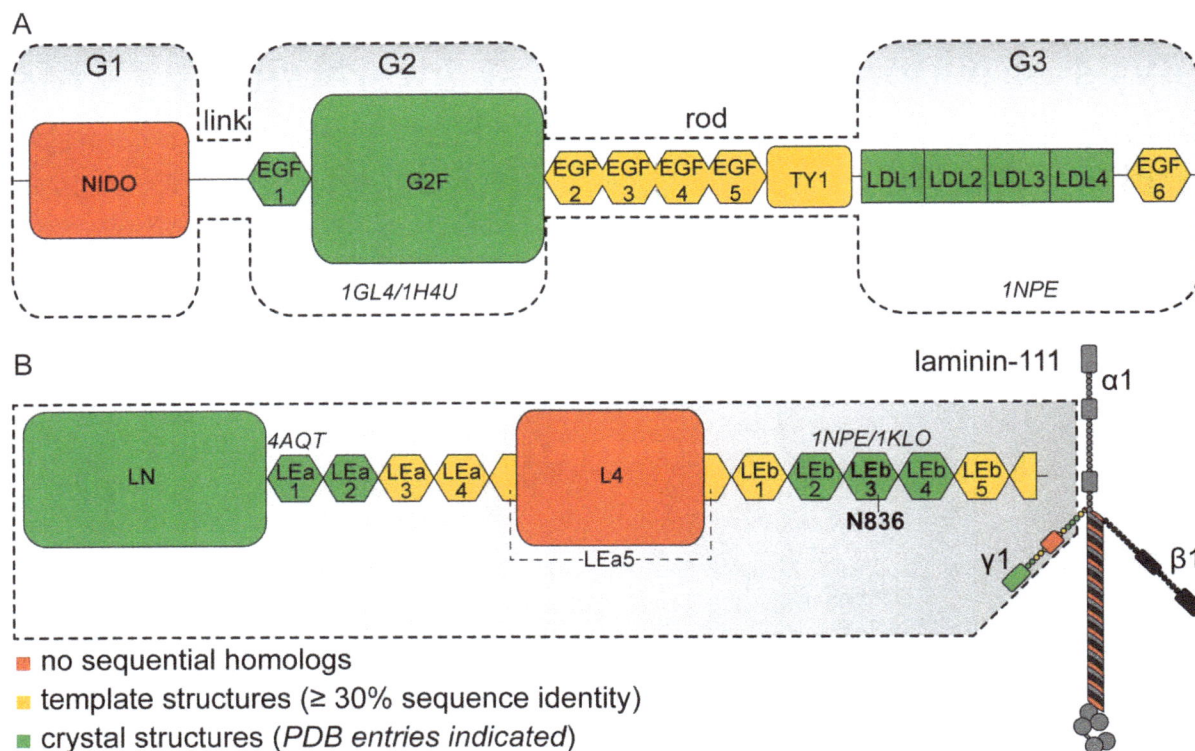

**Figure 1. Arrangement of (A) nidogen-1 and (B) laminin γ1 short arm.** The domains are color-coded with respect to the availability of crystal structures (green), template structures for comparative modeling (yellow) or neither of both (red). PDB IDs of the respective crystal structures are indicated in italics. (A) Nidogen-1 domain assignments are according to the UniProt KB entry P10493. Additionally, the historic domain names (G1–link–G2–rod–G3) are given. (B) Laminin domain designations follow the nomenclature of Aumailley *et al.* [5]. Laminin γ1 short arm is part of the heterotrimeric protein laminin-111, the overall structure of which is schematically depicted.

(ca. 150 kDa), have been identified in vertebrates [8–10]. Both isoforms exhibit a similar repertoire of binding partners [11], resulting in basement formation even after knocking out one nidogen isoform [12]. Electron microscopic studies have revealed nearly identical arrangements of nidogen-1 and -2 [9,13]. Both comprise three globular domains (G1–G3), which are connected by two rod-shaped domains ('link' and 'rod' regions) (Figure 1). Sequence analyses have confirmed the modular structures of nidogens. Identified motifs, such as epidermal growth factor (EGF)-like sequences, are also present in a number of other proteins of the extracellular matrix [14]. X-ray structures are available for G2 and G3 domains of nidogen-1 (PDB entries 1GL4, 1H4U, 1NPE). Nidogen is considered to be a stabilizer and adaptor protein within the basement membrane. In addition to isoform-specific interactions of nidogen-1 with fibulin-1 and -2 as well as of nidogen-2 with tropoelastin and type XVIII collagen, both nidogens bind to the essential basement membrane proteins perlecan, type IV collagen, and laminin [14].

Initial studies of nidogen-1 revealed an extraordinarily strong interaction with laminin [15] and led to the elucidation of a 1:1 stoichiometry between nidogen-1 and laminin in the complex [16]. The *C*-terminal G3 domain of nidogen-1 [16,17] as well as the EGF-like motif LEb3 in laminin γ1 [18–21] were identified as binding regions. A weaker nidogen-1 binding has been observed in laminin γ2 and γ3 [14]. Using radioligand binding assays with variants of this domain, amino acids Asp-834, Asn-836, and Val-838 were identified to be essential for the interaction of the laminin γ1 LEb3 domain with nidogen-1. The exchange of Asn-836 against aspartic acid resulted in a drastic decrease of nidogen-1

binding with a 25,000-fold loss in affinity [22]. The sequence numbering used here follows the amino acid sequences shown in Figure S1 with Asp-834, Asn-836, and Val-838 corresponding to Asp-800, Asn-802, and Val-804 in the classical laminin γ1 numbering [23].

In 2003, the three-dimensional structure of the complex between the G3 domain of nidogen-1 and LEb domains 2–4 of laminin γ1 was elucidated [24]. The G3 domain exhibits a ß-propeller composed of six LDL receptor YWTD modules, creating a concave interface for the amino acids Asp-834, Asn-836, and Val-838 within the loop of laminin γ1 LEb3. Complex formation is enhanced by additional interactions of laminin γ1 LEb2 with the ß-propeller.

Recently, the interaction of full-length nidogen-1 with laminin γ1 short arm has been investigated by size-exclusion chromatography, dynamic light scattering, and small-angle X-ray scattering [25]. These studies indicate that the interaction is mediated solely by the *C*-terminal domains, while the remaining regions of both proteins do not participate in complex formation.

For our studies investigating the interaction between nidogen-1 and laminin γ1 short arm, we chose an alternative approach providing 3D-structural insights into proteins. This strategy relies on chemical cross-linking and a subsequent mass spectrometry (MS)-based analysis of the created products [26,27]. Structural information can be obtained by the insertion of a chemical cross-linker between two functional groups within a protein. The cross-linker has a defined length and is connected via covalent bonds to functional groups of amino acid side chains, allowing the cross-linked amino acids to be identified after enzymatic digestion. This chemical cross-linking approach is also applied to study protein-protein interfaces. The sequence separation of cross-linked amino acids, combined with the cross-linker length, impose a distance constraint on the 3D structure of the protein complex [7,28–31]. Analysis of cross-linked peptides by MS makes use of several advantages: First, the mass of the protein or the protein complex under investigation is theoretically unlimited as the proteolytic peptides of the cross-linked proteins are analyzed – in case a "bottom-up" strategy is employed for MS protein analysis. Second, the experiment is rapid and requires very low $(10^{-14}-10^{-15}$ mol) amounts of protein. Finally, as the cross-linking reaction can be executed in a native-like environment, protein structure and flexibility are accurately reflected. It is possible to study membrane proteins, post-translational modifications as well as splice variants. The broad range of cross-linking reagents with different specificities (primary amines, sulfhydryls, or carboxylic acids) and the wide range of distances (0 Å up to 20 Å) allow a setup of fine-tuned experimental strategies.

However, despite the straightforwardness of the cross-linking approach, the identification of the cross-linked products can be cumbersome due to the complexity of the reaction mixtures. Several strategies have been employed to enrich cross-linker-containing species by affinity chromatography or to facilitate the identification of the cross-linked products, e.g. by using MS/MS cleavable cross-linkers or isotope-labeled cross-linkers or proteins [26,27,32].

By combining sparse distance constraints from disulfide bonds and cross-links imposed by $bis$(sulfosuccinimidyl)glutarate (BS$^2$G), MS identification of the cross-linked products, and computational modeling we predicted a galactose-binding domain-like fold for laminin ß1 and γ1 LN domains [7]. This fold was later confirmed by crystal structures of the α5 LN-LE1–2 [33], the β1 LN-LEa1–4, and the γ1 LN-LEa1–2 fragments [34].

In this work, we extend the cross-linking tool box from the exclusive use of amine-reactive cross-linkers towards the incorporation of photo-reactive amino acids that can deliver valuable short-distance information [35]. Combined with a mass spectrometric analysis of the created cross-links and computational modeling, we were able to gain detailed insights into the interaction mechanisms between full-length nidogen-1 and the laminin variants γ1 LEb2–4, γ1 LEb2–4 N836D, γ1 short arm, and γ1 short arm N836D. Our results suggest the existence of multiple nidogen-1/laminin γ1 interfaces in addition to the known interaction site.

## Materials and Methods

### Materials

The cross-linking reagent BS$^2$G and the photo-amino acids (photo-methionine and photo-leucine) were obtained from Thermo Fisher Scientific. The proteases trypsin (porcine), chymotrypsin, and GluC were obtained from Promega, all other chemicals were purchased from Sigma. Solvents used for nano-high performance liquid chromatography (HPLC) were spectroscopic grade (Uvasol, VWR). Milli-Q water was produced by a TKA Pacific system with X-CAD dispenser from Thermo Electron LED GmbH (part of Thermo Fisher Scientific).

### Expression and Purification of Nidogen-1 and Laminin γ1 Variants

Genes encoding all proteins (murine amino acid sequences, see Figure S1) were expressed in human embryonic kidney (HEK) 293 EBNA cells with $N$-terminal (His)$_6$-tag (nidogen-1) or with $N$- or $C$-terminal double Strep tag II (laminin γ1 variants γ1 LEb2–4, γ1

LEb2–4 N836D, γ1 short arm, and γ1 short arm N836D). Incorporation of photo-reactive amino acids was achieved by growing the cells in a Leu- and Met-depleted medium (DMEM-LM, Thermo Fisher Scientific) to which photo-Met and photo-Leu were added. Strep II-tagged proteins were purified using Strept-Actin sepharose matrix (IBA), (His)$_6$-tagged nidogen-1 was purified with Nickel-NTA Superflow matrix (GE Healthcare) using an ÄKTA FPLC system (GE Healthcare). Amino acid sequences were confirmed by peptide fragment fingerprint analysis.

### Surface Plasmon Resonance Spectroscopy (SPR)

Experiments were conducted at 25°C using a Biacore T100 instrument (GE Healthcare). Nidogen-1 was immobilized by covalent coupling on a Series S Sensor Chip CM 5 (GE Healthcare) using amine-coupling chemistry. A 1:1 mixture of 100 mM $N$-hydroxysulfosuccinimide and 400 mM 1-ethyl-3-(3-dimethylaminopropyl)carbodiimide hydrochloride (Pierce) was passaged over the sensor chip surface for 7 min at a flow rate of 10 μl/min to activate the matrix. For protein immobilization, a solution of 300 nM nidogen-1 in 10 mM sodium acetate (pH 5.5) was injected (contact time 35 min, flow rate 10 μl/min). Remaining activated groups were blocked by injecting 1 M ethanolamine (pH 8.5) for 7 min at a flow rate of 10 μl/min. As reference, a flow cell was prepared in the same manner omitting the injection of protein solution.

Binding assays using all laminin variants (wild type and N836D) as mobile analytes were performed as single-cycle kinetic experiments [36] at a constant flow rate of 30 μl/min. The analyte was diluted in running buffer (20 mM HEPES, 100 mM NaCl, 0.05% ($v/v$) Tween 20, pH 7.5) and 1 min injections of these dilutions were applied with increasing concentrations. Individual analyte injections were followed by a 40.6 min flow of running buffer to allow for partial dissociation. Blank cycles, where running buffer was injected instead of analyte solution, were performed prior to each analyte cycle to facilitate double referencing. After completion of each analysis cycle, the sensor chip surface was extensively washed with running buffer to achieve complete dissociation of the analyte. Response data were doubly referenced and kinetic parameters were determined as described previously [36] using BIAevaluation software 4.1.1 (GE Healthcare) to fit a 1:1 binding model to the data.

### Enyzme-Linked Immunosorbent Assays (ELISA)

ELISA-based binding assays were performed in 96-well plates (Nunc Maxisorp, Thermo Fisher Scientific). Laminin fragments were coated overnight at 4°C with a concentration of 10 μg/ml in 50 μl Tris-buffered saline (TBS) per well (50 mM Tris-HCl, 150 mM NaCl, pH 7.4). The supernatant was discarded after coating and the plates were washed once with a TBS-T solution (TBS with 0.05% ($v/v$) Tween-20) with a volume of 400 μl/well. Unspecific binding sites were blocked with 50 μl/well of 1% ($w/v$) bovine serum albumin (BSA) in TBS-T solution for 1 h at room temperature. Increasing concentrations of nidogen-1 ranging from 0.03 nM to 234 nM in TBS-T/1% BSA were added to the wells until saturation was reached. After ligand incubation for 1 h at room temperature plates were washed three times with TBS-T solution. Nidogen-1 was detected with mouse monoclonal antibodies directed against the (His)$_6$-tag (Qiagen, dilution 1:2000). Antibodies were added in 50 μl TBS-T/1% BSA per well and incubated for 1 h at room temperature. Plates were then washed three times with TBS-T solution and the secondary horseradish peroxidase-conjugated α-mouse IgG antibodies raised in rabbit (Dako, 1:2000 diluted in TBS-T/1% BSA) were added. Plates were washed three times with TBS-T solution and one time with water. For signal detection, 50 µl/well tetramethylbenzidine solution (1-Step ultra TMB-ELISA solution, Thermo Fisher Scientific) was used. The color reaction was stopped with 50 µl/well of 10% $H_2SO_4$ solution and absorption at 450 nm was determined in an ELISA reader (Tecan), subtracting the background (0 nM ligand). As a second control, increasing ligand concentrations were added to uncoated wells. All measurements were performed in triplicates. $K_d$ values were calculated using Origin v6.0.

## Cross-Linking Reactions

Cross-linking reactions were conducted with 1 to 10 µM protein solutions in 20 mM HEPES, 100 mM NaCl, pH 7.5. Freshly prepared stock solutions of the homobifunctional amine-reactive N-hydroxysuccinimide ester $BS^2G$ (in dimethylsulfoxide) were added in 200-fold molar excess to the protein solution. The reactions were conducted at room temperature and were quenched after 30 min by adding $NH_4HCO_3$ to a final concentration of 20 mM.

For photo-cross-linking, nidogen-1 (0.5 µM) was incubated with laminin γ1 short arm (1 µM) or γ1 LEb2–4 (10 µM) for 20 min at room temperature. Then, the samples were irradiated with UV-A light (max. 360 nm) at 8000 $mJ/cm^2$ in a home-built device [37].

## In Gel Digestion

After SDS-PAGE of the cross-linking reaction mixtures, bands of interest were excised, reduced with dithiothreitol, alkylated with iodoacetamide, and digested. For *in situ* digestion, either GluC or Chymotrypsin was added and gel pieces were incubated at 4°C for 1 h before trypsin was added (enzyme:substrate ratio 1:100). The digestion was performed overnight at 37°C. Peptides were extracted and samples were concentrated in a vacuum concentrator to volumes between 60 to 120 µl before LC/MS/MS analysis.

## In Solution Digestion

For *in solution* digestion, proteins were incubated with acetone (−20°C, 1 h) to precipitate them from solution. The pellet was dried, solubilized with 1.6 M urea, reduced, alkylated, and digested with a mixture of GluC and trypsin (enzyme:substrate ratio 1:50).

## Nano-HPLC/Nano-ESI-LTQ-Orbitrap-MS/MS

Fractionation of peptide mixtures was carried out on an Ultimate nano-HPLC system (Thermo Fisher Scientific) using reversed phase C18 columns (precolumn: Acclaim PepMap, 300 µm • 5 mm, 5 µm, 100 Å, separation column: Acclaim PepMap, 75 µm • 250 mm, 3 µm, 100 Å, Thermo Fisher Scientific). After washing the peptides on the precolumn for 15 min with water containing 0.1% trifluoroacetic acid, peptides were eluted and separated using gradients from 0 to 40% B (gradient times varying between 30 to 90 min), 40 to 100% B (1 min), and 100% B (gradient times varying between 11 to 30 min), with solvent A: 5% ACN containing 0.1% formic acid, and solvent B: 80% ACN containing 0.08% formic acid. The nano-HPLC system was directly coupled to the nano-ESI source (Proxeon) of an LTQ-Orbitrap XL hybrid mass spectrometer (Thermo Fisher Scientific). Data were acquired in data-dependent MS/MS mode: Each high-resolution full scan (*m/z* 350 to 2000, R = 60,000) in the orbitrap was followed by five product ion scans in the LTQ (collision-induced dissociation with 35% normalized collision energy) of the five most intense signals in the full-scan mass spectrum (isolation window 1.5 u). Dynamic exclusion

(exclusion duration 120 sec, exclusion window −1 to 2 u) was enabled to allow detection of less abundant ions. Data acquisition was controlled via XCalibur 2.0.7 (Thermo Fisher Scientific) in combination with DCMS link 2.0.

## Identification of Cross-Linked Products

Cross-linked products were analyzed with StavroX v2.0.6 [38]. MS and MS/MS data were automatically analyzed and annotated. All cross-links were manually validated. A maximum mass deviation of 3 ppm between calculated and experimental precursor masses was applied as well as a signal-to-noise ratio of $\geq 2$. Primary amino groups (Lys side chains and N-termini) were considered as cross-linking sites for $BS^2G$; all amino acid residues were regarded as potential sites for UV-A-induced cross-linking of photo-Met and photo-Leu. Oxidation of Met was set as variable modification for all cross-linked proteins. Additionally, carbamidomethylation was included as fixed modification for Cys. Two missed cleavage sites were considered for each amino acid (Lys and Arg for trypsin; Tyr, Trp, and Phe for chymptrypsin; Glu for GluC).

## Identification of Templates for Computational Modeling

The sequences of nidogen-1 and laminin γ1 short arm were split into separate domains as defined by the UniProt Knowledgebase (www.uniprot.org, nidogen-1: entry P10493, laminin γ1: entry P02468) and modeled independently. The domains were subdivided in three classes according to the availability of structural models or templates (Figure 1). The first class comprises the nidogen-1 LDL-receptor class B repeats (G3 domain), EGF-like domain 1 and G2 ß-barrel domain as well as the laminin γ1 LN, LEa1, LEa2, and LEb2–4 domains, all of which have been characterized by X-ray crystallography. The remaining EGF-like domains of nidogen-1 and laminin γ1 short arm as well as the nidogen-1 thyroglobulin type-1 (TY1) domain were assigned to the second class because a DELTA-BLAST [39] search run had led to the identification of homologous domains with existing crystal structures that could serve as templates for comparative modeling. Finally, the third class comprises the nidogen-1 G1 domain and the laminin γ1 L4 domain, for which we did not identify any sequential homologues. Both domains were searched against the threading servers PHYRE2 [40], HHPred [41], PSIPRED (pDomTHREADER and GenTHREADER) [42], and I-TAS-SER [43] for fold recognition.

All following modeling experiments were performed with Rosetta v3.4. The Rosetta total scores reported herein were calculated using the score12 full-atom scoring function. Full command lines for each step are included in Files S1 and S2.

## Comparative Modeling Based on Sequential Homologous Templates

After comparing the DELTA-BLAST sequences of potential template structures with the actual sequences of the corresponding PDB entries, sequence alignments of all target and template sequences were performed with ClustalW 2.1 [44]. Only templates showing $\geq 30\%$ sequence identity were used for comparative modeling. The corresponding alignments are shown in Figure S2.

The LE domain crystal structures (PDB entries 4AQS, 4AQT, 1NPE, 1KLO, 2Y38) were considered as templates for the remaining LE domains, namely LEa3–5, LEb1 and LEb5. LEa5 is split in two parts by the L4 domain (Figure 1). The N-terminal LEa5.1 was not modeled as an individual domain since it comprises only ten residues. Similarly, LEb6 was not modeled as an individual domain since only nine residues are contained in the laminin γ1 short arm fragment used here.

We identified EGF-like domains in 14 X-ray structures (PDB entries 1GL4, 1YO8, 1SZB, 1TOZ, 1UZJ, 2BO2, 3P5B, 1NFU, 3H5C, 3POY, 3QCW, 3S94, 3V64 and 2W86) as templates for nidogen-1 EGF2–6. The nidogen-1 TY1 domain was modeled using TY1 domains within the crystal structures 1ICF and 2DSR as templates. Threading of the primary sequences onto the 3D template structures, modeling of missing loop regions, and clustering of the created models were carried out as reported previously [45]. For each template/target sequence pair, 1000 models were constructed. All structural models for one target sequence were ranked according to their Rosetta total score and the best-scoring 10% were used for clustering with automated detection of the clustering radius by Rosetta. An overview of all clusters is given in Table S1. The best-scoring structures within the top three clusters were considered as final models.

## Comparative Modeling of Laminin γ1 L4

Using the threading servers listed above, 21 potential structural homologues were identified for the L4 domain, 13 of which exhibit a ß-sandwich topology and carbohydrate-binding activity. Structural alignment with MUSTANG [46] revealed that these templates share a common topology that is in accordance with the PSIPRED [47] and JUFO9D [48] secondary structure prediction for L4. Hence, we hypothesize that the laminin γ1 L4 domain shares the galactose-binding domain-like fold and built the comparative model based on these templates. ClustalW 2.1 alignments of the laminin γ1 L4 domain with all 13 templates identified by fold recognition (PDB entries 1GU3, 1GUI, 1K42, 1CX1, 1WKY, 1WMX, 3OEA, 2ZEW, 2ZEZ, 1DYO, 3F95, 3ZXJ, 1D7B) revealed sequence identities of 3–14%. Hence, the alignments had to be adjusted manually to guarantee for correctly aligned secondary structure elements (Figure S3). Adjustments were based on PSIPRED secondary structure prediction for L4, DSSP secondary structure annotations of the template structures [49] and manual inspection of all templates in Pymol v1.5 (Schrödinger LLC). After optimization of the alignments, comparative modeling was performed as described above. For each template, 1000–2000 models were generated (25,296 models in total) and the top 10% were selected for clustering (Table S2). To obtain informative clustering results, long loop regions (>5 residues) were not considered for root-mean square deviation (RMSD) calculation. The clustering radius was set to 2 Å.

## *De Novo* Folding of the Nidogen-1 NIDO Domain

No likely structural homologues of the nidogen-1 NIDO domain were identified by fold recognition. Hence, models were generated by *de novo* folding using Rosetta AbinitioRelax within the Rosetta Topology Broker framework [50]. Initially, Rosetta fragment picker was used to create a fragment library consisting of the primary sequence split into overlapping 3-mers and 9-mers, each of them represented by 200 peptide structures, mimicking the entire distribution of conformations these segments are likely to adopt in a protein structure [45,51]. Full atom refinement during *de novo* modeling resulted in partial unfolding of the created models. Therefore, we generated 11,902 centroid models, omitting full-atom refinement after running AbinitioRelax. To filter for structures that are likely to occur in nature we pursued two complementary strategies.

First, the 10% best-scoring models were compared to a precompiled PISCES library of structurally diverse PDB models (soluble proteins, sequence ID <25%, resolution <2.0 Å) [52] using MAMMOTH [53]. This served to evaluate whether the

domain adopts a known fold. Structural homologues for two of the generated models were identified (>50% of target sequence aligned, MAMMOTH Z-score >5).

Second, we generated 10,000 to 20,000 models for sequences of six homologous NIDO domains present in other organisms. Homologous domains were identified with DELTA-BLAST. Non-redundant sequences with an identity of >75% to the murine NIDO domain were selected and manually compared using Jalview [54]. The sequences of *Sarcophilus harrisii*, *Rattus norvegicus*, *Homo sapiens*, *Bos taurus*, *Cricetulus griseus*, and *Felis catus* were chosen, each of them showing differences in diverse regions that are otherwise conserved (Figure S4). To identify common topologies among the different NIDO domains, models were ranked based on their total centroid score as well as their strand pairing energy score and MAMMOTH was used to compare the best 1% models of each homologue with the best 10% models of the murine NIDO domain. Models of murine NIDO were regarded as representatives of a common topology when significant structural similarities (MAMMOTH Z-score >5) to at least one model of each homologue were found.

As a result, 109 candidate models were identified and selected for full-atom refinement. To prevent distortion of secondary structure element arrangement, side chains and peptide backbone were relaxed sequentially, keeping one of both fixed ('-relax:bb_move false' or '-relax:chi_move false') before a final round of refinement in thorough relax mode was performed on the complete structure. We generated 50 models per input model, which were inspected by visualization with Pymol, ranked according to their Rosetta total score, and clustered with a radius of 1.5 Å. The best-scoring output structure for each template and the best-scoring structures representing the top 20 clusters were kept for further analysis. The two candidate models, identified by comparison with the PISCES library, were processed similarly. For each of them, the five best-scoring models and the best-scoring models representing individual clusters were included in the final list of potential NIDO models, resulting in a total of 132 models. These models were combined and re-scored (Table S3).

## Incorporation of Cross-Linking Constraints into the Nidogen-1 G3/Laminin γ1 LEb2–4 Experimental Structure

We identified eleven $BS^2G$- and two photo-cross-links within the experimental structure of the nidogen-1 G3/laminin γ1 LEb2–4 complex (PDB entry 1NPE), which were used as distance constraints for adapting the 3D structure. The cross-linked residues and their respective Cα–Cα distances are listed in Table 1. The maximum Cα–Cα distance for lysine-lysine cross-links was calculated to 26 Å by adding the lysine side chain length (2×6.3 Å), the distance spanned by $BS^2G$ (7.5 Å [55]) and a tolerance of 5.9 Å to account for structural flexibility [56]. Acceptable Cα–Cα distances for the "zero-length" photo-cross-links varied depending on the side-chain length of the cross-linked residue. Fulfillment of cross-linking restraints was evaluated with a flat harmonic scoring function that renders an energy penalty when the Euclidean distance between cross-linked residues exceeds the allowed Cα–Cα distance [56]. A Rosetta constraint file containing all cross-links was created as described elsewhere [57]. The standard deviation granted for each cross-link was 1 Å. Additionally, distances of the hydrogen bonds formed by Asp-834, Asn-836, and Val-838 in the laminin γ1 LEb3 domain were restrained as they have been reported to be essential for high-affinity interaction [23,24]. Structural refinement was carried out by generating 800 models using the Rosetta Relax application with an atom pair constraint scoring weight of 1.0 ('-constraints:cst_fa_weight 1.0'). Structures ranking among the top 20 in terms of total

**Table 1.** Overview of cross-links within the nidogen-1 G3/laminin γ1 LEb2–4 complex.

| cross-linked lysines | Cα–Cα distances (Å) | | |
| --- | --- | --- | --- |
| | 1NPE | model (best atom-pair constraint score) | model (best total score) |
| K-948 × K-953 | 10.4 | 10.9 | 11.1 |
| K-1128 × K-1165 | 13.3 | 12.3 | 16.0 |
| K-1072 × K-1128 | 16.7 | 19.1 | 16.2 |
| K-948 × K-1144 | 17.9 | 16.4 | 17.6 |
| K-850 (laminin) × K-1072 (nidogen-1) | 20.9 | 17.5 | 16.9 |
| K-948 × K-1152 | 22.2 | 21.2 | 22.2 |
| K-1032 × K-1072 | 27.1 | 27.1 | 27.0 |
| K-961 × K-1072 | 28.7 | 28.0 | 28.2 |
| K-864 (laminin) × K-1152 (nidogen-1) | 32.2 | 22.4 | 27.1 |
| K-850 (laminin) × K-953 (nidogen-1) | 33.0 | 29.5 | 29.4 |
| K-1032 × K-1152 | 35.8 | 35.4 | 35.4 |
| Photo-L-990 × Arg-1038 | 24.7 | 23.4 | 23.5 |
| Photo-L-844 (laminin) × K-1072 (nidogen-1) | 33.8 | 19.4 | 20.8 |

Cα–Cα distances of cross-linked residues were determined for the unmodified crystal structure (PDB entry 1NPE) as well as for the Rosetta models with the best Rosetta total score and atom-pair constraint score, respectively (shown in Figure 6). For intermolecular contacts, residues are assigned to the respective protein. All other cross-links are located within nidogen-1.

score and atom-pair constraint score (reflecting the fulfillment of distance constraints) were selected as potential models of the nidogen-1 G3/laminin γ1 LEb2–4 core complex (Table S4).

## Results

As there are only very limited structural data available for full-length nidogen-1/laminin γ1 complexes, we sought to investigate the complexes created between nidogen-1 and laminin variants γ1 LEb2–4, γ1 LEb2–4 N836D, γ1 short arm, and γ1 short arm N836D by applying a combination of chemical cross-linking and high-resolution nano-HPLC/nano-ESI-LTQ-Orbitrap mass spectrometry. In the laminin γ1 short arm and γ1 LEb2–4 N836D point mutants, the Asn residue, which is crucial for nidogen binding, was exchanged for an acidic Asp. The obtained distance constraints were then used to identify previously unknown nidogen/laminin interfaces, to generate 3D-structural models of all their individual domains, and to generate a refined model of the high-affinity nidogen-1/laminin γ1 binding site based on the known 3D structure. In addition, SPR and ELISA assays allowed us to derive kinetic constants of the nidogen-1/laminin γ1 interaction. Together, the data presented herein shed new light on the mechanisms underlying nidogen/laminin interaction.

### Deriving Kinetic Constants of the Nidogen/Laminin Interaction

The binding affinities between nidogen-1 and laminin γ1 short arm and γ1 LEb2–4 were investigated with surface plasmon resonance spectroscopy by measuring single-cycle kinetics [36]. Representative examples of the obtained sensorgrams are depicted in Figure S5A. Apparent $K_d$ values were determined to be in the (sub-)nanomolar range, confirming high binding affinities between nidogen-1 and laminins (Table S5). However, the nidogen-1 binding activity of laminin γ1 short arm variant was found to be one order of magnitude lower than that of laminin γ1 LEb2–4, which is almost exclusively caused by differences in the association

phase. Nidogen-1 binding was also investigated for N836D variants of laminin γ1 short arm and γ1 LEb2–4. When analyte concentrations were increased ca. 100-fold compared to the laminin γ1 'wild-type' variants, signals were detected for nidogen-1/laminin γ1 LEb2–4 N836D binding (Figure S5A), which, however, did not allow to derive kinetic constants. For nidogen-1 and laminin γ1 short arm N836D, no interaction was detected by SPR.

In addition, the binding affinities between nidogen-1 and laminin γ1 short arm, γ1 LEb2–4 and their respective N836D variants were investigated by ELISA-based binding assays (Figure S5B). These assays revealed no differences in binding affinities of the laminin γ1 short arm and the LEb2–4 fragment to nidogen-1. Both interactions showed strong binding with apparent $K_d$ values of 1.1 and 1.4 nM (Table S5). Binding of the respective N836D variants to nidogen-1 showed a considerable loss of binding affinity with apparent $K_d$-values of 34 and 45 nM (γ1 LE2–4 N836D and γ1 short arm N836D, respectively).

### Cross-Linking of Nidogen-1/Laminin γ1 Complexes

For gaining insights into the interaction between nidogen-1 and laminin γ1 variants on the molecular level, the proteins were cross-linked using the homobifunctional cross-linker $BS^2G$. Additionally, we pursued a complementary approach by incorporating the unnatural diazirine-containing amino acids photo-Met and photo-Leu instead of methionine and leucine [58] into nidogen-1, laminin γ1 LEb2–4, and laminin γ1 short arm. MS/MS analysis of the non-cross-linked proteins revealed 13–25% of all leucines and methionines to be partially replaced by their photo-reactive counterparts. After UV-A-induced or $BS^2G$-mediated cross-linking, the reaction mixtures were separated by SDS-PAGE and analyzed by LC/MS/MS. Experiments were conducted in the presence of varying laminin concentrations to optimize the efficiency of heterodimer formation between nidogen-1 and laminin. The verified cross-links are summarized in Figure 2 and in Tables S6 and S7.

## A

## B

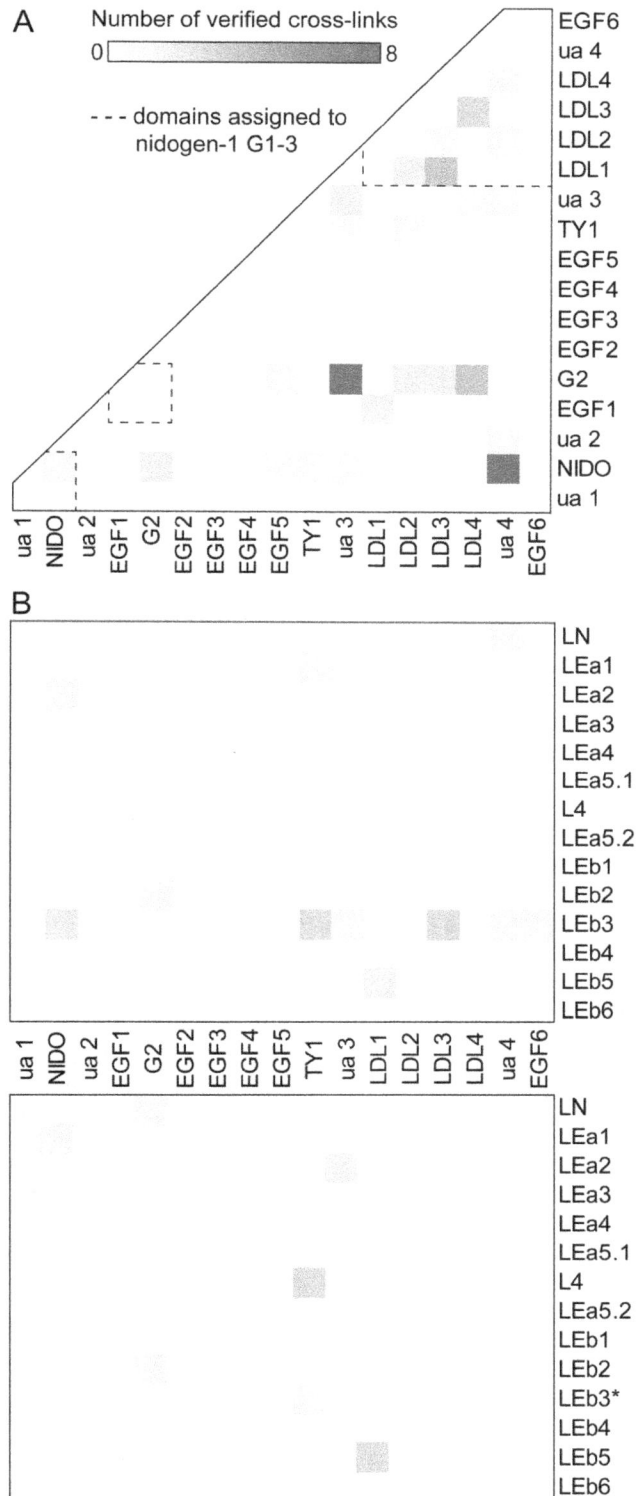

**Figure 2. Diagonal plots of all cross-links identified.** Cross-links are assigned to domains based on the UniProt KB entry P10493 (nidogen-1) and the laminin nomenclature of Aumailley *et al.* [5]. Unannotated areas within the sequences are named 'ua'. Corresponding to the number of inter-domain contacts, areas of intersection are color-coded from white (none) to dark grey (maximum). (A) Intramolecular cross-links within nidogen-1. The globular domains G1, G2, and G3 are denoted by dotted lines. Cross-links located nearby the diagonal

border represent contacts between domains being close to each other in the protein sequence. (B) Cross-links between nidogen-1 and laminin γ1 wild type (upper panel) as well as N836D variants (lower panel). The LEb3 domain, bearing the N836D mutation, is marked with an asterisk.

We confirmed 47 intramolecular BS$^2$G cross-links within nidogen-1, delivering 26 non-redundant distance constraints. Although the efficiency of photo-amino acid incorporation at the partially modified Leu and Met sites was moderate (~35% for photo-Met, ~3% for photo-Leu, see Figure S6), we identified two additional non-redundant photo-cross-links within nidogen-1, one connecting the link region with the G3 domain and one within the G3 domain, highlighting the sensitivity of our cross-linking/MS approach.

More than half of the intermolecular cross-links were inter-domain contacts, in which all globular nidogen-1 domains were connected with each other (Figure 2A). The majority of distance constraints were found in the G3 domain as well as between G2 and G3 domains, while the 'link' region (ua 2) and EGF domains 2–4 within the 'rod' region were not involved in any cross-link. As an example, the fragment ion mass spectrum representing a cross-link between the G2 and the G3 domain of nidogen-1 is shown in Figure 3.

Ten out of 19 non-redundant BS$^2$G cross-links between nidogen-1 and laminin γ1 were exclusively found in experiments with laminin γ1 wild type, six only in experiments with N836D variants, and three for both proteins. Additionally, one intermolecular contact between nidogen-1 G3 and laminin γ1 LEb3 was reproducibly identified in two consecutive photo-cross-linking experiments. With the exception of the link region, all nidogen-1 domains were involved in intermolecular cross-links with the nidogen binding motif of laminin γ1 (LEb2–4) as well as with additional regions within laminin γ1 short arm (Figure 2B). However, when laminin γ1 LEb2–4 N836D variants were used as interaction partners, the distribution of cross-links changed considerably with only two nidogen-1/laminin γ1 LEb2–4 contacts being identified (Figure 2C). Instead, L4 and LEb5 domains were repeatedly found to be cross-linked to nidogen-1.

Examining the intramolecular laminin γ1 cross-links underpins the complementarity of the two cross-linking strategies applied. Whereas only one BS$^2$G cross-link within laminin γ1 variants was identified (LEb2 with LEb5), UV-A irradiation revealed connections of the L4 domain to LEa1 and LEa2 domains as well as one contact between the LEa4 and LEb6 domains. Notably, we exclusively identified cross-links between sequentially non-adjacent laminin domains.

### Structural Characterization of Laminin γ1 and Nidogen-1 Domains by Comparative Modeling

Structures of the laminin γ1 LEa3–5.2, LEb1, and LEb5 domains as well as for the nidogen-1 domains EGF2–6 and TY1 were derived by comparative modeling (Figure S7). Template structures were identified by sequence homology search using DELTA-BLAST. Since we did not observe cross-links within these domains only known disulfide linkages served as distance constraints. 'Laminin-type EGF-like' (LE) domains exhibit a characteristic disulfide linkage pattern that is different from classical EGF-like domains [21,24,33,34]. Hence, only crystal structures of LE domains were used as templates for the remaining LE modules and nidogen-1 EGF domain models were based on the structures of classical EGF-like repeats.

Structural templates for the laminin γ1 L4 domain were identified by fold recognition as described in 'Materials & Methods'. A topology dominated by ß-sheets was predicted by

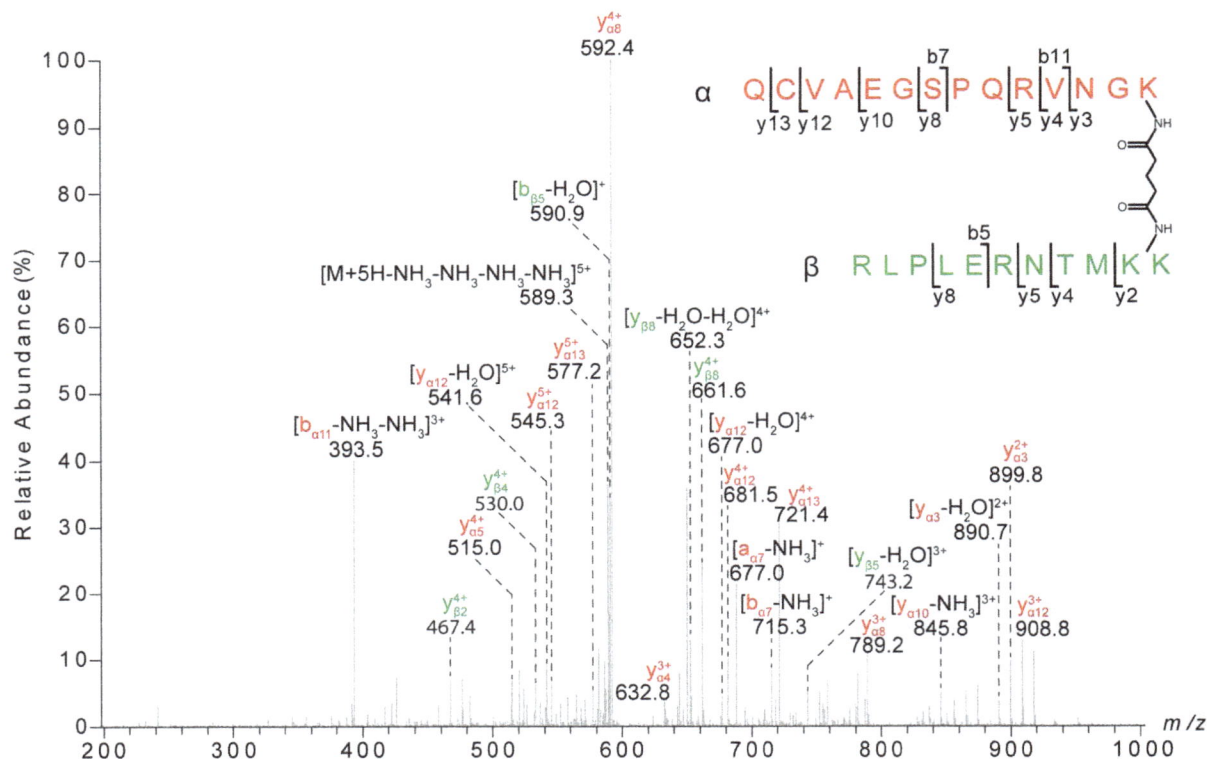

**Figure 3. Nano-HPLC/nano-ESI-LTQ-Orbitrap-MS/MS analysis of cross-linked peptides derived from nidogen-1 G2 and G3 domain.**
The cross-linked product comprises amino acids 407–420 of the G2 domain (α-peptide, red) and 939–949 of the G3 domain (β-peptide, green), in which K-407 is connected to K-948/949.

PSIPRED [47] (Figure S8A) and JUFO9D [48]. In compliance with this prediction, 13 potential templates with a common ß-sandwich topology were identified by the applied threading servers (Figure S8B). Comparative modeling of laminin γ1 L4 was thus based on these structural homologues. The best 10% of all generated models were clustered and further validated by generating score-vs-RMSD plots, assuming each of the best-scoring models of the top five clusters as the native structure (RMSD = 0). Convergence of the models towards a minimum Rosetta total score and RMSD underpins the validity of the structures with the lowest score (Figure S8C). The two final models, both adopting a galactose-binding domain-like fold, match the secondary structure predictions for the L4 domain and exhibit energetically favorable residue conformations throughout the structure (Figure 4).

## Generating an *Ab Initio* Model of the Nidogen-1 G1 Domain

G1, the *N*-terminal globular region of nidogen-1, essentially consists of a NIDO domain (Figure 1). As neither sequential nor structural homologues were identified for nidogen-1 NIDO we decided for a *de novo* folding strategy. This approach is exceptionally challenging because of NIDO's size (156 amino acids). Thus, we could not only rely on the Rosetta scoring function, but had to pursue alternative validation strategies [59]. First, we sought to identify known protein folds among the generated models by comparing them to a PISCES library of diverse PDB structures using MAMMOTH structural alignment (see 'Materials & Methods'). Second, we reasoned that the topologies being sampled during *de novo* folding can be substan-

tially influenced by subtle changes in the protein sequence. However, folding of highly similar sequences should result in identical tertiary structures. Similar topologies, sampled for several closely homologous sequences, are thus more likely to resemble the native structure. Therefore, we performed *de novo* folding not only for the murine NIDO domain, but also for homologues from six additional organisms with sequence identities larger than 75%. Taken together, these strategies resulted in the identification of 132 candidate structures after full-atom refinement and clustering (Table S3) and the ten top-scoring models were found to originate from four initial NIDO centroid models. Three of these models were validated by one BS²G cross-link, which had been identified within the NIDO domain, but was not used as a distance constraint during the modeling process. The determined Cα–Cα distances were within the range of the cross-linker, as shown in Figure 5. Additionally, a coarse clustering (radius = 10 Å) was performed to check whether further final candidate structures can be traced back to common centroid models. In view of the variety of conformations being sampled during the generation of centroid models, clusters were merged, when their best-scoring member structures originated from the same centroid model. We identified six additional centroid models that are represented by more than three full-atom refined models among the 132 final candidate structures (Table S8). For two of these centroid models, structural homologues in the PISCES PDB library have been identified. The best-scoring full-atom structural models for the six centroid models are shown in Figure S9. In all these NIDO models the cross-linking distance constraint is fulfilled. Taken together, our *de novo* approach suggests that NIDO adopts a compact topology containing a ß-sheet with at least four strands and two α-helices.

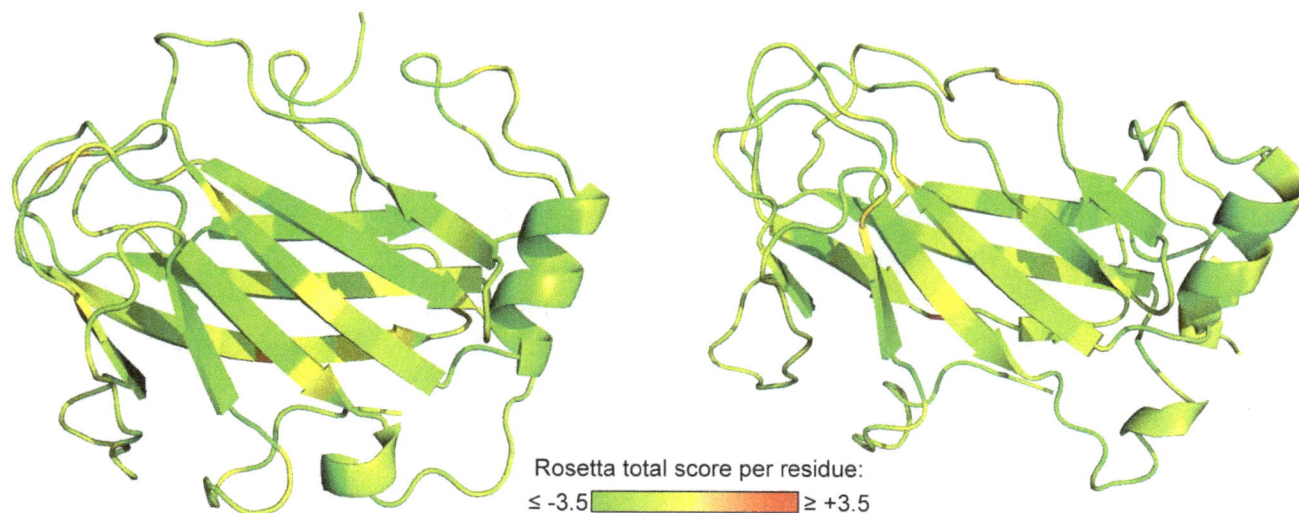

Rosetta total score per residue:
≤ -3.5 ≥ +3.5

**Figure 4. Structural models for laminin γ1 L4.** Shown are the two best-scoring models generated by comparative modeling based on 13 structural homologues that have been identified by fold recognition. The residues are colored according to their Rosetta total score. Scores below zero (yellow-green color) indicate energetically favorable conformations.

## Incorporation of Cross-Linking Distance Constraints into the Nidogen-1 G3/Laminin γ1 LEb2–4 Core Structure

Eleven $BS^2G$ cross-links originate from the known interaction region between the LEb2–4 domains of laminin γ1 and the G3 domain of nidogen-1. The distance constraints obtained by our cross-linking experiments should thus be in agreement with the known 3D structure of the nidogen-1 G3/laminin γ1 LEb2–4 complex (PDB entry 1NPE) [24]. According to the spacer arm length of $BS^2G$ (7.5 Å [55]) and the length of the cross-linked lysine side chains (2×6.3 Å), the maximum Euclidian Cα–Cα distance between the cross-linked residues should be 20.1 Å. However, longer distances are frequently observed, when cross-links are mapped in structural models. Therefore, it is common practice to grant a distance tolerance of 6–7 Å to account for structural flexibility. Recently, a rationale for this approach was presented by studying the lysine–lysine distances within 807 proteins during molecular dynamics simulations [60]. This analysis gave evidence that the ε-amino groups of lysines, with initial Cα–Cα distances of up to 38 Å, will move inside the range of the amine-reactive cross-linker DSS (spacer arm length 11.4 Å) during a 50 ns-simulation. Considering the shorter spacer arm length of $BS^2G$, the maximum Euclidean Cα–Cα distance that is likely to allow cross-linking can, therefore, be estimated to 34 Å. In an earlier study, similar results were obtained by cross-linking seven model proteins and comparing the identified cross-links with their respective crystal structures [61]. Mapping our cross-links into the X-ray structure of the nidogen-1 G3/laminin γ1 LEb2–4 complex resulted in Cα–Cα distances between 10.4 and 35.8 Å. By integrating the crystal structure and our cross-linking data in a Rosetta-based modeling approach, we aimed to derive structural models that better reflect plausible in-solution conformations of the nidogen-1 G3/laminin γ1 LEb2–4 complex, which would be signified by a decrease in observed Cα–Cα distances. The best-scoring models of the complex are depicted in Figure 6. The laminin γ1 fragment is considerably bent compared to the X-ray structure (Figure 6C), which is conceivable as the three LE repeats do not form a compact tertiary structure around a defined hydrophobic core. Both the disulfide bond pattern of the LE domains and the β-propeller structure of the G3 domain are well maintained indicating that the structural rearrangements in our models are reasonable. Notably, the spanned Cα–Cα distances of all except one $BS^2G$ cross-link are significantly reduced compared to the X-ray structure, suggesting that the models give a better picture of the conformations the nidogen-1 G3/laminin γ1 LEb2–4 complex can adopt in solution (Table 1). However, one has to be aware that the Rosetta models, just as the crystal structure, represent conformational samples and do not reflect the entire conformational space of the protein complex.

In this context, it has to be pointed out that the single cross-link that exceeds the 34 Å distance limit of the X-ray structure is in conflict with the models. Visual inspection of the structural models shows that the respective lysines cannot be cross-linked, since the spacer arm would have to traverse through the center of the β-propeller, which is blocked by residues of the G3 domain. We interpret this cross-link as an intermolecular contact between two nidogen-1 molecules. Forcing Rosetta to fulfill this distance constraint resulted in unfolding of the G3 domain, confirming that a sound model cannot be forced to match the experimental data, but complies only with sterically feasible cross-links. In addition to the $BS^2G$ cross-links we obtained two distance constraints by photo-cross-linking. The respective Cα–Cα distances of the cross-links, in which photo-Leu is connected to Arg and Lys, were 24.7 and 33.7 Å within the X-ray structure as well as 23.4–23.5 Å and 19.4–20.8 Å within the models (Table 1). This is well above the expected maximum values of 10.4 Å for a photo-Leu/Lys cross-link and 11.4 Å for a photo-Leu/Arg cross-link, which were determined from the side chain lengths within representative amino acid crystal structures deposited in the Cambridge Structural Database [62]. However, in both cases, one of the cross-linked amino acids is located in a loop region. It is conceivable that the obtained Cα–Cα distances are longer than the maximum expected distances as the loop regions are flexible. Granting a similar distance tolerance as for $BS^2G$ results in maximum allowed Cα–Cα distances of 23.4 Å and 24.4 Å, respectively, both of which are met by our models. Intriguingly, the cross-link between photo-Leu-844 (laminin) and Lys-1072 (nidogen-1) concurs with a $BS^2G$ cross-link pointing to the same region (laminin Lys-850 with nidogen-1 Lys-1072).

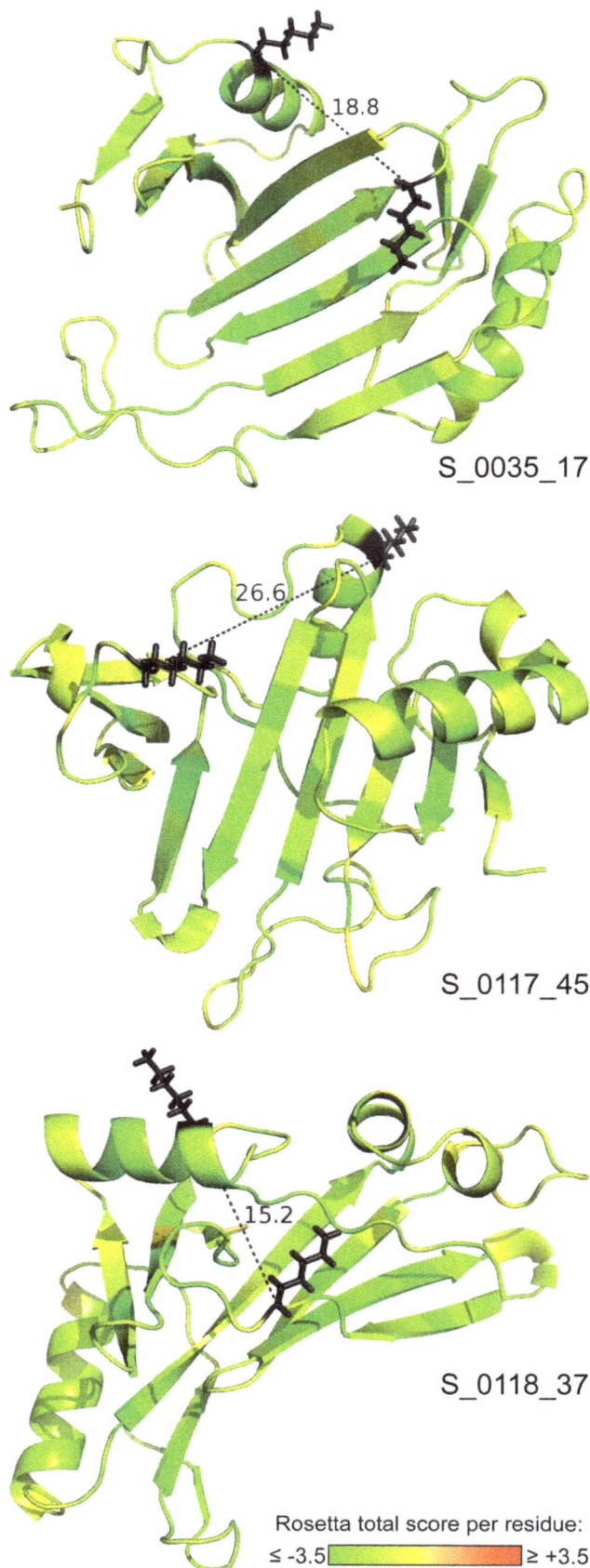

**Figure 5.** *De novo* **folded models of nidogen-1 G1.** The ten best-scoring structures among the final models were all derived from four initial centroid models of the G1 domain. Structures representing three of these centroid models are shown here. These models comply with the single distance constraint in this region that was identified by cross-linking/MS. Cross-linked residues are displayed as black sticks. Cα–Cα distances are given in Å. The residues are colored according to their Rosetta total score. Scores below zero (yellow-green color) indicate energetically favorable conformations. The identifiers of the underlying centroid models are indicated.

## Discussion

The primary goal of this work was to gain novel insights into the nidogen-1/laminin γ1 interaction in solution by combining amine-reactive and photo-chemical cross-linking with high-resolution mass spectrometry and computational modeling. Additionally, we probed the affinity of nidogen-1 to different laminin γ1 variants by means of SPR and ELISA-based binding assays.

So far, quantitative analyses of the nidogen-1/laminin affinity have only been performed using the laminin P1 fragment, which is produced by limited pepsin proteolysis and comprises parts of all three chains (α1, β1, γ1) of laminin-111 [9,11,13,63]. Apparent $K_d$ values determined in those studies ranged between 0.5 nM and 1 nM, complying with the dissociation constants derived from the ELISA assays (Table S5). SPR analysis yielded similar results for the nidogen-1/laminin γ1 LEb2–4 interaction, but a lower apparent $K_d$ value (~12 nM) for laminin γ1 short arm. In the SPR experiments, nidogen-1 was immobilized in random orientations by covalently linking lysine residues to the sensor chip surface, while the laminin variants were used as mobile analytes. Notably, the lower affinity of laminin γ1 short arm is solely caused by a slower association – the dissociation rate constant is similar for both laminin variants (Table S5). Therefore, we conclude that the deviations in the SPR experiments are caused by different *in solution* properties of laminin γ1 short arm compared to laminin γ1 LEb2–4. This hypothesis is supported by the ELISA experiments where nidogen-1 was used as ligand, while the laminin variants were immobilized through passive adsorption in a 96-well plate [64], resulting in similar nidogen-1-binding affinities. These different experimental setups may introduce different degrees of steric hindrance that probably do not affect the binding of the relatively small laminin γ1 LEb2–4 fragment (25 kDa), but hamper the interaction with the laminin γ1 short arm variant, which is considerably bulkier (113 kDa). In other words, the laminin γ1 short arm molecules might not be equally binding-competent due to different *in solution* conformations. This results in a slower association and consequently, a lower apparent $K_d$ value when using laminin γ1 short arm as mobile analyte.

Moreover, interactions between nidogen-1 and laminin γ1 N836D variants were verified by both affinity assays. The ELISA results suggest a ~30-fold loss in affinity upon mutating Asn-836 to aspartic acid. This finding is in contrast to a previous study by Pöschl *et al.* who found this mutation to cause a 25,000-fold loss in affinity, practically abolishing any interaction [23]. Our data suggest a much less dramatic influence of laminin γ1 Asn-836 on nidogen-1 binding.

The large number of inter-domain contacts verified for nidogen-1 implies a globular conformation rather than a linear domain arrangement. This is in agreement with a previous structural investigation of nidogen-1 by electron microscopy showing a wide range of conformations, including both linear as well as globular structures [13]. This variability within nidogen-1 might be caused by high flexibility of the elongated regions connecting the globular domains ('link' and 'rod'), which could likewise be an explanation

A

B

C

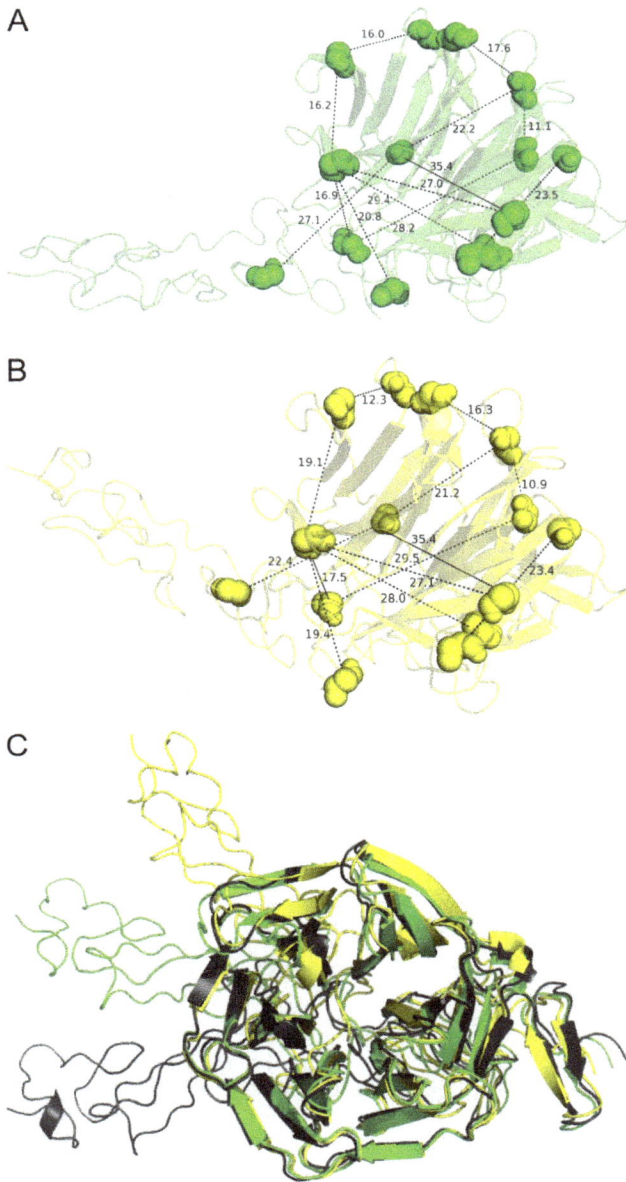

**Figure 6. Refined models of the nidogen-1 G3/laminin γ1 LEb2–4 complex.** Based on PDB entry 1NPE and the identified cross-links, modified structural models for the high-affinity interaction region of laminin γ1 and nidogen-1 were generated. Cross-linked residues are displayed as spheres. Cα–Cα distances are given in Å. (A) Model with the best Rosetta total score and a Rosetta atom-pair constraint score ranking among the top 2.5%. (B) Model with the best Rosetta atom-pair constraint score and a Rosetta total score ranking among the top 2.5%. (C) Alignment of both models and the unmodified crystal structure 1NPE (black). The orientation of LEb2–4 clearly has changed during structural refinement. The β-propeller fold of the G3 domain is still intact.

for the almost complete lack of cross-links in these regions (Figure 2A). Within laminin γ1 short arm variants, intramolecular contacts were exclusively found between non-adjacent domains supporting a compact globular protein architecture as well. The cross-links also give hints on the existence of an ensemble of defined conformations for both nidogen-1 and laminin γ1 short arm.

The intermolecular cross-links between nidogen-1 and the laminin variants strongly indicate additional interaction regions next to the known nidogen-1 G3/laminin γ1 LEb2–4 binding site. Even laminin γ1 LEb2–4 was found to form contacts to all globular nidogen-1 domains and the rod region (Figure 7). We conclude that our cross-linking experiments allowed us to pick up different nidogen-1/laminin γ1 complexes that are present in solution, while the published X-ray structure reflects only one interaction 'snapshot' — most likely the best crystallizable conformation.

The existence of additional interaction patterns was further substantiated by cross-linking experiments using nidogen-1 and laminin γ1 N836D variants, which were unanimously shown to interact regardless of the N836D substitution within the laminin γ1 LEb3 domain. Interestingly, contacts of nidogen-1 to LEb3 were almost completely abolished, while alternative laminin γ1 domains were found to be cross-linked to nidogen-1 (Figure 2). Taken together, these results suggest alternative nidogen-1/laminin γ1 binding modes, conceivably as a result of the conformational flexibility of both proteins, which had already been implied by electron microscopy [65]. When performing cross-linking experiments, one has to keep in mind that amine-reactive cross-linking depends on the reactivity of lysines, which is influenced by solvent exposure of their side chains and local $pK_a$ values [66]. Considering the spacer length of $BS^2G$, the observation of an intermolecular cross-link is not *per se* equivalent to physical binding at exactly that site. While our cross-linking results indeed indicate additional binding sites, it is still imaginable that laminin γ1 LEb2–4 remains the primary anchoring site of nidogen-1. In fact, solid phase binding studies have proven that the nidogen-1 G3 domain is essential for nidogen-1/laminin γ1 interaction [63,67]. Therefore, nidogen-1 G3/laminin γ1 LEb2–4 binding is most likely crucial for a high-affinity interaction and binding at alternative sites proceeds with substantially lower affinity.

This finding seems reasonable also in view of the stabilizing function of nidogen-1 within basement membranes, connecting the laminin and the type IV collagen network [14]. Given the mechanical stress basement membranes have to withstand [2], secondary interactions at alternative binding sites may occur when the basement membrane is in a more relaxed state. In contrast, the high-affinity anchoring interaction between nidogen-1 G3 and laminin γ1 LEb2–4 is likely to be continuously present, thereby ensuring a high mechanical stability. Although this binding region represents one of the smallest high-affinity interfaces known so far [24], our SPR, ELISA, and cross-linking/MS data indicate a certain robustness against the single point mutation N836D, which is plausible in light of the physiological importance of the nidogen/laminin interaction [68].

In line with these findings, we were not able to fit all cross-linking distance constraints for nidogen-1 and laminin γ1 into one single model of the protein complex. Assigning the cross-links to defined conformations within an ensemble of co-existing structural arrangements is currently beyond the bounds of the method. Consequently, a computational model of the entire nidogen-1/laminin γ1 short arm complex based on cross-linking distance constraints would be overly speculative. Comparative modeling and *de novo* folding using the Rosetta modeling suite, however, enabled us to create models of all nidogen-1 and laminin γ1 short arm domains that have not been structurally characterized so far. The structural models of laminin γ1 L4 and nidogen-1 NIDO are of particular interest as there are not any sequential homologues with known structures. We were able to identify structural homologues of laminin γ1 L4 suggesting a galactose-binding

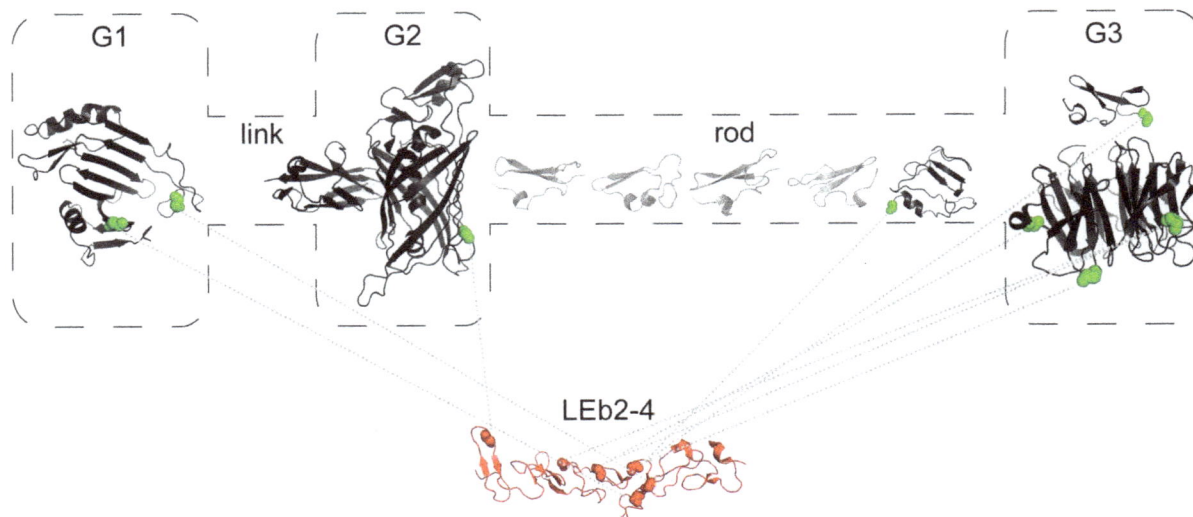

**Figure 7. Contacts of laminin γ1 LEb2–4 wild type with nidogen-1.** The LEb2–4 structure (red) is taken from PDB entry 1NPE. Nidogen-1 (grey) is schematically depicted as a combination of the crystal structures 1GL4 (G2 domain) and 1NPE (G3 domain) and representative models of the remaining domains. Residues involved in intermolecular contacts are shown as spheres. Gray dotted lines represent verified cross-links.

domain-like fold of this domain. Interestingly, this fold is also adopted by the *N*-terminal LN domains of laminin α5, β1, and γ1 [7,33,34]. To date, experimental evidence for carbohydrate-binding activity of laminins only exists for *C*-terminal laminin α LG domains [69].

Structures of the nidogen-1 NIDO domain were obtained by *de novo* folding. The proposed models either exhibit structural similarities to known PDB structures or were consistently modeled based on six NIDO sequences of evolutionary closely related organisms. All but one of the final models were further validated as they fulfill the only cross-linking constraint identified in this region. These models exhibit a compact topology with a β-sheet surrounded by two α-helices as central elements.

Finally, integrating the cross-links found within the nidogen-1 G3/laminin γ1 LEb2–4 complex and the known X-ray structure confirms the main structural features suggested by X-ray crystallography, yet indicating a more bent topology of the LEb2–4 domains (Figure 6C). Again, this finding depicts the flexibility of laminin γ1 resulting in overlying conformations of the nidogen-1/laminin γ1 complex, which can be captured by chemical cross-linking and thus reflect the whole picture of nidogen-1/laminin γ1 interaction in solution.

## Conclusions

With our approach integrating chemical cross-linking, mass spectrometry, and computational modeling, we were able to structurally characterize conformations of nidogen-1/laminin γ1 complexes in solution. We applied two complementary cross-linking approaches, one using a classical homobifunctional amine-reactive cross-linker, the other one relying on the incorporation of unnatural photo-reactive amino acids.

Cross-links between laminin γ1 short arm and nidogen-1 were found in all protein regions. Therefore, it is likely that both proteins exhibit several additional contact regions apart from their known interaction site. In addition, different modes of interaction resulting in several distinct protein-protein interfaces can be imagined. Our results indicate that Asn-836 within laminin γ1 LEb3 domain is not essential for complex formation. Conclusively,

this work describes the first structural insights into the conformational dynamics of the nidogen-1/laminin γ1 complex and provides, for the first time, structural models of all nidogen-1 and laminin γ1 short arm domains. Chemical cross-linking, MS, and computational modeling allowed elucidating different conformations of the nidogen-1/laminin γ1 complex, which exist simultaneously in solution but are not reflected in the X-ray structure.

## Supporting Information

**Figure S1   Amino acid sequences of (A) nidogen-1 and (B) laminin γ1 short arm.**

**Figure S2   Clustal W2.1 sequence alignments for comparative modeling.** Shown are pairwise sequence alignments of all nidogen-1 and laminin γ1 target sequences to template sequences with sequence identities ≥30%. The scheme for template sequences is termed 'PDB-entry_domain-name_chain-identifier'. Annotations comply with the Clustal nomenclature with identical (*), conserved (:) and semi-conserved (.) residues being denoted.

**Figure S3   Manual sequence alignments for modeling of the laminin γ1 L4 domain.** Shown are pairwise sequence alignments to all template sequences. Alignments are manually optimized to obtain maximum overlap of secondary structure elements. The scheme for template sequences is termed 'PDB-entry_chain-identifier'.

**Figure S4   Jalview sequence alignment of nidogen-1 NIDO domains from different organisms.** All NIDO domains share sequence identities >75% but exhibit short sequence stretches that are diverse.

**Figure S5   Probing the nidogen-1/laminin γ1 interaction with SPR and ELISA assays.** (A) Single-cycle kinetic experiments were performed by injecting mobile analyte (laminin) at increasing concentrations followed by partial dissociation. Initially, experiments were carried out with 6.25 nM, 12.5 nM, 25 nM, 50 nM and 100 nM laminin γ1 LEb2–4 wild type and N836D. Binding of laminin γ1 LEb2–4 N836D was additionally probed with 100-fold increased concentrations. System artefact signals (~30 min after each injection) were removed from the sensorgrams. (B) ELISA assays were performed in 96-well plates with immobilized laminin γ1 variants. Nidogen-1 was added in increasing concentrations (0.03–234 nM) until saturation was reached (incubation time: 1 h). Error bars represent standard deviations.

**Figure S6   Incorporation efficiency of photo-amino acids into nidogen-1 and laminin γ1 short arm.** (A) Met and Leu variants that were considered during MS analysis, including the reaction products of the photo-amino acids identified in [35] (1: photo-Leu, 5: photo-Met, 2 and 6: alkene; 3 and 7: alcohol; 4: unmodified Leu; 8 and 9: unmodified and oxidized Met). (B and C) MS-based label-free quantification of photo-amino acid incorporation. The pie charts show the number of leucines (blue) and methionines (red) within nidogen-1 (B) and laminin γ1 short arm (C) that remained unmodified (light shades) or were partially replaced by their photo-reactive counterparts (dark shades). The bars represent the relative abundance of partially modified peptides, containing the Leu (blue) and Met (red) variants listed in (A) [70].

**Figure S7   Homology models of (A) nidogen-1 and (B) laminin γ1 short arm domains.** Alignments of the best-scoring models representing the top three clusters are shown. Disulfide bridges are depicted as black sticks. All models were generated based on X-ray structures sharing more than 30% sequence identity with the respective domains.

**Figure S8   Comparative modeling of laminin γ1 L4.** (A) PSIPRED secondary structure prediction for the L4 domain. A β-sheet-rich fold and one long α-helix are predicted. (B) MUSTANG alignment of 13 potential structural homologs of L4 identified by fold recognition using several threading servers. All template candidates exhibit a β-sandwich topology. The number of β-strands is in line with the predicted secondary structure of L4. Instead of an α-helix, all structures contain a long loop region. (C) Rosetta total score of the top 10% of all generated models plotted against their RMSD from the best-scoring structure. Only α-helices, β-sheets and short loops (≤5 residues) were included in RMSD calculations. The models are converging to a minimum in score and RMSD indicating that the best-scoring models are valid. The two best-scoring models shown in Figure 4 are marked with red circles.

**Figure S9   Best-scoring nidogen-1 NIDO models originating from common centroid models.** The full-atom candidate structures of the NIDO domain were examined for common initial centroid models. Next to the centroid models underlying the structures depicted in Figure 5, we identified six centroid models that form the basis for more than three full-atom refined candidate structures. Shown are the best-scoring final candidate structures representing these initial centroid models. The Cα–Cα distances corresponding to the cross-link located within the models are given in Å. The residues are colored according to their Rosetta total score. Scores below zero (yellow-

green color) indicate energetically favorable conformations. The identifiers of the underlying centroid models are given.

**Table S1   Results of Rosetta clustering of comparative nidogen-1 and laminin γ1 domain models.** The best 10% of all generated models were clustered. The ideal clustering radius was automatically determined by the Rosetta algorithm. Shown are clusters with a size >1.

**Table S2   Results of Rosetta clustering of the laminin γ1 L4 domain models.** The best 10% of all generated models were clustered using a clustering radius of 1 Å. Shown are clusters with a size >1. The best-scoring models of clusters 1 and 2 were chosen as final models of the L4 domain.

**Table S3   Scores of the final nidogen-1 NIDO domain models.** Models sharing the 'centroid model identifier' originate from the same initial low-resolution centroid model. Models, for which structural homologues within the PDB have been identified, are listed in italics. The remaining models share a similar topology to models generated based on highly homologous sequences of NIDO domains derived from related organisms.

**Table S4   Scores of the final models of the nidogen-1 G3/laminin γ1 LEb2–4 complex.** The listed models rank among the top 20 of 800 generated models considering both Rosetta total score and atom pair constraint score, which reflects their compliance with the cross-linking distance constraints.

**Table S5   Affinities and kinetic parameters of the nidogen-1/laminin γ1 interaction.** For SPR measurements, given values for $k_a$ and $k_d$ are the weighted mean from two individual measurements and $K_d$ was calculated from these values as $K_d = k_d/k_a$. All ELISA-based measurements were performed in triplicates and $K_d$ values were determined by non-linear regression of the saturation binding curves. The values in parentheses represent standard deviations.

**Table S6   Verified products of BS²G-mediated cross-linking.** Peptide sequences written in parentheses are part of the protein affinity tags and do thus not belong to the native amino acid sequences of the proteins studied. Oxidized methionines within the peptide sequences are denoted with 'm'. Loss of water or ammonia is indicated by addition of '−H₂O' or '−NH₃' to the fragment ion.

**Table S7   Verified products of UV A-induced cross-linking.** Peptide sequences written in parentheses are part of the protein affinity tags and do thus not belong to the native amino acid sequences of the proteins studied. For ambiguous cross-links, all potential cross-linked amino acids are listed. Within the peptide sequences, photo-leucine and photo-methionine are assigned with 'z' and 'o', respectively. Oxidized methionines are denoted with 'm'. Loss of water or ammonium is indicated by addition of '−H₂O' or '−NH₃' to the fragment ion.

**Table S8   Rosetta clustering results of the final nidogen-1 NIDO domain models.** The clustering radius was set to 10 Å. Clusters represented by models originating from the same initial low-resolution centroid model were merged. Shown are clusters with more than three member structures. Models, for which structural homologues within the PDB have been identified, are listed in italics. The remaining models share a similar topology to models generated based on highly homologous sequences of NIDO domains derived from related organisms.

**File S1    Command line execution commands and flags used for computational modeling with Rosetta.**

**File S2    Rosetta loops files generated for comparative modeling.** The position of the loops was determined based on the sequence alignments of the target sequences (listed in italics) to the respective templates. The scheme for template sequences is termed 'PDB-entry_domain-name_chain-identifier'.

## References

1. Yurchenco PD, O'Rear JJ (1994) Basal lamina assembly. Curr Opin Cell Biol 6 (5): 674–681.
2. Yurchenco PD (2011) Basement membranes: cell scaffoldings and signaling platforms. Cold Spring Harb Perspect Biol 3 (2): a004911.
3. Timpl R, Rohde H, Robey PG, Rennard SI, Foidart JM, et al. (1979) Laminin - a glycoprotein from basement membranes. J Biol Chem 254 (19): 9933–9937.
4. Tunggal P, Smyth N, Paulsson M, Ott MC (2000) Laminins: structure and genetic regulation. Microsc Res Techniq 51 (3): 214–227.
5. Aumailley M, Bruckner-Tuderman L, Carter WG, Deutzmann R, Edgar D, et al. (2005) A simplified laminin nomenclature. Matrix Biol 24 (5): 326–332.
6. Beck K, Hunter I, Engel J (1990) Structure and function of laminin: anatomy of a multidomain glycoprotein. FASEB J 4 (2): 148–160.
7. Kalkhof S, Haehn S, Paulsson M, Smyth N, Meiler J, et al. (2010) Computational modeling of laminin N-terminal domains using sparse distance constraints from disulfide bonds and chemical cross-linking. Proteins 78 (16): 3409–3427.
8. Carlin B, Jaffe R, Bender B, Chung AE (1981) Entactin, a novel basal lamina-associated sulfated glycoprotein. J Biol Chem 256 (10): 5209–5214.
9. Kohfeldt E, Sasaki T, Göhring W, Timpl R (1998) Nidogen-2: a new basement membrane protein with diverse binding properties. J Mol Biol 282 (1): 99–109.
10. Timpl R, Dziadek M, Fujiwara S, Nowack H, Wick G (1983) Nidogen: a new, self-aggregating basement membrane protein. Eur J Biochem 137 (3): 455–465.
11. Salmivirta K (2002) Binding of mouse nidogen-2 to basement membrane components and cells and its expression in embryonic and adult tissues suggest complementary functions of the two nidogens. Exp Cell Res 279 (2): 188–201.
12. Miosge N, Sasaki T, Timpl R (2002) Evidence of nidogen-2 compensation for nidogen-1 deficiency in transgenic mice. Matrix Biol 21 (7): 611–621.
13. Fox JW, Mayer U, Nischt R, Aumailley M, Reinhardt D, et al. (1991) Recombinant nidogen consists of three globular domains and mediates binding of laminin to collagen type IV. EMBO J 10 (11): 3137–3146.
14. Ho MSP, Boese K, Mokkapati S, Nischt R, Smyth N (2008) Nidogens - extracellular matrix linker molecules. Microsc Res Techniq 71 (5): 387–395.
15. Dziadek M, Paulsson M, Timpl R (1985) Identification and interaction repertoire of large forms of the basement membrane protein nidogen. EMBO J 4 (10): 2513–2518.
16. Paulsson M, Aumailley M, Deutzmann R, Timpl R, Beck K, et al. (1987) Laminin-nidogen complex. Extraction with chelating agents and structural characterization. Eur J Biochem 166 (1): 11–19.
17. Mann K, Deutzmann R, Timpl R (1988) Characterization of proteolytic fragments of the laminin-nidogen complex and their activity in ligand-binding assays. Eur J Biochem 178 (1): 71–80.
18. Mayer U, Nischt R, Pöschl E, Mann K, Fukuda K, et al. (1993) A single EGF-like motif of laminin is responsible for high affinity nidogen binding. EMBO J 12 (5): 1879–1885.
19. Gerl M, Mann K, Aumailley M, Timpl R (1991) Localization of a major nidogen-binding site to domain III of laminin B2 chain. Eur J Biochem 202 (1): 167–174.
20. Baumgartner R, Czisch M, Mayer U, Pöschl E, Huber R, et al. (1996) Structure of the nidogen binding LE module of the laminin gamma 1 chain in solution. J Mol Biol 257 (3): 658–668.
21. Stetefeld J, Mayer U, Timpl R, Huber R (1996) Crystal structure of three consecutive laminin-type epidermal growth factor-like (LE) modules of laminin gamma 1 chain harboring the nidogen binding site. J Mol Biol 257 (3): 644–657.
22. Pöschl E, Mayer U, Stetefeld J, Baumgartner R, Holak TA, et al. (1996) Site-directed mutagenesis and structural interpretation of the nidogen binding site of the laminin gamma 1 chain. EMBO J 15 (19): 5154–5159.
23. Sasaki M, Yamada Y (1987) The laminin B2 chain has a multidomain structure homologous to the B1 chain. J Biol Chem 262 (35): 17111–17117.
24. Takagi J, Yang YT, Liu JH, Wang JH, Springer TA (2003) Complex between nidogen and laminin fragments reveals a paradigmatic beta-propeller interface. Nature 424 (6951): 969–974.

## Acknowledgments

Prof. Gunter Fischer and Dr. Cornelia Schiene-Fischer are acknowledged for generously providing their cell culture facilities. The authors are indebted to Prof. Mats Paulsson for continuous support and valuable discussions.

## Author Contributions

Conceived and designed the experiments: PL KK FZ JM AS. Performed the experiments: PL DT DN CHI MVK MS. Analyzed the data: PL KK CHI MS MVK AS. Contributed reagents/materials/analysis tools: DT CHI FZ JM AS. Contributed to the writing of the manuscript: PL DN MS FZ JM AS.

25. Patel TR, Bernards C, Meier M, McEleney K, Winzor DJ, et al. (2014) Structural elucidation of full-length nidogen and the laminin-nidogen complex in solution. Matrix Biol 33: 60–67.
26. Sinz A (2014) The advancement of chemical cross-linking and mass spectrometry for structural proteomics: from single proteins to protein interaction networks. Expert Rev Proteomics Sept. 16: 1–11, online available.
27. Sinz A (2006) Chemical cross-linking and mass spectrometry to map three-dimensional protein structures and protein–protein interactions. Mass Spectrom Rev 25 (4): 663–682.
28. Kalisman N, Adams CM, Levitt M (2012) Subunit order of eukaryotic TRiC/CCT chaperonin by cross-linking, mass spectrometry, and combinatorial homology modeling. Proc Natl Acad Sci USA 109 (8): 2884–2889.
29. Leitner A, Joachimiak LA, Bracher A, Mönkemeyer L, Walzthoeni T, et al. (2012) The molecular architecture of the eukaryotic chaperonin TRiC/CCT. Structure 20 (5): 814–825.
30. Rappsilber J (2011) The beginning of a beautiful friendship: cross-linking/mass spectrometry and modelling of proteins and multi-protein complexes. J Struct Biol 173 (3): 530–540.
31. Rinner O, Seebacher J, Walzthoeni T, Mueller LN, Beck M, et al. (2008) Identification of cross-linked peptides from large sequence databases. Nat Meth 5 (4): 315–318.
32. Müller MQ, Dreiocker F, Ihling CH, Schäfer M, Sinz A (2010) Cleavable cross-linker for protein structure analysis: reliable identification of cross-linking products by tandem MS. Anal Chem 82 (16): 6958–6968.
33. Hussain S, Carafoli F, Hohenester E (2011) Determinants of laminin polymerization revealed by the structure of the α5 chain amino-terminal region. EMBO Rep 12 (3): 276–282.
34. Carafoli F, Hussain S, Hohenester E (2012) Crystal structures of the network-forming short-arm tips of the laminin β1 and γ1 chains. PLoS ONE 7 (7): e42473.
35. Kölbel K, Ihling CH, Sinz A (2012) Analysis of peptide secondary structures by photoactivatable amino acid analogues. Angew Chem Int Ed Engl 51 (50): 12602–12605.
36. Karlsson R, Katsamba PS, Nordin H, Pol E, Myszka DG (2006) Analyzing a kinetic titration series using affinity biosensors. Anal Biochem 349 (1): 136–147.
37. Schaks S, Maucher D, Ihling CH, Sinz A (2012) Investigation of a calmodulin/peptide complex by chemical cross-linking and high-resolution mass spectrometry. In: König S, editor. Biomacromolecular mass spectrometry. Tips from the bench. Hauppauge: Nova Science Publishers. 1–18.
38. Götze M, Pettelkau J, Schaks S, Bosse K, Ihling CH, et al. (2012) StavroX - a software for analyzing crosslinked products in protein interaction studies. J Am Soc Mass Spectrom 23 (1): 76–87.
39. Boratyn GM, Schäffer AA, Agarwala R, Altschul SF, Lipman DJ, et al. (2012) Domain enhanced lookup time accelerated BLAST. Biol Direct 7: 12.
40. Kelley LA, Sternberg MJE (2009) Protein structure prediction on the Web: a case study using the Phyre server. Nat Protoc 4 (3): 363–371.
41. Söding J, Biegert A, Lupas AN (2005) The HHpred interactive server for protein homology detection and structure prediction. Nucleic Acids Res 33 (Web Server issue): W244–8.
42. Buchan DWA, Minneci F, Nugent TCO, Bryson K, Jones DT (2013) Scalable web services for the PSIPRED Protein Analysis Workbench. Nucleic Acids Res 41 (Web Server issue): W349–57.
43. Zhang Y (2008) I-TASSER server for protein 3D structure prediction. BMC Bioinformatics 9: 40.
44. Larkin MA, Blackshields G, Brown NP, Chenna R, McGettigan PA, et al. (2007) Clustal W and Clustal X version 2.0. Bioinformatics 23 (21): 2947–2948.
45. Combs SA, DeLuca SL, DeLuca SH, Lemmon GH, Nannemann DP, et al. (2013) Small-molecule ligand docking into comparative models with Rosetta. Nat Protoc 8 (7): 1277–1298.
46. Konagurthu AS, Whisstock JC, Stuckey PJ, Lesk AM (2006) MUSTANG: a multiple structural alignment algorithm. Proteins 64 (3): 559–574.

47. Jones DT (1999) Protein secondary structure prediction based on position-specific scoring matrices. J Mol Biol 292 (2): 195–202.

48. Leman JK, Mueller R, Karakas M, Woetzel N, Meiler J (2013) Simultaneous prediction of protein secondary structure and transmembrane spans. Proteins 81 (7): 1127–1140.

49. Joosten RP, te Beek TAH, Krieger E, Hekkelman ML, Hooft RWW, et al. (2010) A series of PDB related databases for everyday needs. Nucleic Acids Res 39 (Database): D411.

50. Lange OF, Baker D (2012) Resolution-adapted recombination of structural features significantly improves sampling in restraint-guided structure calculation. Proteins 80 (3): 884–895.

51. Gront D, Kulp DW, Vernon RM, Strauss CEM, Baker D, et al. (2011) Generalized Fragment Picking in Rosetta: Design, Protocols and Applications. PLoS ONE 6 (8): e23294.

52. Wang G, Dunbrack RL (2003) PISCES: a protein sequence culling server. Bioinformatics 19 (12): 1589–1591.

53. Ortiz AR, Strauss CE, Olmea O (2002) MAMMOTH (matching molecular models obtained from theory): An automated method for model comparison. Protein Sci 11 (11): 2606–2621.

54. Waterhouse AM, Procter JB, Martin DMA, Clamp M, Barton GJ (2009) Jalview Version 2–a multiple sequence alignment editor and analysis workbench. Bioinformatics 25 (9): 1189–1191.

55. Green NS, Reisler E, Houk KN (2001) Quantitative evaluation of the lengths of homobifunctional protein cross-linking reagents used as molecular rulers. Protein Sci 10 (7): 1293–1304.

56. Herzog F, Kahraman A, Boehringer D, Mak R, Bracher A, et al. (2012) Structural probing of a protein phosphatase 2A network by chemical cross-linking and mass spectrometry. Science 337 (6100): 1348–1352.

57. Kahraman A, Herzog F, Leitner A, Rosenberger G, Aebersold R, et al. (2013) Cross-link guided molecular modeling with ROSETTA. PLoS ONE 8 (9): e73411.

58. Suchanek M, Radzikowska A, Thiele C (2005) Photo-leucine and photo-methionine allow identification of protein-protein interactions in living cells. Nat Meth 2 (4): 261–267.

59. Borek F (1961) A new two-stage method for cross-linking proteins. Nature 191 (4795): 1293–1294.

60. Merkley ED, Rysavy S, Kahraman A, Hafen RP, Daggett V, et al. (2014) Distance restraints from crosslinking mass spectrometry: mining a molecular dynamics simulation database to evaluate lysine-lysine distances. Protein Sci. 23 (6): 747–759.

61. Leitner A, Walzthoeni T, Kahraman A, Herzog F, Rinner O, et al. (2010) Probing native protein structures by chemical cross-linking, mass spectrometry, and bioinformatics. Mol Cell Proteomics 9 (8): 1634–1649.

62. Allen FH (2002) The Cambridge Structural Database: a quarter of a million crystal structures and rising. Acta Crystallogr., B 58 (Pt 3 Pt 1): 380–388.

63. Ries A, Göhring W, Fox JW, Timpl R, Sasaki T (2001) Recombinant domains of mouse nidogen-1 and their binding to basement membrane proteins and monoclonal antibodies. Eur J Biochem 268 (19): 5119–5128.

64. Butler JE (2000) Solid supports in enzyme-linked immunosorbent assay and other solid-phase immunoassays. Methods 22 (1): 4–23.

65. Aumailley M, Wiedemann H, Mann K, Timpl R (1989) Binding of nidogen and the laminin-nidogen complex to basement membrane collagen type IV. Eur J Biochem 184 (1): 241–248.

66. Guo X, Bandyopadhyay P, Schilling B, Young MM, Fujii N, et al. (2008) Partial acetylation of lysine residues improves intraprotein cross-linking. Anal Chem 80 (4): 951–960.

67. Bechtel M, Keller MV, Bloch W, Sasaki T, Boukamp P, et al. (2012) Different domains in nidogen-1 and nidogen-2 drive basement membrane formation in skin organotypic cocultures. FASEB J 26 (9): 3637–3648.

68. Sasaki T (2004) Laminin: the crux of basement membrane assembly. J Cell Biol 164 (7): 959–963.

69. Hohenester E, Yurchenco PD (2013) Laminins in basement membrane assembly. Cell Adh Migr 7 (1): 56–63.

70. Lössl P, Sinz A (2014) Combining amine-reactive cross-linkers and photo-reactive amino acids for 3D-structure analysis of proteins and protein complexes. Meth Mol Biol (submitted).

# Identification of Functionally Important Residues of the Rat P2X4 Receptor by Alanine Scanning Mutagenesis of the Dorsal Fin and Left Flipper Domains

**Vendula Tvrdonova**[1,2], **Milos B. Rokic**[1,3], **Stanko S. Stojilkovic**[3], **Hana Zemkova**[1]*

1 Department of Cellular and Molecular Neuroendocrinology, Institute of Physiology Acacemy of Sciences of the Czech Republic, Prague, Czech Republic, 2 Department of Physiology of Animals, Faculty of Science, Charles University, Prague, Czech Republic, 3 Section on Cellular Signaling, Program in Developmental Neuroscience, National Institute of Child Health and Human Development, National Institutes of Health, Bethesda, Maryland, United States of America

## Abstract

Crystallization of the zebrafish P2X4 receptor in both open and closed states revealed conformational differences in the ectodomain structures, including the dorsal fin and left flipper domains. Here, we focused on the role of these domains in receptor activation, responsiveness to orthosteric ATP analogue agonists, and desensitization. Alanine scanning mutagenesis of the R203-L214 (dorsal fin) and the D280-N293 (left flipper) sequences of the rat P2X4 receptor showed that ATP potency/efficacy was reduced in 15 out of 26 alanine mutants. The R203A, N204A, and N293A mutants were essentially non-functional, but receptor function was restored by ivermectin, an allosteric modulator. The I205A, T210A, L214A, P290A, G291A, and Y292A mutants exhibited significant changes in the responsiveness to orthosteric analog agonists 2-(methylthio)adenosine 5'-triphosphate, adenosine 5'-(γ-thio)triphosphate, 2'(3'-O-(4-benzoylbenzoyl)adenosine 5'-triphosphate, and α,β-methyleneadenosine 5'-triphosphate. In contrast, the responsiveness of L206A, N208A, D280A, T281A, R282A, and H286A mutants to analog agonists was comparable to that of the wild type receptor. Among these mutants, D280A, T281A, R282A, H286A, G291A, and Y292A also exhibited increased time-constant of the desensitizing current response. These experiments, together with homology modeling, indicate that residues located in the upper part of the dorsal fin and left flipper domains, relative to distance from the channel pore, contribute to the organization of the ATP binding pocket and to the initiation of signal transmission towards residues in the lower part of both domains. The R203 and N204 residues, deeply buried in the protein, may integrate the output signal from these two domains towards the gate. In addition, the left flipper residues predominantly account for the control of transition of channels from an open to a desensitized state.

**Editor:** Jon Brown, University of Exeter, United Kingdom

**Funding:** This work was supported by the Grant Agency of the Czech Republic (P304/12/G069), the Centrum of Biomedicine Research (CZ.1.07/2.3.00/30.0025), the "BIOCEV" project with the Biotechnology and Biomedicine Centre of the Academy of Sciences and Charles University in Vestec (CZ.1.05/1.1.00/02.0109) from the European Regional Development Fund, the Grant Agency of Charles University in Prague (3446/2011), the Academy of Sciences of the Czech Republic (Research Project No. RVO 67985823, and the Intramural Research Program of the NICHD, NIH. The funders had no role in study design, data collection and analysis, decision to publish, or preparation of the manuscript.

**Competing Interests:** The authors declare that no competing interests exist.

* Email: zemkova@biomed.cas.cz

## Introduction

The purinergic P2X receptors (P2XRs) are ATP-gated ion channels that are permeable to $Na^+$, $K^+$, $Ca^{2+}$, and small organic cations. Seven subunits of P2XRs have been identified in mammals [1], and functional receptors are composed of three homologous or heterologous subunits [2]. Each subunit consists of a large, glycosylated, and cystine-rich extracellular domain that contributes to the formation of the intersubunit ATP binding sites, two transmembrane domains that form the pore of the channel, and intracellular N- and C- termini that contribute to gating specificity [3]. Previous studies using single-point mutagenesis have identified most of conserved amino acid residues involved in ATP binding and have shown that ATP binding occurs at the interface between adjacent receptor subunits, assuming that ATP stabilizes the P2X trimer [4–10]. In contrast to the large number of studies

using the native ligand, ATP, there are very few studies providing structural information derived from the use of orthosteric ATP analog agonists. Understanding receptor interactions with these analog agonists may provide significant insights aiding the design of drugs that compete with the native ligand.

The recent crystallization of the zebrafish P2X4R receptor (zfP2X4R) in the absence (closed state; PDB entry codes: 3H9V and 4DW0) and presence (open state; 4DW1) of ATP has confirmed the predicted topology and locations of the ATP binding sites in P2XRs. The authors suggest that the architecture of the P2XRs resembles a dolphin, with a rigid central extracellular body domain, a flexible head, a left flipper (LF), a right flipper, and a dorsal fin (DF). The crystal structure of zfP2XR in the apo-closed state and the ATP-bound open state has also provided structural insights into the mechanisms of ATP binding, the opening of ion channel pore, and a series of conformational changes associated with channel gating [11,12]. These insights have enabled a better understanding of precrystallization studies focused on the structural-functional characterization of P2XR transmembrane domains [13–21] and facilitated further studies focused on extracellular vestibule function [22–24] and molecular dynamics to model conformation transitions [25].

Following ATP binding, the head, upper body, and LF domains of one subunit and the lower body and DF domains of another subunit undergo marked movement that results in the closing of the ATP binding site jaw [11]. During this movement, the LF and DF domains remain in close proximity (Fig. 1B). This promotes expansion of the upper vestibule, leading to the activation of P2XRs [26]. The P2X6R receptor lacks most of the LF domain (Fig. 1A) and is incapable of forming functional homomeric channels [27]. However, it can form functional heteromeric channels with P2X2 and P2X4 subunits [28,29], which may indicate that one or two complete LF domains per receptor are needed to activate the channel after ATP binding.

Crystallographic data also indicate that the DF and LF domains are intrinsically unfolded and lack secondary structures. These regions have significant conformational flexibility due to higher Debye-Waller factors (B-factors; Fig. 1B). This reflects the thermal fluctuation of atoms in zfP2X4R crystals, as assessed by X-ray scattering techniques, around their average positions and provide important information about protein dynamics [30]. Most importantly, the specific role(s) of non-structural and low-conserved DF and LF regions (Fig. 1A) is not well understood. In particular, we do not know the physiological relevance of having these domains positioned between an ATP binding site and the downstream K313-I333 β-sheet that has been previously identified in rat P2X4R (rP2X4R) as important for transmission of signal from the binding site to the channel gate [31].

We examined the hypothesis that the DF and LF domains may influence the organization of the ATP binding pocket, transmission of ATP-induced signal from ATP binding pocket to the gate, and receptor desensitization. To do this, we used 26 mutants generated by alanine scanning mutagenesis of the R203-L214 (DF) and D280-N293 (LF) sequences. These regions are highly variable between P2XRs, and only few of these residue mutants have been previously characterized electrophysiologically (Table 1). We expressed the wild type (WT) rP2X4R and alanine mutants in HEK293 cells and studied the current responses induced by the application and withdrawal of ATP or its analog agonists 2-(Methylthio)adenosine 5′-triphosphate (2-MeS-ATP), Adenosine 5′-(γ-thio)triphosphate (ATPγS), 2′(3′-O-(4-Benzoylbenzoyl)adenosine 5′-triphosphate (BzATP), and α,β-methyleneadenosine 5′-triphosphate (α,β-meATP), both in the presence and absence of ivermectin (IVM), an allosteric regulator of P2X4R [32–34].

## Methods

### Cells culture and transfection

To express the recombinant channels, we used human embryonic kidney (HEK) 293T cells (American Type Culture Collection, Rockville, MD, USA) grown in Dulbecco modified Eagle's medium (Thermo Fisher Scientific, Waltham, MA) supplemented with 10% fetal bovine serum (Sigma-Aldrich, St Louis, MO), 50 U/ml penicillin and 50 μg/ml streptomycin (both Thermo Fisher Scientific, Waltham, MA) in a humidified 5% $CO_2$ and 95% air at 37°C. Cells were cultured in 75 cm² plastic culture flasks (NUNC, Rochester, NY) for 36–72 hours until reaching 80–95% confluence. Before the day of transfection, the cells were plated on 35 mm culture dishes (Sarstedt, Newton, NC) and incubated at 37°C for at least 24 h. Transfection was done using 2 μg of either WT or mutant receptor DNA with 2 μl of JetPrime reagent in 2 ml of Dulbecco modified Eagle's medium, according to manufacturer's instructions (PolyPlus-transfection, Illkirch, France). Transfected cells were identified by the fluorescence signal of EGFP using the Olympus IX71 inverted research microscope with fluorescence illuminators (Model IX71; Olympus, Melville, NY).

### DNA constructs

cDNAs encoding the sequences of the rP2X4 and mutated subunits were subcloned into the pIRES2-EGFP vector (Clontech, Mountain View, CA, USA). To generate the mutants, oligonucleotides (synthesized and provided by VBC-Genomics, Vienna, Austria or Sigma Aldrich) containing specific mutagenesis mismatches were introduced into the rP2X4/pIRES2-EGFP template using PfU Ultra DNA polymerase (Thermo Fisher Scientific). A High-Speed Plasmid Mini Kit (Geneaid, Taipei City, Taiwan) was used to isolate the plasmids for transfection. Dye terminator cycle sequencing (ABI PRISM 3100, Applied Biosystems, Foster City, CA) was used to identify and verify the presence of the mutations. The sequencing was performed by the DNA Sequencing Laboratory, Institute of Microbiology, ASCR, Prague.

### Patch clamp recordings

ATP-induced currents were recorded from whole cells clamped to −60 mV using an Axopatch 200B patch-clamp amplifier (Axon Instruments, Union City, CA). The recordings were captured and stored using the Digidata 1322A and pClamp9 software package (Axon Instruments). During the experiments, the cell culture was perfused with a bath solution containing: 142 mM NaCl, 3 mM KCl, 2 mM $CaCl_2$, 1 mM $MgCl_2$, 10 mM 4-(2-Hydroxyethyl)piperazine-1-ethanesulfonic acid (HEPES) and 10 mM D-glucose, adjusted to pH 7.3 with 1 M NaOH. The patch electrodes were filled with a solution containing: 154 mM CsCl, 11 mM EGTA and 10 mM HEPES, adjusted to pH 7.2 with 1.6 M CsOH. The whole-cell configuration was used to abolish the influence of natively present metabotropic receptors for ATP and we used intracellular cesium to block any kind of possible background potassium conductance. Potency of ATP was measured based on the activation of naïve (not previously stimulated) receptors using a short (2–5 s) application of various concentrations of ATP. The results are expressed as molar concentration of ATP required to produce 50% of the maximal response ($EC_{50}$). One or two responses were recorded from one cell, if not otherwise stated, and responses from different cells were pooled. The maximum current amplitude ($I_{max}$) was measured in response to application of supramaximal concentrations (100–1000 μM) of ATP. Responsiveness to P2XR agonists 2-MeS-ATP, ATPγS, BzATP, and α,β-meATP, all applied in 100 μM concentration, was expressed as percentage of response in comparison to 100 μM ATP treatment for selected mutants. In all mutants, the whole cell currents were also measured in the presence of 3 μM IVM, which was dissolved in dimethyl sulfoxide, stored in stock solutions at 10 mM, and diluted to required concentrations in bath solution in the day of experiment. The control and drug containing solutions were applied via a rapid (exchange time 30–40 ms) perfusion system (RSC-200, BIOLOGIC, Claix, France). All other chemicals are from Sigma-Aldrich.

### Calculations

The concentration-response data points were fitted with the equation $y = I_{max}/[1 + (EC_{50}/x)^h]$, where $y$ is the amplitude of the current evoked by ATP, $I_{max}$ is the maximum current amplitude

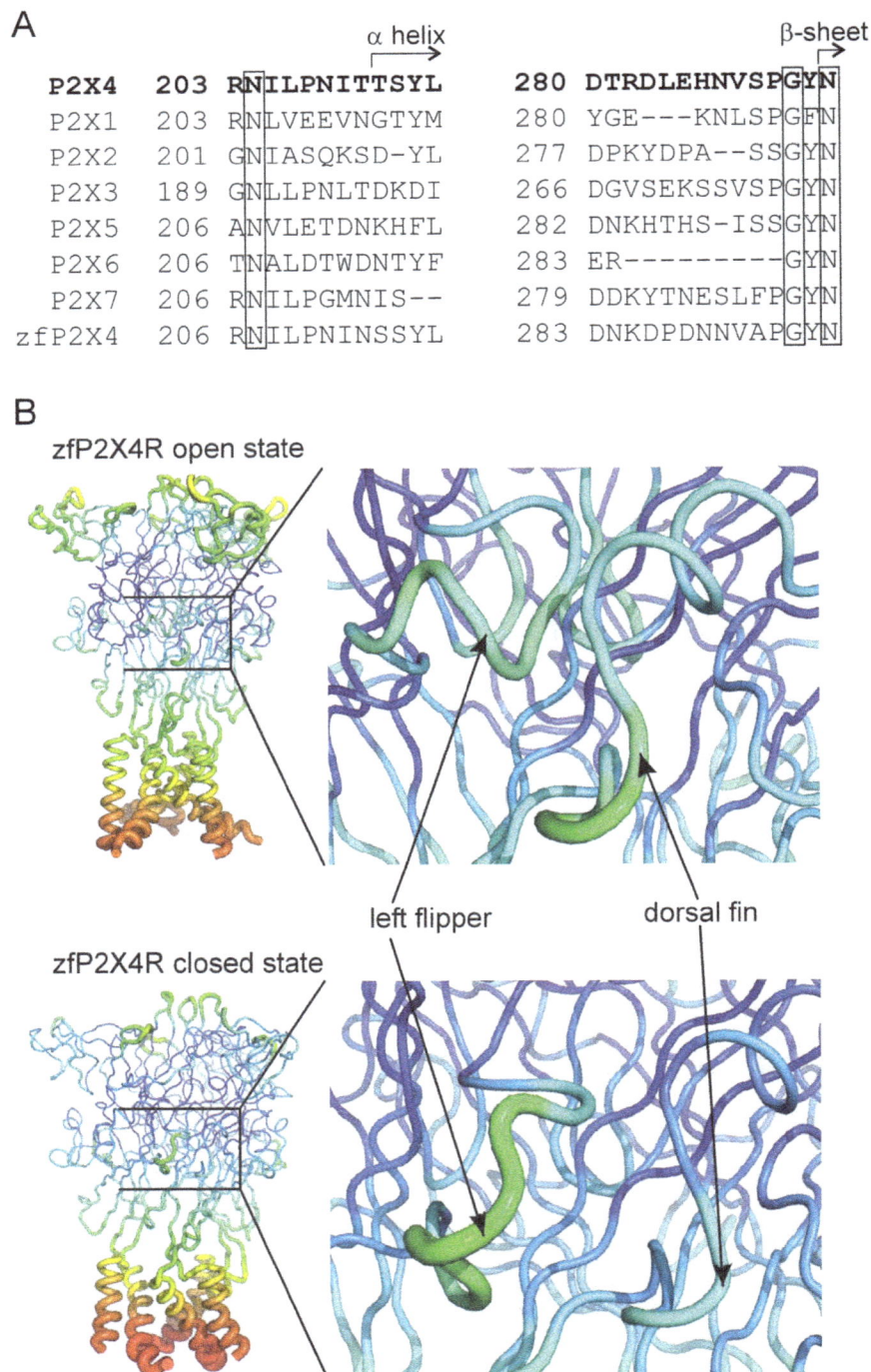

**Figure 1. Structural and tridimensional organization of the DF and LF domains.** (A) Alignment of amino acid sequences from R203-L214 (DF domain) and D280-N293 (LF domain) using seven rP2X and zfP2X subunits. Structurally, these regions are composed of random coils that terminate with short α-helix and β-sheet structures (indicated by arrows). Conserved amino acid residues are shown in boxes. (B) The models of the zfP2X4.1R are shown in the open (4DW1) or closed (4DW0) state, and the Debye-Waller factor (B-factor) indicates the degree to which the electron density is spread (miniatures). The model shows the elevated B-factor values within the region of intersubunit interaction (magnified segments). Higher B-factors are indicated with thicker cylinders and a red-shifted color, while the lowest B-factors are represented with the thinnest cylinders and a blue-shifted color.

induced by 100–1000 µM ATP, $EC_{50}$ is the agonist concentration producing 50% of the maximal response, $h$ is the Hill coefficient, and $x$ is the concentration of ATP (SigmaPlot 2000 v9.01; SPSS

Inc., Chicago, IL). Hill coefficient was fixed to 1.3 in all experiments, a value obtained for the WT receptor by fitting. The kinetics of deactivation (current decay evoked by washout of

**Table 1.** Summary of the changes in estimated $EC_{50}$ values for ATP and changes in desensitization at the DF and LF alanine/cysteine mutants of P2X1-4R residues from published data.

| P2X1R | P2X2R | P2X3R | P2X4R |
|---|---|---|---|
| *DF* | | | |
| N204A: ↑ 3,5x[45] | N202A: n.f.[6] | - | N204A: n.i. |
| - | - | - | I205A: ↑ 2,5x[49] |
| - | - | - | L206A: ≈[49] |
| - | - | T196A: ≈[54] | T210A: n.i. |
| - | - | E197A: D[53] | T211A: n.i. |
| - | - | M200C: ↑ 8x[58] | L214A: ↑ 8x[41], 3x[49] |
| *LF* | | | |
| - | D277A:: ≈[56] | D266A: D[53] | D280A ↑ >100x[10,52] |
| - | - | S267A: ≈[59] | T281A: n.i. |
| - | - | S269A: D[54] | D283A: n.i. |
| - | - | - | H286A: ↑ 2–4x[10,52,57,60] |
| - | - | - | N287A: ≈[10,52] |
| - | - | - | V288A: ↑ 2x[49] |
| - | - | S275A: D[55] | S289A: n.i. |
| P287C: ≈[9] | - | - | P290A: n.i. |
| G288A/C: ≈[47], ↑ 10x[9] | - | - | G291A: n.i. |
| N290A: ↑ >50x[45] | N288A: ↑ >100x[6] | N279A: ↑ 20x[4] | N293A: n.i. |

n. f., non-functional mutants; ↑ mutant with significant increased $EC_{50}$ in comparison to WT (values represent fold increase); ≈ close to WT receptor; D, affected time-constant of the desensitizing current response; -, non-investigated position; n.i., non-investigated P2X4R mutants that were analyzed in this study.

agonist) and desensitization (current decay in the continuous presence of agonist) were fitted by a single exponential function ($y = A_1 \exp(-t/\tau_1)$) or by the sum of two exponentials ($y = A_1 \exp(-t/\tau_2) + A_2 \exp(-t/\tau_2)$), respectively, using the program Clampfit 10 (Axon Instruments), where $A_1$ and $A_2$ are the relative amplitudes of the first and second exponentials, and $\tau_1$ and $\tau_2$ are the time constants. The derived time constants for deactivation and desensitization were labeled as $\tau_{off}$ and $\tau_{des}$, respectively. Weight desensitization constant was calculated as $y = [(A_1\tau_{des1}) + (A_2\tau_{des2})]/(A_1 + A_2)$. Correlation coefficient was calculated using linear regression wizard (SigmaPlot 2000 v9.01). Data points are presented as mean ± SEM values. Significant differences (**$p < 0.01$ and *$p < 0.05$) between means were determined using SigmaStat 2000 v9.01. The data for alanine mutants were analyzed by an ANOVA and Tukey's post hoc test.

## Homology modeling

The rP2X4R (P51577) and zfP2X4.1R (Q6NYR1) share 61% identity at the amino-acid level, measured with the Basic Local Alignment Search Tool (BLAST; The UniProt Knowledgebase), a value sufficient to build a homology model of the rP2X4R using the automated mode of the SWISS-MODEL server [35]. We extracted a tertiary structure template from the Brookhaven Protein Data Bank under the accession number 4DW0 for the receptor in the apo-closed state and 4DW1 for the zfP2X4.1R in the ATP-bound open state. Model quality was estimated by a SWISSMODEL through a Qualitative Model Energy Analysis (QMEAN) score, which represents a composite scoring function describing the major geometrical aspects of protein structures by taking into consideration five different structural descriptors [36], and which was 0.593. The graphical representations of the protein

structure were prepared using PyMOL software (DeLano Scientific LLC, USA).

## Results

### 1. Identification of DF and LF mutants with affected ATP potency and efficacy

To address the structure-function relationship between the LF and DF regions of rP2X4R, we performed single-point mutagenesis on sequences encompassing the LF and DF regions R203-L214 (DF) and D280-N293 (LF) (Fig. 1A). The crystal structure of the zfP2X4R showed elevated B-factor values in these regions, indicating conformational flexibility (Fig. 1B). For the initial electrophysiological characterization, we examined the $EC_{50}$ and $I_{max}$ values to determine ATP potency and efficacy, respectively. The results from experiments on both mutant and WT receptors are summarized in Figs. 2A and 3 and Table S1.

There were no significant effects on the $EC_{50}$ or $I_{max}$ values for P207A, I209A, T211A, S212A, Y213A, D283A, L284A, E285A, N287A, V288A, and S289A mutant receptors. The $EC_{50}$ values for mutants R203A, N204A, and N293A could not be determined because they displayed a very low ATP-induced current ($I_{max} \leq$ 0.2 nA). Mutants I205A, L214A, D280A, R282A, and P290A showed a significant reduction ($p < 0.01$) in $I_{max}$, and with exception of P290A, their $EC_{50}$ values were approximately 10-fold rightward shifted when compared to the WT receptor. The $EC_{50}$ values for T210A, H286A, and G291A mutants were 6- to 10-fold rightward shifted but these receptors showed no significant difference in $I_{max}$ values when compared to the WT receptor (Fig. 3). Slightly less significant increases ($p < 0.05$) in $EC_{50}$ values were observed in the L206A, N208A, T281A, P290A, and Y292A mutants. With the exception of P290A, none of these mutants

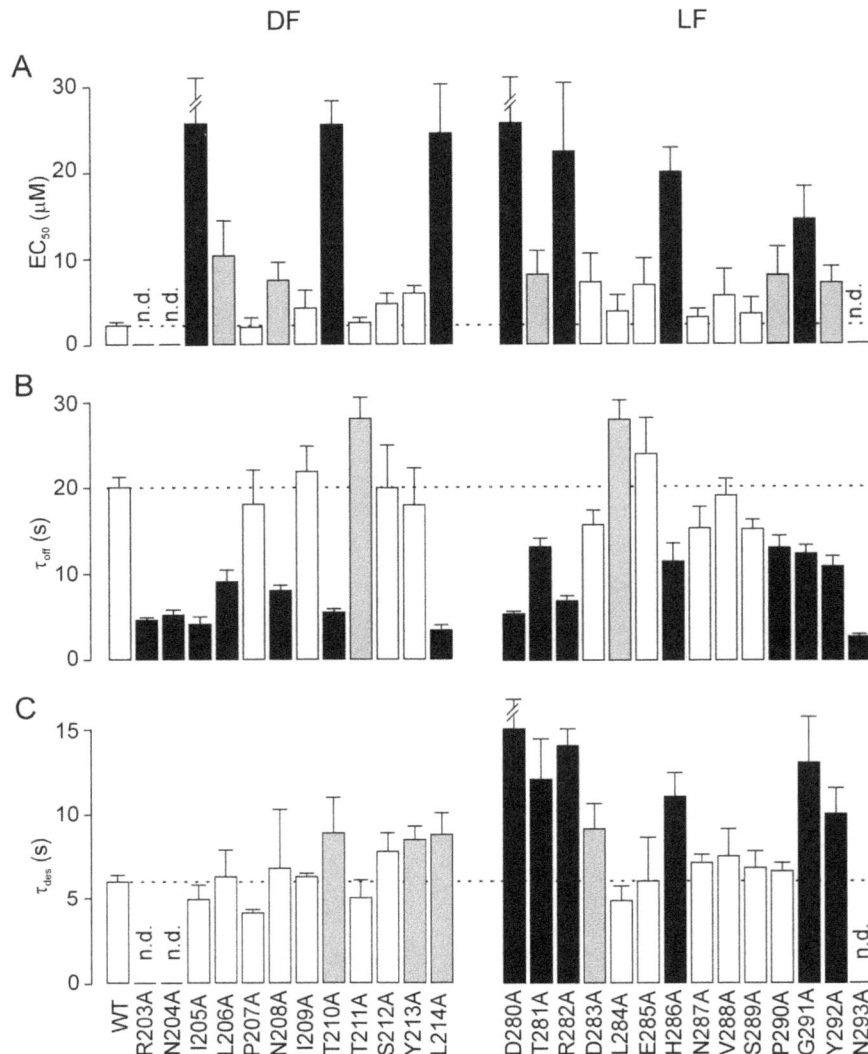

**Figure 2. Characterization of rP2X4R–DF and -LF residue mutants.** Effect of alanine substitutions on the potency of ATP (A), deactivation (B), and desensitization (C) kinetics. Summary histograms show the concentration of ATP producing a half-maximal current (EC$_{50}$), deactivation time constants ($\tau_{off}$) were estimated by the monoexponential fit of the decay of current in response to 2 s of stimulation with 1–3 μM ATP after 4–6 min of preincubation with 3 μM IVM and desensitization time constants ($\tau_{des}$) were derived from the biexponential fit of the response to 60 s of stimulation with 100 μM ATP for WT and alanine mutants of the dorsal fin (DF) and the left flipper (LF) domains. Values shown (and also given in Table S1) are the mean ± SEM of 21–63 measurements per mutant and 267 measurements for the WT. Significant differences between the WT and the mutant receptors are shown in gray (p<0.05) or black (p<0.01) columns. Horizontal dotted lines illustrate the values for WT receptor and n.d. indicates that the value could not be determined.

exhibited significant changes in their I$_{max}$ value. Thus, in 15 of 26 alanine mutants located at the interface of the LF and DF domains the potency and/or efficacy of ATP was significantly reduced, indicating the relevance of these residues in receptor functions.

## 2. IVM rescues the low-functioning DF- and LF-rP2X4R mutants

The effect of IVM on I$_{max}$ was tested initially during ongoing responses to 100 μM ATP to determine whether the low current amplitudes observed in R203A, N204A, I205A, L214A, D280A, R282A, P290A, and N293A mutants could be rescued. The application of 3 μM IVM increased immediately the amplitude of ATP-induced responses in all low-functioning mutants (Fig. 4A). Next, we performed quantitative analysis of I$_{max}$ in WT and all

alanine mutants before and after 4–6 min pretreatment with IVM (Table S1). The WT receptor was potentiated 1.5-fold by IVM, while the low-functioning mutants were potentiated 3.7- to 16-fold. In the presence of IVM, the I$_{max}$ values of all low-functioning mutants were comparable to those of WT receptors, except for N293A (Fig. 4B). These experiments indicate that R203, N204, I205, L214, D280, R282, R290, and N293 residues play a critical role in agonist binding and/or channel gating.

The IVM-induced rescue of I$_{max}$ values made it possible to examine the deactivation time constant ($\tau_{off}$) for these mutants, which inversely correlates with EC$_{50}$ values [37,38]. As a result, we were able to more precisely characterize the potency of ATP under comparable conditions. A prolongation of current decay comparable to that observed in WT receptor would suggest that normal ATP potency has been maintained. Alternatively, a

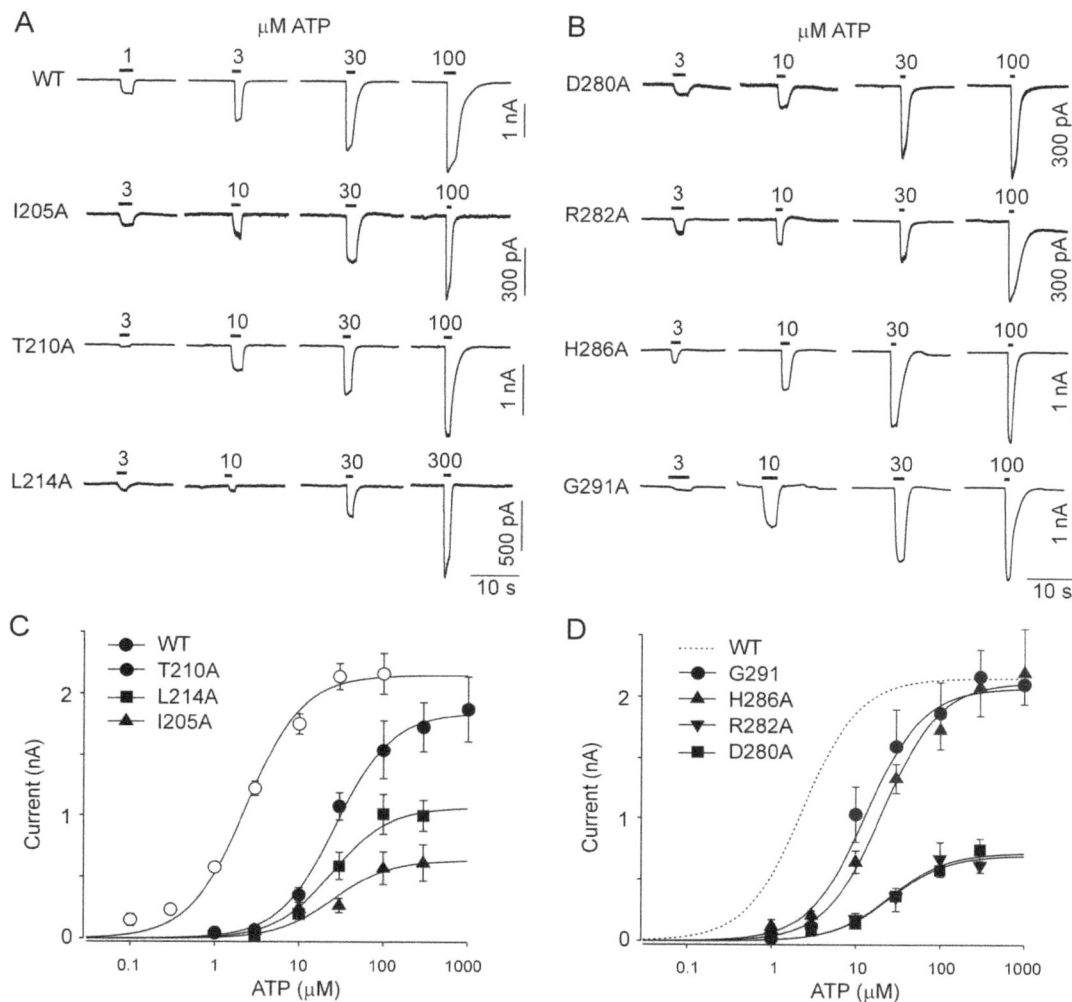

**Figure 3. DF and LF mutants exhibit a rightward shift in $EC_{50}$.** (A, B) Example records of ATP-induced currents from cells expressing the WT receptor and I205A, T210A, and L214A DF mutants (A) and D280A, R282A, H286A, and G291A LF mutants (B). Currents were stimulated by a short (2–5 s) application of different concentrations of ATP (1–1000 μM), indicated by horizontal bars above the traces. Experiments were performed on naïve receptors, and traces from different cells are shown. (C, D) Concentration response curves for WT, I205A, T210A, and L214A DF mutants (C) and D280A, R282A, H286A, and G291A LF mutants (D). Data points are presented as the mean ± SEM from 7–35 measurements per mutant, per concentration and 78 measurements for WT.

decrease or increase in the rate of decay would argue for reduced or enhanced ATP potency, respectively [10]. The deactivation time constant was examined in all alanine mutants by mono-exponential fitting of the decay of current after washout of a non-desensitizing concentration of agonist (1 or 3 μM ATP) in the presence of 3 μM IVM. Example traces from WT and mutant receptors with changed deactivations are shown in Fig. S1. The results of $\tau_{off}$ measurements are summarized in Fig. 2B and Table S1.

In parallel with the rightward shift changes in $EC_{50}$ values, we observed a significantly ($p < 0.01$) accelerated rate of deactivation in non-responding mutants (R203A, N204A, and N293A) and all rightward shifted mutants (I205A, L206A, N208A, T210A, L214A, D280A, T281A, R282A, H286A, P290A, G291A, and Y292A). Less significant ($p < 0.05$) prolonged deactivation times were found in the L284A and T211A mutants. There was a significant correlation, with highly comparable slopes, between the $EC_{50}$ vs. $\tau_{off}$ values for both DF and LF mutants (Fig. 5A). These data confirmed that residues in both domains contribute

significantly to receptor activation as well as to receptor deactivation, i.e., that deactivation is a reverse process occurring through the same signal transmission pathway.

## 3. Desensitization kinetics of LF- and DF-rP2X4R mutants

Next, we determined the desensitization kinetics of alanine mutants at the interface of the LF and DF domains. In the presence of 100 μM ATP for 60 s, the WT receptor current declined biexponentially with $\tau_{des1} = 1.3 \pm 0.2$ s and $\tau_{des2} = 9.0 \pm 0.7$; the slow component contributed to the decay with $63 \pm 3.0\%$ (Table S2; Fig. S1). The decay of current was also biexponential in all mutants, but in some cases monoexponential fit was the best, and we used a weighted desensitization time constant ($\tau_{des}$) for comparison between the mutants and the P2X4R–WT (WT, $\tau_{des} = 6.0 \pm 0.4$ s; Table S1 and Fig. 2C). The LF mutants D280A, T281A, R282A, D283A, H286A, G291A, and Y292A exhibited 1.5- to 2.6-fold slower desensitization kinetics when compared to the WT receptor. Less significantly (1.3- to 1.5-fold; $p < 0.05$) prolonged $\tau_{des}$ were observed in DF

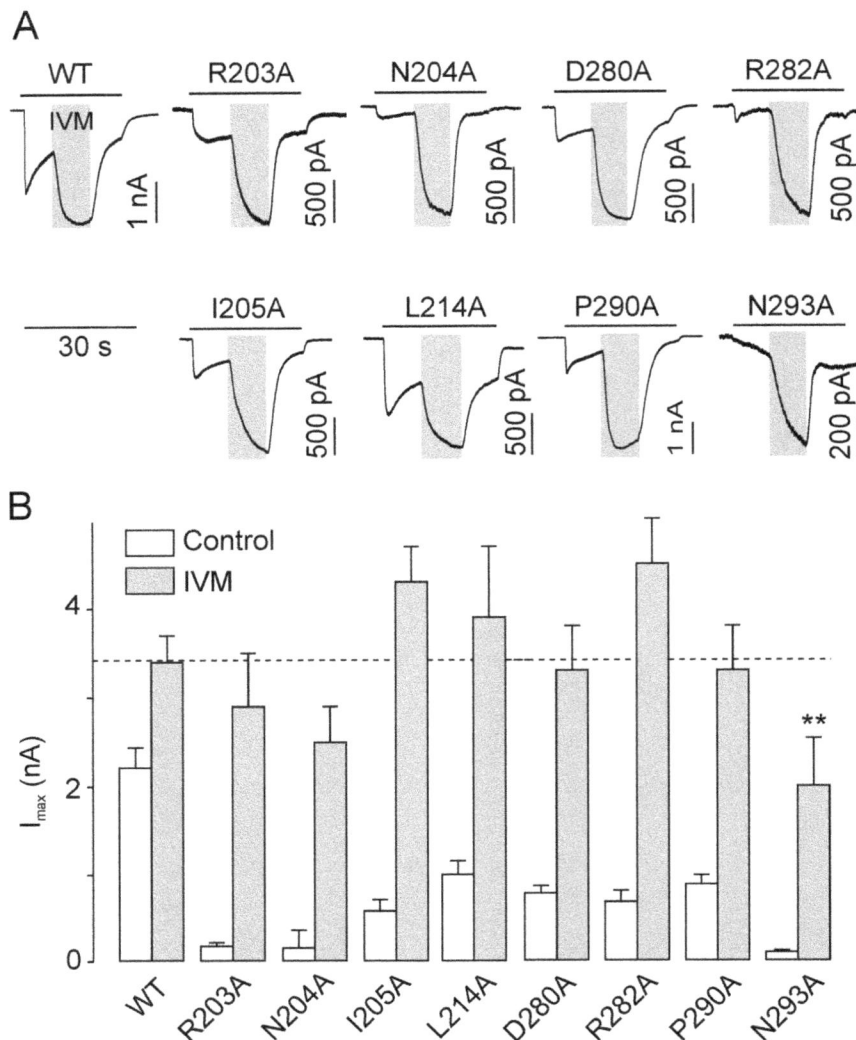

**Figure 4. Ivermectin rescues the $I_{max}$ of low-functioning mutants.** (A) Acute effect of 3 µM ivermectin (IVM) applied for 10 s (gray areas) during ongoing stimulation with 100 µM ATP for 30 s (horizontal bars) in cells expressing the WT, the DF mutants (R203A, N204A, I205A, and L214A), or the LF mutants (D280A, R282A, P290A and N293A). Recordings are examples of traces similar to 3–5 traces per mutant and 30 per WT receptor. (B) Summary data showing the potentiating effect of IVM preapplication (for 4–6 min) on $I_{max}$ in WT and alanine mutant receptors. The $I_{max}$ values were derived from measurements taken in the absence (open bars) or in the presence (filled bars) of IVM. Values are presented as the mean ± SEM from 5–8 measurements per mutant and 15 measurements per WT. IVM treatment rescued the $I_{max}$ of all low-functioning receptors, except in the case of N293A, which is an ATP binding mutant. The statistical significance was determined by an ANOVA comparing the WT $I_{max}$ and the $I_{max}$ of mutant receptors in the presence of IVM. **, $p < 0.01$.

mutants T210A, Y213A, and L214A. The remaining mutants (I205A, L206A, P207A, N208A, I209A, T211A, S212A, L284A, E285A, N287A, V288A, S289A, and P290A) displayed no changes in the desensitization rate. Plotting $\tau_{des}$ versus $EC_{50}$ (Fig. 5B) revealed a significant correlation for LF, but not for DF mutants. These results indicate that clusters of residues rather than individual amino acids, are responsible for the desensitization rate of P2X4R and that the LF domain plays a dominant role in this process.

## 4. The influence of the LF and DF domains of rP2X4R on agonist selectivity

To examine whether mutations in the LF and DF domains alter the responsiveness to orthosteric ligands, we compared the efficacy of ATP with four partial agonists for P2X4R. In WT receptor,

100 µM was maximal concentration for all analogue agonists, except α,β-meATP (2-MeS-ATP, $EC_{50} = 7.9 \pm 1.0$ µM; ATPγS, $EC_{50} = 8.4 \pm 1.8$ µM; BzATP, $EC_{50} = 11.1 \pm 2.9$ µM; α,β-meATP, $EC_{50} = 62 \pm 18$ µM; Fig. S2A, *upper panel*) and the agonist efficacy profile of the WT receptor was ATP (100%) >2-MeS-ATP (67%) > ATPγS (50%) > BzATP (38%) > α,β-meATP (32%). We examined all of the functional mutants that displayed significant changes in ATP potency and/or deactivation kinetics (I205A, L206A, N208A, T210A, T211A, L214A, D280A, T281A, R282A, L284A, H286A, P290A, G291A, and Y292A) and two substitution-insensitive mutants (S212A, Y213A) (Table 2).

The mutants could be divided into one of three groups. The *Group I* was composed of mutants with changes in ATP potency/ efficacy and deactivation kinetics that also exhibited a significant decrease in the relative responsiveness to some or all of the

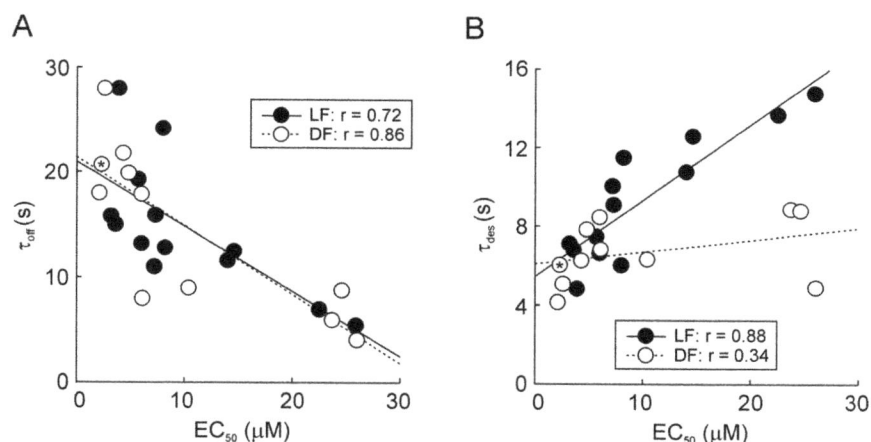

**Figure 5. Deactivation and desensitization properties depend on the potency of ATP.** (A, B) Correlation between $EC_{50}$ and the deactivation time constant $\tau_{off}$ (A) and $EC_{50}$ and desensitization time constant $\tau_{des}$ (B) for alanine DF and LF mutants. DF mutants are shown as open circles and LF mutants as closed circles. WT receptors are shown as an asterisk in an open circle. Values are derived from Table S1. Correlation analysis was performed as described in the Methods.

orthosteric agonists applied in 100 μM concentration (Table 2, Fig. S2B). This group includes I205A, T210A, L214A, P290A, G291A, and Y292A mutants. The I205A and L214A mutants displayed a significantly reduced responsiveness to all agonists when compared to the WT receptor, and the profile (2-MeS-ATP > ATPγS > BzATP ≥ α,β-meATP) was preserved (Table 2). This suggests that these residues also play a role in the mechanism by which an ATP-induced signal is coupled to the channel gating. In contrast, there were changes in the agonist profile for the T210A, P290A, G291A, and Y292A mutants. For example, the

**Table 2.** The relative responsiveness of the wild type (WT) and selected rP2X4R mutants to P2XR agonist analogs.

| Receptor | Domain | 2-MeS-ATP | ATPγS | BzATP | α,β-meATP |
|----------|--------|-----------|-------|-------|-----------|
| **WT** | | 67.3±4.6 | 49.6±3.5 | 38.2±3.4 | 31.8±3.3 |
| *Group I* | | | | | |
| **I205A** | DF | 35.6±1.9** | 11.3±2.2** | 4.4±2.6** | 4.6±1.0** |
| **T210A** | DF | 59.2±6.9 | 34.2±4.1* | 23.9±3.2** | 8.8±2.3** |
| **L214A** | DF | 21.4±1.7** | 7.9±2.0** | 5.9±0.5** | 6.5±1.5** |
| **P290A** | LF | 50.7±3.7 | 37.0±11.0 | 16.9±3.2** | 17.8±2.4** |
| **G291A** | LF | 36.9±8.8** | 31.6±6.1* | 53.0±4.0** | 2.8±0.5** |
| **Y292A** | LF | 49.8±8.6 | 42.3±7.0 | 28.9±8.2 | 16.2±4.4** |
| *Group II* | | | | | |
| **L206A** | DF | 68.3±6.5 | 39.6±8.4 | 33.7±8.0 | 27.5±2.5 |
| **N208A** | DF | 60.0±11.2 | 42.2±5.2 | 33.4±5.4 | 28.5±2.3 |
| **D280A** | LF | 56.3±7.8 | 55.7±3.8 | 25.4±1.5 | 33.1±6.0 |
| **T281A** | LF | 69.6±5.1 | 50.2±5.9 | 26.7±3.9 | 28.2±5.5 |
| **R282A** | LF | 55.1±12.9 | 45.5±5.2 | 25.7±3.7 | 27.3±7.9 |
| **H286A** | LF | 67.9±1.2 | 39.8±6.2 | 25.6±4.0 | 30.0±6.8 |
| *Group III* | | | | | |
| **S212A** | DF | 65.1±19.7 | 61.7±3.9 | 47.5±5.2 | 19.2±6.2 |
| **Y213A** | DF | 63.8±16.4 | 37.1±5.5 | 27.3±5.5 | 31.3±6.7 |
| **T211A** | DF | 79.3±6.7 | 61.8±6.1 | 23.4±4.7* | 39.1±7.2 |
| **L284A** | LF | 69.0±3.0 | 32.4±8.2 | 20.6±0.2** | 22.2±3.9 |

ATP and agonists were applied in 100 μM concentrations for 2 s with a washing interval of 60 s. The data are the mean ± SEM, relative to ATP efficacy (100%), from 26 to 37 measurements for the WT receptor and from 3 to 18 measurements per mutant. *Group I:* Mutants that exhibited changes in ATP potency/efficacy, deactivation kinetics, and changes in the relative responsiveness to orthosteric analog agonists. *Group II:* Mutants that exhibited changes in ATP potency/efficacy and deactivation, but no changes in the relative responsiveness to analog agonists. *Group III:* Mutants that showed no significant changes in ATP potency/efficacy. The statistical significance was determined by an ANOVA comparing the responsiveness to agonists between WT and mutant receptors: **, p<0.01, *, p<0.05. DF, Dorsal Fin; LF, Left Flipper.

profile for the G291A mutant was BzATP > 2-MeS-ATP > ATPγS > α,β-meATP (Fig. S2A, *lower panel*), indicating changes in the folding of the jaw for ATP.

The *Group II* of mutants showed changes in ATP potency/efficacy and deactivation, but not in the relative responsiveness to orthosteric agonists to induce current. This includes the L206A, N208A, D280A, T281A, R282A, and H286A mutants (Table 2). The four members of *Group III*, T211A, S212A, Y213A, and L284A, showed no significant changes in ATP potency/efficacy or gating, and among them, only T211A and L284A displayed slightly (p<0.05) prolonged deactivation times (Fig. 2B). These two mutants also showed a significant decrease in the responsiveness to BzATP but no decrease in the responsiveness to 2-MeS-ATP, ATPγS, and α,β-meATP (Table 2). Therefore, residues T211 and L284 were not considered as residues of interest.

## 5. Model prediction for the positions of residues of interest in the DF and LF domains

We developed the rP2X4R homology model, as described in Materials and Methods, to identify the position of residues in the DF and LF domains in the ATP-bound open state. The data presented in Table 2 are also summarized in Fig. 6 as a receptor structure view. Native residues from the *Group I* mutants are located close to the ATP molecule, near the N293 residue. All of the residues from the *Group II* mutants are located downstream of the ATP binding domain and near the R203 and N204 residues that are burrowed in the protein. This topology suggests that the I205, T210, L214, P290, G291 and Y292 (green spheres) contribute to the organization of the structure of the ATP binding pocket and therefore dictate the specificity of responsiveness to the synthetic orthosteric ligands as well as the transmission of the conformational change induced by ATP binding. Residues L206, N208, D280, T281, R282, and H286 (red spheres) are important for transmitting the signal from the ATP binding cleft. Residues of mutants that have shown significant gating impairment (R203, N204 and N293) are shown as gray spheres.

The N293 residue is in close proximity (less than 5 angstroms) to ATP and, together with Y292, may directly interact with the β-

sheet segment of the K313-I333 sequence, which is responsible for the signal transmission from the ATP binding site to the pore [31] (Fig. S3A). The R203 and N204 residues are situated at the bottom of the ATP binding site (Fig. S3B) and play a crucial role in the transmission of signals towards the pore. The model also indicates that residues D283, H286, V288, and S289, from the adjacent subunit, are in the proximity of R203, while the K190, N191, N204, I205, and L206 residues are in close proximity within the same subunit. Close residues for N204 (I205, L206, and Y274) are also located within the same subunit (Fig. S3B). This suggests that the R203 and N204 residues may integrate the output signal from two neighboring subunits towards the gate and their mutants may display radical conformational misfolding of the DF and LF domains.

## Discussion

The interface between the DF and LF domains, formed by sequences R203-L214 and D280-N293 in rP2X4R, is one of the most variable parts of the P2XRs [12]. Alignments of these regions for seven rat P2X subunits indicate that only three of 26 amino acids are fully conserved (N204, G291, and N293). Four hydrophobic residues, at positions I205, L206, L214, and V288, are partially conserved, and the residual amino acids of this interface are variable (Fig. 1A). Such variability in the structure of the DF and LF domains could indicate that they are not essential for receptor function or that they contribute to receptor subtype specificity in terms of agonist binding and/or gating, and desensitization.

In this study, the physiological relevance of the residues that comprise the DF and LF domains of rP2X4R was systematically analyzed for the first time by substituting each residue with an alanine. This approach enabled us to eliminate interactions between the side chains and to study the effects of that elimination on rP2X4R structure and activity. Alanine was also used as a substituent because its polarity is in the middle of the polarity scale [39] when compared to other residues. Moreover, alanine

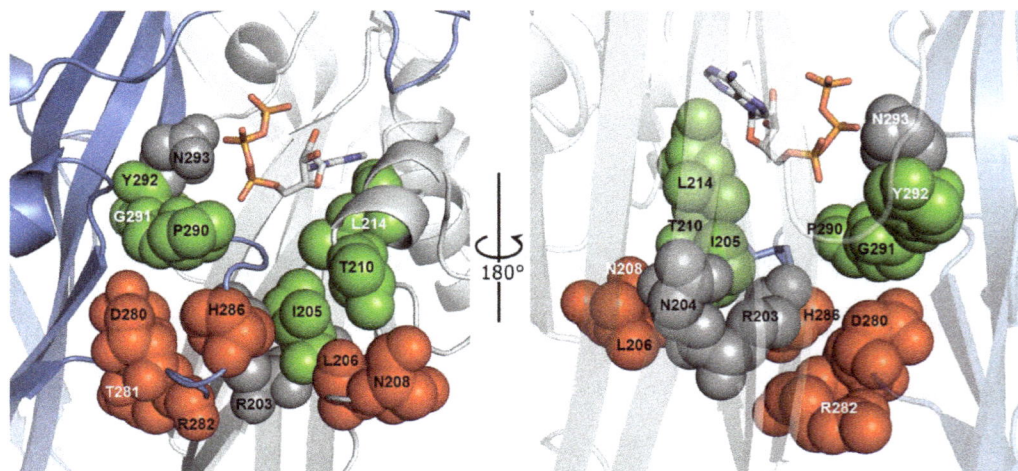

**Figure 6. The structure of the ATP binding site in a rP2X4R homology model.** Two panels show the position of affected residues (rotated 180°) at the interface between the LF and DF domains. The low-response residues without defined EC$_{50}$ values are grey spheres. The amino acid residues presented as green spheres demonstrate the topology of mutants with changes in ATP potency and/or efficacy (EC$_{50}$ and I$_{max}$), and agonist profile (*Group I* from Table 2). The amino acids presented in red spheres illustrate the position of residues whose mutation has affected ATP potency and/or efficacy without changing the action of ATP analogs (*Group II* from Table 2). In both panels, the ATP molecule is situated between two adjacent P2X4R subunits (blue and gray). The ATP molecule is shown in a wireframe model.

scanning mutagenesis has been widely used in research on P2XRs [6,18,23,40].

Electrophysiological and pharmacological characterization of the mutants revealed that substitution of 15 of 26 residues in the R203-L214 and D280-N293 sequences significantly attenuated the receptor function: R203A, N208A, T210A, T281A, R282A, P290A, G291A and Y292A, not previously studied, and N204A, I205A, L206A, L214A, D280A, H286A and N293A, previously studied across different P2XR subtypes (for overview see Table 1). However, the receptor function for all identified mutants, including almost non-functional mutants R203A, N204A and N293A, was rescued by the addition of IVM, an allosteric agonist of P2X4R [41].

In general, IVM allosterically potentiates the $I_{max}$ of P2X4R, causes a leftward shift in the ATP concentration response curve and significantly prolongs deactivation [32,33]. Single channel analysis showed that IVM increases the probability of channel opening [34]. We have recently found that IVM induces dilation of the pore of the P2X4R ion channel and that the IVM-dependent transition from open to dilated state temporally coincides with receptor sensitization, which rescues the receptor from desensitization and subsequent internalization [42]. This suggests that the observed increase in the number of cell surface P2X4Rs after 2–30 min of preincubation with IVM [43] is not due to insertion of new receptors to the plasma membrane, but rather reflects its influence on channel pore dilation. The use of P2X4-pHluorin123 also revealed that IVM does not acutely increase the fraction of P2X4Rs in the plasma membrane [44]. Therefore, it is reasonable to conclude that the trafficking of mutant receptors is not affected, i.e., they are expressed at the plasma membrane, but ATP has reduced binding affinity and/or potency to activate them.

In two of three nonfunctional mutants, N204A and N293A, alanine substitutes conserved asparagine residue. The N293 amino acid was previously identified in the crystal structure of zfP2X4R as an ATP binding residue. This residue is a part of the NFR motif, which is important for the recognition of the triphosphate moiety of ATP [11]. In our experiments, the $I_{max}$ of N293A could not be fully rescued by IVM, similar to previous observations of two other P2X4R ATP binding mutants, K67A and R295A [10]. A decrease in agonist potency was also observed in the corresponding N293A mutants of other receptors, including P2X1R-N290A [45], P2X2R-N288A [6], and P2X3R-N279A [4]. These studies further support the importance of N293 residue in the formation of the ATP binding pocket. In further agreement with our data, mutation of the conserved N204 residue is nonfunctional in P2X1R [6] and causes a 3-fold decrease in ATP potency in P2X2R [45]. An arginine in the position equivalent to 203 is present in P2X1R and P2X7R, and a lysine substitution of residue R206 enhances the sensitivity of P2X7R to activation by ATP [46].

The 12 mutants were functional but exhibited significant changes in the $EC_{50}$, $I_{max}$, $\tau_{off}$, and/or $\tau_{des}$ values. These mutants were divided into two groups based on the relative responsiveness to stimulation with ATP and four P2XR agonist analogs. Group I is composed of mutants with altered relative response to the agonist analogs and includes the I205A, T210A, L214A, P290A, G291A, and Y292A mutants. In contrast, Group II mutants showed no change in the responsiveness to analogs and includes the L206A, N208A, D280A, T281A, R282A, and H286A mutants (Table 2). The model prediction for the positions of these residues supports the conclusions that Group I residues contribute to the formation of the large ATP binding pocket in addition to signal transmission, while Group II residues contribute to signal transmission only. Therefore, both the DF and LF domain residues participate significantly in receptor function.

In our receptor model, in close proximity to the bound ATP molecule and asparagine 293, are the Y292, G291, and P290 residues. The substitution of these residues with alanine altered the ATP potency and/or efficacy, deactivation kinetics, and agonist profile. Among them, the most affected was the G291A mutant that exhibited an approximately 6-fold rightward shifted $EC_{50}$ value and large changes in agonist selectivity profile. Glycine 291 is conserved in all rat P2XRs (Fig. 1A), and the corresponding cysteine mutant of P2X1R showed a 10-fold decrease in ATP potency [9], while the alanine mutant had little effect [47]. However, its role in ligand selectivity and ATP potency in other P2XR subtypes remains to be determined. These data, combined with the topology of residues in the receptor model, suggest that the N293-P290 sequence forms part of a wall in the ATP binding cleft and contributes to signal transmission through downstream LF domain residues (Fig. 6). The model predicts that this segment will also act on the transmission of signals to the gate, possibly by interactions with the K313-I333 β-sheet (Fig. S3A). Consistent with this hypothesis, mutagenesis of the Y315 and G316 residues significantly affects receptor function [31].

The partially conserved hydrophobic residue L214 has also been implicated in the recognition of the ATP ribose ring [11,48], which is fully consistent with our data. We observed that the L214A mutant displays full recovery of $I_{max}$ in the presence of IVM, and has reduced responsiveness to all orthosteric agonists. However, the agonist profile of the WT receptor was preserved. The DF mutants I205A and T210A also exhibited a reduced potency/efficacy for ATP and its analogs. Topologically, these native residues may account for the bottom part of the ATP binding pocket (Fig. 6). A recent study on hydrophobic interactions between the LF and DF domains during receptor activation has identified several non-polar residues, including L214 and I205, that are important for the coordinated relative movements of these domains after ATP binding [49]. Therefore, we suggest that L214 residue plays a dual role in receptor functions: agonist binding and signal transmission.

The topology of T210 in zfP2X4R revealed that the residue is situated nearby the α-helix containing L214 involved in ATP recognition [11], (Fig. 1A). We observed changes in agonist profile for T210A mutant, suggesting that T210 could contribute to coordination of agonist position in the binding cleft. This explanation needs assumption that the T210 side chain position is variable and might be oriented towards the binding pocket, similarly as L214, and that orientation of ATP is different from that predicted by crystal, suggesting the existence of several ATP binding modes [50,51]. Further experiments are needed to explain the role of T210 in receptor function.

The homology model of rP2X4R predicts that the Group II amino acids are clustered into two subgroups: one composed of D280-H286 LF residues and the other composed of L206-N208 DF residues. The position of these residues is consistent with their roles in signal transmission. Fig. 6 suggests that the influence of ATP binding on gating is transmitted downstream through two signal transmission lines. The first is composed of N293, Y292, G291, and P290 towards D280, T281, R282, and H286 (from top to bottom) in the LF domain. The other unit appears to be composed of L214, T210, I205, N204, L206, and N208 (from top to bottom) in the DF domain. The model also suggests that R203 and N204 are positioned to accept the signal from the binding domain through both lines and from two neighboring subunits, and to integrate it towards the gate region.

Finally, seven out of the 13 LF mutants tested showed significantly slower rates of receptor desensitization and our correlation analysis of the relationship between $EC_{50}$ vs. $\tau_{des}$ suggests that that the LF domain plays the major role in the transition from the open to the desensitized state, with signal transmission through the N293-D280 sequence (Fig. 2C and 5B). Alanine substitution of the corresponding positions D266A [53], S269A [54], but also S275A [55], prolongs desensitization of P2X3R, but the P2X2R-D277A mutant was normal [56]. These data indicate that this group of polar and charged residues might play a receptor-specific role in desensitization.

In conclusion, we have shown that the interface between DF and LF domains has dual roles in rP2X4R function. One role is the formation of the ligand-binding pocket and the other is for the transmission of signals from the pocket toward the gate. Both domains contribute to the specificity of binding sites for orthosteric agonists by residues in the upper part of interface, relative to distance from the channel pore, and to the transmission of signals towards the gate by residues in the lower part of the interface. The R203 and N204 may integrate the influence of both lines of transmission. The LF domain appears to have two additional roles: the transmission of signals towards the gate in the second transmembrane domain through the K313-I333 β-sheet and the control of desensitization of receptors.

## Supporting Information

**Figure S1   Deactivation and desensitization responses of WT and selected DF and LF mutants.** (A) An example of the WT response and that of the L214A, D280A, and H286A mutant receptors when stimulated with 3 μM ATP for 2 s in the presence of IVM. Cells were preincubated with 3 μM IVM for 4–6 min, and the deactivation time constants ($\tau_{off}$) were estimated by the monoexponential fit of decay of current after removal of the agonist. (B) The desensitization of WT, L214A, D280A, and H286A receptors when stimulated with 100 μM ATP for 60 s (gray traces), and the curves obtained by fitting (black). Weighted desensitization time constants ($\tau_{des}$) were derived from mono-exponential (D280A) or biexponential fitting.

**Figure S2   Responsiveness of the WT and mutant receptors to ATP analogue agonists.** (A) Concentration

response curves for WT and G291A receptors. Currents were stimulated by a short (2–5 s) application of different (1–300 μM) concentrations of ATP, 2-MeS-ATP (2MeS), ATPγS, BzATP (Bz) and α, βme-ATP (αβme). Even if the full dose response curve for analogue agonists could not be constructed for G291A, these experiments clearly show differences in agonist profile between the WT (*upper panel*) and G291A (*lower panel*) receptor. Experiments were performed on naïve receptors, and data points are presented as the mean ± SEM from 3–27 measurements per agonist, per concentration for both WT and G291A. (B) Example responses to 100 μM of ATP and several P2XR agonists, including 2-MeS-ATP (2MeS), ATPγS, BzATP (Bz) and α, βme-ATP (αβme), recorded from cells expressing the WT receptor and selected mutant receptors from from *Group I* (I205A, T210A, L214A, and G291A) and *Group II* (D280A and H286A). Each trace represents a continuous response from a single cell.

**Figure S3   The structure of the ATP binding site in the rP2X4R homology model.** (A) The possible interaction of N293 and Y292 residues with Y315 and G316 residues from the β-sheet segment from the K313-I333 sequence. (B) The multiple interactions of residues R203 (yellow spheres) and N204 (green spheres) with partners (all in cyan wireframes) from the same (K190, N191, N204, I205, L206, Y274) and adjacent (D283, H236, V288, S289) subunits. Two adjacent rP2X4R subunits are represented in blue and gray.

**Table S1**   Characterization of the DF and LF alanine mutants of rP2X4R.

**Table S2**   Desensitization parameters for the DF and LF mutants of the rP2X4R.

## Author Contributions

Conceived and designed the experiments: VT SS HZ. Performed the experiments: VT MR HZ. Analyzed the data: VT MR SS HZ. Contributed reagents/materials/analysis tools: VT SS HZ. Contributed to the writing of the manuscript: VT MR SS HZ.

## References

1.  North RA (2002) Molecular physiology of P2X receptors. Physiol Rev. 82: 1013–67.
2.  Nicke A, Baumert HG, Rettinger J, Eichele A, Lambrecht G, et al. (1998) P2X1 and P2X3 receptors form stable trimers: a novel structural motif of ligand-gated ion channels. Embo J. 17: 3016–28.
3.  Coddou C, Yan Z, Obsil T, Huidobro-Toro JP, Stojilkovic SS (2011) Activation and regulation of purinergic P2X receptor channels. Pharmacol Rev. 63: 641–83.
4.  Bodnar M, Wang H, Riedel T, Hintze S, Kato E, et al. (2011) Amino acid residues constituting the agonist binding site of the human P2X3 receptor. J Biol Chem. 286: 2739–49.
5.  Ennion S, Hagan S, Evans RJ (2000) The role of positively charged amino acids in ATP recognition by human P2X(1) receptors. J Biol Chem. 275: 29361–7.
6.  Jiang LH, Rassendren F, Surprenant A, North RA (2000) Identification of amino acid residues contributing to the ATP-binding site of a purinergic P2X receptor. J Biol Chem. 275: 34190–6.
7.  Marquez-Klaka B, Rettinger J, Bhargava Y, Eisele T, Nicke A (2007) Identification of an intersubunit cross-link between substituted cysteine residues located in the putative ATP binding site of the P2X1 receptor. J Neurosci. 27: 1456–66.
8.  Roberts JA, Digby HR, Kara M, El Ajouz S, Sutcliffe MJ, et al. (2008) Cysteine substitution mutagenesis and the effects of methanethiosulfonate reagents at P2X2 and P2X4 receptors support a core common mode of ATP action at P2X receptors. J Biol Chem. 283: 20126–36.
9.  Roberts JA, Evans RJ (2007) Cysteine substitution mutants give structural insight and identify ATP binding and activation sites at P2X receptors. J Neurosci. 27: 4072–82.
10. Zemkova H, Yan Z, Liang Z, Jelinkova I, Tomic M, et al. (2007) Role of aromatic and charged ectodomain residues in the P2X4 receptor functions. J Neurochem. 102: 1139–50.
11. Hattori M, Gouaux E (2012) Molecular mechanism of ATP binding and ion channel activation in P2X receptors. Nature. 485: 207–12.
12. Kawate T, Michel JC, Birdsong WT, Gouaux E (2009) Crystal structure of the ATP-gated P2X(4) ion channel in the closed state. Nature. 460: 592–8.
13. Egan TM, Haines WR, Voigt MM (1998) A domain contributing to the ion channel of ATP-gated P2X2 receptors identified by the substituted cysteine accessibility method. J Neurosci. 18: 2350–9.
14. Haines WR, Voigt MM, Migita K, Torres GE, Egan TM (2001) On the contribution of the first transmembrane domain to whole-cell current through an ATP-gated ionotropic P2X receptor. J Neurosci. 21: 5885–92.
15. Jelinkova I, Vavra V, Jindrichova M, Obsil T, Zemkova HW, et al. (2008) Identification of P2X(4) receptor transmembrane residues contributing to channel gating and interaction with ivermectin. Pflugers Arch. 456: 939–50.
16. Jiang LH, Rassendren F, Spelta V, Surprenant A, North RA (2001) Amino acid residues involved in gating identified in the first membrane-spanning domain of the rat P2X(2) receptor. J Biol Chem. 276: 14902–8.
17. Khakh BS, Egan TM (2005) Contribution of transmembrane regions to ATP-gated P2X2 channel permeability dynamics. J Biol Chem. 280: 6118–29.

18. Li Z, Migita K, Samways DS, Voigt MM, Egan TM (2004) Gain and loss of channel function by alanine substitutions in the transmembrane segments of the rat ATP-gated P2X2 receptor. J Neurosci. 24: 7378–86.

19. Migita K, Haines WR, Voigt MM, Egan TM (2001) Polar residues of the second transmembrane domain influence cation permeability of the ATP-gated P2X(2) receptor. J Biol Chem. 276: 30934–41.

20. Rassendren F, Buell G, Newbolt A, North RA, Surprenant A (1997) Identification of amino acid residues contributing to the pore of a P2X receptor. Embo J. 16: 3446–54.

21. Silberberg SD, Chang TH, Swartz KJ (2005) Secondary structure and gating rearrangements of transmembrane segments in rat P2X4 receptor channels. J Gen Physiol. 125: 347–59.

22. Allsopp RC, El Ajouz S, Schmid R, Evans RJ (2011) Cysteine scanning mutagenesis (residues Glu52-Gly96) of the human P2X1 receptor for ATP: mapping agonist binding and channel gating. J Biol Chem. 286: 29207–17.

23. Rokic MB, Stojilkovic SS, Vavra V, Kuzyk P, Tvrdonova V, et al. (2013) Multiple roles of the extracellular vestibule amino acid residues in the function of the rat P2X4 receptor. PLoS One. 8: e59411.

24. Samways DS, Khakh BS, Dutertre S, Egan TM (2011) Preferential use of unobstructed lateral portals as the access route to the pore of human ATP-gated ion channels (P2X receptors). Proc Natl Acad Sci U S A. 108: 13800–5.

25. Du J, Dong H, Zhou HX (2012) Gating mechanism of a P2X4 receptor developed from normal mode analysis and molecular dynamics simulations. Proc Natl Acad Sci U S A. 109: 4140–5.

26. Lorinczi E, Bhargava Y, Marino SF, Taly A, Kaczmarek-Hajek K, et al. (2012) Involvement of the cysteine-rich head domain in activation and desensitization of the P2X1 receptor. Proc Natl Acad Sci U S A. 109: 11396–401.

27. Collo G, North RA, Kawashima E, Merlo-Pich E, Neidhart S, et al. (1996) Cloning of P2X5 and P2X6 receptors and the distribution and properties of an extended family of ATP-gated ion channels. J Neurosci. 16: 2495–507.

28. Le KT, Babinski K, Seguela P (1998) Central P2X4 and P2X6 channel subunits coassemble into a novel heteromeric ATP receptor. J Neurosci. 18: 7152–9.

29. Ormond SJ, Barrera NP, Qureshi OS, Henderson RM, Edwardson JM, et al. (2006) An uncharged region within the N terminus of the P2X6 receptor inhibits its assembly and exit from the endoplasmic reticulum. Mol Pharmacol. 69: 1692–700.

30. Carugo O (1999) Correlation between occupancy and B factor of water molecules in protein crystal structures. Protein Eng. 12: 1021–4.

31. Yan Z, Liang Z, Obsil T, Stojilkovic SS (2006) Participation of the Lys313-Ile333 sequence of the purinergic P2X4 receptor in agonist binding and transduction of signals to the channel gate. J Biol Chem. 281: 32649–59.

32. Jelinkova I, Yan Z, Liang Z, Moonat S, Teisinger J, et al. (2006) Identification of P2X(4) receptor-specific residues contributing to the ivermectin effects on channel deactivation. Biochem Biophys Res Commun. 349: 619–25.

33. Khakh BS, Proctor WR, Dunwiddie TV, Labarca C, Lester HA (1999) Allosteric control of gating and kinetics at P2X(4) receptor channels. J Neurosci. 19: 7289–99.

34. Priel A, Silberberg SD (2004) Mechanism of ivermectin facilitation of human P2X4 receptor channels. J Gen Physiol. 123: 281–93.

35. Schwede T, Kopp J, Guex N, Peitsch MC (2003) SWISS-MODEL: An automated protein homology-modeling server. Nucleic Acids Res. 31: 3381–5.

36. Benkert P, Tosatto SC, Schomburg D (2008) QMEAN: A comprehensive scoring function for model quality assessment. Proteins. 71: 261–77.

37. Rettinger J, Schmalzing G (2004) Desensitization masks nanomolar potency of ATP for the P2X1 receptor. J Biol Chem. 279: 6426–33.

38. Zemkova H, He ML, Koshimizu TA, Stojilkovic SS (2004) Identification of ectodomain regions contributing to gating, deactivation, and resensitization of purinergic P2X receptors. J Neurosci. 24: 6968–78.

39. Karplus PA (1997) Hydrophobicity regained. Protein Sci. 6: 1302–7.

40. Samways DS, Migita K, Li Z, Egan TM (2008) On the role of the first transmembrane domain in cation permeability and flux of the ATP-gated P2X2 receptor. J Biol Chem. 283: 5110–7.

41. Coddou C, Stojilkovic SS, Huidobro-Toro JP (2011) Allosteric modulation of ATP-gated P2X receptor channels. Rev Neurosci. 22: 335–54.

42. Zemkova H, Khadra A, Rokic MB, Tvrdonova V, Sherman A, et al. (2014) Allosteric regulation of the P2X4 receptor channel pore dilation. Pflugers Arch. 2014 Jun 11. [Epub ahead of print].

43. Toulme E, Soto F, Garret M, Boue-Grabot E (2006) Functional properties of internalization-deficient P2X4 receptors reveal a novel mechanism of ligand-gated channel facilitation by ivermectin. Mol Pharmacol. 69: 576–87.

44. Xu J, Chai H, Ehinger K, Egan TM, Srinivasan R, et al. (2014) Imaging P2X4 receptor subcellular distribution, trafficking, and regulation using P2X4-pHluorin. J Gen Physiol. 144: 81–104.

45. Roberts JA, Evans RJ (2006) Contribution of conserved polar glutamine, asparagine and threonine residues and glycosylation to agonist action at human P2X1 receptors for ATP. J Neurochem. 96: 843–52.

46. Adriouch S, Bannas P, Schwarz N, Fliegert R, Guse AH, et al. (2008) ADP-ribosylation at R125 gates the P2X7 ion channel by presenting a covalent ligand to its nucleotide binding site. Faseb J. 22: 861–9.

47. Digby HR, Roberts JA, Sutcliffe MJ, Evans RJ (2005) Contribution of conserved glycine residues to ATP action at human P2X1 receptors: mutagenesis indicates that the glycine at position 250 is important for channel function. J Neurochem. 95: 1746–54.

48. Zhang L, Xu H, Jie Y, Gao C, Chen W, et al. (2014) Involvement of ectodomain Leu 214 in ATP binding and channel desensitization of the P2X4 receptor. Biochemistry. 53: 3012–9.

49. Zhao WS, Wang J, Ma XJ, Yang Y, Liu Y, et al. (2014) Relative motions between left flipper and dorsal fin domains favour P2X4 receptor activation. Nat Commun. 5: 4189.

50. Jiang R, Lemoine D, Martz A, Taly A, Gonin S, et al. (2011) Agonist trapped in ATP-binding sites of the P2X2 receptor. Proc Natl Acad Sci U S A. 108: 9066–71.

51. Huang LD, Fan YZ, Tian Y, Yang Y, Liu Y, et al. (2014) Inherent Dynamics of Head Domain Correlates with ATP-Recognition of P2X4 Receptors: Insights Gained from Molecular Simulations. PLoS One. 9: e97528.

52. Yan Z, Liang Z, Tomic M, Obsil T, Stojilkovic SS (2005) Molecular Determinants of the Agonist Binding Domain of a P2X Receptor Channel. Mol Pharmacol. 67: 1078–88.

53. Fabbretti E, Sokolova E, Masten L, D'Arco M, Fabbro A, et al. (2004) Identification of negative residues in the P2X3 ATP receptor ectodomain as structural determinants for desensitization and the Ca2+ sensing modulatory sites. J Biol Chem. 279: 53109–115.

54. Stanchev D, Flehmig G, Gerevich Z, Norenberg W, Dihazi H, et al. (2006) Decrease of current responses at human recombinant P2X3 receptors after substitution by Asp of Ser/Thr residues in protein kinase C phosphorylation sites of their ecto-domains. Neurosci Lett. 393: 78–83.

55. Petrenko N, Khafizov K, Tvrdonova V, Skorinkin A, Giniatullin R (2011) Role of the ectodomain serine 275 in shaping the binding pocket of the ATP-gated P2X3 receptor. Biochemistry. 50: 8427–36.

56. Friday SC, Hume RI (2008) Contribution of extracellular negatively charged residues to ATP action and zinc modulation of rat P2X2 receptors. J Neurochem. 105: 1264–75.

57. Coddou C, Morales B, Gonzalez J, Grauso M, Gordillo F, et al. (2003) Histidine 140 plays a key role in the inhibitory modulation of the P2X4 nucleotide receptor by copper but not zinc. J Biol Chem. 278: 36777–85.

58. Kowalski M, Hausmann R, Dopychai A, Grohmann M, Franke H, et al. (2014) Conformational flexibility of the agonist binding jaw of the human P2X3 receptor is a prerequisite for channel opening. Br J Pharmacol. 2014 Jul 2. doi: 10.1111/bph.12830. [Epub ahead of print].

59. Wirkner K, Stanchev D, Koles L, Klebingat M, Dihazi H, et al. (2005) Regulation of human recombinant P2X3 receptors by ecto-protein kinase C. J Neurosci. 25: 7734–42.

60. Xiong K, Stewart RR, Hu XQ, Werby E, Peoples RW, et al. (2004) Role of extracellular histidines in agonist sensitivity of the rat P2X4 receptor. Neurosci Lett. 365: 195–9.

# Transcriptome-Based Identification of ABC Transporters in the Western Tarnished Plant Bug *Lygus hesperus*

**J. Joe Hull**[1]*, **Kendrick Chaney**[1], **Scott M. Geib**[2], **Jeffrey A. Fabrick**[1], **Colin S. Brent**[1], **Douglas Walsh**[3], **Laura Corley Lavine**[3]

1 USDA-ARS, Arid Land Agricultural Research Center, Maricopa, Arizona, United States of America, 2 USDA-ARS, Daniel K. Inouye Pacific Basin Agricultural Research Center, Hilo, Hawaii, United States of America, 3 Dept. of Entomology, Washington State University, Pullman, Washington, United States of America

## Abstract

ATP-binding cassette (ABC) transporters are a large superfamily of proteins that mediate diverse physiological functions by coupling ATP hydrolysis with substrate transport across lipid membranes. In insects, these proteins play roles in metabolism, development, eye pigmentation, and xenobiotic clearance. While ABC transporters have been extensively studied in vertebrates, less is known concerning this superfamily in insects, particularly hemipteran pests. We used RNA-Seq transcriptome sequencing to identify 65 putative ABC transporter sequences (including 36 full-length sequences) from the eight ABC subfamilies in the western tarnished plant bug (*Lygus hesperus*), a polyphagous agricultural pest. Phylogenetic analyses revealed clear orthologous relationships with ABC transporters linked to insecticide/xenobiotic clearance and indicated lineage specific expansion of the *L. hesperus* ABCG and ABCH subfamilies. The transcriptional profile of 13 LhABCs representative of the ABCA, ABCB, ABCC, ABCG, and ABCH subfamilies was examined across *L. hesperus* development and within sex-specific adult tissues. All of the transcripts were amplified from both reproductively immature and mature adults and all but LhABCA8 were expressed to some degree in eggs. Expression of LhABCA8 was spatially localized to the testis and temporally timed with male reproductive development, suggesting a potential role in sexual maturation and/or spermatozoa protection. Elevated expression of LhABCC5 in Malpighian tubules suggests a possible role in xenobiotic clearance. Our results provide the first transcriptome-wide analysis of ABC transporters in an agriculturally important hemipteran pest and, because ABC transporters are known to be important mediators of insecticidal resistance, will provide the basis for future biochemical and toxicological studies on the role of this protein family in insecticide resistance in *Lygus* species.

**Editor:** Youjun Zhang, Institute of Vegetables and Flowers, Chinese Academy of Agricultural Science, China

**Funding:** This work was supported by CSB - Cotton Incorporated (project no. 12-373), by KC - Agriculture and Food Research Initiative Competitive Grant from the USDA National Institute of Food and Agriculture (no. 2011-38422-30955), and by SMG - National Science Foundation (no. OCI-1053575XSEDE under allocation TG-MCB140032). The funders had no role in study design, data collection and analysis, decision to publish, or preparation of the manuscript.

**Competing Interests:** The research described in the manuscript was partially funded by a research grant from Cotton Inc. (project no. 12-373) to CSB, JAF, and JJH.

* Email: joe.hull@ars.usda.gov

## Introduction

ATP-binding cassette (ABC) proteins are an extensive family of transmembrane proteins that are ubiquitous to all organisms. The defining characteristic for most members of this superfamily is ATP hydrolysis driven unidirectional translocation of substrates (either import or export) across lipid membranes, typically in a thermodynamically unfavorable direction. However, ABC proteins also function as ion channels, regulators of ion channels, receptors, and in ribosome assembly and translation. They are structurally characterized by two highly conserved cytosolic nucleotide-binding domains (NBD) and two variable transmembrane domains (TMD) [1–4]. The NBDs, which are critical for ATP-binding and hydrolysis, provide the energy necessary for

driving a substrate across the membrane. They are characterized by a catalytic core comprised of a Walker A motif (GXXGXGKS/T) and a Walker B motif ($\varphi\varphi\varphi\varphi$D; where $\varphi$ represents a hydrophobic residue) separated by a conserved Q-loop and a Walker C motif. This latter component is a structurally diverse helical segment encompassing the ABC signature sequence (LSGGQ) that distinguishes ABC transporter family members from other ATP-binding proteins. Unlike the NBDs, TMDs vary in sequence, length, and helix number and are thought to provide initial substrate contact points. In eukaryotic organisms, the ABC transporter domains are organized as either full-transporters (FT) containing four domains (2 TMDs and 2 NBDs) or half-transporters (HT) comprised of only two domains (1 TMD and

1 NBD) that require homo- or heterodimerization for full functionality [2,3].

Based on conserved domain homology and organization, the eukaryotic ABC family can be divided into eight distinct subfamilies (A-H) with most family members facilitating the movement of a diverse array of substrates (sugars, lipids, peptides, polysaccharides, metals, inorganic ions, amino acids, and xenobiotics) across membranes. The first characterized eukaryotic ABC transporter (P-glycoprotein, HsABCB1) was identified based on its multidrug efflux pump functionality in mammalian cancer cell lines [5]. Since then members of the ABCB, ABCC, and ABCG subfamilies have been reported to play roles in drug resistance and detoxification in a number of species across multiple phyla [6–8]. Unlike the other subfamilies, the ABCE and ABCF transporters lack TMDs and function in ribosome assembly and protein translation rather than substrate transport [9–13]. The ABCH subfamily, which so far has only been identified in arthropod genomes and zebrafish [8,14,15], has not been as extensively characterized as the other subfamilies.

Insect ABC transporters mediate diverse functions with critical roles in molting, cuticle differentiation, and egg development [16], eye pigmentation [17–19], uric acid uptake [20], germ cell migration [21], 20-hydroxyecdysone mediated circadian rhythmicity [22], phytochemical sequestration [23], and biogenic amine transport [24]. Insect ABC transporters also function in the clearance of xenobiotics including plant defensive compounds and numerous insecticides representing disparate chemical classes and modes of action [8,25]. Resistance of some lepidopteran species to *Bacillus thuringiensis* (Bt) toxins has also been shown to involve novel interactions with ABC transporters independent of substrate transport [26–31].

Detailed genome/transcriptome wide analyses of ABC subfamilies have been conducted for a number of arthropods including water flea (*Daphnia pulex*) [32], spider mite (*Tetranychus urticae*) [33], fruitfly (*Drosophila melanogaster*) [34], olive fruit fly (*Bactrocera oleae*) [35], African malaria mosquito (*Anopheles gambiae*) [36], red flour beetle (*Tribolium castaneum*) [16], honeybee (*Apis mellifera*) [37], silkmoth (*Bombyx mori*) [37,38], and poplar leaf beetle (*Chrysomela populi*) [39]. Among hemipterans, knowledge of ABC transporters is limited to unverified genome annotations for the pea aphid (*Acyrthosiphon pisum*) [40] and partial transcriptomic/EST analyses in whiteflies (*Bemisia tabaci*) [38,41–43], bed bugs (*Cimex lectularius*) [44], and the brown planthopper (*Nilaparvata lugens*) [45].

The western tarnished plant bug (*Lygus hesperus* Knight) is a highly polyphagous hemipteran agricultural pest that primarily feeds on and injures the buds, flowers and seeds of many crops in the western United States and Canada [46–48]. Control strategies have traditionally relied on broad-spectrum insecticides. However, decreased efficacy has been reported for many of these compounds against field populations of *Lygus* [49–51]. Further compounding its pest status, our understanding of the molecular basis underlying *Lygus* resistance to insecticides is limited. To begin to address this issue, we performed a transcriptome-wide survey of ABC transporters by supplementing currently available transcriptomic data [52] with RNA-Seq analyses. We provide detailed sequence comparisons of the ABC subfamilies (ABCA-ABCH) in *L. hesperus* with those from four other arthropods and humans. In addition, the transcriptional profile of a subset of the assembled sequences was examined across developmental stages and in various sex-specific tissues. These findings will facilitate future exploration of the biological and physiological functions mediated by *L. hesperus* ABC transporters and their potential role in insecticide resistance.

## Results and Discussion

### RNA-Seq assembly and annotation

An earlier 454-based transcriptome of *L. hesperus* adults [52] contains 44 putative ABC transporter sequences. Here, we used Illumina RNA-Seq to obtain a comprehensive transcriptome that more extensively reflects ABC transporter transcription in *Lygus* adults. Furthermore, because some ABC transporters are associated with cellular stress [53–57], we combined the transcriptomes of *L. hesperus* exposed to cold and heat stress as well as non-stressed cohorts. Illumina HiSeq generated 144,898,116 raw 100 bp read pairs across nine libraries representing the three treatment groups. After quality filtering and *in silico* normalization, *de novo* assembly of 16,191,383 read pairs generated a raw assembly containing 132,802 isoforms across 77,246 unigenes with an N50 isoform length of 2,228 bp. Filtering this assembly by retaining only transcripts that have a predicted open reading frame reduced the assembly to 45,723 isoforms across 21,049 unigenes with an N50 isoform length of 2,989 bp.

### Identification of L. hesperus ABC transporter transcripts

The *L. hesperus* RNA-Seq database was searched using protein sequences corresponding to the full complement of ABC transporters from seven metazoans as well as the 44 putative ABC transporter sequences from the previous transcriptome [52]. We identified 65 putative *L. hesperus* ABC transporter-like (LhABC) transcripts. Based on Transdecoder predictions, significant matches with the Pfam-A database, and manual inspection of sequences spanning the first in-frame Met and stop codons, 36 of the sequences are predicted to encompass complete open reading frames. The number of LhABCs identified is comparable to that reported for other arthropods (Table 1). However, this number may under-represent (exclusion of temporally or spatially restricted transcripts not expressed) or over-represent (multiple partial transcripts corresponding to different portions or isoforms of the same gene) the actual number of LhABC transporters.

BLASTx (Table S1) and tBLASTn (Table S2) analyses revealed that the 65 putative LhABC transcripts represent all eight ABC transporter subfamilies (A-H). The most similar sequences from both BLAST analyses were putative transporters from *A. pisum* and *T. castaneum* (Figure S1). These results are consistent with previous transcriptome comparisons and not unexpected given the shared hemipteran lineage with *A. pisum* and the extensive genomic annotation in *T. castaneum*. The putative LhABC transcripts encode proteins or protein fragments ranging in size from 144 to 2237 amino acids. Although additional partial transcripts were identified in the dataset, only transcripts with coding sequences of >100 amino acids were further analyzed. Consistent with substrate translocation across the plasma membrane, 30 of the 36 full-length sequences identified are predicted to localize to the cell surface (Table 2). The six exceptions include sequences with highest similarity to ABCDs, which localize to the peroxisome, and ABCE/F subfamily members, which do not function in substrate transport. Multiple TMD prediction algorithms indicated the presence of numerous helical segments in 32 of the full-length sequences and 25 of the partial sequences (Table 2). No helices are predicted for the *L. hesperus* ABCE and ABCF transporters. ABC transporter motifs and/or NBDs are present in 62 of the 65 LhABC sequences (Table S3), and even for those lacking these domains there was significant sequence similarity (E-value $< 10^{-12}$) with genes annotated as ABC transporters (Table S1 and S2). LhABCC1A and LhABCC1B share the highest amino acid identity (96%), whereas sequence identity across the other LhABC transporters varies from 1% to 51% with

**Table 1.** ABC subfamilies in *L. hesperus* and ten other species.

| phylum | Chordata | Nematoda | Arthropoda | | | | | | | | | |
|---|---|---|---|---|---|---|---|---|---|---|---|---|
| class | | | | | Insecta | | | | | | | |
| order | | | Branchiopoda | Arachnida | Diptera | | Coleoptera | Hymenoptera | | Lepidoptera | Hemiptera | |
| ABC subfamily | *H. sapiens*[a] | *C. elegans*[b] | *D. pulex*[c] | *T. urticae*[d] | *D. melanogaster*[a] | *A. gambiae*[e] | *T. castaneum*[f] | *C. poul*[g] | *A. mellifera*[h] | *B. mori*[h,i] | *L. hesperus* |
| ABCA | 12 | 7 | 4 | 9 | 10 | 9 | 10 | 5 | 3 | 6, 9 | 11 |
| ABCB | 11 | 24 | 7 | 4 | 8 | 5 | 6 | 8 | 5 | 8, 9 | 6 |
| ABCC | 12 | 9 | 7 | 39 | 14 | 13 | 35 | 29 | 9 | 15, 15 | 12 |
| ABCD | 4 | 5 | 3 | 2 | 2 | 2 | 2 | 2 | 2 | 2, 2 | 2 |
| ABCE | 1 | 1 | 1 | 1 | 1 | 1 | 1 | 1 | 1 | 1, 1 | 1 |
| ABCF | 3 | 3 | 4 | 3 | 3 | 3 | 3 | 3 | 3 | 3, 3 | 3 |
| ABCG | 5 | 9 | 24 | 23 | 15 | 16 | 13 | 14 | 15 | 13, 12 | 19 |
| ABCH | 0 | 0 | 15 | 22 | 3 | 3 | 3 | 3 | 3 | 3, 2 | 11 |
| TOTAL | 48 | 58 | 65 | 103 | 56 | 52 | 73 | 65 | 41 | 51, 53 | 65 |

Subfamily numbers are from: [a][34]; [b][151]; [c][32]; [d][33]; [e][36]; [f][16]; [g][39]; [h][37]; [i][38].

highest levels of conservation observed within the respective subfamilies (Table S4).

## ABCA subfamily

ABCA transporters are among the largest known ABCs and typically exhibit an extended extracellular domain between the TMDs, a dipolar diacidic motif downstream of the NBD region, and a conserved amino terminal sequence (XLXXKN) involved in post-Golgi trafficking [58,59]. We found 11 putative ABCA transporter sequences (Table S1 and S2) exhibiting 3–50% amino acid identity (Table S5). LhABCA4 and LhABCA5 encompass full-length coding sequences with extended extracellular domains (653 aa in LhABCA4, 189 aa in LhABCA5) and the dipolar diacidic motif. The amino terminal post-Golgi trafficking motif is only found in LhABCA4. This motif is present in 11 of the 12 HsABCAs but only 6 of 38 ABCAs from *T. urticae*, *D. melanogaster*, *B. mori*, and *T. castaneum*, suggesting that arthropods may use an alternative post-Golgi targeting mechanism.

Alignment and phylogenetic analyses of putative LhABCAs with ABCAs from *T. urticae*, *D. melanogaster*, *B. mori*, *T. castaneum* and humans are consistent with those reported by other groups [8,33,37,38] and indicate conserved subfamily clustering within six central clades that we have designated ABCA.1-ABCA.6 (Figure 1). LhABCA1, LhABCA2, and LhABCA4 aligned within clade ABCA.3, which is comprised of multiple HsABCAs that function in the transport of membrane lipids [60,61]. LhABCA5, LhABCA6, LhABCA3, and LhABCA9 aligned with two BmAB-CAs to form one branch in clade ABCA.6. The first two LhABCAs are most similar to BmABCA3, while the other two share similarity with BmABCA7. Consistent with previous phylogenetic analyses [33], a second branch of clade ABCA.6 is comprised solely of *T. urticae* ABCAs. LhABCA7 and LhABCA11 form a separate branch in clade ABCA.5, which is dominated by *D. melanogaster* and human ABCAs. While no function has been assigned to the arthropod ABCAs in this clade, the HsABCAs function in lipid homeostasis [60,61]. LhABCA10 aligned with sequences in clade ABCA.2, whereas LhABCA8 clustered with a group of BmABCAs to form a branch in clade ABCA.4. In addition to the LhABCA8 branch, the ABCA.4 clade is also characterized by gene expansion in *T. castaneum*. HsABCA3, a transporter involved in pulmonary surfactant secretion [60,61], also sorted to ABCA.4. No arthropod sequences sorted to clade ABCA.1, which is comprised of HsABCA12, a keratinocyte lipid transporter [62], and HsABCA13, the biological function of which is not known.

## ABCB subfamily

ABCBs are structurally organized as either HTs characterized by two domains (1 TMD, 1 NBD) or FTs that contain four domains (2 TMDs, 2 NBDs). Mammalian ABCBs transport diverse hydrophobic substrates including bile acids, peptides, steroids, drugs, and other xenobiotics. This broad substrate range of ABCBs likely contributes to their involvement in multidrug resistance phenotypes [63–65] and insecticide resistance [8,25]. Based on BLAST analyses, we identified six LhABCB transcripts, which is similar to the number reported for other arthropods (Table 1). The LhABCB1 and LhABCB4 transcripts include full-length ORFs, whereas the other LhABCB transcripts encode protein fragments ranging in size from 469 to 830 amino acids (Table 2) that encompass predicted ABC transporter-like domains (Table S3). As a group, the LhABCBs, exhibit 12–51% sequence identity (Table S6). Consistent with the phylogenetic analyses of Dermauw et al.[8], the HTs aligned into five clades (ABCB.1-

**Table 2.** Bioinformatics analysis of putative LhABC transporters.

| L. hesperus id | Illumina assembly id | Size (aa) | Full/Partial CDS | Localization WoLF PSORT[a] | Number of helical domains | | |
|---|---|---|---|---|---|---|---|
| | | | | | TMHMM[b] | TopPredII[c] | TopCons[d] |
| LhABCA1 | comp10339_c0_seq1 | 474 | partial | nd[e] | 5 | 6 | 5 |
| LhABCA2 | comp9970_c0_seq1 | 663 | partial | nd | 5 | 5(7)[f] | 6 |
| LhABCA3 | comp35546_c0_seq1 | 637 | partial | nd | 5 | 6(8) | 6 |
| LhABCA4 | comp34679_c0_seq2 | 2237 | full | PM | 12 | 14(16) | 12 |
| LhABCA5 | comp28522_c0_seq4 | 1326 | full | PM | 14 | 16(19) | 12 |
| LhABCA6 | comp12312_c0_seq2 | 416 | partial | nd | 1 | 2 | 1 |
| LhABCA7 | comp33391_c0_seq1 | 585 | partial | nd | 5 | 6 | 5 |
| LhABCA8 | comp9781_c0_seq1 | 415 | partial | nd | 0 | 1 | 1 |
| LhABCA9 | comp38669_c0_seq1 | 275 | partial | nd | 0 | 0(1) | 0 |
| LhABCA10 | comp39633_c0_seq1 | 243 | partial | nd | 0 | 0 | 0 |
| LhABCA11 | comp28530_c0_seq1 | 795 | partial | nd | 5 | 5(6) | 6 |
| LhABCB1 | comp37357_c0_seq3 | 1191 | full | PM | 9 | 10(11) | 11 |
| LhABCB2 | comp30116_c0_seq1 | 830 | partial | nd | 6 | 6 | 6 |
| LhABCB3 | comp31442_c0_seq2 | 687 | partial | nd | 5 | 6(7) | 6 |
| LhABCB4 | comp35640_c0_seq1 | 834 | full | PM | 10 | 8(11) | 11 |
| LhABCB5 | comp37353_c1_seq7 | 735 | partial | nd | 5 | 7 | 6 |
| LhABCB6 | comp21601_c0_seq1 | 469 | partial | nd | 6 | 6(7) | 6 |
| LhABCC1A | comp37257_c0_seq2 | 944 | full | PM | 9 | 10(12) | 11 |
| LhABCC1B | comp19288_c0_seq1 | 944 | full | PM | 7 | 9(12) | 11 |
| LhABCC2 | comp36363_c3_seq3 | 589 | partial | nd | 5 | 5 | 6 |
| LhABCC3 | comp29144_c0_seq2 | 1407 | partial | nd | 9 | 11(16) | 12 |
| LhABCC4 | comp26101_c0_seq1 | 1226 | partial | nd | 9 | 7(10) | 12 |
| LhABCC5 | comp28277_c0_seq1 | 1316 | full | PM | 9 | 10(12) | 12 |
| LhABCC6 | comp28206_c0_seq2 | 633 | partial | nd | 4 | 4 | 6 |
| LhABCC7 | comp1105_c0_seq1/comp20506_c0_seq1 | 601 | partial | nd | 5 | 5(6) | 6 |
| LhABCC8 | comp36620_c0_seq1 | 1495 | full | PM | 16 | 14(17) | 17 |
| LhABCC9 | comp33891_c0_seq2 | 1415 | full | PM | 9 | 10(11) | 12 |
| LhABCC10 | comp14659_c0_seq1 | 256 | partial | nd | 0 | 0(1) | 0 |
| LhABCC11 | comp37326_c0_seq3 | 856 | full | PM | 10 | 9 | 10 |
| LhABCD1 | comp30627_c0_seq1 | 676 | full | cytosol | 3 | 4(5) | 5 |
| LhABCD2 | comp32676_c0_seq1 | 656 | full | mito | 5 | 4 | 5 |
| LhABCE1 | comp29836_c0_seq1 | 608 | full | cytosol | 0 | 1 | 0 |
| LhABCF1 | comp37052_c0_seq9 | 589 | full | cyto/nucleus | 0 | 0(1) | 0 |
| LhABCF2 | comp34354_c0_seq1 | 629 | full | nucleus | 0 | 1 | 0 |

**Table 2.** Cont.

| L. hesperus id | Illumina assembly id | Size (aa) | Full/Partial CDS | Localization WoLF PSORT[a] | Number of helical domains | | |
| --- | --- | --- | --- | --- | --- | --- | --- |
| | | | | | TMHMM[b] | TopPredII[c] | TopCons[d] |
| LhABCF3 | comp24795_c0_seq2 | 712 | full | cyto/nucleus | 0 | 0(1) | 0 |
| LhABCG1 | comp33145_c0_seq2 | 685 | full | PM | 6 | 6(7) | 6 |
| LhABCG2 | comp28267_c0_seq1 | 583 | partial | nd | 6 | 6(8) | 6 |
| LhABCG3 | comp36162_c0_seq6 | 622 | full | PM | 6 | 6(7) | 6 |
| LhABCG4 | comp3947_c0_seq1 | 226 | partial | nd | 0 | 0(1) | 0 |
| LhABCG5 | comp33057_c0_seq1 | 608 | full | PM | 5 | 5(8) | 6 |
| LhABCG6 | comp32204_c0_seq5 | 617 | full | PM | 6 | 6(7) | 6 |
| LhABCG7 | comp30007_c0_seq1 | 608 | full | PM | 5 | 4(6) | 6 |
| LhABCG8 | comp26890_c0_seq1 | 651 | partial | nd | 7 | 6(9) | 6 |
| LhABCG9 | comp31063_c0_seq2 | 479 | partial | nd | 4 | 4(7) | 6 |
| LhABCG10 | comp32700_c0_seq1 | 703 | full | PM | 5 | 6(7) | 6 |
| LhABCG11 | comp35069_c0_seq5 | 623 | full | PM | 6 | 4(8) | 6 |
| LhABCG12 | comp28984_c1_seq1 | 654 | full | PM | 5 | 9(11) | 6 |
| LhABCG13 | comp35811_c1_seq7 | 614 | full | PM | 7 | 7(11) | 6 |
| LhABCG14 | comp26007_c1_seq1 | 655 | partial | nd | 6 | 5(6) | 6 |
| LhABCG15 | comp34164_c0_seq8 | 601 | full | PM | 5 | 6(8) | 5 |
| LhABCG16 | comp32376_c3_seq1 | 606 | full | PM | 7 | 6(7) | 6 |
| LhABCG17 | comp32660_c1_seq2 | 611 | full | PM | 7 | 7(8) | 6 |
| LhABCG18 | comp9896_c0_seq1 | 319 | partial | nd | 0 | 1 | 1 |
| LhABCG19 | comp20817_c0_seq1 | 227 | partial | nd | 4 | 4(5) | 4 |
| LhABCH1 | comp36633_c0_seq2 | 795 | full | PM | 7 | 6(8) | 6 |
| LhABCH2 | comp37335_c0_seq4 | 748 | full | PM | 5 | 6(7) | 6 |
| LhABCH3 | comp34118_c0_seq2 | 683 | full | PM | 5 | 6(7) | 6 |
| LhABCH4 | comp32271_c0_seq13 | 768 | full | PM | 8 | 8(9) | 7 |
| LhABCH5 | comp31632_c0_seq1 | 680 | full | PM | 6 | 6(8) | 6 |
| LhABCH6 | comp6701_c0_seq1 | 673 | full | PM | 5 | 6(7) | 6 |
| LhABCH7 | comp35902_c0_seq6 | 685 | full | PM | 5 | 6(7) | 6 |
| LhABCH8 | comp37338_c0_seq18 | 144 | partial | nd | 3 | 3 | 4 |
| LhABCH9 | comp22796_c0_seq1 | 526 | partial | nd | 0 | 2(3) | 2 |
| LhABCH10 | comp31498_c1_seq1 | 693 | partial | nd | 7 | 7(9) | 6 |
| LhABCH11 | comp27033_c0_seq3 | 682 | full | PM | 6 | 6(7) | 6 |

[a][155]; [b][156]; [c][158]; [d][157]; [e] nd - not determined;
first number indicates certain TMS (score potential >1), number in parenthesis indicates number of putative TMs.

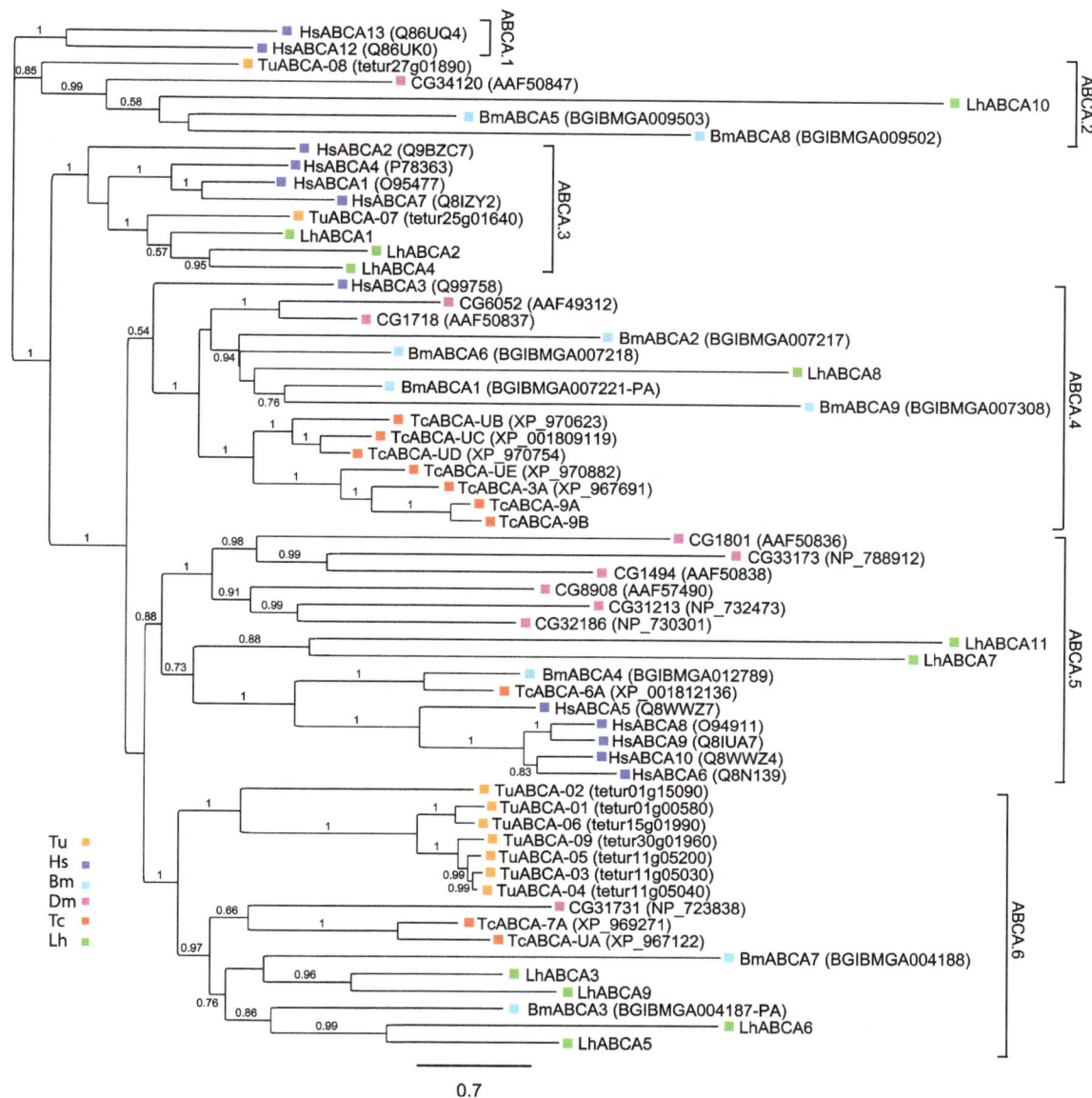

**Figure 1. Phylogenetic analysis of ABCA transporters from *L. hesperus* and five metazoan species.** Putative *L. hesperus* ABCA sequences and full-length ABCA proteins from five additional species were aligned using MUSCLE [159] and analyzed using the FastTree2 approximate likelihood method [160]. Numbers at the branch point of each node represent support values. Species abbreviations and color coding are: Bm, *Bombyx mori* (teal); Dm, *Drosophila melanogaster* (pink); Hs, *Homo sapiens* (blue); Lh, *Lygus hesperus* (green); Tc, *Tribolium castaneum* (red); and Tu, *Tetranychus urticae* (orange). Accession numbers are indicated in parentheses. The scale bar represents 0.7 amino acid substitutions per site. A full listing of the accession numbers for the five metazoan sequences is available in Table S11. LhABC transporter sequences are available in Table S12.

ABCB.5), whereas the FTs clustered into four clades (ABCB.6-ABCB.9) largely composed of lineage specific branches (Figure 2). The branching pattern in four of the HT clades (ABCB.1-ABCB.4) is suggestive of orthologous relationships. LhABCB4, LhABCB5, and LhABCB3 aligned with proteins in clades ABCB.1, ABCB.2, and ABCB.4 respectively. The expression of HsABCB6 (clade ABCB.1) has been correlated with increased drug resistance [65,66] and arthropod transporters in clades ABCB.1 and ABCB.2 are reported to play roles in heavy metal detoxification [67], insecticide resistance [68], cold stress tolerance [69], and pupal-adult development in *T. castaneum* [16]. Despite orthologous sequences in the other arthropods examined, no LhABCB sequences aligned with proteins comprising clade

ABCB.3. The three human transporters (HsABCB2, HsABCB3, HsABCB9) that function in antigen processing [70,71] form a human specific clade (ABCB.5).

In contrast to the HTs, the FT sequences separated into clades with species-specific branches. Two LhABCBs (LhABCB1 and LhABCB2) and two *T. urticae* ABCBs form separate branches of the ABCB.6 clade, which is consistent with diversification arising from lineage specific gene duplication events [32]. Similar to the ABCB.5 clade, ABCB.7 is human specific with no arthropod sequences aligning with the four HsABCBs. The other two FT clades (ABCB.8 and ABCB.9) comprise sequences similar to *D. melanogaster* multiple drug resistance (MDR) proteins (i.e., DmMDR49, DmMDR50, and DmMDR65) that have been

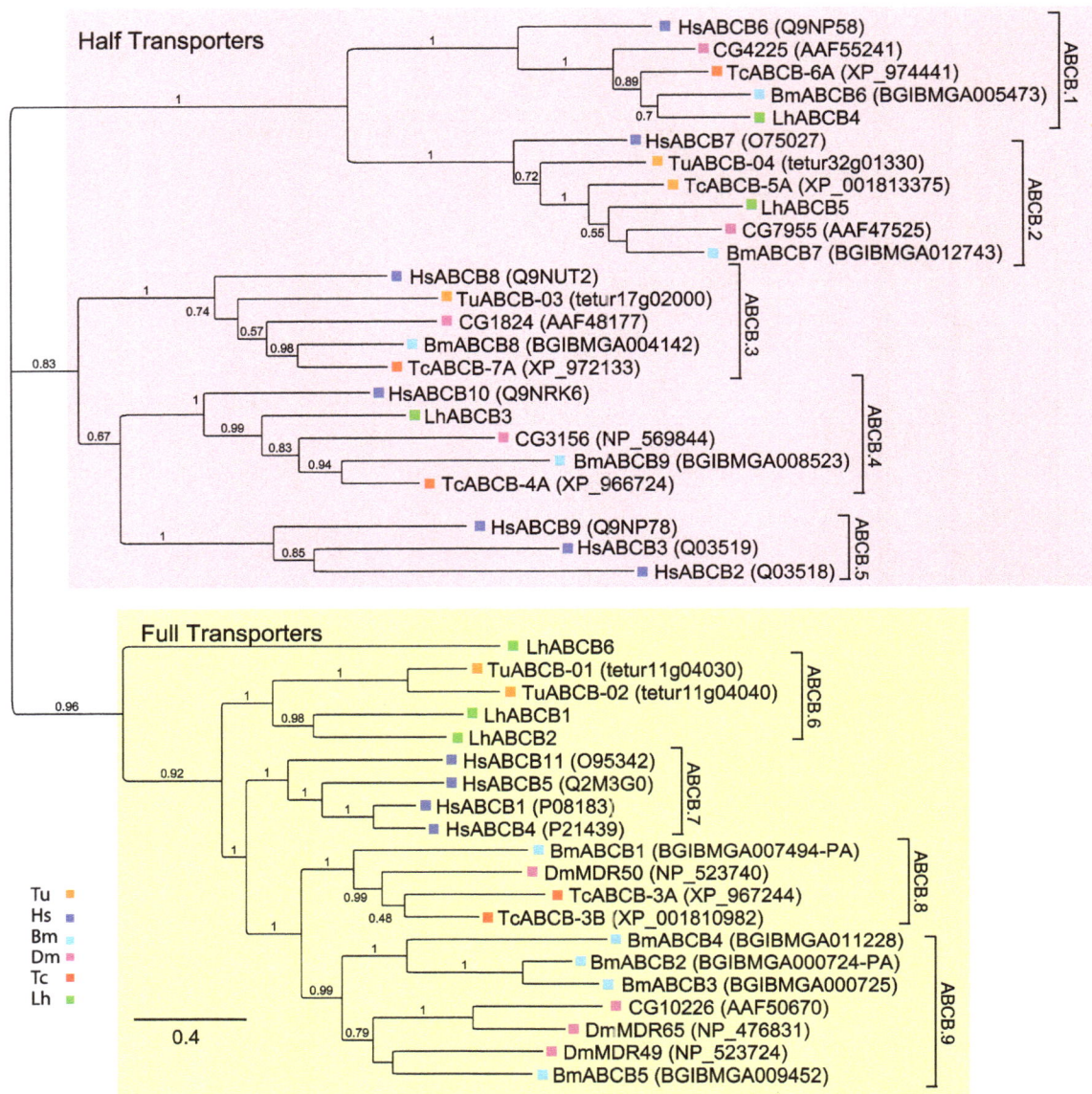

**Figure 2. Phylogenetic analysis of ABCB transporters from *L. hesperus* and five metazoan species.** Clades corresponding to full-transporters and half-transporters are shaded light yellow and light red, respectively. The scale bar represents 0.4 amino acid substitutions per site. Analyses, abbreviations, and color-coding are as in Figure 1.

identified in abiotic stress and insecticide resistance [45,53,72–78]. Surprisingly, no *L. hesperus* sequences nor any *T. urticae* sequences clustered within either of these two clades. LhABCB6, which is a partial sequence, aligned with the FTs but did not sort with any of the clades.

## ABCC subfamily

The ABCC subfamily consists of three functionally distinct classes of transporters: broad-specificity multidrug resistance-associated proteins (MRPs), sulfonylurea receptors (SUR), and the cystic fibrosis transmembrane conductance regulator (CFTR) [34,79–81]. MRPs interact with a diverse array of substrates including a number of endogenous metabolites, xenobiotics, and various conjugated (glutathione, sulfate, and glucuronate) anions. Nine members of the mammalian ABCC subfamily are classified

as MRPs: ABCC1–6 and ABCC10–12. In contrast to the xenobiotic transport activities of MRP-ABCCs, SURs (i.e. ABCC8 and ABCC9) function as regulators of specific potassium channels, and CFTR (i.e. ABCC7) functions as an ATP-gated chloride ion channel [34].

Based on sequence similarities, we identified 12 ABCC-like transcripts in *L. hesperus*, a number comparable to that reported for other arthropods with the exception of *T. urticae*, *T. castaneum*, and *C. populi*, all of which have undergone significant expansion of the ABCC subfamily (Table 1). Sequence identity among the LhABCC transcripts ranges from 2–96% (Table S7). Half of the transcripts comprise complete coding sequences, and the rest encode protein fragments (256 to 1407 amino acids) containing ABC transporter-like domains (Table S3). Our phylogenetic analysis generated five major clades (designated ABCC.1-ABCC.5) with 11 nested clades (designated ABCC.5A-

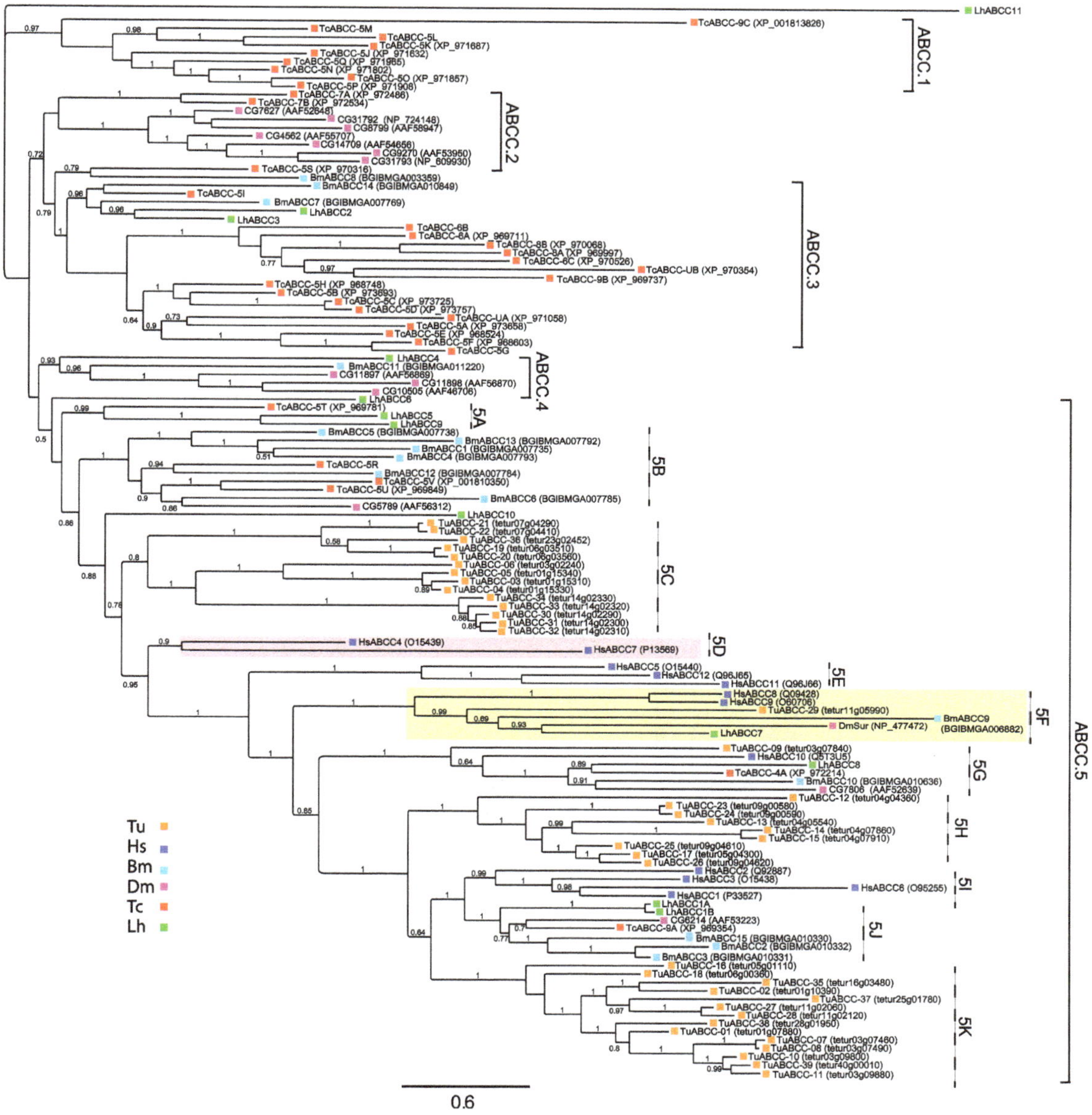

**Figure 3. Phylogenetic analysis of ABCC transporters from *L. hesperus* and five metazoan species.** The scale bar represents 0.6 amino acid substitutions per site. Clades corresponding to SUR-like sequences and CFTR are shaded light yellow and light red, respectively. Analyses, abbreviations, and color-coding are as in Figure 1.

5 K) branching from ABCC.5 (Figure 3). Consistent with other studies supporting gene duplication-based expansion of the ABCC subfamily [33,82], our analysis clustered many of the *T. castaneum* and *T. urticae* sequences into lineage specific clades/branches (*T. castaneum* – clade ABCC.1 and two nested branches in clade ABCC.3; *T. urticae* – clades ABCC.5C, ABCC.5 H, and ABCC.5 K). As before, clear relationships in sequence alignment were seen for a number of the LhABCCs.

LhABCC1A and LhABCC1B sorted to nested clade ABCC.5J. The two transcripts encode full-length sequences that are 96% identical (Table S7) with sequence variation (86% pairwise sequence identity) primarily in the first 260 amino acids (Figure S2). Sanger sequencing of multiple clones confirmed the respective sequences. The sequence variation could reflect the heterogeneity of our *L. hesperus* colony. Alternatively, they could represent splice

variants similar to that reported for *D. melanogaster* ABCC CG6214 [83] and CG6214 orthologs in *A. gambiae* [36] and *Trichoplusia ni* [84]. CG6214 is most similar with the two LhABCCs in ABCC.5J and is upregulated in response to xenobiotic feeding [75,85]. The HsABCC specific clade (ABCC.5I) that branches from the same node as ABCC.5J is comprised of classic MRPs with broad substrate specificities and diverse resistance phenotypes [80,86].

LhABCC8 aligned to clade ABCC.5G with HsABCC10 and four other arthropod ABCCs (Figure 3). HsABCC10 transports a wide range of substrates and has been linked with multiple multidrug resistance phenotypes [87]. While the biological function of the arthropod ABCCs in this grouping is unknown, ecdysone treatment has been shown to elevate expression of the *D. melanogaster* CG7806 transporter [88]. LhABCC4 sorts to clade

ABCC.4 with BmABCC11 and three *D. melanogaster* ABCCs (CG11897, CG11898, and CG10505). The expression of CG11897 is strongly induced following immune challenge with an entomopathogenic bacterium [89], whereas that of CG10505 has been linked to heavy metal homeostasis [90] and alcohol exposure [91].

LhABCC2 and LhABCC3 aligned with two BmABCCs and numerous *T. castaneum* transporters in clade ABCC.3 (Figure 3), which is a sister clade to ABCC.2. The ABCC.2 clade is characterized by a cluster of six *D. melanogaster* ABCCs, two of which (CG14709 and CG8799) are expressed and/or upregulated in response to cellular stress [92–94]. A third ABCC (CG4562) in that clade was recently shown to be upregulated following knockdown of a detoxifying cytochrome P450 [95], suggesting potential compensatory cross talk occurs between the two detoxification mechanisms.

ABCC2 MRP-like transporters have been linked with resistance to *Bacillus thuringiensis* (Bt) Cry1 toxins in a number of lepidopterans [26–31]. This resistance does not appear to be linked to ATP-dependent transport but rather to the cell surface transporter functioning as a putative Bt Cry1A toxin receptor [96]. In our phylogenetic analysis, BmABCC13 (BGIBMGA007792) and BmABCC4 (BGIBMGA007793), which represent a single gene [28], correspond to the ABCC2 transporter linked to Bt resistance. Both *B. mori* sequences align within a lineage-specific branch of clade ABCC.5B, which also included several *T. castaneum* ABCCs and a *D. melanogaster* sequence. No LhABCC sequence aligned within that clade. Although a transgenic cotton plant expressing a hemipteran-active Bt toxin has been developed [97], it remains to be determined if LhABCCs will also function as Bt toxin receptors.

Well-supported alignment of LhABCC7 with the SUR ABCC transporters in clade ABCC.5F suggests possible conservation of function. These transporters assemble with other proteins to form ATP-sensitive potassium channels that function in a number of physiological processes [34,98]. In insects, SUR is important for glucose homeostasis [99], protection against hypoxic stress [100], and has been proposed as the putative binding site for benzoylphenylureas, a class of insecticides that inhibit chitin synthesis [16,101]. However, recent findings suggest that in some species SUR is dispensable for benzoylphenylurea action [8,16,102,103].

The third functional class of ABCC transporters is CFTR, an ABC transporter that has undergone functional divergence to exhibit ATP-gated chloride channel activity [34]. In our phylogenetic analyses, HsABCC7 (i.e., CFTR) sorted to clade ABCC.5D with HsABCC4, the closest mammalian homolog of HsABCC7 [104]. While structurally similar, HsABCC4 is functionally differentiated by a broad substrate range that includes diverse xenobiotics [80]. Consistent with other studies, we did not find any *L. hesperus* sequences orthologous with the two HsABCCs. The top BLAST hits for LhABCC11, however, are with mammalian MRP4/ABCC4 sequences (Table S1 and S2). The LhABCC11 transcript encodes an 856 aa protein (Table 1) containing characteristic ABC transporter domains (Table S3), but it did not align within any of the ABCC clades. A *C. populi* ABCC transporter, CpMRP, that functions in plant-derived phenolglucoside sequestration also shares similarity with MRP4 proteins [23]. Low sequence identity (~15%) between LhABCC11 and CpMRP, however, suggests that the two transporters may interact with different substrates.

## ABCD, ABCE, and ABCF subfamilies

ABCD transporters function in peroxisomal import of long chain fatty acids and/or fatty acyl CoAs [105,106]. We found that, like most other arthropods, *L. hesperus* has two ABCD transcripts

(Table 1). LhABCD1 and LhABCD2 share 36% sequence identity (Table S8), which is comparable to the homology (27–35%) shared between the ABCDs in our phylogenetic analyses. Both LhABCDs have transporter domains (Table S3), as well as the two motifs, EEA-like (EEIAFYGG) and loop1 (LXXRT), that are considered essential for ABCD function [107]. LhABCD1 and LhABCD2 aligned with clades ABCD.1 and ABCD.2, respectively (Figure 4). Although little is known about arthropod ABCDs, the *D. melanogaster* ABCD transporter CG2316 (clade ABCD.1) is overexpressed in a cell line resistant to the insect growth regulator methoxyfenozide [108].

ABCE and ABCF proteins contain the characteristic NBD but lack TMDs and do not transport substrates. Instead, ABCEs function in ribosome recycling and regulation of protein translation [13,109], whereas ABCFs are involved in translation [10,12]. Like other arthropods [8], *L. hesperus* has a single LhABCE and three LhABCF transcripts (Table 2). LhABCE shares ~80% sequence identity with the other ABCEs examined in our analyses, supporting a possible evolutionarily conserved role. The three LhABCFs, which share 32–35% sequence identity (Table S8), sorted into distinct but well-supported clades (Figure 4). Sequence identity within each clade ranged from 45–84%, with highest identities in the LhABCF2 clade.

## ABCG subfamily

ABCG transporters have a HT motif in which the lone NBD is localized on the amino terminal side of the TMD, in contrast to localization on the carboxyl terminal side as in other ABC transporters. In comparison with the ABCG subfamily in other metazoans, which frequently have 5–10 members [110], the arthropod lineage has undergone extensive gene expansion (Table 1). Consistent with this, we identified 19 LhABCG transcripts, including 12 full-length coding sequences (Table 2). The LhABCGs share low (10%) to moderate (58%) sequence identity (Table S9), and all but LhABCG19 have ABC transporter domains (Table S3). Despite the lack of known ABC domains, LhABCG19 shares sequence similarity (BLASTx E value$<10^{-10}$) with other putative ABCG transporters (Table S1 and S2).

Our phylogenetic analysis resulted in four major clades (ABCG.1-ABCG.4) with a number of minor clades branching from ABCG.3 and ABCG.4 (Figure 5). As previously reported [33], *T. urticae* has multiple lineage-specific clades (ABCG.1 and ABCG.4G) indicating expansion by multiple gene duplication events [32]. Aside from these sequences, the clustering pattern of the other ABCGs within the respective clades suggests clear orthologous relationships (Figure 5). LhABCG2 aligned to clade ABCG.3A along with *D. melanogaster* ABCG CG2969 (i.e., DmAtet), reportedly a target of the transcriptional regulator gene, *clock* [111], and two HsABCGs involved in sterol homeostasis [112,113]. LhABCG1 sorted to clade ABCG.3B with arthropod ABCGs that have been implicated in cuticular lipid transport [16]. LhABCG3, LhABCG11, and LhABCG15 sorted to separate branches of clade ABCG.3C with LhABCG15 aligning to the same branch as the *D. melanogaster* ABCG CG9663 and LhABCG11 aligning with *D. melanogaster* ABCG CG17646. CG9663 is a putative target of *clock* [111] that has also been linked to decreased susceptibility to oxidative stress [114]. CG17646 functions in triglyceride storage [115] and ethanol sensitivity [116]. LhABCG6, LhABCG13, LhABCG16, and LhABCG17 aligned to separate branches of clade ABCG.3D with arthropod ABCGs of unknown function. LhABCG4 and LhABCG5 aligned with potential orthologs of HsABCG8 and HsABCG5 in clades ABCG.4B and ABCG.4A respectively. The obligate heterodimerization of the two HsABCGs in sterol homeostasis [112,113]

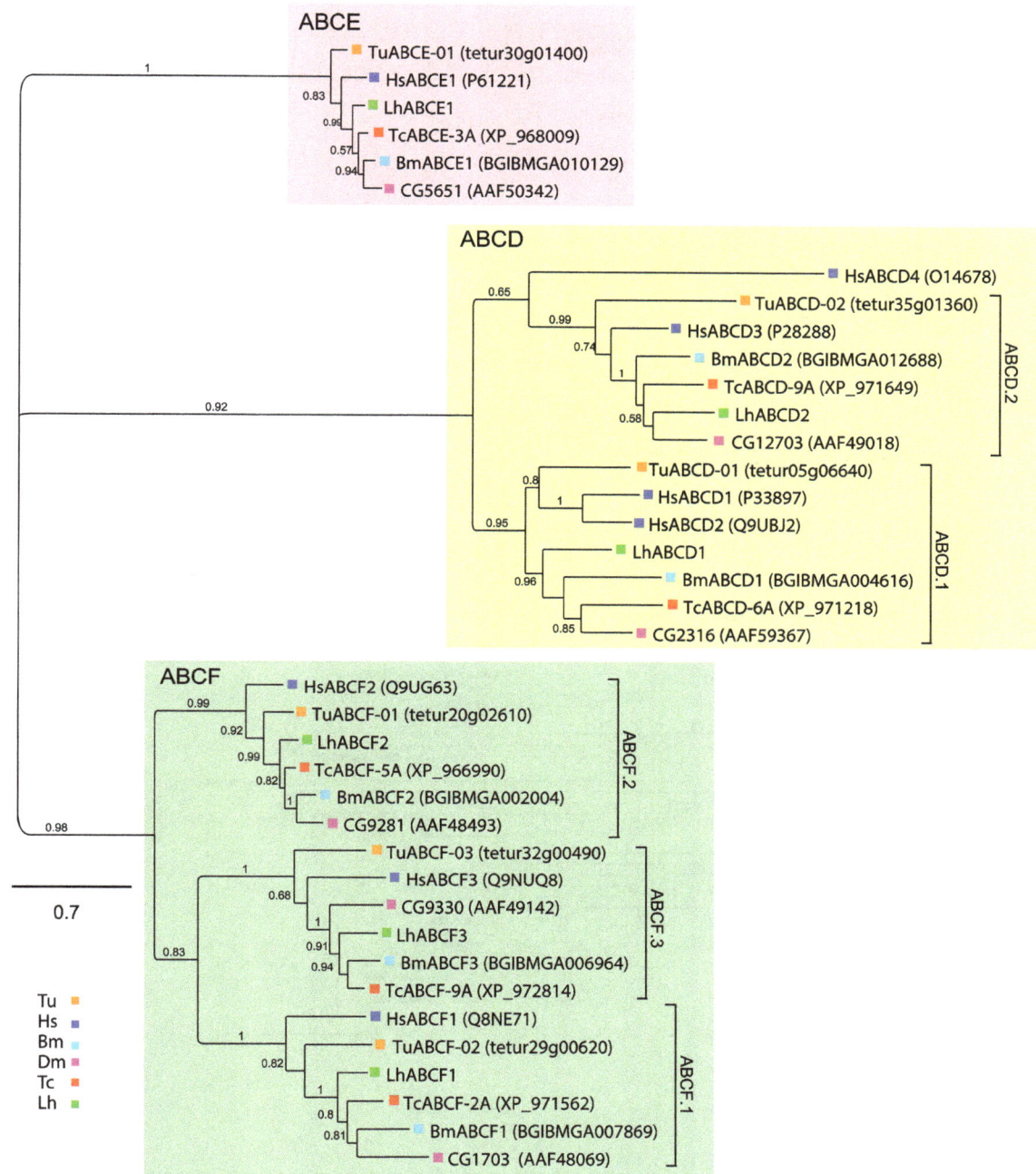

**Figure 4. Phylogenetic analysis of ABCD, ABCE, and ABCF transporters from *L. hesperus* and five metazoan species.** Clades corresponding to the three subfamilies are shaded light red (ABCE), light yellow (ABCD), and green (ABCF). The scale bar represents 0.7 amino acid substitutions per site. Analyses, abbreviations, and color-coding are as in Figure 1.

suggests similar dimerization of LhABCG5 and LhABCG4 could be important for their functionality. LhABCG14 aligned to clade ABCG.4C with arthropod ABCGs potentially involved in ecdysteroid signaling [16,22,117–119].

Six LhABCGs sorted to three clades (ABCG.4D, ABCG.4E, and ABCG.4F) characterized by *D. melanogaster* ABC transporter genes (*white, brown,* and *scarlet*) that function in the import of eye pigment precursors [17,18]. Heterodimers of the *white* and *brown* gene products transport red-pigmented pteridine precursors,

whereas heterodimers of the *white* and *scarlet* gene products are crucial for the import of brown-pigmented monochrome precursors. Consequently, *white* mutants are characterized by white eyes (complete loss of pigmentation), *brown* mutants by dark brown eyes (loss of red pigments), and *scarlet* mutants by bright red eyes (loss of brown pigments). Similar roles in pigment transport have been described for homologs of the three genes in *B. mori* [19,20,120,121]. These ABCGs also function in biogenic amine transport [24], uric acid uptake [20,121], and courtship behavior

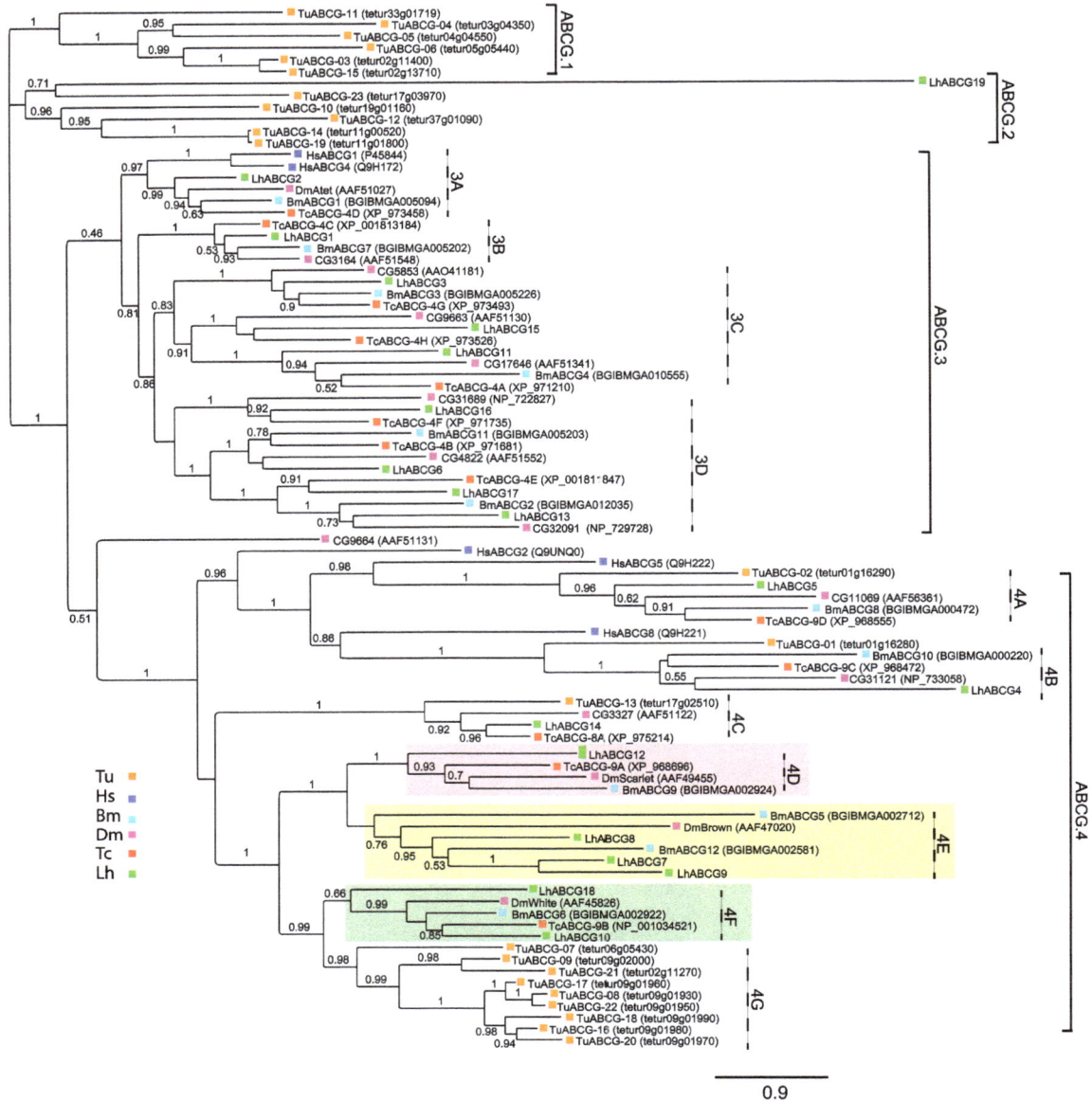

Figure 5

**Figure 5. Phylogenetic analysis of ABCG transporters from *L. hesperus* and five metazoan species.** Clades corresponding to transporters involved in eye pigmentation have been shaded light red (*scarlet*), light yellow (*brown*), and green (*white*). The scale bar represents 0.9 amino acid substitutions per site. Analyses, abbreviations, and color-coding are as in Figure 1.

in *D. melanogaster* [122]. In our analyses, LhABCG12 clustered within the *scarlet* clade (ABCG.4D), LhABCG10 and LhABCG18 within the *white* clade (ABCG.4F), and three LhABCGs (LhABCG7, LhABCG8, and LhABCG9) that share 40–58% sequence identity (Table S9) aligned to the *brown* clade (ABCG.4E). While red eye mutants of various plant bugs, including a species sympatric to *L. hesperus* (*Lygus lineolaris*), have been reported [123–126], the functional importance of *scarlet* in these phenotypes is unknown. The LhABCG19 partial sequence aligned with a group from *T. urticae* in clade ABCG.2.

## ABCH subfamily

Similar to ABCGs, ABCH transporters also have the inverted NBD-TMD configuration. ABCHs though, with the exception of zebrafish [15,127], are specific to arthropods [8,34]. While most arthropods have 3 ABCH transporters (Table 1), lineage-specific gene duplications have resulted in 22 and 14 ABCH genes in *T. urticae* and *D. pulex* respectively [32,33]. We found similar expansion in *L. hesperus* with 11 LhABCH transcripts, 8 of which are full-length coding sequences. Although BLASTx analyses indicate that these transcripts share sequence similarity with

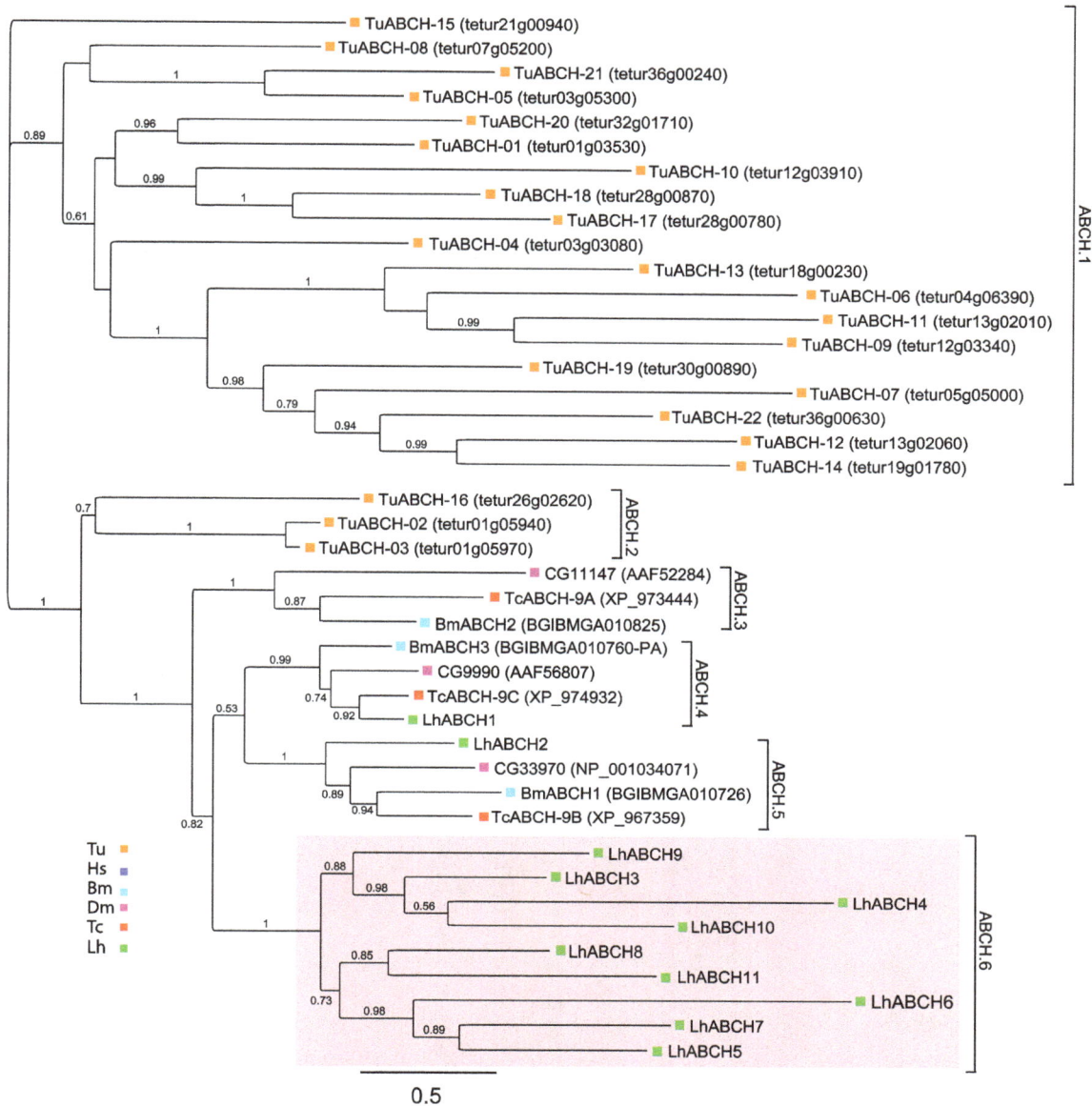

**Figure 6. Phylogenetic analysis of ABCH transporters from *L. hesperus* and four metazoan species.** The scale bar represents 0.5 amino acid substitutions per site. Analyses, abbreviations, and color-coding are as in Figure 1. Because the ABCH subfamily is largely restricted to the arthropod lineage, no representative sequences for *H. sapiens* are available and thus the analyses were performed using only arthropod sequences. *L. hesperus* ABCH sequences that have undergone gene expansion are shaded light red.

ABCGs (Table S1 and S2), phylogenetic analyses clustered these transcripts in the ABCH clade (Figure S3). As reported previously [33], *T. urticae* sequences are unique and do not align well with other ABCHs (Figure 6). LhABCH1 aligned with transporters in clade ABCH.4 that likely function in cuticular lipid transport [16,128,129]. LhABCH2 sorted to the sister ABCH.5 clade along with the *D. melanogaster* ABCH CG33970, which is upregulated in response to cold hardening [130]. The remaining LhABCH sequences formed a separate clade, implying that, like *T. urticae* and *D. pulex*, independent lineage-specific gene duplication events have contributed to the expansion of the LhABCH subfamily. While the physiological functions of the ABCH subfamily remain largely uncharacterized, differential expression of ABCHs has been reported for *T. urticae* females in diapause [119] and some

ABCH transcript levels are elevated in insecticide resistant strains of *T. urticae* [33] and *P. xylostella* [131].

## Expression profile of LhABC transcripts

Many ABC transporters function in development [16,21]. Consequently, we used end-point PCR to examine the developmental expression profile of a subset of 13 LhABCs representative of the ABCA, ABCB, ABCC, ABCG, and ABCH subfamilies. All of the transcripts were amplified from both reproductively immature and mature adults and all but LhABCA8 were expressed to some degree in eggs (Figure 7A). Five LhABCs had limited nymphal expression; LhABCA8 and LhABCC3 were detected in early and late stadium fifth instars, LhABCB2 and LhABCB6 in late stadium fifth instars, and LhABCC5 in first,

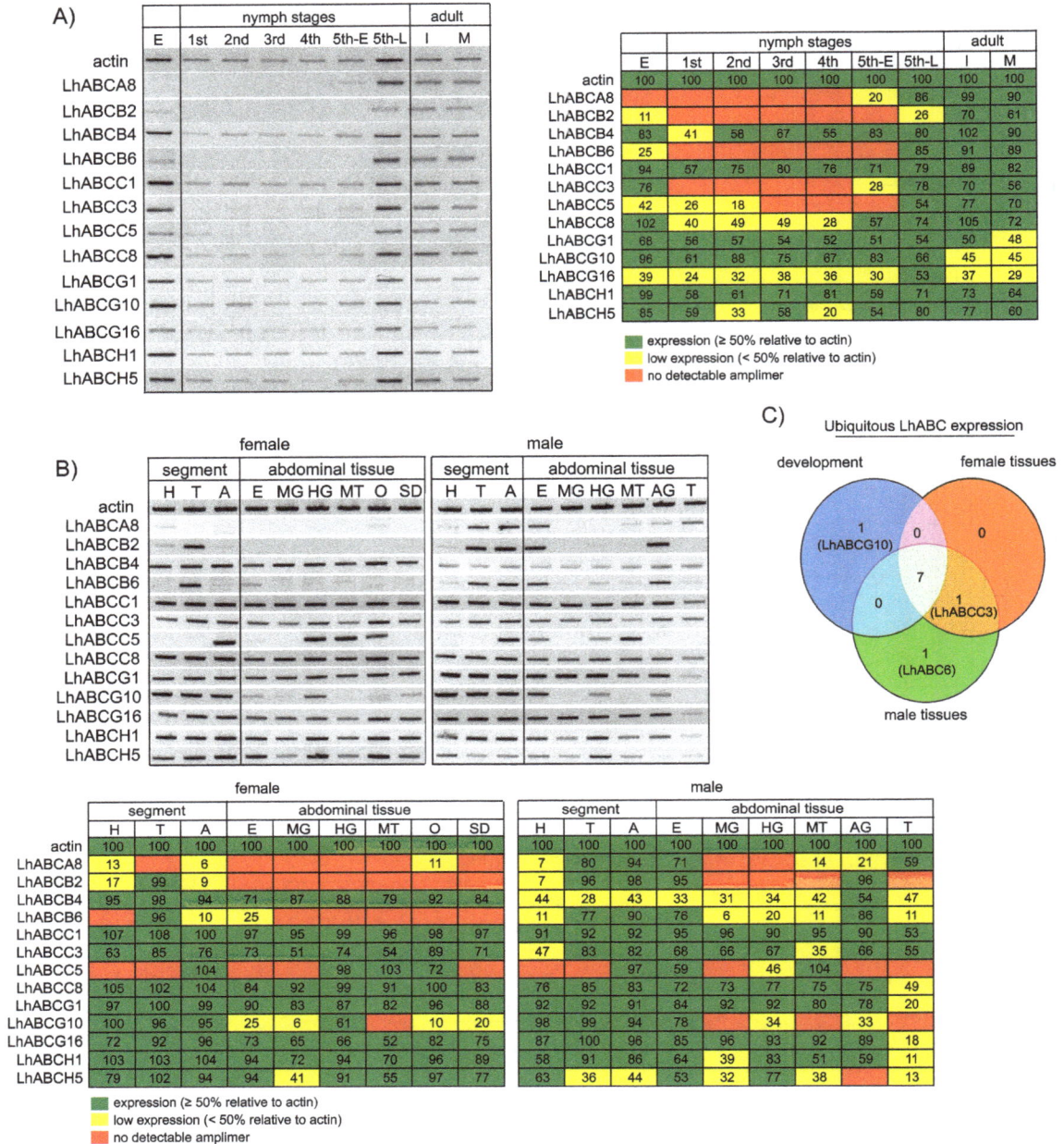

**Figure 7. Transcriptional expression profile of 13 LhABC transcripts.** A) Developmental profile. Expression profile of 13 LhABC transporters in eggs through 12-day-old adults was examined by end-point PCR using primers designed to amplify a ~500 bp fragment of each transcript. Abbreviations: E, eggs; 1st, first instars; 2nd, second instars; 3rd, third instars; 4th, fourth instars; 5th-E, early stadium fifth instars; 5th-L, late stadium fifth instars; I, reproductively immature 1-day-old mixed sex adults; M, mature 12-day-old mixed sex adults. Products were analyzed on 1.5% agarose gels and stained with SYBR Safe. Actin was used as a positive control. Leftmost image - Representative gel image. For clarity, the negative image is shown. Rightmost image – Visual aid depicting semi-quantitative analysis of amplimers of interest compared to actin in the representative gel. Relative to the actin amplimer of each developmental stage, cells with amplimer intensity ≥50% are indicated in green, while cells <50% are indicated in yellow. Red cells indicate no detectable amplimer. Numbers inside individual cells denote the percentage of amplimer intensity compared to actin. The primer set used for LhABCC1 profiling amplifies a shared region of LhABCC1A and LhABCC1B. B) Adult tissue profile. The expression profile of the 13 LhABC transcripts was examined as above in adult body segments and various abdominal tissues prepared from 7-day-old adults. Abbreviations: H, head; T, thorax; A, abdomen; E, epidermis; MG, midgut; HG, hindgut; MT, Malpighian tubules; O, ovary; SD, seminal depository; AG, accessory glands (lateral and medial); T, testis. Top image - Representative gel image. For clarity, the negative image is shown. Lower image – Visual aid depicting semi-quantitative analysis of amplimers of interest compared to actin in the representative gel. Color shading is as above but relative to the intensity of actin amplimers within each respective tissue. C) Three-way Venn diagram comparing the transcriptional expression profile of LhABC transcripts ubiquitously expressed throughout *L. hesperus* development (blue) with those expressed in female (red) and male (green) tissues.

second (albeit weakly) and late stadium fifth instars (Figure 7A). Orthologs of the eight LhABCs expressed throughout *L. hesperus* nymphal development have been reported to have similar expression profiles [16], suggesting potential roles in basic physiological functions.

We next examined the transcription profiles of the 13 LhABC transporters in specific body segments (head, thorax, abdomen) and abdominal tissues (epidermis/cuticle, midgut, hindgut, Malpighian tubule, ovary, seminal depository, male medial and lateral accessory glands, and testis) from 7-day-old virgins of each sex (Figure 7B). Eight of the transcripts were amplified from all tissues/segments in both sexes with seven of them also constitutively transcribed throughout development (Figure 7C). In contrast, LhABCC3, which is present in all adult tissues/segments, was only amplified from eggs, fifth instar nymphs and adults (Figure 7A), suggesting a potential role in reproductive development. The inverse was observed with LhABCG10, a *white*-like gene that is transcribed throughout development but which exhibits tissue specific transcription in adults (Figure 7B). The low abundance of LhABCG10 transcripts in Malpighian tubules differs from that reported for *white* genes in *D. melanogaster* and *B. mori* [18,120,132], which play key roles in the uptake and concentration of excess tryptophan [133]. The undetectable transcript levels of LhABCG10 in this tissue suggests tryptophan transport in *L. hesperus* is mediated by some other mechanism or involves a different transporter such as LhABCG18, which also sorts with *white*-like genes in clade ABCG.4F (Figure 5).

Malpighian tubules are the main excretory and osmoregulatory organs in insects and are thus crucial in clearing toxic endogenous compounds and xenobiotics [134,135]. In support of this role, ABC transporters and detoxification enzyme levels are frequently present at relatively high levels in Malpighian tubules [75,85,136–138]. LhABCC5 expression was specific to the abdomen of both sexes where it predominantly localized to the hindgut and Malpighian tubules (Figure 7B), suggesting a potential role in xenobiotic excretion.

The expression of LhABCA8, LhABCB2, and LhABCB6 was sex-biased with higher levels of the three transcripts in male abdomen compared to female abdomen (Figure 7B). Among the male abdominal tissues, the three transporters were enriched in reproductive tissues (LhABCA8– testis; LhABCB2 and LhABCB6– accessory gland). The developmental profile of the three transcripts is likewise similar with expression limited to eggs and fifth instars (Figure 7A). This latter period coincides with the development of male reproductive organs (Figure S4), suggesting an association with sexual maturation. A number of *D. melanogaster* ABC transporters are highly expressed in male reproductive tissues [84,132,139] and elevated testicular expression of ABC transporters has been reported for *B. mori* [37,38,77]. The elevated levels of ABC transporters in reproductive tissues may be critical for protection of spermatozoa [140]. Additionally, the ABC transporters might function in the loading of accessory glands and other male secretory reproductive tissues with seminal fluid components (e.g., prostaglandins, lipids, peptides, hormones, etc.) [141,142].

The presence of LhABC transcripts in the abdominal epidermis of both males and females (Figure 7B) could indicate potential functions in integument coloration [19,20,143] or in the transport of cuticular lipids to prevent water loss [16]. Alternatively, the transporters may be expressed in oenocytes, polyploid insect cells found in close association with the epidermis that have been reported to function in xenobiotic detoxification, the synthesis of cuticle components, and innate immunity [144]. Elevated levels of ABC transporter transcripts in epidermis have been reported for some insecticide-resistant bed bug populations [145], suggesting a potential role in xenobiotic transport at the cuticular layer.

## Conclusions

The genus *Lygus* encompasses more than 30 different species of polyphagous pests that attack crops worldwide. However, reports of insecticide resistance in field populations threaten the sustainability of insecticide-based management strategies. Consequently, there is growing interest in elucidating the molecular basis of resistance. While a number of studies have focused on identifying detoxification enzymes, the role of ABC transporters in insecticide clearance in hemipteran pests has been largely neglected. To address this, we used RNA-Seq to identify the ABC transporter superfamily in *L. hesperus*. Defining the functional relevance and substrate specificity of the 65 LhABC-like transcripts will be a future research priority. Initial efforts will focus on assessing the effects of insecticide exposure on expression levels of the LhABCs, in particular LhABCC5, the tissue localization of which is consistent with a role in insecticide/xenobiotic clearance. Furthermore, targeting ABC transporters by RNAi may facilitate the development of novel control strategies for *L. hesperus* and other hemipteran pests.

## Materials and Methods

### Insects

*L. hesperus* were obtained from an established laboratory colony. Stock insects were maintained at 27.5–29.0°C under 40% humidity with a L14:D10 photoperiod, and fed artificial diet packaged in Parafilm M [146]. Experimental nymphs and adults were generated from eggs deposited in oviposition packets (agarose gel packaged within Parafilm M) and maintained as described previously [147].

### RNA isolation and Illumina sequencing

To induce expression of potential stress-related genes, 10-day-old *L. hesperus* adults from a single cohort were placed individually in covered glass Petri dishes (60×15 mm) along with a section of green bean. Dishes were transferred to environmental chambers and exposed to one of three temperatures (4°C, 25°C, or 39°C) for 4 hr. Insects were stored in RNALater (Ambion, Life Technologies, Carlsbad, CA) at −80°C. Total RNA from frozen samples was isolated by the University of Arizona Genomics Center (http://uagc.arl.arizona.edu; Tucson, AZ) using an RNeasy Mini Kit (Qiagen, Valencia, CA) followed by on-column DNAse digestion according to the manufacturer's instructions. RNA samples were eluted in 30 μL RNAse-free $H_2O$. RNA quality was assessed on a Fragment Analyzer Automated CE System (Advanced Analytical Technologies, Ames, IA) and RNA was quantified using RiboGreen (Molecular Probes, Eugene, OR). Triplicate RNA libraries for each of the three temperature regimens were constructed using a TruSeq RNA Sample Preparation Kit v2 (Illumina Inc., San Diego, USA) and sequenced on an Illumina HiSeq2500 in rapid run mode. CASAVA version 2.8 was used for base calling and de-multiplexing.

### Transcriptome assembly

Raw de-multiplexed reads for each sample were assembled into single files and then trimmed and quality filtered with Trimmomatic version 0.30 [148] using the parameters LEADING:20, TRAILING:20, WINSIZE:5, WINCUTOFF:25, MINLEN:50, and ILLUMINACLIP:TruSeq3-PE.fa:2:30:10. Quality metrics were calculated for the unfiltered and filtered data using FASTQC version 0.10.1. After quality filtering, orphaned pairs were discarded while reads still having a read pair were used for assembly. Reads were normalized *in silico* using the "normalize_by_kmer_coverage.pl" script distributed with the Trinity transcriptome assembly pipeline (r2013_08_14) [149] and a kmer size of 25 and maximum read coverage of 40. Normalized reads were used to create a *de novo* transcriptome assembly with Trinity (Inchworm, Chrysalis, and Butterfly) using default parameters except with the jaccard_clip option used to compare paired-read consistency to reduce the creation of fused transcripts from non-strand specific data. The initial Trinity assembly was further filtered to maintain only transcripts exhibiting evidence of a coding region. Open reading frames (ORFs) were predicted using Transdecoder (r2012–08–15) with training against the 500 longest ORFs in the transcriptome. ORF transcripts were also identified based on significant matches to the Pfam-A database using a HMMER search [150]. Transcripts were only retained if they had an ORF predicted by Transdecoder with a length longer than 100 amino acids. The filtered transcriptome was annotated using InterProScan 5, and gene names assigned via BLASTp alignment to the UniProtKB/SwissProt database. The raw data was deposited in the NCBI sequence read archive under BioProject PRJNA238835, BioSamples SAMN02679940 - SAMN02679948, SRA Submission ID "PBARC: *Lygus hesperus* Heat Experiment", SRA Study Accession SRP039607. To facilitate submission to the NCBI TSA database, transcript sequences were modified to put all coding sequence on the positive strand by reverse complementing when appropriate and the longest coding sequence for each transcript was submitted to TSA using an open source transcriptome preparation software package (http://genomeannotation.github.io/transvestigator/). The annotated assembly with putative gene name and functional annotations was submitted to NCBI under TSA submission GBHO00000000. The version described in this paper is the first version, GBHO01000000.

## Annotation and bioinformatic analysis of the L. hesperus ABC transporter superfamily

ABC transporter annotations were performed on the longest isoform for each unigene. Putative ABC transporter sequences were identified by initially performing a BLASTn search (E value $\leq 10^{-10}$) of the assembly described above with queries consisting of sequences annotated previously as ABC transporters [52]. A BLASTx search (E value $\leq 10^{-10}$) of the RNA-Seq data was also performed using the full complement of ABC transporters from *H. sapiens* [34], *C. elegans* [151], *D. pulex* [32], *T. urticae* [33], *D. melanogaster* [34], *T. castaneum* [16], and *B. mori* [37,38]. The longest isoform of the resulting *L. hesperus* sequence hits were then re-evaluated against the NCBI nr database using BLASTx and tBLASTn (E value $\leq 10^{-10}$). The sequence list of positive hits was curated to remove duplicates.

Identification of putative ABC transporter domains was performed using ScanProsite [152,153] and the HMMscan module on the HMMER webserver [154]. Subcellular localization prediction of LhABC transporters was performed with WoLF PSORT [155]. Transmembrane domain predictions were performed using TMHMM v2.0 [156], TOPCONS [157], and TopPred II [158].

Putative LhABC transporters were initially assigned to the respective subfamilies (A-H) based on BLAST analyses, then assignments were refined based on phylogenetic inferences. Multiple sequence alignments consisting of the putative *L.*

*hesperus* sequences and ABCs from five metazoans (*H. sapiens*, *T. urticae*, *D. melanogaster*, *T. castaneum*, and *B. mori*) were performed using default settings for MUSCLE [159]. Phylogenetic trees were constructed utilizing FastTree2 [160] implemented in Geneious 7.0.6 using default settings augmented by the Whelan-Goldman model and optimization for Gamma20 likelihood. Parallel analyses were also performed in MEGA5 [161] with bootstrap testing of 500 replicates using both the maximum parsimony method and the maximum likelihood method based on the JTT matrix-based model [162]. Phylogenetic analyses were performed using available sequences (both complete coding sequences and partial fragments) rather than specific domains as reported for other species [16,33]. The clustering of subfamilies and orthologous genes was compared amongst the three phylogenetic methods and with previous analyses of ABC transporters [8,16,33,37,38]. Heat identity maps for the LhABC transporters were generated using Geneious 7.1.7 (Biomatters Ltd., Auckland, New Zealand) and MUSCLE-based sequence alignments. See Table S11 for the accession numbers of the proteins used in the phylogenetic analyses, and Table S12 for LhABC transporter amino acid sequences. *T. castaneum* sequences lacking accession numbers were downloaded as FASTA protein files directly from BeetleBase (http://beetlebase.org/) based on the reported GLEAN accessions [16].

## End point PCR expression analyses

The expression profile of a subset of LhABC transporters was examined across three biological replicates throughout *L. hesperus* development and among sex-specific adult body segments/tissues. TRI Reagent Solution (Ambion) was used to isolate total RNA from pooled samples of eggs, first - fourth instars, early stadium fifth instars, late stadium fifth instars, reproductively immature adults (1-day-old) of each sex, and mature virgin adults (12-day-old) of each sex. Total RNA was also isolated from pooled 7-day-old adult virgin male and female body segments and abdominal tissues: 10×head, 3×thorax, 5×abdomen, 15×abdominal carcass, 5×midgut, 20×hindgut, 20×Malpighian tubules, 20×seminal depository, and 5 pairs each of ovaries, male medial and lateral accessory glands, and testes. First-strand cDNAs were generated using a Superscript III first-strand cDNA synthesis kit (Invitrogen) with custom-made random pentadecamers (IDT, San Diego, CA) and 500 ng of DNase I-treated total RNAs. PCR expression profiling was performed using 0.4 µL of the prepared cDNAs with Sapphire Amp Fast PCR Master Mix (Takara Bio Inc./Clontech, Madison, WI) and sequence-specific primers (Table 3) designed to amplify ~500 bp fragments of the LhABC transcripts. Thermocycler conditions consisted of 95°C for 2 min followed by 35 cycles at 94°C for 20 s, 56°C for 20 s, and 72°C for 20 s, and finished with a 1 min incubation at 72°C. PCR products were analyzed by gel electrophoresis on 1.5% agarose gels stained with SYBR Safe (Life Tech.) and a Tris/acetate/EDTA buffer system. Representative amplimers of the expected sizes were gel excised using an EZNA Gel Extraction kit (Omega Bio-Tek Inc., Norcross, GA), sub-cloned into the pCR2.1TOPO TA cloning vector (Invitrogen), and sequenced at the Arizona State University DNA Core Lab (Tempe, AZ). In all cases, sequence variation of the cloned LhABC transporter fragments was minimal (>98% nucleotide identity) compared to the transcriptomic data, indicating that the assembled data accurately represent the sequences. The minor variations in sequence were likely attributable to the allelic heterogeneity of the *L. hesperus* laboratory colony, or to rare errors introduced during amplification.

**Table 3.** Oligonucleotide primers used in expression profiling and cloning.

| gene | sequence (5'-3') | Amplimer size |
|---|---|---|
| Lygus actin F | ATGTGCGACGAAGAAGTTG | 555 |
| Lygus actin R1 | GTCACGGCCAGCCAAATC | |
| LhABCA8 8 F | AAGGCTGGTTTGCTGTGGCT | 533 |
| LhABCA8 541 R | AGGAGCTGGATCGATAGCTCG | |
| LhABCB2 444 F | CACCGCTCAGCAATGCAACC | 491 |
| LhABCB2 935 R | TGAGGACCTCGCCTGCGATA | |
| LhABCB4 1848 F | CTGGCTACGTCGTCCAGCTC | 511 |
| LhABCB4 2359 R | GTCAGCATTTCTCGCAGCGG | |
| LhABCB6 51 F | CCTCGCCTGGCTGAGAAGTC | 481 |
| LhABCB6 532 R | AATACAATCCGCCCGCCTCC | |
| LhABCC1 1240 F | GGCTGCACATACGGACTGCT | 468 |
| LhABCC1 1708 R | GCAGCACCGAAGTAGGCACT | |
| LhABCC3 1755 F | AGATGGCTGGGACCTTCCGT | 492 |
| LhABCC3 2247 R | TGGTTCAGAGCTGACACGGC | |
| LhABCC5 1143 F | GGTGGCCGAATGTCGAGTGT | 523 |
| LhABCC5 1666 R | GCGATGCTCCCCTTTCACCA | |
| LhABCC8 1210 F | CGTCGCAGGATTTTGCACCC | 496 |
| LhABCC8 1706 R | TGCGGTTGTCCTTGTGTTGC | |
| LhABCG1 220 F | TGACCATCACGCCGTGTCAG | 467 |
| LhABCG1 687 R | TCGTCTTCACGTGCTCCTGG | |
| LhABCG10 788 F | AAGCGTCTTGGTGGGCTCAG | 517 |
| LhABCG10 1305 R | AGGGCCCACTGACAAGGCTA | |
| LhABCG16 850 F | TCTGCACGATTCACCAGCCC | 483 |
| LhABCG16 1333 R | TTGCTACCGTCTTGTCCCGC | |
| LhABCH1 927 F | AGCTCCCCAAGTCCTGCTGA | 473 |
| LhABCH1 1400 R | CCTGTAGGGTCCCGACCGAT | |
| LhABCH5 934 F | ATAGGCACAGCAGTGCACCC | 495 |
| LhABCH5 1429 R | TATCCCCGGGACGTGATGCT | |
| LhABCC1A start F | ATGGCCGAGGATACGCTT | 767 |
| LhABCC1B start F | ATGGCAGAGGAAACACTTC | 767 |
| LhABCC1A/B 767 R | TCGCCAATGTGGCTTTCC | |

## Supporting Information

**Figure S1**  Distribution of the most highly represented species in BLASTx and tBLASTn analyses of *L. hesperus* ABC transporter sequences. BLAST analyses were performed using the NCBI non-redundant database with an E value $\leq 10^{-10}$.

**Figure S2**  Amino acid sequence alignment of LhABCC1A and LhABCC1B. Alignment was performed using the default settings in MUSCLE [159]. Pairwise sequence identity for the full-length transporters is 96%. Pairwise sequence identity for the first 260 amino acids is 86% (223/260 aa), whereas identity over the remaining 684 amino acids is 99.6% (681/684 aa). Black shading is indicative of 100% amino acid sequence identity.

**Figure S3**  Phylogenetic analysis of ABCG and ABCH transporters from *L. hesperus* and four other species. The scale bar represents 1.0 amino acid substitutions per site. Analyses,

abbreviations, and color-coding are as in Figure 1. Clades corresponding to the two subfamilies are indicated by tan shading (ABCH) or yellow shading (ABCG). As before, because the ABCH subfamily is restricted to the arthropod lineage, no representative sequences for *H. sapiens* were included in the analysis.

**Figure S4**  Length of male *L. hesperus* accessory glands and testes in fifth instar nymphs and adults. Testis length was measured from the base to the apical tip of the longest lobe. Accessory gland length was measured from the insertion at the common duct to its anterior end where the accessory gland folds over on itself. It should be noted that while primordial reproductive tissues are present in fourth instar nymphs they are smaller than that seen in early fifths and very poorly developed. Stage selection criteria were: early stadium fifth instars - small green abdomen and thin wing buds with light pigmentation; late stadium fifth instars - enlarged abdomen with yellow color and significant fatty deposits, thickened wing bugs with heavy pigmentation; adults – light body pigmentation, minimal body fat, wings not hardened, sampled within 12 h of eclosion. All specimens sampled were from the same cohort. Error bars represent standard deviation (n = 20 for each group).

**Table S1**   Top five BLASTx hits from a search against the non-redundant protein database using the 65 putative LhABC transporter sequences as a query. Analysis performed with an E value $\leq 10^{-10}$.

**Table S2**   Top five tBLASTn hits from a search against the non-redundant database using the 65 putative LhABC transporter sequences as a query. Analysis performed with an E value $\leq 10^{-10}$.

**Table S3**   Identification of potential protein domains in the putative LhABC transporter sequences. Analyses were performed using default settings for ScanProsite [152] and HMMScan on the HMMER webserver [154] using default settings with protein databases set to Pfam, Gene3D, and Superfamily.

**Table S4**   MUSCLE based multiple sequence alignment heat map of the percent amino acid identities among the LhABC transporters. The matrix, which includes partial sequences, was generated from a MUSCLE alignment and indicates the percent identity across the predicted protein sequences. Cell shading is based on a sliding three color scale with lowest percent identities in red and highest percent identities in blue.

**Table S5**   MUSCLE based multiple sequence alignment heat map of the percent amino acid identities among the LhABCA transporters. The matrix and cell shading are as described in Table S4.

**Table S6**   MUSCLE based multiple sequence alignment heat map of the percent amino acid identities among the LhABCB transporters. The matrix and cell shading are as described in Table S4.

**Table S7**   MUSCLE based multiple sequence alignment heat map of the percent amino acid identities among the LhABCC

transporters. The matrix and cell shading are as described in Table S4.

**Table S8**   MUSCLE based multiple sequence alignment heat map of the percent amino acid identities among the LhABCD, LhABCE, and LhABCF transporters. The matrix and cell shading are as described in Table S4.

**Table S9**   MUSCLE based multiple sequence alignment heat map of the percent amino acid identities among the LhABCG transporters. The matrix and cell shading are as described in Table S4.

**Table S10**   MUSCLE based multiple sequence alignment heat map of the percent amino acid identities among the LhABCH transporters. The matrix and cell shading are as described in Table S4.

**Table S11**   Gene accession/model numbers of ABC transporter protein sequences used in phylogenetics analyses.

**Table S12**   LhABC transporter protein sequences.

## Acknowledgments

The authors thank Dr. David J. Hawthorne (University of Maryland) for initial discussions and critical reading of the manuscript. The authors also thank Daniel Langhorst and Lynn Forlow Jech (both from USDA-ARS ALARC) for maintaining the *L. hesperus* colony and assistance with tissue dissections, and Brian Hall (USDA-ARS Daniel K. Inouye Pacific Basin Agricultural Research Center) for assistance with NCBI data deposition. Mention of trade names or commercial products in this article is solely for the purpose of providing specific information and does not imply recommendation or endorsement by the U. S. Department of Agriculture. USDA is an equal opportunity provider and employer.

## Author Contributions

Conceived and designed the experiments: JJH JAF CSB LCL. Performed the experiments: JJH KC CSB. Analyzed the data: JJH KC SMG CSB. Contributed reagents/materials/analysis tools: JJH SMB CSB DW LCL. Contributed to the writing of the manuscript: JJH SMG CSB JAF DW LCL.

## References

1. Oldham ML, Davidson AL, Chen J (2008) Structural insights into ABC transporter mechanism. Curr Opin Struct Biol 18: 726–733.
2. Rees DC, Johnson E, Lewinson O (2009) ABC transporters: the power to change. Nature 10: 218–227.
3. Jones PM, O'Mara ML, George AM (2009) ABC transporters: a riddle wrapped in a mystery inside an enigma. Trends Biochem Sci 34: 520–531.
4. George AM, Jones PM (2012) Perspectives on the structure-function of ABC transporters: the Switch and Constant Contact models. Prog Biophys Mol Biol 109: 95–107.
5. Kartner N, Riordan JR, Ling V (1983) Cell surface P-glycoprotein associated with multidrug resistance in mammalian cell lines. Science 221: 1285–1288.
6. Lage H (2003) ABC-transporters: implications on drug resistance from microorganisms to human cancers. Int J Antimicrob Agents 22: 188–199.
7. Leprohon P, Légaré D, Ouellette M (2011) ABC transporters involved in drug resistance in human parasites. Essays Biochem 50: 121–144.
8. Dermauw W, Van Leeuwen T (2014) The ABC gene family in arthropods: Comparative genomics and role in insecticide transport and resistance. Insect Biochem Mol Biol 45: 89–110.
9. Bisbal C, Martinand C, Silhol M, Lebleu B, Salehzada T (1995) Cloning and characterization of a RNAse L inhibitor. A new component of the interferon-regulated 2–5A pathway. J Biol Chem 270: 13308–13317.
10. Tyzack JK, Wang X, Belsham GJ, Proud CG (2000) ABC50 interacts with eukaryotic initiation factor 2 and associates with the ribosome in an ATP-dependent manner. J Biol Chem 275: 34131–34139.
11. Dong J, Lai R, Nielsen K, Fekete CA, Qiu H, et al. (2004) The essential ATP-binding cassette protein RLI1 functions in translation by promoting preinitiation complex assembly. J Biol Chem 279: 42157–42168.
12. Paytubi S, Wang X, Lam YW, Izquierdo L, Hunter MJ, et al. (2009) ABC50 promotes translation initiation in mammalian cells. J Biol Chem 284: 24061–24073.
13. Barthelme D, Dinkelaker S, Albers S-V, Londei P, Ermler U, et al. (2011) Ribosome recycling depends on a mechanistic link between the FeS cluster domain and a conformational switch of the twin-ATPase ABCE1. Proc Natl Acad Sci USA 108: 3228–3233.
14. Dean M, Annilo T (2005) Evolution of the ATP-binding cassette (ABC) transporter superfamily in vertebrates. Annu Rev Genomics Hum Genet 6: 123–142.
15. Popovic M, Zaja R, Loncar J, Smital T (2010) A novel ABC transporter: the first insight into zebrafish (*Danio rerio*) ABCH1. Mar Environ Res 69 Suppl: S11–S13.
16. Broehan G, Kroeger T, Lorenzen M, Merzendorfer H (2013) Functional analysis of the ATP-binding cassette (ABC) transporter gene family of *Tribolium castaneum*. BMC Genomics 14: 6.

17. Ewart GD, Cannell D, Cox GB, Howells AJ (1994) Mutational analysis of the traffic ATPase (ABC) transporters involved in uptake of eye pigment precursors in *Drosophila melanogaster*. Implications for structure-function relationships. J Biol Chem 269: 10370–10377.

18. Mackenzie SM, Brooker MR, Gill TR, Cox GB, Howells AJ, et al. (1999) Mutations in the white gene of *Drosophila melanogaster* affecting ABC transporters that determine eye colouration. Biochim Biophys Acta 1419: 173–185.

19. Kômoto N, Quan G-X, Sezutsu H, Tamura T (2009) A single-base deletion in an ABC transporter gene causes white eyes, white eggs, and translucent larval skin in the silkworm w-3(oe) mutant. Insect Biochem Mol Biol 39: 152–156.

20. Wang L, Kiuchi T, Fujii T, Daimon T, Li M, et al. (2013) Mutation of a novel ABC transporter gene is responsible for the failure to incorporate uric acid in the epidermis of ok mutants of the silkworm, *Bombyx mori*. Insect Biochem Mol Biol 43: 562–571.

21. Ricardo S, Lehmann R (2009) An ABC transporter controls export of a Drosophila germ cell attractant. Science 323: 943–946.

22. Itoh TQ, Tanimura T, Matsumoto A (2011) Membrane-bound transporter controls the circadian transcription of clock genes in Drosophila. Genes Cells 16: 1159–1167.

23. Strauss AS, Peters S, Boland W, Burse A (2013) ABC transporter functions as a pacemaker for the sequestration of plant glucosides in leaf beetles. eLife 2: e01096.

24. Borycz J, Borycz JA, Kubow A, Lloyd V, Meinertzhagen IA (2008) Drosophila ABC transporter mutants white, brown and scarlet have altered contents and distribution of biogenic amines in the brain. J Exp Biol 211: 3454–3466.

25. Buss DS, Callaghan A (2008) Interaction of pesticides with p-glycoprotein and other ABC proteins: A survey of the possible importance to insecticide, herbicide and fungicide resistance. Pest Biochem Physiol 90: 141–153.

26. Gahan LJ, Pauchet Y, Vogel H, Heckel DG (2010) An ABC transporter mutation is correlated with insect resistance to *Bacillus thuringiensis* Cry1Ac toxin. PLoS Genet 6: e1001248.

27. Baxter SW, Badenes-Perez FR, Morrison A, Vogel H, Crickmore N, et al. (2011) Parallel evolution of *Bacillus thuringiensis* toxin resistance in Lepidoptera. Genetics 189: 675–679.

28. Atsumi S, Miyamoto K, Yamamoto K, Narukawa J, Kawai S, et al. (2012) Single amino acid mutation in an ATP-binding cassette transporter gene causes resistance to Bt toxin Cry1Ab in the silkworm, *Bombyx mori*. Proc Natl Acad Sci USA 109: E1591–E1598.

29. Heckel DG (2012) Learning the ABCs of Bt: ABC transporters and insect resistance to *Bacillus thuringiensis* provide clues to a crucial step in toxin mode of action. Pest Biochem Physiol 104: 103–110.

30. Lei Y, Zhu X, Xie W, Wu Q, Wang S, et al. (2014) Midgut transcriptome response to a Cry toxin in the diamondback moth, *Plutella xylostella* (Lepidoptera: Plutellidae). Gene 533: 180–187.

31. Park Y, González-Martínez RM, Navarro-Cerrillo G, Chakroun M, Kim Y, et al. (2014) ABCC transporters mediate insect resistance to multiple Bt toxins revealed by bulk segregant analysis. BMC Biology 12: 46.

32. Sturm A, Cunningham P, Dean M (2009) The ABC transporter gene family of *Daphnia pulex*. BMC Genomics 10: 170.

33. Dermauw W, Osborne EJ, Clark RM, Grbić M, Tirry L, et al. (2013) A burst of ABC genes in the genome of the polyphagous spider mite *Tetranychus urticae*. BMC Genomics 14: 317.

34. Dean M, Rzhetsky A, Allikmets R (2001) The human ATP-binding cassette (ABC) transporter superfamily. Genome Res 11: 1156–1166.

35. Pavlidi N, Dermauw W, Rombauts S, Chrisargiris A, Van Leeuwen T, et al. (2013) Analysis of the olive fruit fly *Bactrocera oleae* transcriptome and phylogenetic classification of the major detoxification gene families. PLoS ONE 8: e66533.

36. Roth CW, Holm I, Graille M, Dehoux P, Rzhetsky A, et al. (2003) Identification of the *Anopheles gambiae* ATP-binding cassette transporter superfamily genes. Mol Cells 15: 150–158.

37. Liu S, Zhou S, Tian L, Guo E, Luan Y, et al. (2011) Genome-wide identification and characterization of ATP-binding cassette transporters in the silkworm, *Bombyx mori*. BMC Genomics 12: 491.

38. Xie X, Cheng T, Wang G, Duan J, Niu W, et al. (2012) Genome-wide analysis of the ATP-binding cassette (ABC) transporter gene family in the silkworm, *Bombyx mori*. Mol Biol Rep 39: 7281–7291.

39. Strauss AS, Wang D, Stock M, Gretscher RR, Groth M, et al. (2014) Tissue-specific transcript profiling for ABC transporters in the sequestering larvae of the phytophagous leaf beetle *Chrysomela populi*. PLoS ONE 9: e98637.

40. International Aphid Genomics Consortium (2010) Genome sequence of the pea aphid *Acyrthosiphon pisum*. PLoS Biol 8: e1000313.

41. Ye X-D, Su Y-L, Zhao Q-Y, Xia W-Q, Liu S-S, et al. (2014) Transcriptomic analyses reveal the adaptive features and biological differences of guts from two invasive whitefly species. BMC Genomics 15: 370.

42. Yang N, Xie W, Jones CM, Bass C, Jiao X, et al. (2013) Transcriptome profiling of the whitefly *Bemisia tabaci* reveals stage-specific gene expression signatures for thiamethoxam resistance. Insect Mol Biol 22: 485–496.

43. Xia J, Zhang C-R, Zhang S, Li F-F, Feng M-G, et al. (2013) Analysis of whitefly transcriptional responses to *Beauveria bassiana* infection reveals new insights into insect-fungus interactions. PLoS ONE 8: e68185.

44. Mamidala P, Wijeratne AJ, Wijeratne S, Kornacker K, Sudhamalla B, et al. (2012) RNA-Seq and molecular docking reveal multi-level pesticide resistance in the bed bug. BMC Genomics 13: 6.

45. Bao Y-Y, Li B-L, Liu Z-B, Xue J, Zhu Z-R, et al. (2010) Triazophos up-regulated gene expression in the female brown planthopper, *Nilaparvata lugens*. J Insect Physiol 56: 1087–1094.

46. Scott DR (1977) An annotated listing of host plants of *Lygus hesperus* Knight. Entomol Soc Am Bull 23: 19–22.

47. Young OP (1986) Host plants of the tarnished plant bug, *Lygus lineolaris* (Heteroptera: Miridae). Ann Entomol Soc Am 79: 747–762.

48. Wheeler AG (2001) Biology of the plant bugs (Hemiptera: Miridae): pests, predators, opportunists. Ithaca, NY: Comstock Publishing Associates. 528 p.

49. Snodgrass GL (1996) Insecticide resistance in field populations of the tarnished plant bug (Heteroptera: Miridae) in cotton in the Mississippi Delta. J Econ Entomol 89: 783–790.

50. Snodgrass G, Scott W (2002) Tolerance to acephate in tarnished plant bug (Heteroptera: Miridae) populations in the Mississippi river delta. Southwestern Entomologist 27: 191–199.

51. Snodgrass G, Gore J, Abel C, Jackson R (2009) Acephate resistance in populations of the tarnished plant bug (Heteroptera: Miridae) from the Mississippi River Delta. J Econ Entomol 102: 699–707.

52. Hull JJ, Geib SM, Fabrick JA, Brent CS (2013) Sequencing and de novo assembly of the western tarnished plant bug (*Lygus hesperus*) transcriptome. PLoS ONE 8: e55105.

53. Tapadia MG, Lakhotia SC (2005) Expression of mdr49 and mdr65 multidrug resistance genes in larval tissues of *Drosophila melanogaster* under normal and stress conditions. Cell Stress Chaperones 10: 7–11.

54. de Boussac H, Orbán TI, Várady G, Tihanyi B, Bacquet C, et al. (2012) Stimulus-induced expression of the ABCG2 multidrug transporter in HepG2 hepatocarcinoma model cells involves the ERK1/2 cascade and alternative promoters. Biochem Biophys Res Commun 426: 172–176.

55. de Araujo Leite JC, de Vasconcelos RB, da Silva SG, de Siqueira-Junior JP, Marques-Santos LF (2013) ATP-binding cassette transporters protect sea urchin gametes and embryonic cells against the harmful effects of ultraviolet light. Mol Reprod Dev 81: 66–83.

56. Kim Y, Park S-Y, Kim D, Choi J, Lee Y-H, et al. (2013) Genome-scale analysis of ABC transporter genes and characterization of the ABCC type transporter genes in *Magnaporthe oryzae*. Genomics 101: 354–361.

57. Kulkarni SR, Donepudi AC, Xu J, Wei W, Cheng QC, et al. (2014) Fasting induces nuclear factor E2-related factor 2 and ATP-binding cassette transporters via protein kinase A and sirtuin-1 in mouse and human. Antioxid Redox Signal 20: 15–30.

58. Peelman F, Labeur C, Vanloo B, Roosbeek S, Devaud C, et al. (2003) Characterization of the ABCA transporter subfamily: identification of prokaryotic and eukaryotic members, phylogeny and topology. J Mol Biol 325: 259–274.

59. Beers MF, Hawkins A, Shuman H, Zhao M, Newitt JL, et al. (2011) A novel conserved targeting motif found in ABCA transporters mediates trafficking to early post-Golgi compartments. J Lipid Res 52: 1471–1482.

60. Kaminski WE, Piehler A, Wenzel JJ (2006) ABC A-subfamily transporters: Structure, function and disease. Biochim Biophys Acta 1762: 510–524.

61. Albrecht C, Viturro E (2007) The ABCA subfamily–gene and protein structures, functions and associated hereditary diseases. Pflugers Arch 453: 581–589.

62. Akiyama M, Sugiyama-Nakagiri Y, Sakai K, McMillan JR, Goto M, et al. (2005) Mutations in lipid transporter ABCA12 in harlequin ichthyosis and functional recovery by corrective gene transfer. J Clin Invest 115: 1777–1784.

63. Bain LJ, LeBlanc GA (1996) Interaction of structurally diverse pesticides with the human MDR1 gene product P-glycoprotein. Toxicol Appl Pharmacol 141: 288–298.

64. Gottesman MM, Fojo T, Bates SE (2002) Multidrug resistance in cancer: role of ATP-dependent transporters. Nat Rev Cancer 2: 48–58.

65. Szakács G, Annereau J-P, Lababidi S, Shankavaram U, Arciello A, et al. (2004) Predicting drug sensitivity and resistance: profiling ABC transporter genes in cancer cells. Cancer Cell 6: 129–137.

66. Yasui K, Mihara S, Zhao C, Okamoto H, Saito-Ohara F, et al. (2004) Alteration in copy numbers of genes as a mechanism for acquired drug resistance. Cancer Res 64: 1403–1410.

67. Sooksa-Nguan T, Yakubov B, Kozlovskyy VI, Barkume CM, Howe KJ, et al. (2009) Drosophila ABC transporter, DmHMT-1, confers tolerance to cadmium. DmHMT-1 and its yeast homolog, SpHMT-1, are not essential for vacuolar phytochelatin sequestration. J Biol Chem 284: 354–362.

68. Bariami V, Jones CM, Poupardin R, Vontas J, Ranson H (2012) Gene amplification, ABC transporters and cytochrome P450s: unraveling the molecular basis of pyrethroid resistance in the dengue vector, *Aedes aegypti*. PLoS Negl Trop Dis 6: e1692.

69. Telonis-Scott M, Hallas R, McKechnie SW, Wee CW, Hoffmann AA (2009) Selection for cold resistance alters gene transcript levels in *Drosophila melanogaster*. J Insect Physiol 55: 549–555.

70. Parcej D, Tampé R (2010) ABC proteins in antigen translocation and viral inhibition. Nature Chemical Biology 6: 572–580.

71. Abele R, Tampé R (2011) The TAP translocation machinery in adaptive immunity and viral escape mechanisms. Essays Biochem 50: 249–264.

72. Wu CT, Budding M, Griffin MS, Croop JM (1991) Isolation and characterization of Drosophila multidrug resistance gene homologs. Mol Cell Biol 11: 3940–3948.

73. Vache C, Camares O, Cardoso-Ferreira M-C, Dastugue B, Creveaux I, et al. (2007) A potential genomic biomarker for the detection of polycyclic aromatic hydrocarbon pollutants: multidrug resistance gene 49 in Drosophila melanogaster. Environ Toxicol Chem 26: 1418–1424.

74. Azad P, Zhou D, Russo E, Haddad GG (2009) Distinct mechanisms underlying tolerance to intermittent and constant hypoxia in Drosophila melanogaster. PLoS ONE 4: e5371.

75. Chahine S, O'Donnell MJ (2009) Physiological and molecular characterization of methotrexate transport by Malpighian tubules of adult Drosophila melanogaster. J Insect Physiol 55: 927–935.

76. Figueira-Mansur J, Ferreira-Pereira A, Mansur JF, Franco TA, Alvarenga ESL, et al. (2013) Silencing of P-glycoprotein increases mortality in temephos-treated Aedes aegypti larvae. Insect Mol Biol 22: 648–658.

77. Tian L, Yang J, Hou W, Xu B, Xie W, et al. (2013) Molecular cloning and characterization of a P-glycoprotein from the diamondback moth, Plutella xylostella (Lepidoptera: Plutellidae). Int J Mol Sci 14: 22891–22905.

78. Luo L, Sun Y-J, Wu Y-J (2013) Abamectin resistance in Drosophila is related to increased expression of P-glycoprotein via the dEGFR and dAkt pathways. Insect Biochem Mol Biol 43: 627–634.

79. Toyoda Y, Hagiya Y, Adachi T, Hoshijima K, Kuo MT, et al. (2008) MRP class of human ATP binding cassette (ABC) transporters: historical background and new research directions. Xenobiotica 38: 833–862.

80. Keppler D (2010) Multidrug Resistance Proteins (MRPs, ABCCs): importance for pathophysiology and drug therapy. Drug Transporters. Handbook of Experimental Pharmacology. Berlin, Heidelberg: Springer Berlin Heidelberg, Vol. 201. 299–323.

81. Slot AJ, Molinski SV, Cole SPC (2011) Mammalian multidrug-resistance proteins (MRPs). Essays Biochem 50: 179–207.

82. Dassa E, Bouige P (2001) The ABC of ABCs: a phylogenetic and functional classification of ABC systems in living organisms. Res Microbiol 152: 211–229.

83. Grailles M, Brey PT, Roth CW (2003) The Drosophila melanogaster multidrug-resistance protein 1 (MRP1) homolog has a novel gene structure containing two variable internal exons. Gene 307: 41–50.

84. Labbé R, Caveney S, Donly C (2011) Genetic analysis of the xenobiotic resistance-associated ABC gene subfamilies of the Lepidoptera. Insect Mol Biol 20: 243–256.

85. Chahine S, O'Donnell MJ (2011) Interactions between detoxification mechanisms and excretion in Malpighian tubules of Drosophila melanogaster. J Exp Biol 214: 462–468.

86. Luckenbach T, Epel D (2008) ABCB- and ABCC-type transporters confer multixenobiotic resistance and form an environment-tissue barrier in bivalve gills. Am J Physiol Regul Integr Comp Physiol 294: R1919–R1929.

87. Kruh GD, Guo Y, Hopper-Borge E, Belinsky MG, Chen Z-S (2006) ABCC10, ABCC11, and ABCC12. Pflugers Arch 453: 675–684.

88. Beckstead RB, Lam G, Thummel CS (2007) Specific transcriptional responses to juvenile hormone and ecdysone in Drosophila. Insect Biochem Mol Biol 37: 570–578.

89. Vodovar N, Vinals M, Liehl P, Basset A, Degrouard J, et al. (2005) Drosophila host defense after oral infection by an entomopathogenic Pseudomonas species. Proc Natl Acad Sci USA 102: 11414–11419.

90. Yepiskoposyan H, Egli D, Fergestad T, Selvaraj A, Treiber C, et al. (2006) Transcriptome response to heavy metal stress in Drosophila reveals a new zinc transporter that confers resistance to zinc. Nucleic Acids Res 34: 4866–4877.

91. Morozova TV, Anholt RRH, Mackay TFC (2006) Transcriptional response to alcohol exposure in Drosophila melanogaster. Genome Biol 7: R95.

92. Monnier V, Girardot F, Cheret C, Andres O, Tricoire H (2002) Modulation of oxidative stress resistance in Drosophila melanogaster by gene overexpression. Genesis 34: 76–79.

93. Huang H, Haddad GG (2007) Drosophila dMRP4 regulates responsiveness to O2 deprivation and development under hypoxia. Physiol Genomics 29: 260–266.

94. Fernández-Ayala DJ, Chen S, Kemppainen E, O'Dell KMC, Jacobs HT (2010) Gene expression in a Drosophila model of mitochondrial disease. PLoS ONE 5: e8549.

95. Shah S, Yarrow C, Dunning R, Cheek B, Vass S, et al. (2011) Insecticide detoxification indicator strains as tools for enhancing chemical discovery screens. Pest Manag Sci 68: 38–48.

96. Tanaka S, Miyamoto K, Noda H, Jurat-Fuentes JL, Yoshizawa Y, et al. (2013) The ATP-binding cassette transporter subfamily C member 2 in Bombyx mori larvae is a functional receptor for Cry toxins from Bacillus thuringiensis. FEBS J 280: 1782–1794.

97. Baum JA, Sukuru UR, Penn SR, Meyer SE, Subbarao S, et al. (2012) Cotton plants expressing a hemipteran-active Bacillus thuringiensis crystal protein impact the development and survival of Lygus hesperus (Hemiptera: Miridae) nymphs. J Econ Entomol 105: 616–624.

98. Aittoniemi J, Fotinou C, Craig TJ, de Wet H, Proks P, et al. (2009) SUR1: a unique ATP-binding cassette protein that functions as an ion channel regulator. Philos Trans R Soc Lond B Biol Sci 364: 257–267.

99. Kim SK, Rulifson EJ (2004) Conserved mechanisms of glucose sensing and regulation by Drosophila corpora cardiaca cells. Nature 431: 316–320.

100. Akasaka T, Klinedinst S, Ocorr K, Bustamante EL, Kim SK, et al. (2006) The ATP-sensitive potassium (KATP) channel-encoded dSUR gene is required for Drosophila heart function and is regulated by tinman. Proc Natl Acad Sci USA 103: 11999–12004.

101. Li Y, Qin Y, Yang N, Sun Y, Yang X, et al. (2013) Studies on insecticidal activities and action mechanism of novel benzoylphenylurea candidate NK-17. PLoS ONE 8: e66251.

102. Merzendorfer H, Kim HS, Chaudhari SS, Kumari M, Specht CA, et al. (2012) Genomic and proteomic studies on the effects of the insect growth regulator diflubenzuron in the model beetle species Tribolium castaneum. Insect Biochem Mol Biol 42: 264–276.

103. Meyer F, Flötenmeyer M, Moussian B (2013) The sulfonylurea receptor Sur is dispensable for chitin synthesis in Drosophila melanogaster embryos. Pest Manag Sci 69: 1136–1140.

104. Jordan IK, Kota KC, Cui G, Thompson CH, McCarty NA (2008) Evolutionary and functional divergence between the cystic fibrosis transmembrane conductance regulator and related ATP-binding cassette transporters. Proc Natl Acad Sci USA 105: 18865–18870.

105. Wanders RJ, Visser WF, van Roermund CWT, Kemp S, Waterham HR (2007) The peroxisomal ABC transporter family. Pflugers Arch 453: 719–734.

106. Morita M, Imanaka T (2012) Peroxisomal ABC transporters: structure, function and role in disease. Biochim Biophys Acta 1822: 1387–1396.

107. Shani N, Sapag A, Valle D (1996) Characterization and analysis of conserved motifs in a peroxisomal ATP-binding cassette transporter. J Biol Chem 271: 8725–8730.

108. Mosallanejad H, Badisco L, Swevers L, Soin T, Knapen D, et al. (2010) Ecdysone signaling and transcript signature in Drosophila cells resistant against methoxyfenozide. J Insect Physiol 56: 1973–1985.

109. Khoshnevis S, Gross T, Rotte C, Baierlein C, Ficner R, et al. (2010) The iron-sulphur protein RNase L inhibitor functions in translation termination. EMBO Reports 11: 214–219.

110. Liu S, Li Q, Liu Z (2013) Genome-wide identification, characterization and phylogenetic analysis of 50 catfish ATP-binding cassette (ABC) transporter genes. PLoS ONE 8: e63895.

111. Abruzzi KC, Rodriguez J, Menet JS, Desrochers J, Zadina A, et al. (2011) Drosophila CLOCK target gene characterization: implications for circadian tissue-specific gene expression. Genes Dev 25: 2374–2386.

112. Tarr PT, Tarling EJ, Bojanic DD, Edwards PA, Baldán A (2009) Emerging new paradigms for ABCG transporters. Biochim Biophys Acta 1791: 584–593.

113. Kerr ID, Haider AJ, Gelissen IC (2011) The ABCG family of membrane-associated transporters: you don't have to be big to be mighty. Br J Pharmacol 164: 1767–1779.

114. Weber AL, Khan GF, Magwire MM, Tabor CL, Mackay TFC, et al. (2012) Genome-wide association analysis of oxidative stress resistance in Drosophila melanogaster. PLoS ONE 7: e34745.

115. Buchmann J, Meyer C, Neschen S, Augustin R, Schmolz K, et al. (2007) Ablation of the cholesterol transporter adenosine triphosphate-binding cassette transporter G1 reduces adipose cell size and protects against diet-induced obesity. Endocrinology 148: 1561–1573.

116. Morozova TV, Anholt RRH, Mackay TFC (2007) Phenotypic and transcriptional response to selection for alcohol sensitivity in Drosophila melanogaster. Genome Biol 8: R231.

117. Hock T, Cottrill T, Keegan J, Garza D (2000) The E23 early gene of Drosophila encodes an ecdysone-inducible ATP-binding transporter capable of repressing ecdysone-mediated gene activation. Proc Natl Acad Sci USA 97: 9519–9524.

118. Tan A, Palli SR (2008) Edysone receptor isoforms play distinct roles in controlling molting and metamorphosis in the red flour beetle, Tribolium castaneum. Mol Cell Endocrinol 291: 42–49.

119. Bryon A, Wybouw N, Dermauw W, Tirry L, Van Leeuwen T (2013) Genome wide gene-expression analysis of facultative reproductive diapause in the two-spotted spider mite Tetranychus urticae. BMC Genomics 14: 815.

120. Abraham EG, Sezutsu H, Kanda T, Sugasaki T, Shimada T, et al. (2000) Identification and characterisation of a silkworm ABC transporter gene homologous to Drosophila white. Mol Gen Genet 264: 11–19.

121. Tatematsu K-I, Yamamoto K, Uchino K, Narukawa J, Iizuka T, et al. (2011) Positional cloning of silkworm white egg 2 (w-2) locus shows functional conservation and diversification of ABC transporters for pigmentation in insects. Genes Cells 16: 331–342.

122. Anaka M, MacDonald CD, Barkova E, Simon K, Rostom R, et al. (2008) The white gene of Drosophila melanogaster encodes a protein with a role in courtship behavior. J Neurogenet 22: 243–276.

123. Snodgrass GL (2002) Characteristics of a red-eye mutant of the tarnished plant bug (Heteroptera: Miridae). Ann Entomol Soc Am 95: 366–369.

124. Allen ML (2013) Genetics of a sex-linked recessive red eye color mutant of the tarnished plant bug, Lygus lineolaris. Open J Anim Sci 3: 1–9.

125. Liu S-H, Yao J, Yao H-W, Jiang P-L, Yang B-J, et al. (2014) Biological and biochemical characterization of a red-eye mutant in Nilaparvata lugens (Hemiptera: Delphacidae). Insect Science 21: 469–476.

126. Seo BY, Jung JK, Kim Y (2011) An orange-eye mutant of the brown planthopper, Nilaparvata lugens (Hemiptera: Delphacidae). J Asia-Pacific Entomol 14: 469–472.

127. Annilo T, Chen Z-Q, Shulenin S, Costantino J, Thomas L, et al. (2006) Evolution of the vertebrate ABC gene family: analysis of gene birth and death. Genomics 88: 1–11.

128. Mummery-Widmer JL, Yamazaki M, Stoeger T, Novatchkova M, Bhalerao S, et al. (2009) Genome-wide analysis of Notch signalling in Drosophila by transgenic RNAi. Nature 458: 987–992.

129. Zhang S, Feany MB, Saraswati S, Littleton JT, Perrimon N (2009) Inactivation of Drosophila Huntingtin affects long-term adult functioning and the pathogenesis of a Huntington's disease model. Dis Model Mech 2: 247–266.

130. Qin W, Neal SJ, Robertson RM, Westwood JT, Walker VK (2005) Cold hardening and transcriptional change in Drosophila melanogaster. Insect Mol Biol 14: 607–613.

131. You M, Yue Z, He W, Yang X, Yang G, et al. (2013) A heterozygous moth genome provides insights into herbivory and detoxification. Nat Genet 45: 220–225.

132. Robinson SW, Herzyk P, Dow JAT, Leader DP (2012) FlyAtlas: database of gene expression in the tissues of Drosophila melanogaster. Nucleic Acids Res 41: D744–D750.

133. Sullivan DT, Bell LA, Paton DR, Sullivan MC (1980) Genetic and functional analysis of tryptophan transport in Malpighian tubules of Drosophila. Biochem Genet 18: 1109–1130.

134. Dow JAT, Davies S-A (2006) The Malpighian tubule: Rapid insights from post-genomic biology. J Insect Physiol 52: 365–378.

135. Dow JAT (2009) Insights into the Malpighian tubule from functional genomics. J Exp Biol 212: 435–445.

136. Wang J, Kean L, Yang J, Allan AK, Davies S-A, et al. (2004) Function-informed transcriptome analysis of Drosophila renal tubule. Genome Biol 5: R69.

137. Yang J, McCart C, Woods DJ, Terhzaz S, Greenwood KG, et al. (2007) A Drosophila systems approach to xenobiotic metabolism. Physiol Genomics 30: 223–231.

138. Labbé R, Caveney S, Donly C (2011) Expression of multidrug resistance proteins is localized principally to the Malpighian tubules in larvae of the cabbage looper moth, Trichoplusia ni. J Exp Biol 214: 937–944.

139. Wasbrough ER, Dorus S, Hester S, Howard-Murkin J, Lilley K, et al. (2010) The Drosophila melanogaster sperm proteome-II (DmSP-II). J Proteomics 73: 2171–2185.

140. Jones SR, Cyr DG (2011) Regulation and characterization of the ATP-binding cassette transporter-B1 in the epididymis and epididymal spermatozoa of the rat. Toxicol Sci 119: 369–379.

141. Gillott C (2003) Male accessory gland secretions: modulators of female reproductive physiology and behavior. Annu Rev Entomol 48: 163–184.

142. Avila FW, Sirot LK, LaFlamme BA, Rubinstein CD, Wolfner MF (2011) Insect seminal fluid proteins: identification and function. Annu Rev Entomol 56: 21–40.

143. Quan GX, Kanda T, Tamura T (2002) Induction of the white egg 3 mutant phenotype by injection of the double-stranded RNA of the silkworm white gene. Insect Mol Biol 11: 217–222.

144. Martins GF, Ramalho-Ortigao JM (2012) Oenocytes in insects. Inver Surv J 9: 139–152.

145. Zhu F, Gujar H, Gordon JR, Haynes KF, Potter MF, et al. (2013) Bed bugs evolved unique adaptive strategy to resist pyrethroid insecticides. Sci Rep 3: 1456.

146. Debolt JW (1982) Meridic diet for rearing successive generations of Lygus hesperus. Ann Entomol Soc Am 75: 119–122.

147. Brent CS, Hull JJ (2014) Characterization of male-derived factors inhibiting female sexual receptivity in Lygus hesperus. J Insect Physiol 60: 104–110.

148. Lohse M, Bolger AM, Nagel A, Fernie AR, Lunn JE, et al. (2012) RobiNA: a user-friendly, integrated software solution for RNA-Seq-based transcriptomics. Nucleic Acids Res 40: W622–W627.

149. Haas BJ, Papanicolaou A, Yassour M, Grabherr M, Blood PD, et al. (2013) De novo transcript sequence reconstruction from RNA-seq using the Trinity platform for reference generation and analysis. Nature Protocols 8: 1494–1512.

150. Eddy SR (2011) Accelerated profile HMM searches. PLoS Comput Biol 7: e1002195.

151. Sheps JA, Ralph S, Zhao Z, Baillie DL, Ling V (2004) The ABC transporter gene family of Caenorhabditis elegans has implications for the evolutionary dynamics of multidrug resistance in eukaryotes. Genome Biol 5: R15.

152. de Castro E, Sigrist CJA, Gattiker A, Bulliard V, Langendijk-Genevaux PS, et al. (2006) ScanProsite: detection of PROSITE signature matches and ProRule-associated functional and structural residues in proteins. Nucleic Acids Res 34: W362–W365.

153. Sigrist CJA, Cerutti L, de Castro E, Langendijk-Genevaux PS, Bulliard V, et al. (2010) PROSITE, a protein domain database for functional characterization and annotation. Nucleic Acids Res 38: D161–D166.

154. Finn RD, Clements J, Eddy SR (2011) HMMER web server: interactive sequence similarity searching. Nucleic Acids Res 39: W29–W37.

155. Horton P, Park K-J, Obayashi T, Fujita N, Harada H, et al. (2007) WoLF PSORT: protein localization predictor. Nucleic Acids Res 35: W585–W587.

156. Krogh A, Larsson B, von Heijne G, Sonnhammer EL (2001) Predicting transmembrane protein topology with a hidden Markov model: application to complete genomes. J Mol Biol 305: 567–580.

157. Bernsel A, Viklund H, Hennerdal A, Elofsson A (2009) TOPCONS: consensus prediction of membrane protein topology. Nucleic Acids Res 37: W465–W468.

158. von Heijne G (1992) Membrane protein structure prediction. Hydrophobicity analysis and the positive-inside rule. J Mol Biol 225: 487–494.

159. Edgar RC (2004) MUSCLE: a multiple sequence alignment method with reduced time and space complexity. BMC Bioinformatics 5: 113.

160. Price MN, Dehal PS, Arkin AP (2010) FastTree 2– approximately maximum-likelihood trees for large alignments. PLoS ONE 5: e9490.

161. Tamura K, Peterson D, Peterson N, Stecher G, Nei M, et al. (2011) MEGA5: molecular evolutionary genetics analysis using maximum likelihood, evolutionary distance, and maximum parsimony methods. Mol Biol Evol 28: 2731–2739.

162. Jones DT, Taylor WR, Thornton JM (1992) The rapid generation of mutation data matrices from protein sequences. Comput Appl Biosci 8: 275–282.

# Pilon: An Integrated Tool for Comprehensive Microbial Variant Detection and Genome Assembly Improvement

Bruce J. Walker[1][*][9][¤], Thomas Abeel[1,2][9], Terrance Shea[1], Margaret Priest[1], Amr Abouelliel[1], Sharadha Sakthikumar[1], Christina A. Cuomo[1], Qiandong Zeng[1], Jennifer Wortman[1], Sarah K. Young[1], Ashlee M. Earl[1][*]

1 Broad Institute of MIT and Harvard, Cambridge, Massachusetts, United States of America, 2 VIB Department of Plant Systems Biology, Ghent University, Ghent, Belgium

## Abstract

Advances in modern sequencing technologies allow us to generate sufficient data to analyze hundreds of bacterial genomes from a single machine in a single day. This potential for sequencing massive numbers of genomes calls for fully automated methods to produce high-quality assemblies and variant calls. We introduce Pilon, a fully automated, all-in-one tool for correcting draft assemblies and calling sequence variants of multiple sizes, including very large insertions and deletions. Pilon works with many types of sequence data, but is particularly strong when supplied with paired end data from two Illumina libraries with small e.g., 180 bp and large e.g., 3–5 Kb inserts. Pilon significantly improves draft genome assemblies by correcting bases, fixing mis-assemblies and filling gaps. For both haploid and diploid genomes, Pilon produces more contiguous genomes with fewer errors, enabling identification of more biologically relevant genes. Furthermore, Pilon identifies small variants with high accuracy as compared to state-of-the-art tools and is unique in its ability to accurately identify large sequence variants including duplications and resolve large insertions. Pilon is being used to improve the assemblies of thousands of new genomes and to identify variants from thousands of clinically relevant bacterial strains. Pilon is freely available as open source software.

Editor: Junwen Wang, The University of Hong Kong, Hong Kong

Funding: This project has been funded in part with Federal funds from the National Institute of Allergy and Infectious Diseases, National Institutes of Health, Department of Health and Human Services, under Contract No.:HHSN272200900018C. This project has been also been funded in part with Federal funds from the National Human Genome Research Institute, National Institutes of Health, Department of Health and Human Services, under grant U54HG003067. TA is a postdoctoral fellow of the Research Foundation-Flanders. The funders played no role collection, analysis, and interpretation of data; in the writing of the manuscript; and in the decision to submit the manuscript for publication.

Competing Interests: The authors have declared that no competing interests exist.

* Email: bruce@broadinstitute.org (BJW); aearl@broadinstitute.org (AME)

9 These authors contributed equally to this work.

¤ Current address: Applied Minds, LLC, Boston, Massachusetts, United States of America

## Introduction

Massively parallel sequencing technology has dramatically reduced the cost of genome sequencing, making the generation of large numbers of microbial genomes accessible to a wide range of biological researchers. For example, a single Illumina HiSeq2500 has the ability to generate the equivalent of 300 bacterial genomes of sequencing data in a single day using only one flow cell. Comparisons of whole genome sequence data from hundreds of microorganisms have provided unprecedented views on all aspects of microbial diversity, and there is growing recognition that 'hundreds' of genomes is the minimum scale needed to address pressing questions related to microbial evolution, diversity, pathogenicity and resistance to antimicrobial drugs [1–4]. As such, the methods needed to analyze these large volumes of data — including assembling and calling variants relative to a reference — must be robust, accurate, scalable, and able to operate without human intervention.

Several computational methods exist that make improvements to the quality of draft assemblies by recognizing and correcting errors involving (i) single bases and small insertion/deletion events (indels) [5], (ii) gaps [6], (iii) read alignment discontinuities [7] or by reconciling multiple de novo assemblies into an improved consensus assembly [8]. However, no single tool performs integrated assembly improvement of all error types. Computational tools for identifying sequence polymorphisms also exist [9,10], but focus primarily on identifying variants in the human genome [11], and particularly small events (SNPs and small indels) or structural rearrangements (chromosomal rearrangements) [11]. Furthermore, many of these tools require multiple steps to identify and subsequently filter variants to remove noise and false calls. In addition, for tools able to identify variants that exceed the length of the sequence reads (read-length) being evaluated, they generally indicate the approximate chromosomal location and estimated size of the predicted variant relative to the reference, but often do not provide exact coordinates [11]. For insertions that are longer than the read-length - particularly common in the microbial world - current tools do not assemble and report the inserted sequence.

We introduce Pilon, an integrated software tool for comprehensive microbial genome assembly improvement and variant detection, including detection of variants that exceed sequence read-length. Conceptually, Pilon treats assembly improvement and variant detection as the same process (Figure 1). Both start with an input genome — either an existing draft assembly or a reference assembly from another strain — and use evidence from read alignments to identify specific differences from the input genome supported by the sequencing data. Applying those changes to a draft genome assembly yields an improved assembly, while reporting the changes with respect to a reference genome yields variant calls.

In genomic regions where read alignments are poor, Pilon is capable of filling out and correcting sequence through an internal local reassembly process. This capability allows Pilon to further improve assemblies by filling gaps and correcting local mis-assemblies, and it also enables Pilon to capture many large insertion, deletion, and block substitution variants in their entirety. These larger events are often completely missed or inaccurately characterized by conventional variant calling tools that rely solely on read alignments. Pilon has built-in heuristics to determine which corrections and calls are of high confidence, so no separate filtering criteria need be specified. This allows for the automated processing of hundreds or thousands of data sets representing different microbial species with minimal human intervention.

We benchmarked Pilon both as an assembly refinement tool and variant caller. For assembly refinement, we used finished reference genome sequences from *Mycobacterium tuberculosis* F11, *Streptococcus pneumoniae* TIGR4 and *Candida albicans* SC5314 as benchmarks to evaluate the accuracy of Pilon in improving draft assemblies. Pilon-improved assemblies were more contiguous and complete than non-Pilon-improved assemblies and contained

improved sequences for genes implicated in pathogen-host interaction and virulence. We also evaluated Pilon's performance against tools specializing in assembly base quality improvement and gap filling, and, in each case, Pilon made more correct improvements while making far fewer mistakes than the other tools. For variant calling, we used read data from *M. tuberculosis* F11 to call polymorphisms against the finished *M. tuberculosis* H37Rv genome to evaluate Pilon's ability to accurately call polymorphisms. Pilon performed as well or better when compared with two state-of-the-art variant detection tools in calling small variants, and Pilon differentiated itself in its ability to identify large-scale variants.

## Results

### Assembly improvement evaluation

**Assessing accuracy on bacterial assemblies.** To test the accuracy of Pilon's improvements on bacterial assemblies, we sequenced and created draft assemblies for two bacterial strains with finished references: *S. pneumoniae* TIGR4 and *M. tuberculosis* F11 (see Methods). These strains were chosen because they represent different GC content (40% and 66% GC content, respectively) and both possess genomic features that are known to confound assemblers, leading to mis-assembled and/or incomplete genome sequences [12–14]. Sequence reads from both libraries were aligned back to their respective draft assemblies using *BWA* [15], and Pilon was run with those alignments.

To assess the benefits of running Pilon, we compared the original draft and Pilon-improved assemblies to each other and to their respective finished genome sequence. Pilon made significant improvements to the contiguity of both draft assemblies, increasing the contig N50 size by 443 Kbp for *S. pneumoniae* TIGR4 (see Table 1) and 196 Kbp for *M. tuberculosis* F11, even though the F11 draft assembly had been generated with assistance from a

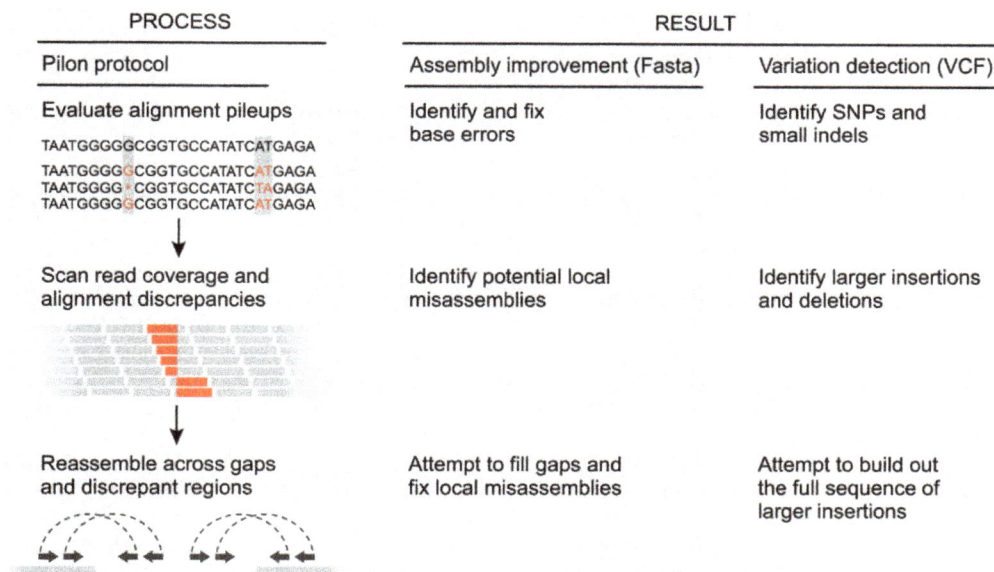

**Figure 1. Simplified overview of the Pilon workflow for assembly improvement and variant detection.** The left column depicts the conceptual steps of the Pilon process, and the center and right columns describe what Pilon does at each step while in assembly improvement and variant detection modes, respectively. During the first step (top row), Pilon scans the read alignments for evidence where the sequencing data disagree with the input genome and makes corrections to small errors and detects small variants. During the second step (second row), Pilon looks for coverage and alignment discrepancies to identify potential mis-assemblies and larger variants. Finally (bottom row), Pilon uses reads and mate pairs which are anchored to the flanks of discrepant regions and gaps in the input genome to reassemble the area, attempting to fill in the true sequence including large insertions. The resulting output is an improved assembly and/or a VCF file of variants.

**Table 1.** Summary assembly statistics before and after Pilon improvement.

| Genome | M. tuberculosis F11 | S. pneumoniae TIGR4 | C. albicans SC5314 |
|---|---|---|---|
| Contig N50 Increase | 196 kb | 443 kb | 56 kb |
| Bases Added | 11,516 | 9,608 | 33,804 |
| Gaps Closed | 9 | 9 | 54 |
| Gaps Shrunk | 7 | 7 | 102 |
| Single-base Modifications | 20 | 27 | 26,939 |
| Mis-assembly fixes | 3 | 1 | 44 |

In all cases the assemblies were more contiguous, contained more bases, and had fewer gaps and errors after Pilon improvement.

close reference. In addition, Pilon assemblies were more complete, with the *M. tuberculosis* F11 and *S. pneumoniae* TIGR4 Pilon-improved assemblies containing an additional 11,516 bp and 9,608 bp, respectively.

Observed gains in genome coverage and contig N50 were principally due to Pilon's ability to recognize and fill (or partially fill) by local assembly "captured gaps", *i.e.*, missing sequence between contigs within a scaffold. When run with default settings, Pilon does not introduce ambiguous bases or additional Ns during this process. Across the two draft assemblies, Pilon completely and accurately filled 17 of the 44 captured gaps (39% closure rate) including 8 gaps that represented more than 1 Kbp in sequence length (see Table S1). None of Pilon's gap closures were incorrect, though one was judged to be "no worse": the sequence used to bridge the gap was correct, but an error in the original assembly in one of the gap flanks was not detected by Pilon. An additional 14 gaps (32% of total captured gaps) were partially filled by Pilon, and 13 (93%) of those extensions were error-free. The one partial fill judged to be "Incorrect" involved a repetitive structure that Pilon extended into flanking sequence belonging to a different copy of the repeat.

We compared Pilon's ability to close gaps in these assemblies with two other tools commonly used for this purpose, IMAGE [16] and GapFiller [17] (see Table S1b). Pilon's overall gap closure rate was only somewhat higher than that of the other tools, but its accuracy was dramatically better. Across the two assemblies, IMAGE closed 13 captured gaps (30% closure rate) but only two of those closures were found to be correct by alignment with the reference (15% precision). Similarly, GapFiller closed 16 of captured gaps (36% closure rate) in the two assemblies, but only four of its closures were correct (25% precision). In addition to filling captured gaps, Pilon also corrected 43 single-bases and 4 small indels across both genomes, and all 47 changes were found to be correct by alignment against the reference (100% accuracy; see Table S2). By comparison, iCORN [18] made 47 single-base changes and 2 single-base deletions, but only 35 of the 49 (71%) of its changes were correct.

Optionally, Pilon can also make changes to genomic locations at which it finds significant evidence for more than one alternative, choosing the allele with the most support even where the evidence is insufficient to make a confident call. When run with this option on these assemblies, Pilon made 10 changes to ambiguous bases, but only 3 were verified to be correct. This option is turned off by default starting with Pilon version 1.8.

Pilon also detected and attempted to fix local mis-assemblies by reassembling contig regions that were suspected to be incorrectly assembled. Three of these regions were correctly fixed (see Table 1 and Table S1) and a fourth we classified as "No worse". For the latter, Pilon correctly identified a repetitive region within the

original *M. tuberculosis* F11 draft assembly that contained extra sequence with respect to the *M. tuberculosis* F11 reference. However, Pilon's change introduced a deletion with respect to the reference, underscoring the difficulty of accurately assembling repetitive regions with short read data [13].

For the *M. tuberculosis* F11 and *S. pneumoniae* TIGR4 Pilon improved assemblies, there were 13 and 4 regions, respectively, where Pilon detected a problem in the draft assembly, but was unable to provide solutions. In each of these cases, Pilon flagged the coordinates of the problematic region, and, in 10 of these cases, it also reported the length of the detected tandem repeat confounding resolution of the region. For example, Figure 2 shows scaffold00001 coordinates 3159800–3159898 of the *M. tuberculosis* F11 draft assembly, along with Pilon-generated genome browser tracks representing some of the internal metrics it used to identify this region as problematic. In this case, Pilon noted that it was unable to resolve a 57 bp tandem repeat, which enabled an experienced analyst to confirm the presence of a mis-assembly and accurately narrow the bounds of the unresolvable region. Manual comparison of the draft assembly with the reference revealed that there should have been three full and one partial copies of the 57 bp repeat in tandem, whereas the draft assembly only contained one full and one partial copy of the repeat.

**Effect of assembly improvements on gene calls.** To assess the impact of Pilon-improvement on gene calls (*i.e.*, functional interpretation of the genome), we examined Pilon improvements with respect to genes by investigating the regions that were affected by Pilon modifications and the effect of these modifications on coding sequences. Thirty-two genes and seven intergenic regions were impacted by Pilon changes to the *M. tuberculosis* F11 Pilon-improved assembly; of these, nearly all (95%; 37 of 39) were correct improvements. Nearly half (13) of the genes that were affected by a fix involved transposases that were completely or partially filled with sequence that perfectly matched the reference genome (see Table S3). One additional transposase had a single base pair corrected with perfect match to the reference. A particularly complex 13 Kbp region in *M. tuberculosis* F11 is highlighted in Figure 3. This region harbors three sets of transposases in close proximity that were not captured in the draft *M. tuberculosis* F11 assembly, but were accurately filled in by Pilon. Two of the gaps were completely closed, and the third transposase set was completely captured along with an additional gene. However, due to Pilon's conservative overlap requirement for closure (95 bp), that gap was not closed despite a 42 bp overlap in the extended flank sequences.

Of the remaining 19 genes, 6 were PE/PPE family protein encoding genes. Five corrections were perfect and, in one case (TBFG_11946), Pilon identified the problematic region, but could not completely resolve the problem. However, the correction that

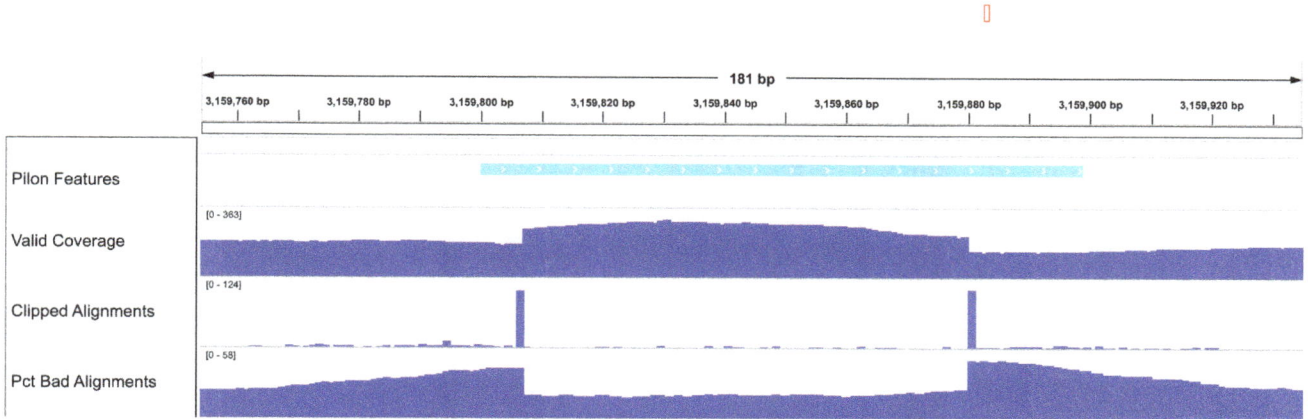

**Figure 2. Example Pilon generated genome browser tracks.** This region was flagged by Pilon as containing a possible local mis-assembly, but Pilon was unable to determine a fix due to a tandem repeat sequence. The tracks shown here include: *Pilon Features* track indicating the extent of the region flagged by Pilon as containing a potential mis-assembly, *Valid Coverage* track indicating the sequence coverage of valid read pair alignments excluding the clipped portions of the alignments, *Clipped Alignments* track indicating the number of reads soft-clipped at each location, *Pct Bad Alignments* track indicating the percentage of the total reads aligned to each location which are not part of *Valid Coverage*. These tracks are created with the '—tracks' command-line option. Together, these tracks reveal the true bounds of the mis-assembly, and indicate that there are likely missing copies of the tandem repeat in the draft assembly. In this case, manual analysis revealed the draft assembly was missing two of three full copies of a 57-base tandem repeat.

Pilon applied did not make the situation worse. Pilon also identified and accurately corrected a mis-assembly (highlighted in Figure S1) in which a gene had been truncated due to a collapsed repeat in the draft assembly.

In *S. pneumoniae* TIGR4, 20 genes and 12 intergenic regions were affected by fixes from Pilon. A majority (15 of 20) of the improved genes were transposases, of which Pilon was able to completely or partially fill 8 that matched completely and perfectly with the reference; the remaining 7 were individual base pair corrections. Pilon was also able to partially fill other genes encoding repetitive cell wall surface proteins - including choline binding protein A [19] and pneumococcal surface protein A [20] - both implicated in adhesion and virulence in *S. pneumoniae*.

**Assessing accuracy on the assembly of the larger, polymorphic genome of *C. albicans*.** To evaluate Pilon's

ability to accurately improve assemblies of diploid genomes containing a high level of heterozygosity, we ran Pilon on an Illumina ALLPATHS-LG assembly of the SC5314 strain of *C. albicans* (Methods), for which there is a high quality reference curated by the Candida Genome Database (www.candidagenome. org). At 14.3 Mb, the *C. albicans* genome is 3- to 7-fold larger than the bacterial genomes evaluated here. It consists of 8 chromosomes that are present at diploid levels with an average of one SNP found at every 330–390 bases, although large regions of most chromosomes display loss of heterozygosity [21,22].

Pilon was capable of improving the assembly and added > 33 Kb of sequence (see Table 1). While the increase in contig N50 was relatively small (56 Kb), Pilon completely or partially filled 61% (156 of 256) of the total captured gaps including in both homo- and hetero-zygous regions of the genome. Homozygous

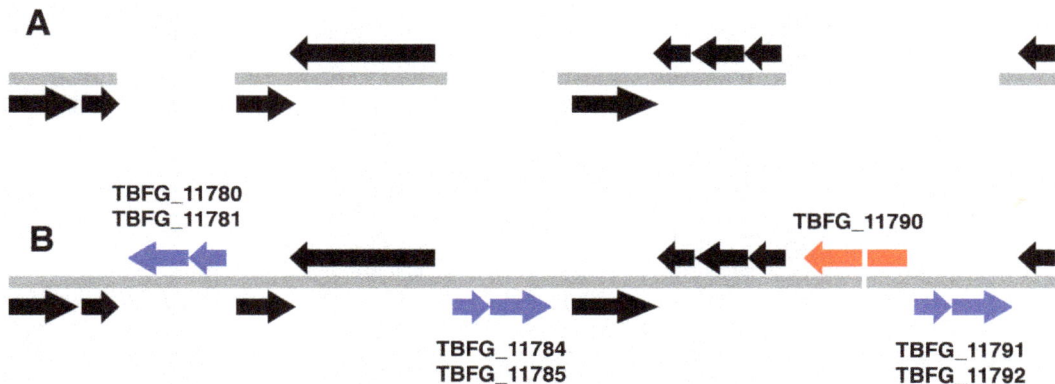

**Figure 3. Comparative view of a transposase-rich region of the *M. tuberculosis* F11 genome (coordinates 1,991,000 to 2,006,300) obtained from the draft (A) and Pilon-improved (B) assemblies.** In the draft assembly, three regions containing transposases (shown in blue) remained unassembled resulting in gaps. In the Pilon-improved assembly, all three sets of transposases were successfully assembled. The Pilon-improved assembly also contained a hypothetical gene, *TBFG_11790* (shown in red), missing from the draft assembly. Though *TBFG_11790* was not fully closed in the Pilon-improved version, closer inspection revealed that there was a 42 bp overlap in assembled sequence at this site. By default, Pilon will not close gaps unless there is at least 95 bp overlapping sequence to minimize spurious joins.

regions had a slightly higher fraction of completely closed gaps (33%; 8 of 24) as compared to heterozygous regions (20%; 46 of 232). Of completely filled gaps, 93% had full length alignment to the reference (including 300 bp of their flanking sequences) at 94% sequence identity or higher. Less than 100% identity is to be expected when comparing a heterozygous genome assembly against a flat reference. In several of the lower-identity cases, most of the base differences were in the flanks present in the draft assembly rather than the filled gap itself, suggesting that the gaps may have been caused by the original assembler's inability to assemble sequence through a highly polymorphic region.

Pilon also identified and corrected regions in the reference assembly where the read alignment evidence disagreed, including 44 regions that were likely mis-assembled. The nearly 27,000 corrected single-bases were mostly at heterozygous sites; Pilon identified these as potential bases to fix, as the majority of read-evidence favored an alternate allele from the reference base in the draft. These positions represented about half of the ~70,000 heterozygous SNP positions in this *Candida* genome [23]. While we did not investigate every change that Pilon made to this assembly, our results indicate that Pilon is suitable to be run on larger diploid genomes and can improve the quality of a draft assembly, resulting in fewer and longer contigs and an improved gene set.

**Assembly improvements in a production environment.** Given promising results from the benchmarking experiments, we implemented Pilon in the Broad Institute's *de novo* genome assembly production pipeline and assessed its performance by comparing assembly metrics from Pilon-improved assemblies of 50 representatives of the *Enterobacteriaceae* (including *Escherichia, Klebsiella, and Enterobacter*) to non-Pilon improved versions. Pilon reduced the mean number of contigs in the 50 assemblies from 33.7 to 20.9 (see Figure S2), a 38% reduction in total contigs representing closure of 47% of captured gaps. As a result, Pilon nearly doubled the contig N50 from 392 Kbp to 780 Kbp (99% increase; see Figure S3), capturing, on average, an additional 14,681 bp of sequence per assembly. This increase in genome size equates roughly to the addition of ~14 genes per genome (based on the average bacterial gene size of ~1 kb). Scaffold numbers were unchanged since, currently, Pilon does not attempt to join or break scaffold structures.

## Variant detection evaluation

**Assessing accuracy of polymorphism calls.** To test the accuracy of Pilon's variant detection capability, we used BWA to align approximately 200-fold coverage of reads from the same *M. tuberculosis* F11 fragment and long insert libraries used in the assembly improvement assessment to the *M. tuberculosis* H37Rv finished reference genome. We generated two sets of variant calls with Pilon, one using both fragment and long insert reads as input, and one using only fragment reads. We also ran two popular variant detection tools, GATK UnifiedGenotyper (GATK-UG) and SAMtools/BCFtools (SAMtools), starting with the same aligned fragment BAM. All variant sites, including substitutions, deletions or insertions, were identified and two categories of variants were assessed: single nucleotide polymorphisms (SNPs) and multi nucleotide polymorphisms (MNPs) greater than 1 bp. Predicted polymorphisms were compared to a curated truth set of variants produced by comparing the *M. tuberculosis* F11 finished genome to the finished *M. tuberculosis* H37Rv genome (see Methods), resulting in a list of 1,325 events (summarized in Table 2) of which the majority were SNPs. We then compared Pilon's performance to that of the other two variant detection algorithms.

Overall, Pilon performed better in identifying SNPs, including single nucleotide insertions and deletions than did GATK-UG or SAMtools (Table 3). Pilon identified 8 to 11 percentage points (pp) more single nucleotide substitutions, 4 pp more single nucleotide deletions, and 4 to 8 pp more single nucleotide insertions from the curation set than did GATK-UG or SAMtools, respectively. Pilon's ability to precisely call single nucleotide substitutions was also high - only 3% of calls were not accounted for in the curation set - which was on par with the other two tools. Similarly, Pilon had perfect precision in calling single nucleotide insertions and only 5% of single nucleotide deletion calls were not accounted for in the curation set.

For MNPs, we allowed for a combination of two or more smaller events in the prediction set to contribute to a larger variant since there may be equivalent ways of representing the changes as a series of smaller edits (see Methods). Pilon greatly outperformed the other two variant callers in accurately identifying variants that involved more than one nucleotide (see Table 3; bottom three rows). Pilon identified three times as many multi nucleotide insertions as either GATK-UG or SAMtools (63% versus 17 or 21% of curated events), but made slightly more false predictions. For multi nucleotide deletions, Pilon identified two times as many events from the curation set than did GATK-UG or SAMtools and made fewer unsupported calls. In addition, Pilon identified all six curated multi-substitution events while the other two tools missed at least one, even when multiple SNPs were accounted for in these regions.

We next examined how overlapping the three tools were in either missing or overcalling variants. Panel A of Figure 4 summarizes the total number of variants appearing in the curation set that could not be detected by one or more of the variant callers. Pilon uniquely missed only one curated variant, while SAMtools and GATK-UG missed many more (32 and 13, respectively). The majority of variants that were missed by Pilon were also missed by SAMtools and GATK-UG (52 events). In addition, all three tools made predictions that were not supported by the curation set (summarized in Panel B of Figure 4), but, among unique unsupported events, Pilon and GATK-UG had ~3-fold fewer than SAMtools. Altogether, there were only 21 predictions where two or more of the tools agreed that there should be a variant called, most of which were SNPs, although four of the seven events shared by Pilon and GATK-UG were multi-nucleotide indels (5–15 nt in length).

Given the broad definition of 'multi' in Table 3 (>1 bp), we also evaluated how well Pilon performed for variants that were larger than 50 bp (see Table 4). We chose 50 bp since it is a length that is larger than the size of events for which short-read aligners are typically able to align, but shorter than the individual read length of the data used (101 bp). Overall, Pilon was able to accurately identify 74% of these large variants, including 100% of substitutions, 68% of insertions and 77% of deletions from the curated list. Of the eleven insertions that were missed by Pilon, eight involved a repetitive element (5 tandem insertions and 3 IS6110 insertions). Similarly, the six deletions not detected by Pilon involved deletion of one or more copies of a tandem repeat. Four of these tandem repeat regions were correctly reported by Pilon as possible tandem repeat variants in its standard output, but Pilon currently makes no attempt to provide a definitive copy number call in the presence of significant tandem repeat structures. Pilon also identified three events >50 bp that did not match variants in the curation set. These unsupported calls occurred within complex variable regions of the genome in which multiple nearby repeat structures prevented Pilon from correctly identifying the precise correct location or form of the events.

**Table 2.** Summary of variant types curated in the *M. tuberculosis* H37Rv and *M. tuberculosis* F11 finished genome comparison.

| Type of variation between F11 and H37RV | Total found |
| --- | --- |
| Single substitution | 1012 |
| Single insertion | 26 |
| Single deletion | 31 |
| Multi substitution | 13 |
| Multi insertion | 56 |
| Multi deletion | 47 |

The full list can be found in Table S8.

Since Pilon performed well in identifying and resolving MNPs and since GATK-UG and SAMtools were not explicitly designed to call these large variants, we sought to compare Pilon's MNP calls to that of methods specifically designed for MNP detection [11]. Though neither was described for use on microbial data, we evaluated how well BreakDancer [24] and CLEVER [25], two algorithms developed to detect large variants in eukaryotes, performed in calling MNPs on the *M. tuberculosis* test set. BreakDancer was unable to identify any MNP found in the curation set and CLEVER identified 21 multi nucleotide deletions, of which only 1 corresponded to a variant in the curated list. No large insertions or substitutions were predicted (data not shown).

**Evaluating Pilon variant calls without long insert data.** It is unsurprising that Pilon was better able to call larger variants since it is optimized to use both fragment (or small) and long (or large) insert libraries. Since many sequencing projects do not have access to long insert data and to also make a more direct comparison to existing variant callers that are not optimized to accept these data, we evaluated Pilon's performance using data from fragment insert libraries alone. To do this, we ran Pilon using the aligned fragment paired end reads from the *M. tuberculosis* F11 genome to the *M. tuberculosis* H37Rv finished reference genome. We then compared this output ("Pilon-frags") to the previously analyzed output from GATK and SAMtools and to Pilon output using data from both library types ("Pilon").

Pilon-frags performed well in identifying both single and multi nucleotide variants (see Table 3). Pilon-frags identified only 2 pp fewer single nucleotide substitutions, 4 pp fewer single insertions and 4 pp fewer single deletions as compared to the original Pilon output. Pilon-frag performance in calling SNPs was better or on par with both GATK-UG and SAMtools. Remarkably, Pilon-frags was also able to identify a large fraction of the MNPs, with nearly identical performance to Pilon with long insert read data. Pilon-frags also performed very well in calling variants larger than 50 bp (see Table 4), with one less insertion call and 4 fewer deletions calls as compared to Pilon.

To better understand the qualitative differences in what Pilon and Pilon-frags reported, we examined the concordance between results for each variant type. For SNP calls, we observed high concordance in the outputs from Pilon-frags and Pilon (95.2%; 871 of 915 events) (see Table S5). Discordance in SNPs often involved a position where a variant was found in both Pilon runs, but was considered high quality in one and low quality in the other. In fact, only 7 of 915 SNPs (0.8% of total) were confidently predicted to differ between the two Pilon run conditions, suggesting that the value of long insert library data when calling SNPs is small. However, for SNPs within repetitive regions of the genome, long insert data appeared to be very helpful in disambiguating these events (Table S6). Small indel variant calls

were also highly concordant for the two Pilon runs (93.3%; 56 of 60 events), and 78.5% concordance (73 of 93) for large indels.

For larger variants, the discordance between Pilon with and without long insert data was larger (see Table S5), particularly in regions of the genome encoding transposable IS6110 repeat elements. While Pilon-frags detected many of these events, the sequences that were assembled and reported at these sites were often incomplete, as illustrated in Table S7. Given the length of the IS6110 repeat (~1.3 Kbp), the fragment pairs — only ~180 bp apart — were unable to span the entire length of the IS6110 elements, leading to two large indels being called, one coming in from each side of the IS6110 *e.g.* Table S7, position 1,541,957. Pilon's improved ability to capture the full sequence of larger insertions is the primary value of including long insert read data for variant calling applications.

**Assessing large-scale genome duplications.** In addition to identifying substitutions and indels of various sizes, Pilon is able to identify areas in which the read evidence suggests additional copies of large genomic regions (>10 Kbp) compared with the input draft assembly or reference genome. These regions could indicate large collapsed repeats in an assembly improvement application or large genomic duplications in a variant detection application. To evaluate Pilon's ability to identify large duplications, we resequenced *M. tuberculosis* T67, a strain previously reported to harbor a large-scale duplication [26], using fragment and long insert libraries, and aligned the reads to the *M. tuberculosis* H37Rv finished reference. Pilon was then run to detect variants in T67 using H37Rv as a reference.

Pilon identified two duplication events that were >10 Kbp in size and separated by ~3 Kbp at *M. tuberculosis* H37Rv coordinates 3,494,063–3,551,070 (57 Kbp) and 3,554,192–3,712,284 (158 Kbp) resulting in a combined duplication of ~215 Kbp. The left gene boundary (Rv3128c) of the first predicted duplication and right gene boundary (Rv3427c) of the second predicted duplication corresponded to the upstream and downstream boundaries in the previously reported *M. tuberculosis* T67 duplication [26]. Upon closer inspection, the 3 Kbp intervening region contained two copies of the IS6110 element, which are routinely found in multiple copies within the *M. tuberculosis* genome (16 copies in H37Rv) [27]. Because these elements occur so frequently in the genome, the incremental coverage from the duplication was not sufficient for Pilon to identify them as part of the duplication event, breaking a true large duplication into two reported pieces.

## Discussion

Pilon is an assembly improvement algorithm and variant caller that identifies differences between a draft assembly or closely

**Table 3.** Recall and precision metrics for *M. tuberculosis* F11 variants called against *M. tuberculosis* H37Rv by Pilon (with and without long insert library data), GATK UnifiedGenotyper and SAMtools.

| | Pilon | | | GATK | | | SAMtools | | | Pilon-frags | | |
|---|---|---|---|---|---|---|---|---|---|---|---|---|
| | R | P | F | R | P | F | R | P | F | R | P | F |
| Single substitution | 0.96 | 0.98 | 0.97 | 0.85 | 0.98 | 0.91 | 0.88 | 0.93 | 0.90 | 0.94 | 0.98 | 0.96 |
| Single insertion | 0.83 | 1 | 0.91 | 0.75 | 1 | 0.86 | 0.79 | 1 | 0.88 | 0.79 | 1 | 0.88 |
| Single deletion | 0.91 | 0.95 | 0.93 | 0.87 | 0.9 | 0.86 | 0.87 | 1 | 0.93 | 0.87 | 0.95 | 0.91 |
| Multi substitution | 1 | 0.95 | 0.97 | 0.67 | N/A | N/A | 1 | 0.98 | 0.99 | 1 | 0.95 | 0.97 |
| Multi insertion | 0.63 | 0.73 | 0.68 | 0.17 | 0.79 | 0.28 | 0.21 | 0.5 | 0.30 | 0.63 | 0.76 | 0.69 |
| Multi deletion | 0.73 | 0.9 | 0.81 | 0.27 | 0.75 | 0.4 | 0.39 | N/A | N/A | 0.71 | 0.87 | 0.78 |

The three rows marked with 'Single' indicate single nucleotide variants. The three rows marked with 'Multi' indicate variants involving two or more nucleotides, which also include very large events that span several Kb. Recall (R) is the fraction of curated events that were called by the program. Precision (P) is the fraction of calls that the program made that were also described in the curation. The F-measure is the harmonic mean of recall and precision and provides measure of the trade-off between recall and precision. "N/A" indicates that all events of this type were captured in another variant category.

related reference assembly using evidence supported by the sequencing data, resulting in either an improved assembly or a list of variants in VCF format. We have demonstrated Pilon's performance on several microbial genomes with varying GC-content and different ploidy. Our results indicate that applying Pilon yields more contiguous and accurate assemblies. Furthermore, variant calls made by Pilon are of high quality when compared with other state-of-the-art tools, and Pilon's ability to find insertion and deletion events considerably larger than read-lengths sets it apart from traditional variant calling tools.

For many of the sub tasks that Pilon performs, there are a variety of existing tools that might be used in sequence to achieve similar output as Pilon. However, using existing tools to achieve the full complement of analyses performed by Pilon would require implementation of a complicated workflow that hands data around to various tools, and development of post-processing algorithms to ensure that results from the various tools are in agreement. Pilon is a single, benchmarked tool that performs comparably well, if not better, than other tools that do only a fraction of the work. In addition, Pilon is easy to use and is successfully being utilized for assembly and variant detection of thousands of data sets in Broad Institute's production pipelines.

We showed that the assembly improvements that were introduced by Pilon were both accurate and biologically relevant. In particular, several highly repetitive genes that were captured by Pilon are known to play a role in virulence and host-pathogen interactions [28,29]. To date, it has been difficult to study these genes comparatively, because they are often not captured or are only partially captured in draft assemblies. Furthermore, we were able to place more genes with repetitive features accurately in the genome. In the *M. tuberculosis* case, Pilon improved the sequence accuracy and placement of genes encoding transposases, which have an important role in genome reorganization in this species [27] and are used in strain typing schemes [30]. In addition, for *M. tuberculosis*, Pilon had a significant impact on the repetitive, GC-rich and not well understood PE and PPE genes, an expanded family of highly repetitive genes that account for about 10% of the gene repertoire in this species [31], and are implicated in pathogen-host interactions and virulence [32,33]. Pilon was able to resolve these repetitive structures because of its ability to use the long-distance mate pair information afforded by long insert libraries. In addition, data from long insert libraries often enable Pilon to completely fill in large sequence insertions and assemble across gaps.

For variant calling, the primary benefit of Pilon over other variant callers is its ability to capture large sequence polymorphisms and highly polymorphic local regions by performing a local assembly to generate complete sequences. By capturing intervening sequences in large variants, Pilon enables a more comprehensive view of the biological differences between strains *e.g.*, new genes that confer antibiotic resistance or virulence. In addition, by integrating the SNP and large sequence variant detection in a single tool, Pilon is less likely to erroneously call SNPs in regions that are affected by a large variant. Long insert libraries provide information that resolves larger, multi-nucleotide events, in particular by allowing Pilon to completely assemble inserted sequences. However, even without long insert libraries, most of these events are identified, albeit often with an incomplete alternative sequence.

Pilon was also able to perform comparably well in calling small variants, including SNPs and small indels. While all three variant callers benchmarked in this study had similar precision in their calls, Pilon demonstrated better recall (fewer false negatives) on single-base polymorphisms. There are two reasons for this.

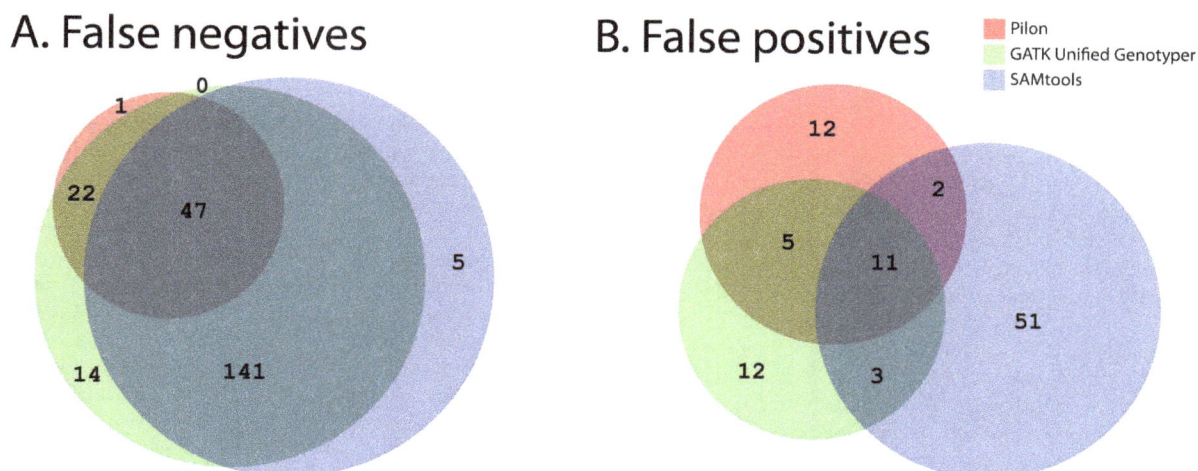

**A. False negatives**

**B. False positives**

Pilon
GATK Unified Genotyper
SAMtools

**Figure 4. Venn diagram of the overlap in false negative (A) and false positive (B) calls by the three variant detection tools, Pilon, GATK UnifiedGenotyper and SAMtools.** False negative calls are the number of unique events from the curation set that was missed by each tool. Overlaps in the Venn diagram show the number of variants that were missed by multiple tools. False positive calls are the number of predictions from *M. tuberculosis* F11 that were not supported by the curation set. Overlaps indicate predictions that were shared among tools.

First, there are a few local areas of the genome with very high polymorphism rates between F11 and H37Rv. When the local polymorphism rate is very high, the short-read aligners are unable to produce enough good alignments for any of the tools to call the base differences from the resulting pileups. However, Pilon was able to detect some of these problem areas and reassemble them into block substitutions, correctly capturing dozens of polymorphic locations the other tools were unable to resolve. Second, the use of long insert libraries allowed Pilon to make definitive calls inside some repeat regions by heavily weighting long insert reads which were unambiguously anchored to the flanks of the repeat area, capturing additional true SNPs.

We note that GATK contains a highly sophisticated collection of tools for variant calling. However, several of its tools and their associated best practices for human variant calling are not applicable to many microbial projects, because they rely upon a database of known variant locations (such as those found at dbSNP) to perform recalibration. For microbial variant calling, the extent of variation across the microbial species under investigation is typically unknown, so no such catalog of variation is available *a priori*. There is also a fundamental difference between GATK and Pilon's approach to variant calling. GATK's UnifiedGenotyper is designed to be aggressive in detecting possible variants, relying on a user-controlled VariantFiltration step to filter out calls of lower confidence or quality to minimize false positives. Pilon, on the other hand, relies exclusively on internal heuristics to make a determination of which calls are confident. This makes Pilon easy to use "out of the box" in a highly automated environment, though it is less configurable than GATK for custom applications.

There are several areas where improvements could be made in future versions of Pilon with respect to assembly improvement and variant calling. First, it seems likely that there will be some benefit in iteratively applying Pilon in the assembly improvement and insertion variant calling process. Currently, Pilon builds out the gap/inserted sequence without re-aligning reads to build off the newly extended sequence. Using an iterative strategy has recently been used with some success by PAGIT [6]. Second, Pilon currently does not attempt to make fixes to larger structural issues within assemblies or make changes to scaffold architecture. With data from long insert libraries, it should be feasible to break and/ or join scaffolds accurately. Third, tandem repeats continue to be challenging and may require a more specific approach. These regions are inherently difficult with short read data because there is no unambiguous information in the data to determine how many copies are present. This challenge is true for any *de novo* assembly; in order to resolve a tandem repeat, reads are needed that anchor into unique sequence on either side and read through the entire tandem repeat sequence. Lacking this, tandem copy numbers can only be speculated from mate-pairing information, depth of coverage calculations and library insert sizes.

Currently, Pilon is rather conservative in its corrections: (i) it uses a large cut-off to merge overlaps, and (ii) it will not attempt to

**Table 4.** Pilon's performance in calling variants in *M. tuberculosis* F11 that were larger than 50 nt.

| | Pilon | | Pilon-frags | |
|---|---|---|---|---|
| | **Missed** | **Called** | **Missed** | **Called** |
| Insertion | 11 | 23 | 12 | 22 |
| Substitution | 0 | 5 | 0 | 5 |
| Deletion | 6 | 20 | 10 | 16 |

Variants are divided by type across the rows. Missed variants are those that were annotated in the curation, but were not identified by Pilon. The called variants are those that were annotated in the curation that Pilon accurately identified.

resolve significant tandem repeats structures definitively. Notwithstanding the challenges encountered with tandem repeats, Pilon does an excellent job with other repetitive sequences and is able to fix many genes of known repetitive gene families and is able to fill in many transposable elements.

While we have evaluated Pilon's assembly improvements on both haploid and diploid genomes and obtained positive results for both, we acknowledge that there is still significant opportunity for future improvement in Pilon's handling of diploid genomes. Pilon could be enhanced to understand IUPAC ambiguity codes in its input genome and generate them in its output, and Pilon's heuristics for identifying insertions and deletions in diploid genomes could be improved, including its ability to recognize and report heterozygous indels. Finally, the local reassembly process could be improved to perform better in heterozygous regions. Even so, our results indicate that in its current form, Pilon is able to make valuable improvements to diploid genomes.

We have evaluated Pilon's performance using microbial genomes with finished references. However, there is no inherent limitation on the size of genomes to which Pilon can be applied. For example, we have used Pilon to improve assemblies of larger genomes, including 16 strains of the *Anopheles* genus (~200 Mbp diploid genome), but we were unable to verify the accuracy of Pilon's improvements since these genomes have not been finished. Pilon runs within minutes on small microbial genomes and will complete overnight on larger eukaryote genomes, such as *Anopheles*, which is similar to the tools included in our benchmarking.

## Conclusion

Ultimately, Pilon has great utility and addresses an urgent need for better and more efficient methods to deal with the thousands of microbial genomes that are being produced. We have shown that Pilon performs well as compared to the state-of-the-art for both assembly improvement and variant detection, often outperforming these tools. Pilon is also unique in its user-friendly integrated approach to assembly improvement and is unique in its ability to identify large variants accurately in microbial genomes. As a recent addition to the production process for microbial genomes at Broad Institute, Pilon has been used to automatically improve the quality of over 8,000 prokaryote and eukaryote genomes prior to their submission to Genbank, and it has been used to call variants on over 6,000 genomes.

## Material and Methods

### Detailed algorithm description

**Input requirements.** Pilon requires an input genome in FASTA format and one or more BAM files containing sequencing reads aligned to the input genome. The BAM files must be sorted in coordinate order and indexed. For Illumina data, these BAM files are usually produced by an aligner such as BWA [15] or Bowtie 2 [34]. It is recommended that single best hit or random selection among equal best alignments is used as input into Pilon. Pilon can use three types of BAM files:

1. *Fragments*: paired read data of short insert size, typically < 1 Kbp. Reads should be in forward-reverse (FR) orientation;
2. *Long inserts*: paired read data of longer insert size, typically > 1 Kbp. Reads should be in forward-reverse (FR) orientation. Sequencing of long insert libraries that are generated using the standard Illumina mate-pair library preparation protocol

typically result in reverse-forward (RF) read orientation, so they will need to be reversed in the BAM file.
3. *Unpaired*: unpaired sequencing read data.

To use Pilon with default arguments, read length should be 75 bases or longer and total sequence coverage should be 50x or greater, though deeper total coverage of >100x is beneficial. Pilon can also make use of longer reads, such as those from Sanger capillary sequencing and circular-consensus or error-corrected reads from Pacific Biosciences (PacBio) sequencing. However, Pilon is not currently tuned to the error model of raw PacBio reads, and their use may introduce false corrections.

Pilon makes extensive use of pairing information when it is available, so paired libraries are highly recommended. Pilon is capable of using paired libraries of any insert size, as it scans the BAMs to compute statistics, including insert size distribution.

**Improving local base accuracy and identifying SNPs.** Pilon improves the local base accuracy of the contigs through analysis of the read alignment information. First, Pilon parses the alignment information from the input data and summarizes the evidence from all the reads covering each base position. Alignments can be less trustworthy near the ends of reads, especially in differentiating between indels and base changes, so Pilon ignores the alignments from a small number of bases at each end of the read, which is configurable at run time. For each base position in the genome, Pilon builds a pileup structure which records both a count and a measure of the weighted evidence for each possible base (A,C,G,T) from the read alignments. The contribution of base information from each read is weighted by the base quality reported by the sequencing instrument as well as the mapping quality computed by the aligner.

When Pilon is building pileups from paired alignment data, only reads from "valid" pairs (*i.e.*, those with the *PROPER_PAIR* flag set in the BAM by the aligner, indicating the reads of a pair align in proper orientation with a plausible separation) contribute evidence to the pileups. It is crucial that the PROPER_PAIR flag is set accurately by the tool that produced the BAM file. A count of non-valid alignments covering each position is also kept to help identify areas of possible mis-assembly. Pilon also keeps track of "soft clipping" in the alignments, which exclude sub-sections of a read which aligned poorly. A tally of soft-clip transitions is kept at each genomic location as another aid in identifying possible local misassemblies.

From the pileup evidence, Pilon classifies each base in the input genome into one of four categories:

1. *Confirmed*: the vast majority of evidence supports the base in the input genome;
2. *Changed*: the vast majority of evidence supports a change of the base in the input genome to another allele;
3. *Ambiguous*: the evidence supports more than one alternative at this position;
4. *Unconfirmed*: there is insufficient evidence to make a determination at this position due to insufficient depth of coverage by valid reads.

Ambiguous bases can occur for several reasons. If the genome is diploid, this is expected at heterozygous polymorphic locations. Difficult-to-sequence regions may result in a large enough fraction of sequencing errors to result in an ambiguous call. Finally, if the input genome has a smaller number of copies of a repeated genomic structure than occurs in the true genome (a "collapsed repeat"), the aligned reads may have originated from more than one instance of the repeat structure; where there are differences in the true instances of the repeat, the alignments can show mixed evidence.

Paired read information, especially information from long insert libraries which span a longer distance, is extremely valuable in helping resolve ambiguous locations due to collapsed repeats. Pairs for which one read lands inside a repeat element, but the other lands in unique anchoring sequence on the flanks of the repeat help to resolve the true base content of the repeat structure. Data from long inserts will typically have a higher alignment mapping quality than short-range fragment pairs that lie completely within the repeat because the fragment pairs may not be able to be placed uniquely among the repeats. Since Pilon uses mapping quality to weigh the evidence from each read, the long inserts can often pick the correct haplotype variations of the repeat structure.

Pilon includes corrections to single-base errors in its output genome, and optionally, it can also change ambiguous bases to the allele with the preponderance of evidence.

**Finding and fixing small indels.** While recording the base-by-base pileup evidence, Pilon also records the location and content of indels present in the alignments. Indel alignments which represent equivalent edits to the input genome may appear at different coordinates in the alignments. For instance, if the input genome has the sequence *ACCCCT*, but the read evidence suggests one of the Cs should be deleted (*ACCCT*), each individual read alignment might show a deletion at any of the four *C* coordinates. Pilon shifts alignment indels to their leftmost equivalent edit in the input genome, so that the evidence from all the equivalent edits is combined into evidence for a single event at a one location.

Pilon makes an insertion or deletion call if a majority of the valid reads support the change, though that threshold is lowered somewhat for longer events, as it is typically more difficult for aligners to identify longer indels in short read data. Called indels from the input genome are fixed in Pilon's output genome.

**Fixing mis-assemblies, detecting large indels, and filling gaps.** Pilon is capable of reassembling local regions of the genome when there is sufficient evidence from the alignments that the contiguity of the input genome does not match the sequencing data. For assembly improvement applications, this could be an indication of a local mis-assembly. For variant calling applications, this could be caused by insertions or deletions too large to be reflected in the short read alignments.

Pilon tries to identify areas of potential local read alignment discontinuity in the contigs of the input genome by employing four heuristics: (i) a large percentage of reads containing a soft-clipped alignment at a given base position, (ii) a large ratio of invalid pairs to valid pairs spanning a location, (iii) areas of extremely low coverage and (iv) rapid drops in alignment coverage over a distance on the order of a read length. Once Pilon has identified an area for local reassembly, it treats the suspicious region (which may be a single base or a larger region) as untrusted, using alignments to the trusted flanks on both sides to identify a collection of reads that might contribute evidence for the true sequence in the suspicious region.

Unpaired reads with partial alignments to the flanks are included in the collection. For paired data, Pilon identifies pairs in which one of the reads is anchored by proper alignment to one of the flanks (*e.g.*, with forward orientation on the left flank, or reverse orientation on the right flank), but whose mate is either unmapped or improperly mapped (*e.g.*, to a remote location in the genome). For fragment pairs, both reads of such pairs are included in the collection; for long inserts, only the unanchored read is included in the collection.

From the collected reads, Pilon builds a De Bruijn assembly graph (default K = 47). For each k-mer in the reads, it uses the same pileup structure to record the bases which follow that k-mer, including weighting by base quality. Then, the pileups are evaluated to determine the link(s) to the next k-mer(s); this results in either a single base call, resulting in one forward link to the next k-mer, or an ambiguous call, resulting in two links forward and a branch in the assembly graph. This process automatically prunes the assembly graph of most sequencing errors, as infrequent base differences are unlikely to present enough evidence to affect the forward links. A minimum coverage cutoff of five for each forward link also prunes the assembly graph of many false links that could appear because of sequencing errors.

Pilon then tries k-mers from the trusted flanks as starting points to walk into the untrusted region from each side, building all possible extensions with up to five branching points ($2^5$ possible extensions). Tandem repeats with combined length $>K$ cause loops in the local assembly graph, and they are detected by noting when the assembly walk reaches an already-incorporated k-mer. Pilon currently does not attempt to determine the copy number of such tandem repeats; instead, it will report the length of the repeat structure encountered in its standard output, and it will not attempt to close the two sides.

When no tandem repeat is detected, the resulting extensions from each side are combinatorially matched for possible perfect overlaps of sufficient length (2K+1) to be considered for closure. If there is exactly one such closure and it differs from the input genome, the assembled flank-to-flank sequence will replace the corresponding sequence in the input genome. Since the default k-mer size is 47, an overlap of 95 bases is required for closure.

If there are no closures or more than one possible closure, Pilon will identify a consensus extension from each flank. If an optional argument is set to allow opening of new gaps, Pilon will replace the suspicious region with the consensus extensions from each flank and create a gap between them; otherwise, it simply reports that it was unable to find a solution. These reports identify areas that an assembly analyst might wish to investigate manually.

Pilon also attempts to fill gaps between contigs in a scaffold ("captured gaps") in the input genome. In order to fill captured gaps, Pilon employs the same local reassembly technique described above, treating the gap itself as the "untrusted" region. If there is a unique closure, the gap is filled; otherwise, consensus extensions from each flank are used to reduce the size of the gap. Pilon does not currently attempt to join or break scaffolds.

**Large collapsed repeat (segmental duplication) detection.** Pilon includes heuristics that attempt to flag areas indicative of large (>10 Kbp) collapsed repeats with respect to the input genome. These are characteristically large contiguous areas that appear to have double (or higher) read coverage compared to the rest of the genomic element being analyzed. Long insert data are excluded from this computation, as we have found long insert coverage to be far more variable across some genomes. Pilon does not attempt to fix these potentially collapsed regions, but it does report them in its standard output for further investigation.

In variant calling applications, large segmental duplications in the sequenced strain with respect to the reference have the same signature as large collapsed repeats in a draft assembly; a duplicated region of the genome will result in double the number of reads covering that sequence. Pilon's reporting of large collapsed repeat regions can be used to identify candidate segmental duplications.

**Output files.** Pilon generates a modified genome as a FASTA file, including all single-base, small indel, gap filling, mis-assembly and large-event corrections from the input genome. In the assembly improvement case, this is the improved assembly consensus. In variant detection mode, this is the reference sequence which has been edited to represent the consensus of the given sample more closely.

Pilon can optionally generate a Variant Call Format (VCF) [http://vcftools.sourceforge.net/specs.html] file, which lists copious detailed information about the base and indel evidence at every base position in the genome, including two scores regarding variant quality: the QUAL column, and a depth-normalized call quality (QD) field in the INFO column. For additional details on the VCF format, we refer to the VCF specification referred above. Changes generated by local reassembly, often triggered by larger polymorphisms in variant calling applications, are included as structural variant records (SVTYPE = INS and SVTYPE = DEL). Pilon can also, optionally, generate a "changes" file which lists the edits applied from input to output genome in tabular form, including source and destination coordinates and source and destination sequence. Finally, Pilon will optionally (with the — tracks option) output a series of visualization tracks ("bed" and "wig" files) suitable for viewing in genome browsers such as IGV [35] and GenomeView [36]. Tracks include basic metrics across the genome, such as sequence coverage and physical coverage, as well as some of the calculated metrics Pilon uses in its heuristics for finding potential areas of mis-assembly, such as percentage of valid read pairs covering every location.

Pilon's standard output also contains useful information, including coverage levels, percentage of the input genome confirmed, a summary of the changes made, as well as some specifically flagged issues which were not corrected, such as potentially large collapsed repeat regions, potential regions of mis-assembly which were not able to be corrected, and detected tandem repeats that were not resolved.

## Data generation

All sequencing data used for these experiments were generated from an Illumina HiSeq 2000 machine. For sequencing *M. tuberculosis* F11 and T67, two libraries were generated: one PCR-free 180 bp insert paired fragment library [37] and large insert 3–5 Kbp long insert library [38]. *S. pneumoniae* TIGR4 data also consisted of two libraries: one robotically size selected 180 bp insert paired fragment library [37] and a large insert 3–5 Kbp long insert library [38]. The sequencing data for *C. albicans* SC5314 was generated from three libraries: a robotically size-selected 180 bp insert paired fragment library [37], a gel-cut 4 Kbp long insert library [39], and a 40 Kb Fosill library [40]. Sequencing data were submitted to the Sequence Read Archive with identifiers: SRX347313, SRX347312, SRX105400, SRX110130, SRX347317 and SRX347316.

## Evaluation methods

**Assembly improvement.** All draft assemblies were generated using ALLPATHS-LG [41]. The draft assembly for *Mycobacterium tuberculosis* F11 utilized 100x of the 180 bp insert fragment library and 50x of the 3–5 Kb long insert library and was executed using ALLPATHS-LG v45395 utilizing the ASSISTED_PATCHING = 2.1 parameter and the *M. tuberculosis* H37RV reference genome for assisting (GenBank accession: CP003248). The draft assembly for *S. pneumoniae* TIGR4 was created using ALLPATHS-LG v45925 with default parameters and using 100x of the 180 bp insert fragment library and 50x of the 3–5 Kb long insert library. The *C. albicans* SC5314 utilized 100x of the 180 bp insert fragment library, 100x of the gel-cut 4 Kb long insert library and 50x of the Fosill library, and was assembled with ALLPATHS-LG v39846 utilizing the ASSISTED_PATCHING and HAPLOI-

DIFY options with the *C. albicans* SC5314 reference sequence as a reference for assisting.

We benchmarked Pilon's ability to close gaps in the draft bacterial assemblies against two tools built for this purpose, IMAGE v2.4.1 [16] and GapFiller v1.10 [17]. The same sets of sequencing reads used as input to Pilon were used for IMAGE (fragment library only) and GapFiller (fragment and long insert libraries). IMAGE was run in the manner implemented in the PAGIT [6] example scripts: 6 iterations, one with a kmer size of 61, three with a kmer size of 49, and two with a kmer size of 41. GapFiller was run for 10 iterations with a libraries.txt file specifying a ratio $r = 0.5$ and library insert sizes computed by Pilon from the aligned bams.

To evaluate the quality of Pilon's single base and small indel corrections to the draft assemblies, we also ran iCORN v0.97 [18], the consensus sequence improvement tool in PAGIT, on the same draft assemblies using the same sets of fragment reads. iCORN was run in the manner implemented in the PAGIT example scripts, only changing the library insert size mean and range parameters. For TIGR4, we used a mean of 180 and a range of 120–300. For F11, we used a mean of 226 and a range of 100–500, since the PCR-free library preparation resulted in a wider range of insert sizes.

Fixes to the assemblies (Table S1) made by Pilon and the other assembly improvement tools were assessed by extracting the changed region of sequence in the output genome along with 300 bp flanks on each side. These extracted sequences were aligned to their respective finished reference genomes with BLASTN [42], and the accuracy of the changes was assessed by manually inspecting the alignments for accuracy, judging each fix as "Correct" or "Incorrect". For larger block changes which resulted from local reassembly (gap filling and fixing of local mis-assemblies), a third category of "No worse" was established for situations in which: (i) the draft assembly contained a mis-assembly in the changed region, (ii) Pilon made a change attempting to fix the mis-assembly, and (iii) the fix was not entirely correct, but was no worse than the original problem.

For the assembly improvement statistics, *Bases added* was calculated by tallying bases added in locations where resulting fixes resulted in a net gain of bases during gap filling and mis-assembly correction processes, as reported in the Pilon standard output indicated by the "fix gap" or "fix break" lines.

**Variant calling.** Variant calls were made using *M. tuberculosis* H37Rv (GenBank accession: CP003248) as the reference and the *M. tuberculosis* F11 aligned read and long insert fragments as input data. From the sequenced fragment and long insert libraries, a random subset of read pairs was selected from each library to obtain an estimated 200x coverage of the *M. tuberculosis* H37Rv reference genome. Each library's reads were aligned to the *M. tuberculosis* H37Rv reference genome using BWA (version 0.5.9-r16) to generate BAM files suitable for input to the variant calling processes.

Pilon: Pilon was run with the —variant command line option, specifying the *M. tuberculosis* H37Rv reference genome and the above BAM file(s) as inputs. We evaluated two Pilon variant calling sets, one generated using both fragment and long insert library BAMs, and one using only the fragment library BAM.

GATK UnifiedGenotyper: Reads in the fragment library BAM were realigned by applying the Genome Analysis Toolkit (GATK version v3.2.2) RealignerTargetCreator and IndelRealigner tools on the fragment library aligned BAM file. Variants were then called from the realigned BAM file using UnifiedGenotyper run with the following settings: -nt 32 -A AlleleBalance -ploidy 1 -pnrm EXACT_GENERAL_PLOIDY -glm BOTH —output_mode

EMIT_ALL_SITES. Low confidence variants were then filtered using VariantFiltration (VF) run with the following settings: −filterExpression "((DP-MQ0)<10) || ((MQ0/(1.0*DP))>=0.8) || (ABHom <0.8) || (Dels >0.5)" −filterName LowConfidence.

These VariantFiltration settings filtered out variant calls at locations with less than 10 unambiguous read alignments, where 80% or more of the read depth had ambiguous mappings, where fewer than 80% of the reads supported the alternate allele, or more than half of the reads contained spanning deletions. This filter expression was based on one previously used to call variants from the European *Escherichia coli* O104:H4 outbreak [38], adjusting depth and allele balance thresholds to yield the best performance tradeoff between false negative and false positive results on these data.

SAMtools/BCFtools: The same aligned fragment library BAM file described above was used as input for variant calling using SAMtool/BCFtools v0.1.19 according to recommendations found on the SAMtools webpage (http://samtools.sourceforge.net/mpileup.shtml). samtools mpileup was used to generate pileups in bcf format, and variants were called using bcftools using the −bcg option. Finally, variants were filtered using vcfutils.pl varFilter -d 10 to filter out calls at locations where the aligned coverage was less than 10 reads. We chose the minimum depth of 10 to be consistent with the filtering used for GATK Unified-Genotyper.

CLEVER and BreakDancer: The aligned fragment and combined fragment and long insert library described above were used as input for CLEVER v2.0rc3 and BreakDancer 1.3.6. clever −sorted −use_xa was used to generate calls for CLEVER. bam2cfg.pl -g -h was used to generate the Break-Dancer config file, which was then used with breakdancer-max.

**Curating differences between F11 and H37Rv.** Differences between the finished *M. tuberculosis* F11 (GenBank accession: CP000717) and *M. tuberculosis* H37RV (GenBank accession: CP003248) references were curated by employing a banded Smith-Waterman algorithm to align syntenic regions of the two genomes. Alignments were run, separately, for each syntenic portion of the two sequences. When the alignment diverged significantly, the program was run again to pick up at the next syntenic block. The resulting alignments over syntenic regions identified coordinates of small blocks of mismatches, typically only a single-base long, but in some cases up to 289 bp. Areas where there was a significant break in synteny or where the banded Smith-Waterman alignment produced questionable results were analyzed using either Nucmer [43], ClustalW [44] or Blast2 [45] to verify the nature of the difference and to obtain more accurate coordinates. In some cases, the alignments proved too difficult to get accurate coordinates, but approximate definitions of these differences were obtained. The resulting table of differences between the two references (Table S8) has each difference annotated with most likely coordinates, with two exceptions where the variation between the strains was so high that it was impossible to know whether each difference was captured individually. The two highly variable regions corresponded to coordinates 1636857–1639600 and 3928967–3949709, which, together, account for less than 0.5% of the *M. tuberculosis* H37Rv genome. These regions were excluded from all variant analyses.

**Variant Assessment.** The resulting variant calls were compared to a manually curated set of differences between *M. tuberculosis* F11 and *M. tuberculosis* H37Rv as described above. Based on this comparison, recall and precision were calculated according to the strategy described in [46]. Briefly, recall is a measure of completeness of calls against the curated truth set; false negatives lower the recall score. Precision is a measure of the accuracy of the calls made; false positives lower the precision score. Specifically, recall = $tp\_c/(tp\_c+fn)$ and precision = $tp\_p/(tp\_p+fp)$, where $tp\_c$ is the number of true positive calls based on the curation set, $tp\_p$ is the number of true positive calls from the set of predicted variants, fp is the number of false calls from the set of predicted variants, and fn is the number of missed calls from the predicted variants based on the curation set. The F-measure is the harmonic mean of the recall and precision rates, providing an "overall" metric that captures tradeoff between recall and precision.

True positives in the prediction set had at least one variant site called in the curation set. For variants that affected more than a single base in the curation set (*i.e.*, multi nucleotide polymorphisms), we allowed for a combination of two or more smaller events in the prediction set to be marked as correct, since tools may call a densely polymorphic region as a block substitution rather than a series of equivalent single-base changes. For example, for the multi nucleotide substitution in the curation set, ACCGT => CCTGA, three SNP calls at the same location, A>C, C>T and T>A, would be counted as a true positive. In addition, predicted variants that affected more than 20 bases were allowed to match only partially with the curation set because there can be different ways to manually curate sites that vary among the *M. tuberculosis* F11 and H37Rv finished reference genomes. In particular, resolution of tandem repeats was challenging for both prediction and curation of variants since it was difficult to determine which copy of the repeat was inserted or deleted. In these cases, we counted the variant as correct if a similar event was predicted within 100 nucleotides.

## Availability

Pilon is open source software available under the GNU General Public License Version 2 (GPLv2). Pilon is written in the Scala programming language, and it makes extensive use of the open source Picard Java libraries (http://picard.sourceforge.net) for parsing BAM and FASTA files. Pilon is compiled into a single Java Archive (JAR) file which runs inside a 64-bit Java Virtual Machine environment. Binary and source distributions can be obtained from GitHub (http://github.com/broadinstitute/pilon/releases/). Results in this paper were obtained with Pilon version 1.5. A summary of all command-line options is available in Table S9.

Online documentation, as well as two example data sets to test Pilon on the same data as was used in this manuscript, are available from the web site http://broadinstitute.org/software/pilon/. We provide the *Streptococcus pneumoniae* TIGR4 data set as an assembly improvement example and the *Mtb* F11 data set as a variant calling example.

## Supporting Information

**Figure S1 Muscle alignment of TB F11 gene TFBG_12611.**

**Figure S2 Contig count reduction in production.**

**Figure S3 Contig N50 increase in 50 production assemblies.**

**Table S1 Assessment of gap filling and local reassembly fixes. b**: Comparison of assembly gap closures among Pilon, IMAGE, and GapFiller.

**Table S2   Assessment of base corrections by Pilon and iCORN.**

**Table S3   Detailed information regarding the gene based assessment of F11 assemblies.**

**Table S4   Detailed information regarding the gene based assessment of TIGR4.**

**Table S5   Summary of SNP, small in-dels, and large in-dels in M. tuberculosis F11 relative to H37Rv.**

**Table S6   Example SNPs only found with regular Pilon.**

**Table S7   Example IS6110 insertion element variation.**

**Table S8   Curation of Mtb F11.**

**Table S9   Pilon Command Line Arguments.**

## Acknowledgments

We acknowledge the Broad Institute Genomics Platform for generating the Illumina libraries and sequence used here, and the Genome Assembly and Analysis Group and the A2E Group for their assistance in configuring and maintaining Pilon within the Broad Institute's production assembly infrastructure. We also thank Gustavo Cerqueira, Christopher Desjardins, Abigail McGuire, Jonathan Livny and, especially, Geraldine Van der Auwera for their helpful comments.

## Author Contributions

Conceived and designed the experiments: BJW TA CAC QZ JW SKY AME. Performed the experiments: BJW TA TS SKY. Analyzed the data: BJW TA TS MP AA SS CAC QZ JW SKY AME. Wrote the paper: BJW TA TS MP AA SS CAC SKY AME.

## References

1. Chewapreecha C, Harris SR, Croucher NJ, Turner C, Marttinen P, et al. (2014) Dense genomic sampling identifies highways of pneumococcal recombination. Nat Genet 46: 305–309. Available: http://www.ncbi.nlm.nih.gov/pubmed/24509479. Accessed 21 March 2014.

2. Comas I, Coscolla M, Luo T, Borrell S, Holt KE, et al. (2013) Out-of-Africa migration and Neolithic coexpansion of Mycobacterium tuberculosis with modern humans. Nat Genet 45: 1176–1182. Available: http://www.ncbi.nlm.nih.gov/pubmed/23995134. Accessed 19 March 2014.

3. Croucher NJ, Finkelstein J a, Pelton SI, Mitchell PK, Lee GM, et al. (2013) Population genomics of post-vaccine changes in pneumococcal epidemiology. Nat Genet 45: 656–663. Available: http://www.pubmedcentral.nih.gov/articlerender.fcgi?artid=3725542&tool=pmcentrez&rendertype=abstract. Accessed 21 March 2014.

4. Grad YH, Kirkcaldy RD, Trees D, Dordel J, Harris SR, et al. (2014) Genomic epidemiology of Neisseria gonorrhoeae with reduced susceptibility to cefixime in the USA: a retrospective observational study. Lancet Infect Dis 14: 220–226. Available: http://www.ncbi.nlm.nih.gov/pubmed/24462211. Accessed 21 March 2014.

5. Ronen R, Boucher C, Chitsaz H, Pevzner P (2012) SEQuel: improving the accuracy of genome assemblies. Bioinformatics 28: i188–96. Available: http://www.pubmedcentral.nih.gov/articlerender.fcgi?artid=3371851&tool=pmcentrez&rendertype=abstract. Accessed 20 January 2014.

6. Swain MT, Tsai IJ, Assefa S a, Newbold C, Berriman M, et al. (2012) A post-assembly genome-improvement toolkit (PAGIT) to obtain annotated genomes from contigs. Nat Protoc 7: 1260–1284. Available: http://www.pubmedcentral.nih.gov/articlerender.fcgi?artid=3648784&tool=pmcentrez&rendertype=abstract. Accessed 24 January 2014.

7. Hunt M, Kikuchi T, Sanders M, Newbold C, Berriman M, et al. (2013) REAPR: a universal tool for genome assembly evaluation. Genome Biol 14: R47. Available: http://www.pubmedcentral.nih.gov/articlerender.fcgi?artid=3798757&tool=pmcentrez&rendertype=abstract. Accessed 22 January 2014.

8. Vicedomini R, Vezzi F, Scalabrin S, Arvestad L, Policriti A (2013) GAM-NGS: genomic assemblies merger for next generation sequencing. BMC Bioinformatics 14 Suppl 7: S6. Available: http://www.pubmedcentral.nih.gov/articlerender.fcgi?artid=3633056&tool=pmcentrez&rendertype=abstract. Accessed 28 January 2014.

9. Li H, Handsaker B, Wysoker A, Fennell T, Ruan J, et al. (2009) The Sequence Alignment/Map format and SAMtools. Bioinformatics 25: 2078–2079. Available: http://www.pubmedcentral.nih.gov/articlerender.fcgi?artid=2723002&tool=pmcentrez&rendertype=abstract. Accessed 20 January 2014.

10. McKenna A, Hanna M, Banks E, Sivachenko A, Cibulskis K, et al. (2010) The Genome Analysis Toolkit: a MapReduce framework for analyzing next-generation DNA sequencing data. Genome Res 20: 1297–1303. Available: http://www.pubmedcentral.nih.gov/articlerender.fcgi?artid=2928508&tool=pmcentrez&rendertype=abstract. Accessed 21 January 2014.

11. Pabinger S, Dander A, Fischer M, Snajder R, Sperk M, et al. (2013) A survey of tools for variant analysis of next-generation genome sequencing data. Brief Bioinform 15: 256–278. Available: http://www.ncbi.nlm.nih.gov/pubmed/23341494. Accessed 19 March 2014.

12. Cubillos-Ruiz A, Morales J, Zambrano MM (2008) Analysis of the genetic variation in Mycobacterium tuberculosis strains by multiple genome alignments. BMC Res Notes 1: 110. Available: http://www.pubmedcentral.nih.gov/articlerender.fcgi?artid=2590607&tool=pmcentrez&rendertype=abstract. Accessed 28 January 2014.

13. El-Metwally S, Hamza T, Zakaria M, Helmy M (2013) Next-generation sequence assembly: four stages of data processing and computational challenges. PLoS Comput Biol 9: e1003345. Available: http://www.pubmedcentral.nih.gov/articlerender.fcgi?artid=3861042&tool=pmcentrez&rendertype=abstract. Accessed 21 January 2014.

14. Tettelin H, Nelson KE, Paulsen IT, Eisen J a, Read TD, et al. (2001) Complete genome sequence of a virulent isolate of Streptococcus pneumoniae. Science 293: 498–506. Available: http://www.ncbi.nlm.nih.gov/pubmed/11463916. Accessed 21 January 2014.

15. Li H, Durbin R (2009) Fast and accurate short read alignment with Burrows-Wheeler transform. Bioinformatics 25: 1754–1760. Available: http://www.pubmedcentral.nih.gov/articlerender.fcgi?artid=2705234&tool=pmcentrez&rendertype=abstract. Accessed 20 January 2014.

16. Tsai IJ, Otto TD, Berriman M (2010) Improving draft assemblies by iterative mapping and assembly of short reads to eliminate gaps. Genome Biol 11: R41. Available: http://www.pubmedcentral.nih.gov/articlerender.fcgi?artid=2884544&tool=pmcentrez&rendertype=abstract. Accessed 10 July 2014.

17. Nadalin F, Vezzi F, Policriti A (2012) GapFiller: a de novo assembly approach to fill the gap within paired reads. BMC Bioinformatics 13 Suppl 14: S8. Available: http://www.pubmedcentral.nih.gov/articlerender.fcgi?artid=3439727&tool=pmcentrez&rendertype=abstract. Accessed 10 July 2014.

18. Otto TD, Sanders M, Berriman M, Newbold C (2010) Iterative Correction of Reference Nucleotides (iCORN) using second generation sequencing technology. Bioinformatics 26: 1704–1707. Available: http://www.pubmedcentral.nih.gov/articlerender.fcgi?artid=2894513&tool=pmcentrez&rendertype=abstract. Accessed 10 July 2014.

19. Luo R, Mann B, Lewis WS, Rowe A, Heath R, et al. (2005) Solution structure of choline binding protein A, the major adhesin of Streptococcus pneumoniae. EMBO J 24: 34–43. Available: http://www.pubmedcentral.nih.gov/articlerender.fcgi?artid=544903&tool=pmcentrez&rendertype=abstract. Accessed 21 March 2014.

20. Tu AH, Fulgham RL, McCrory MA, Briles DE, Szalai AJ (1999) Pneumococcal surface protein A inhibits complement activation by Streptococcus pneumoniae. Infect Immun 67: 4720–4724. Available: http://www.pubmedcentral.nih.gov/articlerender.fcgi?artid=96800&tool=pmcentrez&rendertype=abstract. Accessed 21 March 2014.

21. Butler G, Rasmussen MD, Lin MF, Santos M a S, Sakthikumar S, et al. (2009) Evolution of pathogenicity and sexual reproduction in eight Candida genomes. Nature 459: 657–662. Available: http://www.pubmedcentral.nih.gov/articlerender.fcgi?artid=2834264&tool=pmcentrez&rendertype=abstract. Accessed 28 January 2014.

22. Jones T, Federspiel N a, Chibana H, Dungan J, Kalman S, et al. (2004) The diploid genome sequence of Candida albicans. Proc Natl Acad Sci U S A 101: 7329–7334. Available: http://www.pubmedcentral.nih.gov/articlerender.fcgi?artid=409918&tool=pmcentrez&rendertype=abstract.

23. Muzzey D, Schwartz K, Weissman JS, Sherlock G (2013) Assembly of a phased diploid Candida albicans genome facilitates allele-specific measurements and provides a simple model for repeat and indel structure. Genome Biol 14: R97. Available: http://www.ncbi.nlm.nih.gov/pubmed/24025428. Accessed 2014 January 28.

24. Chen K, Wallis JW, McLellan MD, Larson DE, Kalicki JM, et al. (2009) BreakDancer: an algorithm for high-resolution mapping of genomic structural variation. Nat Methods 6: 677–681. Available: http://www.pubmedcentral.nih.

gov/articlerender.fcgi?artid=3661775&tool=pmcentrez&rendertype=abstract. Accessed 2014 March 20.

25. Marschall T, Costa IG, Canzar S, Bauer M, Klau GW, et al. (2012) CLEVER: clique-enumerating variant finder. Bioinformatics 28: 2875–2882. Available: http://www.ncbi.nlm.nih.gov/pubmed/23060616. Accessed 10 July 2014.

26. Weiner B, Gomez J, Victor TC, Warren RM, Sloutsky A, et al. (2012) Independent large scale duplications in multiple M. tuberculosis lineages overlapping the same genomic region. PLoS One 7: e26038. Available: http://www.pubmedcentral.nih.gov/articlerender.fcgi?artid=3274525&tool=pmcentrez&rendertype=abstract. Accessed 2012 August 24.

27. Ioerger TR, Feng Y, Ganesula K, Chen X, Dobos KM, et al. (2010) Variation among genome sequences of H37Rv strains of Mycobacterium tuberculosis from multiple laboratories. J Bacteriol 192: 3645–3653. Available: http://www.pubmedcentral.nih.gov/articlerender.fcgi?artid=2897344&tool=pmcentrez&rendertype=abstract. Accessed 2014 January 22.

28. Kohli S, Singh Y, Sharma K, Mittal A, Ehtesham NZ, et al. (2012) Comparative genomic and proteomic analyses of PE/PPE multigene family of Mycobacterium tuberculosis H$_{37}$Rv and H$_{37}$Ra reveal novel and interesting differences with implications in virulence. Nucleic Acids Res 40: 7113–7122. Available: http://www.pubmedcentral.nih.gov/articlerender.fcgi?artid=3424577&tool=pmcentrez&rendertype=abstract. Accessed 2014 March 21.

29. Vordermeier HM, Hewinson RG, Wilkinson RJ, Wilkinson K a, Gideon HP, et al. (2012) Conserved immune recognition hierarchy of mycobacterial PE/PPE proteins during infection in natural hosts. PLoS One 7: e40890. Available: http://www.pubmedcentral.nih.gov/articlerender.fcgi?artid=3411574&tool=pmcentrez&rendertype=abstract. Accessed 2014 March 21.

30. Das S, Paramasivan CN, Lowrie DB, Prabhakar R, Narayanan PR (1995) IS6110 restriction fragment length polymorphism typing of clinical isolates of Mycobacterium tuberculosis from patients with pulmonary tuberculosis in Madras, south India. Tuber Lung Dis 76: 550–554. Available: http://www.ncbi.nlm.nih.gov/pubmed/8593378. Accessed 2014 January 28.

31. Karboul A, Mazza A, Gey van Pittius NC, Ho JL, Brousseau R, et al. (2008) Frequent homologous recombination events in Mycobacterium tuberculosis PE/PPE multigene families: potential role in antigenic variability. J Bacteriol 190: 7838–7846. Available: http://www.pubmedcentral.nih.gov/articlerender.fcgi?artid=2583619&tool=pmcentrez&rendertype=abstract. Accessed 2014 January 28.

32. Ford C, Yusim K, Ioerger T, Feng S, Chase M, et al. (2012) Mycobacterium tuberculosis—heterogeneity revealed through whole genome sequencing. Tuberculosis (Edinb) 92: 194–201. Available: http://www.pubmedcentral.nih.gov/articlerender.fcgi?artid=3323677&tool=pmcentrez&rendertype=abstract. Accessed 28 January 2014.

33. McEvoy CRE, Cloete R, Müller B, Schürch AC, van Helden PD, et al. (2012) Comparative analysis of Mycobacterium tuberculosis pe and ppe genes reveals high sequence variation and an apparent absence of selective constraints. PLoS One 7: e30593. Available: http://www.pubmedcentral.nih.gov/articlerender.fcgi?artid=3319526&tool=pmcentrez&rendertype=abstract. Accessed 2012 July 28.

34. Langmead B, Salzberg SL (2012) Fast gapped-read alignment with Bowtie 2. Nat Methods 9: 357–359. Available: http://www.pubmedcentral.nih.gov/articlerender.fcgi?artid=3322381&tool=pmcentrez&rendertype=abstract. Accessed 2014 January 20.

35. Thorvaldsdóttir H, Robinson JT, Mesirov JP (2013) Integrative Genomics Viewer (IGV): high-performance genomics data visualization and exploration. Brief Bioinform 14: 178–192. Available: http://www.pubmedcentral.nih.gov/articlerender.fcgi?artid=3603213&tool=pmcentrez&rendertype=abstract. Accessed 2014 January 20.

36. Abeel T, Van Parys T, Saeys Y, Galagan J, Van de Peer Y (2012) GenomeView: a next-generation genome browser. Nucleic Acids Res 40: e12. Available: http://www.pubmedcentral.nih.gov/articlerender.fcgi?artid=3258165&tool=pmcentrez&rendertype=abstract. Accessed 2012 March 15.

37. Ross MG, Russ C, Costello M, Hollinger A, Lennon NJ, et al. (2013) Characterizing and measuring bias in sequence data. Genome Biol 14: R51. Available: http://www.ncbi.nlm.nih.gov/pubmed/23718773. Accessed 2014 January 22.

38. Grad YH, Lipsitch M, Feldgarden M, Arachchi HM, Cerqueira GC, et al. (2012) Genomic epidemiology of the Escherichia coli O104:H4 outbreaks in Europe, 2011. Proc Natl Acad Sci U S A 109: 3065–3070. Available: http://www.pubmedcentral.nih.gov/articlerender.fcgi?artid=3286951&tool=pmcentrez&rendertype=abstract. Accessed 2014 January 20.

39. Ribeiro FJ, Przybylski D, Yin S, Sharpe T, Gnerre S, et al. (2012) Finished bacterial genomes from shotgun sequence data. Genome Res 22: 2270–2277. Available: http://www.pubmedcentral.nih.gov/articlerender.fcgi?artid=3483556&tool=pmcentrez&rendertype=abstract. Accessed 2014 January 23.

40. Williams LJS, Tabbaa DG, Li N, Berlin AM, Shea TP, et al. (2012) Paired-end sequencing of Fosmid libraries by Illumina. Genome Res 22: 2241–2249. Available: http://www.pubmedcentral.nih.gov/articlerender.fcgi?artid=3483553&tool=pmcentrez&rendertype=abstract. Accessed 28 January 2014.

41. Gnerre S, Maccallum I, Przybylski D, Ribeiro FJ, Burton JN, et al. (2011) High-quality draft assemblies of mammalian genomes from massively parallel sequence data. Proc Natl Acad Sci U S A 108: 1513–1518. Available: http://www.pubmedcentral.nih.gov/articlerender.fcgi?artid=3029755&tool=pmcentrez&rendertype=abstract. Accessed 21 January 2014.

42. Altschul SF, Gish W, Miller W, Myers EW, Lipman DJ (1990) Basic local alignment search tool. J Mol Biol 215: 403–410. Available: http://www.ncbi.nlm.nih.gov/pubmed/2231712. Accessed 23 January 2014.

43. Delcher AL, Phillippy A, Carlton J, Salzberg SL (2002) Fast algorithms for large-scale genome alignment and comparison. Nucleic Acids Res 30: 2478–2483. Available: http://www.pubmedcentral.nih.gov/articlerender.fcgi?artid=117189&tool=pmcentrez&rendertype=abstract.

44. Larkin M a, Blackshields G, Brown NP, Chenna R, McGettigan P a, et al. (2007) Clustal W and Clustal X version 2.0. Bioinformatics 23: 2947–2948. Available: http://www.ncbi.nlm.nih.gov/pubmed/17846036. Accessed 22 January 2014.

45. Tatusova TA, Madden TL (1999) BLAST 2 Sequences, a new tool for comparing protein and nucleotide sequences. FEMS Microbiol Lett 174: 247–250. Available: http://www.ncbi.nlm.nih.gov/pubmed/10339815. Accessed 28 January 2014.

46. Abeel T, Saeys Y, Rouzé P, Van de Peer Y (2008) ProSOM: core promoter prediction based on unsupervised clustering of DNA physical profiles. Bioinformatics 24: i24–31. Available: http://www.pubmedcentral.nih.gov/articlerender.fcgi?artid=2718650&tool=pmcentrez&rendertype=abstract. Accessed 8 July 2012.

# detectIR: A Novel Program for Detecting Perfect and Imperfect Inverted Repeats Using Complex Numbers and Vector Calculation

**Congting Ye[1,2], Guoli Ji[1,3]\*, Lei Li[1,2], Chun Liang[2,4]\***

1 Department of Automation, Xiamen University, Xiamen, Fujian 361005, China, 2 Department of Biology, Miami University, Oxford, Ohio 45056, United States of America, 3 Innovation Center for Cell Biology, Xiamen University, Xiamen, Fujian 361005, China, 4 State Key Laboratory for Biology of Plant Diseases and Insect Pests, Institute of Plant Protection, Chinese Academy of Agricultural Sciences, Beijing 100193, China

## Abstract

Inverted repeats are present in abundance in both prokaryotic and eukaryotic genomes and can form DNA secondary structures – hairpins and cruciforms that are involved in many important biological processes. Bioinformatics tools for efficient and accurate detection of inverted repeats are desirable, because existing tools are often less accurate and time consuming, sometimes incapable of dealing with genome-scale input data. Here, we present a MATLAB-based program called *detectIR* for the perfect and imperfect inverted repeat detection that utilizes complex numbers and vector calculation and allows genome-scale data inputs. A novel algorithm is adopted in *detectIR* to convert the conventional sequence string comparison in inverted repeat detection into vector calculation of complex numbers, allowing non-complementary pairs (mismatches) in the pairing stem and a non-palindromic spacer (loop or gaps) in the middle of inverted repeats. Compared with existing popular tools, our program performs with significantly higher accuracy and efficiency. Using genome sequence data from HIV-1, *Arabidopsis thaliana*, *Homo sapiens* and *Zea mays* for comparison, *detectIR* can find lots of inverted repeats missed by existing tools whose outputs often contain many invalid cases. *detectIR* is open source and its source code is freely available at: https://sourceforge.net/projects/detectir.

**Editor:** Silvio C. E. Tosatto, Universita' di Padova, Italy

**Funding:** This work was supported partially by National Institutes of Health National Institute of General Medical Sciences grant (1R15GM94732-1 A1 to CL), the National Natural Science Foundation of China (61174161, 61375077 and 61201358), the Natural Science Foundation of Fujian Province of China (2012J01154), the specialized Research Fund for the Doctoral Program of Higher Education of China (20130121130004 and 20120121120038), and the Fundamental Research Funds for the Central Universities in China (2013121025, 201412G009 and CBX2014007). The funders had no role in study design, data collection and analysis, decision to publish, or preparation of the manuscript.

**Competing Interests:** The authors have declared that no competing interests exist.

\* Email: liangc@miamioh.edu (CL); glji@xmu.edu.cn (GJ)

## Introduction

An inverted repeat is a nucleotide sequence fragment that can form self-complementary pairing between its two halves. The perfect inverted repeats are also known as palindrome where one of these two halves is exactly the reverse complement of the other; in contrast, imperfect inverted repeats contain nucleotide pairs that are not reversely complementary (*i.e.*, mismatched), often with a non-palindromic spacer (loop or gaps) in the middle [1] [2] [3]. Abundant inverted repeats are present in both prokaryotic and eukaryotic genomes with nonrandom distributions, and they are involved in many biological processes including DNA replication [4], DNA transition [5] and DNA methylation [6]. In yeast, long inverted repeats were demonstrated to be mitotic recombination hotspots, and quasipalindromes (imperfect inverted repeats) underwent deletion more frequently [7]. In mouse embryonic stem cells, inverted repeats get involved in the generation of unstable chromosomal rearrangements [8]. Inverted repeats of $>=6$ complementary nucleotides, either perfect or imperfect, can form secondary structures – cruciforms in double

stranded DNA [9]. Some DNA-binding proteins have their two binding sites arranged as in an inverted repeat [3] [4] [10]. Using atomic force microscopy images, the DNA-binding protein PARP-1 was shown to bind the cruciform structure generated by a 106-*nt* inverted repeat within an *E. coli* plasmid [5]. PARP-1 was found to participate in chromatin structure coordination and gene expression regulation [11], and it did show a binding preference to cruciform structures than loops or linear DNAs [9]. In humans, a 14-*nt* imperfect inverted repeat sequence located in distal promoter region of human HFE gene can form a cruciform structure that binds PARP-1 protein to repress HFE transcription, and increased ion level can trigger PARP-1 breakdown to release such transcriptional repression [12]. Interestingly, using 2D electrophoretic analysis of DNA replication intermediates, single-stranded hairpins formed by imperfect inverted repeats with a central non-palindromic spacer, rather than double-stranded cruciforms, proved to be responsible for replication stalling that induces genome instability [13]. On the other hand, as a part of gene expression production, hairpin is a common and important secondary structural element in RNA transcripts, and its single-stranded template DNA must have a relevant inverted repeat [3], suggesting that inverted repeats in transcribed DNAs might play a potential regulatory role in resultant RNA transcripts. For instance, in yeast and mammalian pre-mRNAs, inverted repeats located in introns were shown to affect alternative splicing [14] [15] [16]. Moreover, inverted repeats have been utilized for gene silencing in fungi and plants for many years [17] [18] [19]. They also delineate transposon element boundaries. For instance, miniature inverted repeat transposable elements (MITEs) are characterized by their terminal inverted repeats [20]. Genes away from MITEs show higher expression than those that contain or are close to MITEs [21]. Clearly, the role of inverted repeats in gene expression regulation is worthy of further wet-lab experimental investigation and validation. The identification and characterization of inverted repeats at genome-wide scale will offer an important glimpse and survey that will facilitate our understanding of inverted repeats and their biological functions.

Lu et al. examined the distribution of perfect inverted repeats (palindromes) in human genome [22]. They found that palindromes show higher abundance in introns than exons while upstream regions (i.e., 2,000 bp upstream from translational start site) also contain rich palindromes that can serve as binding sites for transcription factors. Interestingly, they also scanned the human genome for imperfect inverted repeats (i.e., near-perfect palindromes of < = 4 mismatches between two halves, with a short spacer in the middle) and found a similar distribution pattern as perfect inverted repeats [22]. In yeast Saccharomyces cerevisiae genome, both palindromes and imperfect inverted repeats (i.e., the pairing stem length $>6$ $nt$ and the spacer length less than 77 $nt$) were significantly richer than randomized genome [3]. In particular, imperfect inverted repeats with short spacers, which have a greater susceptibility to cruciform extrusion than long spacers, were significantly enriched in intergenic regions near 3' gene ends than near 5' gene ends [3]. Using yeast relatives S. paradoxus, S. mikatae and S. bayanus, Humphey-Dixon and coworkers studied the conservation of both perfect and imperfect inverted repeats in yeast S. cerevisiae genome, and they found that both conserved inverted repeats in promoters and inverted repeats in the promoters of highly expressed genes are most frequently located near the transcriptional start sites, indicating their potential function in transcriptional regulation [23]. As more and more genome sequences are available for different species, it is important for us to conduct genome-wide comparative study to determine the distributions, properties, and conservation of inverted repeats among different species, either distantly or closely related, in order to deepen our understanding of inverted repeats and their biological importance.

To better understand the roles of inverted repeats in genome organization and evolution, developing efficient programs to conduct genome-wide detection of inverted repeats is particularly important. Recently, a MATLAB-based tool findIR [24] was created for detecting perfect inverted repeats. In comparison with the existing similar tools that adopt conventional string comparison algorithms, findIR deployed a novel algorithm that uses prime number scoring system and turns sequence search into the calculation, search and comparison of numbers. Consequently, findIR proved to have obviously higher accuracy in detecting perfect inverted repeats than several popular tools, and it was capable of processing genome-scale inputs. Unfortunately, it is difficult to use prime numbers to represent imperfect inverted repeats that contain non-complementary pairs in the pairing stem (i.e., two halves) and a central non-palindromic spacer. The search and validation strategy of findIR is not designed for detecting imperfect inverted repeats. Moreover, findIR is limited to detect

perfect inverted repeats of length shorter than 1,000 $nt$ and the robust vector calculation power of MATLAB has not been utilized at all by findIR to enhance the program efficiency.

Here, we developed a novel program detectIR that also turns sequence search into numerical calculation and manipulation using complex numbers, which can represent both perfect and imperfect IRs accurately and efficiently. In comparison with findIR, the novelty of detectIR lies in a novel mapping schema that utilizes complex numbers, a distinctive and effective strategy of search and validation to evaluate candidates of both perfect and imperfect inverted repeats, and the utilization of MATLAB built-in vector calculation power that enables simultaneous detection of inverted repeats of same length to improve the program efficiency significantly.

## Design and Implementation

Most existing bioinformatics tools in inverted repeat detection rely on string comparison that is often computational resource demanding and less accurate. Using a novel algorithm that employs prime number scoring system and numerical calculation and search to detect perfect inverted repeats (or palindromes), findIR was demonstrated to have much higher accuracy than BioPHP (http://www.biophp.org/minitools/find_palindromes), MATLAB built-in palindromes function and EMBOSS palindrome tool [25] in detecting perfect inverted repeats. However, both EMBOSS palindrome tool and MATLAB built-in palindromes function can detect imperfect inverted repeats that contain a central non-palindromic spacer (loop) and/or non-complementary pairs in the pairing stem. In the core algorithm of findIR, nucleotide bases are first mapped to a prime number scoring system in which scores of reversely complementary bases can cancel each other out, and a cumulative score is computed for each base along the whole target sequence. Then, findIR searches all the pairs of positions whose cumulative scores are the same to construct candidates of perfect inverted repeats. findIR finally validates the candidate based on the principle that if a subsequence is a valid perfect inverted repeat, the number of nested perfect inverted repeats within the subsequence, all of which share the same center, should be equal to the half-length of the subsequence. This search and validation strategy makes findIR impossible to detect imperfect inverted repeats. Moreover, the prime number scoring system has its innate difficulty to represent imperfect inverted repeats that often contains a central non-palindromic spacer and/or non-complementary pairs in the pairing stem. In addition, findIR is limited to detect perfect inverted repeats of length shorter than 1,000 $nt$, presenting a size constraint for large genomes where longer inverted repeats may exist. Here, we developed a novel program called detectIR that maps the nucleotide sequence to a complex number vector so that both perfect and imperfect inverted repeats can be searched (see examples in Figure 1). Compared with findIR, detectIR adopted a totally different algorithm in perfect inverted repeat detection, which had been modified and extended to search for imperfect inverted repeats. In particular, for both perfect and imperfect inverted repeat detection, we have taken advantage of MATLAB built-in vector calculation to search and validate inverted repeats candidates of the same size simultaneously. Our program proves to be much faster and more accurate in detecting both perfect and imperfect inverted repeats than the previously mentioned tools. Moreover, our program can accept a large genome input like chromosome 1 of Homo sapiens and Zea mays that can often result in an execution crash in other tools.

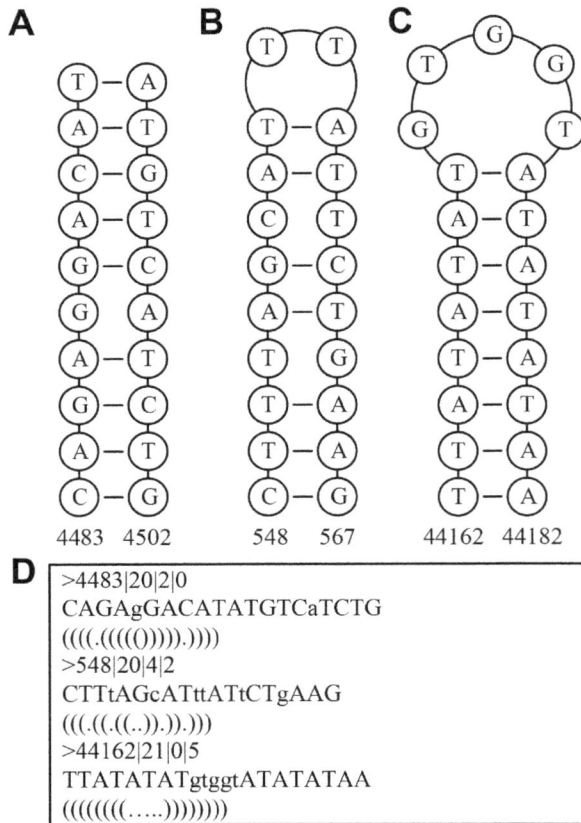

**Figure 1. Examples of imperfect inverted repeats detected in the chromosome 1** *of Arabidopsis thaliana* **by** *detectIR*. Circles highlight the nucleotide bases and numbers represent the start and end genomic coordinates of the imperfect inverted repeats. (A) An imperfect inverted repeat containing only mismatches (un-pairing pair). (B) An imperfect inverted repeat containing both mismatches and a spacer (gaps). (C) An imperfect inverted repeat containing only a spacer (gaps) in the middle. (D) The *detectIR* output for the aforementioned 3 imperfect inverted repeats in the combined format of dot-bracket notation and FASTA. The dots represent the spacer or mismatch nucleotides while brackets indicate the pairing relations. The FASTA description line (*e.g.*, >4483|20|2|0) has the following explanation: > GenomicStartPosition|InvertedRepeatLength|MismatchNumber|Gap-Number.

Within the prime number scoring system (also called cumulative scoring system) implemented in *findIR* [24], a subsequence between two bases whose cumulative scores are identical will be recognized as a perfect inverted repeat candidate. This conclusion is based on the assumption that only the scores between reversely complementary bases can completely eliminate each other in the system. If two bases have identical score, the numbers of reversely complementary bases should be equal in the corresponding subsequence [24]. Unfortunately, this assumption is not correct for some cases, because scores also can be partly eliminated between non-complementary bases. For instance, when using the prime number mapping schema: $A \rightarrow 3$, $T \rightarrow -3$, $G \rightarrow 7$, $C \rightarrow -7$, the scores of seven nucleotides A can be eliminated out by that of three nucleotides C (same to T and G). Therefore, if two bases have an identical accumulative score, the numbers of complementary bases in the corresponding subsequence between them are not always equal. In *findIR*, such a case would be still recognized as a perfect inverted repeat candidate. Although these

cases could be filtered out later by the downstream process, it clearly increases the computational workload. On the other hand, these cases can be avoided by using large prime numbers (*e.g.*, 10007, 10009), which was exactly adopted by the *findIR* [24].

Different from prime numbers, complex numbers can effectively represent both perfect and imperfect inverted repeats that contain non-complementary pairs in the stem and a central non-palindromic spacer. The complex numbers have been utilized in detecting symmetric palindromes (*e.g.*, ACGGCA, the palindromes without reverse complementary) by Gupta et al. [26], using the following mapping schema:

$$A \rightarrow 1+j, T \rightarrow -1-j, C \rightarrow 1-j, G \rightarrow -1+j$$

Firstly, they divided an input nucleotide sequence into subsequences with the length of a desired palindrome, and then utilized this mapping method to convert each subsequence into numerical series and rearranged the subsequence. Secondly, they use periodicity transformation to calculate a periodic sequence that is closest to the rearranged subsequence. Lastly, through calculating a coefficient between the rearranged subsequence and the periodic sequence to determine and verify each subsequence to be a valid symmetric palindrome [26]. Therefore, their method was not designed for both perfect and imperfect inverted repeat detection that requires nucleotide reverse complementary. In contrast, our usage of complex numbers is different from their approach by deploying a novel mapping schema:

$$A \rightarrow 1, T \rightarrow -1, C \rightarrow j, G \rightarrow -j$$

Using this mapping method, only the scores between reversely complementary bases can really cancel each other out within the numeric scoring system. This method makes sure that the numbers of reversely complementary nucleotides A and T (also C and G) are equal within the sequence candidates of perfect inverted repeat. Furthermore, we can precisely define the upper bound of the score of the imperfect inverted repeats using this complex number mapping schema - the sum of the absolute values of the real part and the imaginary part of $C$ is less than or equal to $m$, $|real(C)| + |image(C)| \leq m$ ($C$ denotes the score of the subsequence, $m$ represents the maximal number of mismatches allowed – the sum of all nucleotides within the central non-palindromic spacer and the non-complementary pairs in the pairing stem).

## Algorithm for perfect inverted repeat detection

The algorithm for perfect inverted repeat detection is based on the principle that a perfect inverted repeat of length $h$ ($h>4$) must contain a nested perfect inverted repeat of length $h$-2, which shares the same center. So if we have obtained all the perfect inverted repeats of length $h$-2, extension of one base at both ends will derive inverted repeat candidates with length $h$. Among all these inverted repeat candidates, we then filter out those whose terminal bases are not reversely complementary.

The details of our algorithm are shown in Figures 2 and 3. Here, $l$ is the minimal length, $L$ is the maximal length of perfect inverted repeats to be detected, and $N$ is the length of the input sequence $S$ ($l \leq L \leq N$, $l$ and $L$ should be even number).

**(1). Mapping sequence to vector.** The input sequence $S$ is mapped to a complex number vector $M$. For example, if input sequence $S$ is 'ATCGAACGAATTCGTTAACC', $M$ = [1,-1, j, -j,1,

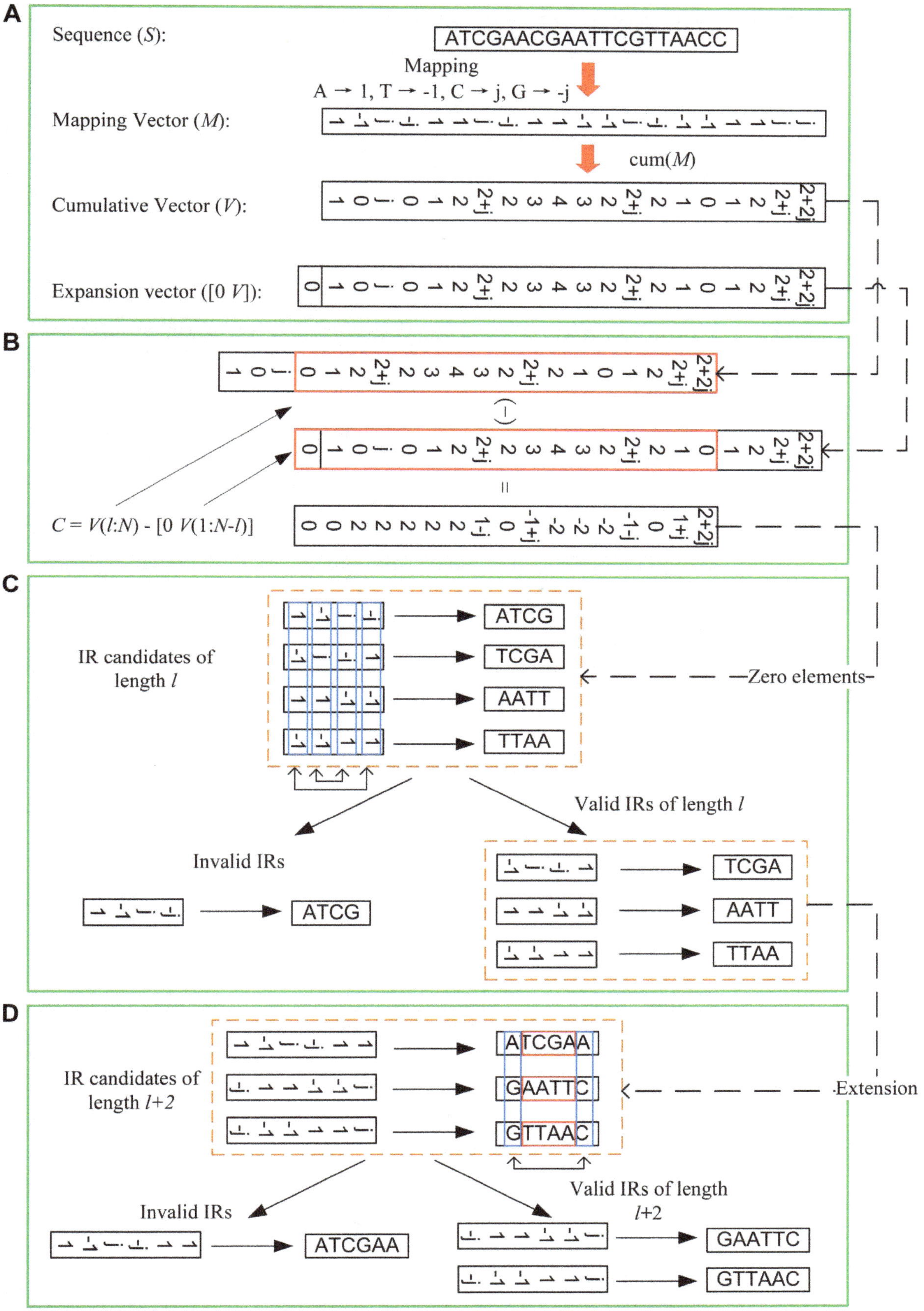
A
Sequence (S):
ATCGAACGAATTCGTTAACC
Mapping
A → 1, T → -1, C → j, G → -j
Mapping Vector (M):
cum(M)
Cumulative Vector (V):
Expansion vector ([0 V]):
B
C = V(l:N) - [0 V(1:N-l)]
C
IR candidates of length l
ATCG
TCGA
AATT
TTAA
Zero elements
Valid IRs of length l
Invalid IRs
ATCG
TCGA
AATT
TTAA
D
IR candidates of length l+2
ATCGAA
GAATTC
GTTAAC
Extension
Valid IRs of length l+2
Invalid IRs
ATCGAA
GAATTC
GTTAAC

**Figure 2. Major steps of the core algorithm for perfect inverted repeat detection.** (A) Map a nucleotide sequence to a complex number vector and calculate its cumulative score value. (B) Calculate the scores of all the subsequences of length $l$ (here $l=4$, $N$ is the length of input sequence). (C) Select out perfect inverted repeat candidates and determine valid perfect inverted repeats. (D) Extend one base at both ends of subsequences to obtain longer perfect inverted repeats.

1, j, -j, 1, 1, -1, -1, j, -j, -1, -1, 1, 1, j, j]. Then the cumulative value $V$ of $M$ is calculated (Figure 2A).

**(2). Calculation of subsequence score.** The scores of all the subsequences of length $l$ are calculated, a subsequence score is defined as the sum of the subsequence's corresponding vector. The scores of all the subsequences of length $l$ can be obtained by the following formula,

$$C = V(l:N) - [0\,V(1:N-l)]$$

The $i$-th element $C(i)$ represent the score of the subsequence $S(i:i+l-1) = s_i s_{i+1} \ldots s_{i+l-2} s_{i+l-1}$ (Figure 2B).

**(3). Identification of candidates.** A subsequence whose score is equal to 0 will be recognized as a candidate of perfect inverted repeat. In this step we will identify all the candidates of length $l$ by finding the zero elements of $C$,

$$P = \mathrm{find}(C == 0)$$

$P$ represents the indices of the zero elements in vector $C$.

**(4). Validation of candidates.** For each perfect inverted repeat candidate, perfect pairing needs to be found between the front half and rear half of bases. We will keep valid perfect inverted repeat candidates, while filter out invalid cases (Figure 2C).

**(5). Extension of inverted repeats.** Base on the aforementioned principle, extend one base at both ends of inverted repeats identified in the previous step, and select out cases whose new

terminal bases are reversely complementary (Figure 2D). For instance, if subsequence $S(i:i+h-3)$ is a perfect inverted repeat of length $h$-2 identified in previous step, the extended subsequence $S(i-1:i+h-2)$ will be a potential inverted repeat of length $h$. If its terminal bases are either G/C or A/T,

$$M(i-1) + M(i+h-2) = 0$$

Then, $S(i-1:i+h-2)$ is a perfect inverted repeat of length $h$.

**(6). Repetition.** Repeat the step 5 until $h=L$, then all the perfect inverted repeats of length ranged between $l$ and $L$ have been identified.

## Algorithm for imperfect inverted repeat detection

Different from the aforementioned algorithm for perfect inverted repeat detection, to find imperfect inverted repeats of length $h$, allowing maximal mismatch number $m$, which is defined as the sum of all nucleotides within the central non-palindromic spacer and the non-complementary pairs in the pairing stem, we will firstly find all subsequences of length $h$-2 with mismatch number $\leq m$, with the assumption that the nested subsequence of an imperfect inverted repeat with mismatch number $\leq m$ must be a sequence with mismatch number $\leq m$ which share the same center. The major difference between the nested subsequence and the imperfect inverted repeat is that the terminal bases of imperfect inverted repeat must be reversely complementary ('AAAAATTTCT'), while the former does not require meeting this condition ('AAAATTTC').

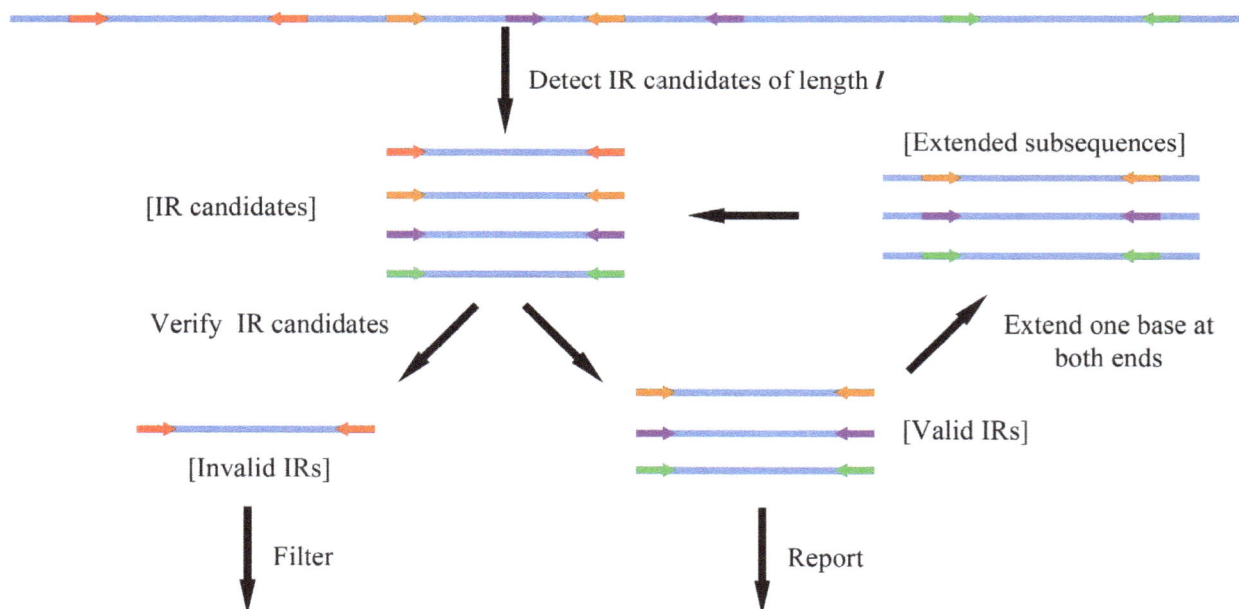

**Figure 3. The flowchart of the core algorithm for perfect inverted repeat detection.** Arrows in the same color represent the borders of inverted repeats.

Figure 4 shows the core algorithm for imperfect inverted repeat detection. $l$ is the minimal length and $L$ is the maximal length of imperfect inverted repeats to be detected, and $N$ is the length of the input sequence ($l \leq L \leq N$), $m$ is the maximal number of mismatches allowing in each imperfect inverted repeat. The first two steps 1 and 2 are same to the detection of perfect inverted repeats.

**(1). Mapping sequence to vector**

**(2). Calculation of subsequence score**

**(3). Identification of candidates.** If the sum of the absolute values of the real part and the imaginary part of a subsequence's score is less than or equal to $m$, the subsequence is recognized as an imperfect inverted repeat candidate of length $l$.

$$|real(C(i))| + |image(C(i))| \leq m$$

**(4). Validation of candidates.** In this step, the mismatch numbers of all the imperfect inverted repeat candidates of *length l* are calculated. Subsequences that satisfy the following three conditions: (1) the mismatch number is less than or equal to $m$; (2) the first and last base of the subsequence are reversely complementary; (3) mismatch number is not zero, will be reported as the detected imperfect inverted repeats. While, all the subsequences meeting (1) with/without (2) and (3) will be picked up for the process in the next step (see Figure 4A).

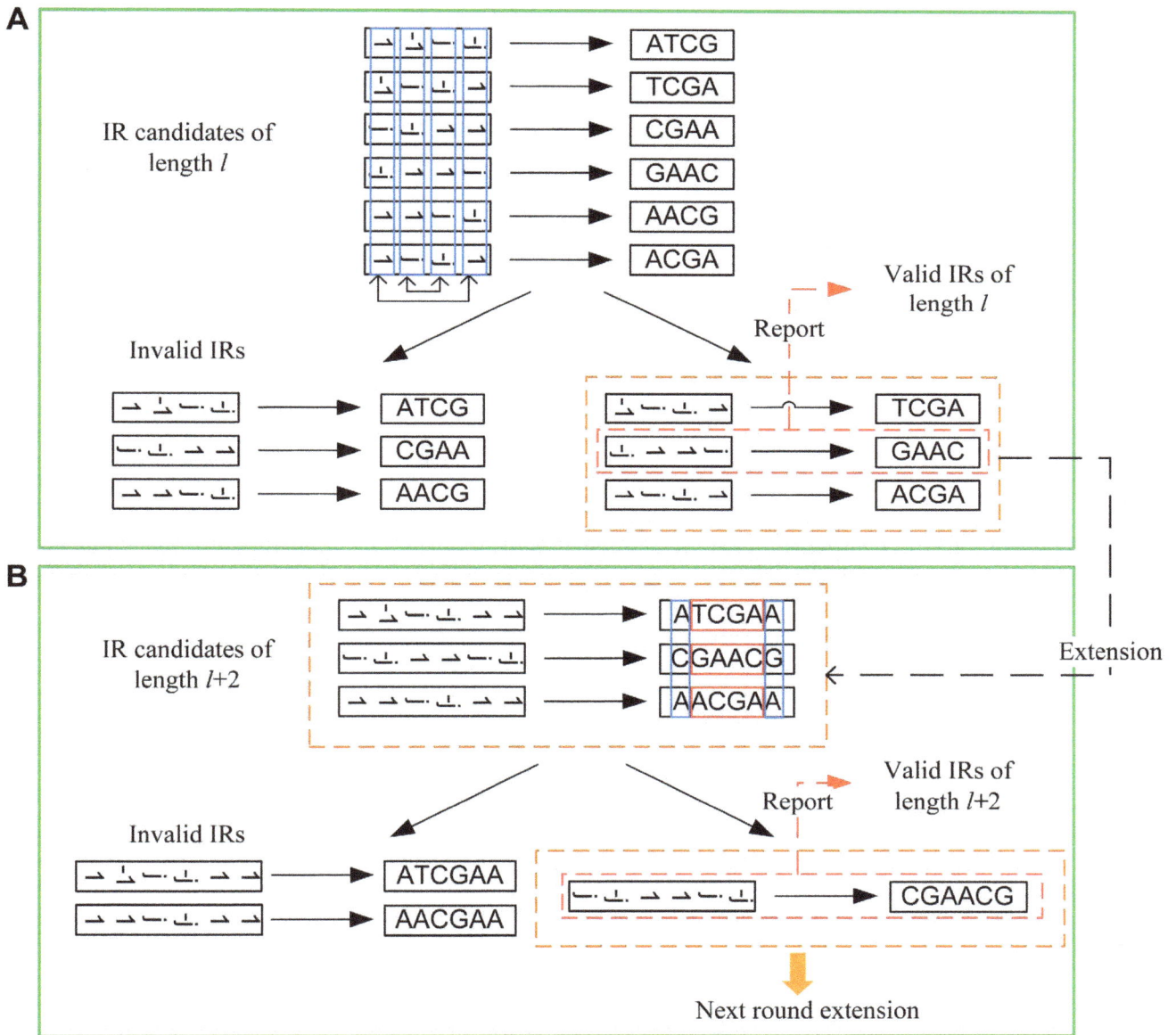

**Figure 4. Major steps of the core algorithm for imperfect inverted repeat detection.** (A) Select out imperfect inverted repeat candidates (here, we just show part of candidates), report valid imperfect inverted repeats and pick up valid subsequences for the downstream process described in the next step (here $l = 4$, $m = 2$). (B) Extend one base at both ends of subsequences to obtain longer imperfect inverted repeats. The subsequences locating at both ends of a sequence (*i.e.*, 'ATCGAA') should not be selected for the next round extension.

**(5). Extension of subsequences.** We will extend one nucleotide at both ends of subsequences generated in the previous step. If the two extended bases are reversely complementary, the corresponding mismatch number of the subsequence keep unchanged, otherwise the mismatch number increases by 2. Then subsequences meeting all three conditions described in step 4 are valid imperfect inverted repeats. Subsequences meeting (1) with/without (2) and (3) will be picked up for the next round extension (see Figure 4B).

For example, when subsequence $S(i{:}i{+}h{-}3)$ of length $h$-2 has a mismatch number $\alpha$ less than or equal to $m$, the subsequence $S(i{-}1{:}i{+}h{-}2)$ is a potential imperfect inverted repeats of length $h$. If,

$$M(i-1) + M(i+h-2) = 0, \alpha > 0$$

Subsequence $S(i{-}1{:}i{+}h{-}2)$ is an imperfect inverted repeat that meets all given conditions.

**(6). Repetition.** Repeat the step 5 until $h = L$ (or $h = L$-1), then all imperfect inverted repeats of odd-length (or even-length) ranged between $l$ and $L$ with a mismatch number $\leq m$ have been identified.

**(7). Detection of even/odd length inverted repeats.** Repeat steps 2–6, starting with calculating the score of subsequences of length $l{+}1$. Finally, all imperfect inverted repeats of even-length (or odd-length) ranged between $l$ and $L$ will been identified.

## Results

The aforementioned algorithms have been implemented into *detectIR*, which was composed of two MATLAB functions *detectPerfectIR* and *detectImperfectIR* respectively. To evaluate the performance of these two functions, they were compared with four inverted repeat detection tools BioPHP (A PHP program obtained from http://www.biophp.org/minitools/find_palindromes), EMBOSS *palindrome* tool (Stable version 6.3.1), MATLAB built-in *palindromes* function (Matlab R2013b) and MATLAB-based *findIR*. HIV-1 genome and chromosome 1 from *Arabidopsis thaliana*, *Homo sapiens* and *Zea mays* were used as the test data. The detailed step-by-step guide for testing and comparison is available on our sourceforge project website (https://sourceforge.net/projects/detectir). Our source codes and relevant documents are also available as in File S1. The result comparison of perfect inverted repeat detection and imperfect inverted repeat detection is presented below separately. For smaller genomes like *Arabidopsis thaliana* and HIV-1, a standard workstation with 8G RAM should be enough for testing, while large genomes like *Homo sapiens* and *Zea mays* will require at least 32G RAM. For the test results described below, all the tests were performed using Ubuntu 12.04 (precise) 64-bit platform with Intel Xeon (2.00 GHz) processer and 125.9 GB RAM.

### Perfect inverted repeat detection

Here, we used each program to search perfect inverted repeats with length between 10 and 1000 *nt*. The detection results are summarized in Table 1. The length distributions of the perfect inverted repeats detected by *detectIR* are shown in Figure S1. It is clear that the numbers of perfect inverted repeats decrease with the increase of inverted repeat length, shorter inverted repeats are much more abundant in genomes (see Figure S1).

To determine the accuracy of *detectPerfectIR*, the outputs from different programs using the same input data were compared against *detectPerfectIR* respectively. The differences discovered between them can be classified into two categories (1) entries only

**Table 1.** The comparison of *detectPerfectIR* with BioPHP, EMBOSS, MATLAB and *findIR* for perfect inverted repeat detection.

|  | HIV-1 | | Arabidopsis thaliana | | Homo sapiens | | Zea mays | |
|---|---|---|---|---|---|---|---|---|
|  | IR count | runtime | IR count | runtime | IR count | runtime | IR count | runtime |
| BioPHP | 7 | 5.56 m | * | * | * | * | * | * |
| EMBOSS(*palindrome*) | 5 | <1 s | 61,705 | 34.53 m | 347,688 | 38.74 h | 448,928 | 4.28 h |
| MATLAB(*palindromes*) | 13 | 0.06 s | 266,187 | 3.27 m | * | * | * | * |
| *findIR* | 7 | 0.23 s | 142,249 | 8.80 m | 746,925 | 1.16 h | 1,147,428 | 1.51 h |
| *detectPerfectIR* | 7 | 0.05 s | 142,249 | 11.85 s | 746,925 | 1.72 m | 1,147,428 | 2.00 m |

*Program crashed in execution or output nothing after 10-day execution.

present in the output of *detectPerfectIR*, (2) entries only present in the output of BioPHP, EMBOSS (*palindrome*), MATLAB (*palindromes*) or *findIR*. As shown in Table 1, consequently, the outputs of *detectPerfectIR* and *findIR* are identical, while *detectPerfectIR* performs much more efficiently.

**Entries only present in the output of detectPerfectIR and findIR.** The entries only detected by *detectPerfectIR* and *findIR* were found to be valid perfect inverted repeats by human validation. In other words, these cases are missed by the other compared tools. For HIV-1 genome, both EMBOSS and MATLAB missed 2 perfect inverted repeats (see File S2 and S3). For chromosome 1 of *Arabidopsis thaliana*, EMBOSS missed 80,548 perfect inverted repeats (see File S4) and MATLAB missed 86,759 perfect inverted repeats (see File S5). For chromosome 1 of *Homo sapiens*, EMBOSS missed 399,237 perfect inverted repeats (see File S6). For chromosome 1 of *Zea mays*, EMBOSS missed 703,018 perfect inverted repeats (see File S7).

**Entries only present in the output of the other compared tools.** The entries only found by the other tools prove to be invalid perfect inverted repeats by human validation, which mean the outputs of these tools contain false positives. For the HIV-1 genome, the output of MATLAB contains 8 cases with a gap in the center, like 'CAAAAATTTTG' (see File S3). For chromosome 1 of *Arabidopsis thaliana*, the output of EMBOSS contains 4 invalid cases like 'A-N(60)-T' etc (see File S4). And the output of MATLAB contains 210,697 invalid cases, 46915 of them are subsequences like 'ATTTTTTAAAAAT' with a gap in the center and 163,782 of them are subsequences like 'NNNNNNNNNN' or 'A-N(27)-T' (see File S5). For chromosome 1 of *Zea mays*, the output of EMBOSS contains 4,518 invalid cases (see File S7).

## Imperfect inverted repeat detection

In imperfect inverted repeat detection, some tools define continuous mismatches in the center of inverted repeat as gaps; mismatches in other locations are still considered as mismatches, while our algorithm does not differentiate these two types of mismatch. For example, EMBOSS *palindrome* tool will recognize sequence 'AACAACTTTCTT' as an inverted repeat with 1 mismatch and 2 gaps, our tool will recognize it as an inverted repeat with 4 mismatches. MATLAB *palindromes* function can only deal with imperfect inverted repeats containing gaps. So, here we let our function *detectImperfectIR* and MATLAB *palindromes* function detect imperfect inverted repeats of length between 20 and 1000 *nt*, mismatch number (or gap number) ≤6, and let EMBOSS palindrome tool search imperfect inverted repeats of length between 20 and 1000 *nt*, gap number ≤2, and mismatch number ≤2 (Using the definition of EMBOSS). BioPHP and *findIR* are not used here, because they are designed to detect only the perfect inverted repeats. Summaries of the detection results are shown in Table 2. The length distributions of the imperfect inverted repeats detected by *detectIR* are shown in Figure S2. Obviously, the length distribution of imperfect inverted repeats is similar to that of perfect inverted repeats. The number of imperfect inverted repeats of even length is more than those of odd length, and the underlying reason may be that the odd-length inverted repeats must contain a spacer in the middle, which means that the mismatches in the stem should be less. To test this assumption, we run detectIR with an odd maximum mismatch number ($m = 7$). As shown in Figure S3, the numbers of odd-length imperfect inverted repeats are approximate to the numbers of even-length imperfect inverted repeats.

In order to make an unbiased comparison, we will filter out entries with a mismatch number larger than 2 or a gap number larger than 2 (Using the definition of EMBOSS) in the output of

**Table 2.** The comparison of *detectImperfectIR* with EMBOSS and MATLAB for imperfect inverted repeat detection.

| | HIV-1 | | Arabidopsis thaliana | | Homo sapiens | | Zea mays | |
|---|---|---|---|---|---|---|---|---|
| | IR count | runtime | IR count | runtime | IR count | runtime | IR count | runtime |
| EMBOSS(*palindrome*) | 8 | <1 s | 65,596 | 43.48 m | 429,662 | 39.96 h | 619,889 | 5.91 h |
| MATLAB(*palindromes*) | 1 | 0.07 s | 173,659 | 11.00 m | * | * | * | * |
| *detectImperfectIR* | 37 | 0.09 s | 311,369 | 1.48 m | 2,166,638 | 11.83 m | 2,878,686 | 16.63 m |

*Program crashed in execution or output nothing after 10-day execution.

*detectImperfectIR* before comparing it with the output of EMBOSS.

**Entries only present in the output of detectImperfectIR.** All these entries are validated by human to be imperfect inverted repeats that meet the defined criteria. For HIV-1, EMBOSS missed 9 cases (see File S8), and MATLAB missed 36 cases (see File S9). For instance, both EMBOSS and MATLAB missed case 'ATCAGATGCTAAAGCATATGAT'. For chromosome 1 of *Arabidopsis thaliana*, EMBOSS missed 78,293 cases (see File S10) and MATLAB missed 304,394 cases (see File S11). Both EMBOSS and MATLAB missed cases like 'CTTTAGCATTTATTCTGAAG' and 'ATAATTTAAAA-TAAAATTAT'. For chromosome 1 of *Homo sapiens*, EMBOSS missed 618,357 cases (see File S12). For chromosome 1 of *Zea mays*, EMBOSS missed 825,455 cases (see File S13).

**Entries only present in the output of the other compared tools.** The entries only present in the output of the other compared tools prove to be false positives again. For chromosome 1 of *Arabidopsis thaliana*, the output of EMBOSS contains 1,765 perfect inverted repeats and 23 entries like 'A-N(27)-TT' (see File S10). The output of MATLAB contains 3,091 perfect inverted repeats and 163,593 entries like 'N(20)' or 'A-N(27)-T' (see File S11). For chromosome 1 of *Homo sapiens*, the output of EMBOSS contains 6,771 perfect inverted repeats (see File S12). For chromosome 1 of *Zea mays*, the output of EMBOSS contains 17,926 perfect inverted repeats and 27,761 entries like 'A-N(100)-T' (see File S13).

### Random nucleotide sequence test

To evaluate the efficiency of *detectIR*, we generated several random nucleotide sequences of varied lengths, use function *detectPerfectIR* to detect perfect inverted repeats of length between 4 and 1000, and use function *detectImperfectIR* to detect imperfect inverted repeats of length between 10 and 1000 with mismatches less than or equal to 6. The average runtimes are shown below in Figure 5.

As shown in Figure 5, average runtime of the *detectIR* increases with the increase of the sequence length. The execution time of

*detectImperfectIR* is ~6 fold of that of *detectPerfectIR* with same input data, because the perfect inverted repeats has more strict requirements than imperfect inverted repeats which reduce the search space. So it is clear that *detectIR* shows good scalability dealing with large genome sequence inputs.

### Availability and Future Directions

*detectIR* is platform independent and can be used in Windows or Linux as long as MATLAB can be run. The source codes, test and comparison scripts, and documents are freely available at: https://sourceforge.net/projects/detectir.

Clearly, inverted repeats are not randomly distributed in both prokaryotic and eukaryotic genomes. Without a doubt, more wet-lab experiments for important genes are needed to clearly understand their biological functions. However, *in silico* genome-wide scan of inverted repeats can effectively help us to determine their overall distributions and characteristics and discover the groups of inverted repeats that are more likely to have important biological functions. As shown in Information S1, we used *detectIR* to conduct comparative study of perfect and imperfect inverted repeats in *Arabidopsis* genome. We found that both perfect and imperfect inverted repeats are not randomly distributed along the genome, and imperfect inverted repeats are much more abundant than perfect inverted repeats. In particular, imperfect inverted repeats are significantly enriched in near intergenic regions than far intergenic regions, while both perfect and imperfect inverted repeats are significantly more abundant in introns than exons. Our results are in line with the findings in human and yeast genomes [3] [22] [23]. Obviously, the inverted repeats in introns and promoter regions are worthy of closer examination in the future.

In conclusion, we developed an accurate and efficient program *detectIR* for detecting both perfect and imperfect inverted repeats in a given nucleotide sequence. *detectIR* is capable of processing large genome sequences, given enough memory in computation. Compared to BioPHP, EMBOSS *palindrome* tool, MATLAB built-in *palindromes* function and MATLAB-based *findIR*, the test

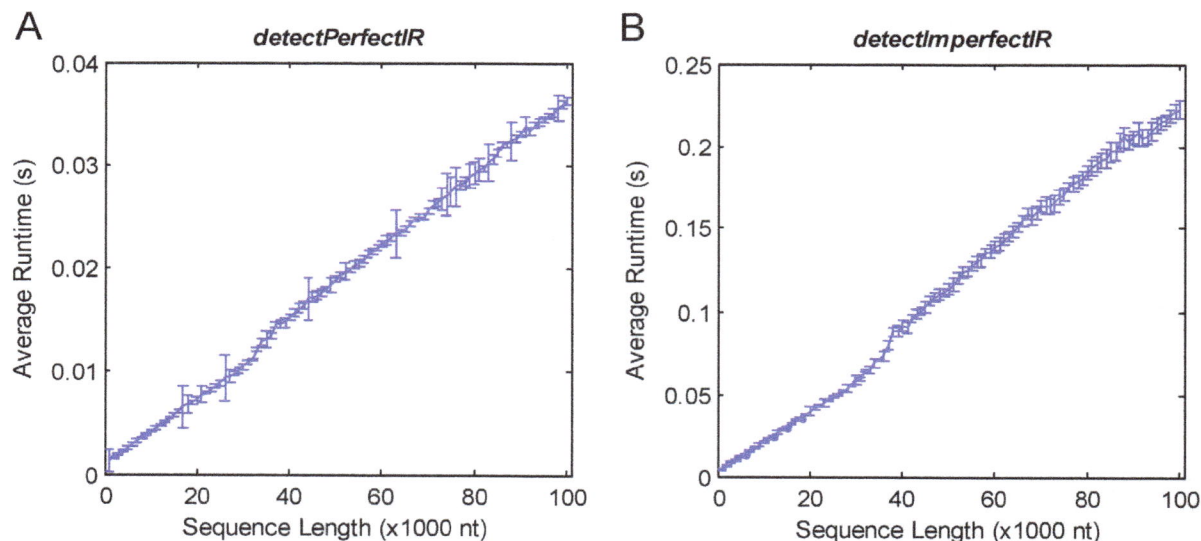

**Figure 5. Average runtimes of *detectPerfectIR* and *detectImperfectIR* using random nucleotide sequence inputs of varied lengths.** (A) The average runtimes of *detectPerfectIR* using random nucleotide sequence inputs of varied lengths. (B) The average runtimes of *detectImperfectIR* using random nucleotide sequence inputs of varied lengths.

results show that our program can more efficiently detect inverted repeats without sacrificing accuracy. Future directions will focus on improving the program to detect imperfect inverted repeats with indels within the paring stem, and reducing the memory consumption of program.

## Supporting Information

**Figure S1** Length distributions of perfect inverted repeats in different species detected by *detectIR*.

**Figure S2** Length distributions of imperfect inverted repeats in different species detected by *detectIR* with the maximum mismatch number of 6 ($m = 6$).

**Figure S3** Length distributions of imperfect inverted repeats in different species detected by *detectIR* with the maximum mismatch number of 7 ($m = 7$).

**File S1** Source codes, test and comparison scripts, and user documents of *detectIR*.

**File S2** The output files of the perfect inverted repeats detected differentially by *detectIR* and EMBOSS *palindrome* tool (Using the parameters described in the manuscript and HIV-1 genome as input data).

**File S3** The output files of the perfect inverted repeats detected differentially by *detectIR* and MATLAB *palindromes* function (Using the parameters described in the manuscript and HIV-1 genome as input data).

**File S4** The output files of the perfect inverted repeats detected differentially by *detectIR* and EMBOSS *palindrome* tool (Using the parameters described in the manuscript and chromosome 1 of *Arabidopsis thaliana* as input data).

**File S5** The output files of the perfect inverted repeats detected differentially by *detectIR* and MATLAB *palindromes* function (Using the parameters described in the manuscript and chromosome 1 of *Arabidopsis thaliana* as input data).

**File S6** The output files of the perfect inverted repeats detected differentially by *detectIR* and EMBOSS *palindrome* tool (Using the parameters described in the manuscript and chromosome 1 of *Homo sapiens* as input data).

**File S7** The output files of the perfect inverted repeats detected differentially by *detectIR* and EMBOSS *palindrome* tool (Using the parameters described in the manuscript and chromosome 1 of *Zea mays* as input data).

**File S8** The output files of the imperfect inverted repeats detected differentially by *detectIR* and EMBOSS *palindrome* tool (Using the parameters described in the manuscript and HIV-1 genome as input data).

**File S9** The output files of the imperfect inverted repeats detected differentially by *detectIR* and MATLAB *palindromes* function (Using the parameters described in the manuscript and HIV-1 genome as input data).

**File S10** The output files of the imperfect inverted repeats detected differentially by *detectIR* and EMBOSS *palindrome* tool (Using the parameters described in the manuscript and chromosome 1 of *Arabidopsis thaliana* as input data).

**File S11** The output files of the imperfect inverted repeats detected differentially by *detectIR* and MATLAB *palindromes* function (Using the parameters described in the manuscript and chromosome 1 of *Arabidopsis thaliana* as input data).

**File S12** The output files of the imperfect inverted repeats detected differentially by *detectIR* and EMBOSS *palindrome* tool (Using the parameters described in the manuscript and chromosome 1 of *Homo sapiens* as input data).

**File S13** The output files of the imperfect inverted repeats detected differentially by *detectIR* and EMBOSS *palindrome* tool (Using the parameters described in the manuscript and chromosome 1 of *Zea mays* as input data).

**Information S1** The genome-wide distribution of perfect and imperfect inverted repeats in *Arabidopsis thaliana*.

## Acknowledgments

We thank Sutharzan Sreeskandarajan for helpful comments on the manuscript. We also want to thank the anonymous reviewers for their valuable comments and suggestions to improve the manuscript.

## Author Contributions

Conceived and designed the experiments: CL GJ CY. Performed the experiments: CY. Analyzed the data: CY LL. Contributed reagents/materials/analysis tools: CY. Contributed to the writing of the manuscript: CY LL GJ CL.

## References

1. Lilley DM (1980) The inverted repeat as a recognizable structural feature in supercoiled DNA molecules. Proc Natl Acad Sci 77: 6468–6472.
2. Smith GR (2008) Meeting DNA palindromes head-to-head. Genes Dev 22: 2612–2620. doi:10.1101/gad.1724708.
3. Strawbridge EM, Benson G, Gelfand Y, Benham CJ (2010) The distribution of inverted repeat sequences in the Saccharomyces cerevisiae genome. Curr Genet 56: 321–340. doi:10.1007/s00294-010-0302-6.
4. Pearson CE, Zorbas H, Price GB, Zannis-Hadjopoulos M (1996) Inverted Repeats, Stem-Loops, and Cruciforms: Significance for Initiation of DNA Replication. J Cell Biochem 63: 1–20.

5.  Chasovskikh S, Dimtchev A, Smulson M, Dritschilo A (2005) DNA transitions induced by binding of PARP-1 to cruciform structures in supercoiled plasmids. Cytometry A 68A: 21–27. doi:10.1002/cyto.a.20187.

6.  Allers T, Leach DR (1995) DNA Palindromes Adopt a Methylation-resistant Conformation that is Consistent with DNA Cruciform or Hairpin Formation in Vivo. J Mol Biol 252: 70–85.

7.  Gordenin DA, Lobachev KS, Degtyareva NP, Malkova AL, Perkins E, et al. (1993) Inverted DNA repeats: a source of eukaryotic genomic instability. Mol Cell Biol 13: 5315–5322.

8.  Hu L, Kim TM, Son MY, Kim S-A, Holland CL, et al. (2013) Two replication fork maintenance pathways fuse inverted repeats to rearrange chromosomes. Nature 501: 569–572. doi:10.1038/nature12500.

9.  Brázda V, Laister RC, Jagelská EB, Arrowsmith C (2011) Cruciform structures are a common DNA feature important for regulating biological processes. BMC Mol Biol 12: 33.

10. Bhattacharya M, Das AK (2011) Inverted repeats in the promoter as an autoregulatory sequence for TcrX in Mycobacterium tuberculosis. Biochem Biophys Res Commun 415: 17–23. doi:10.1016/j.bbrc.2011.09.143.

11. Tulin A, Chinenov Y, Spradling A (2003) Regulation of Chromatin Structure and Gene Activity by Poly(ADP-Ribose) Polymerases. Curr Top Dev Biol Volume 56: 55–83.

12. Pelham C, Jimenez T, Rodova M, Rudolph A, Chipps E, et al. (2013) Regulation of HFE expression by poly(ADP-ribose) polymerase-1 (PARP1) through an inverted repeat DNA sequence in the distal promoter. Biochim Biophys Acta BBA - Gene Regul Mech 1829: 1257–1265. doi:10.1016/j.bbagrm.2013.10.002.

13. Voineagu I, Narayanan V, Lobachev KS, Mirkin SM (2008) Replication stalling at unstable inverted repeats: interplay between DNA hairpins and fork stabilizing proteins. Proc Natl Acad Sci 105: 9936–9941.

14. Martinez-Contreras R, Fisette J-F, Nasim FH, Madden R, Cordeau M, et al. (2006) Intronic Binding Sites for hnRNP A/B and hnRNP F/H Proteins Stimulate Pre-mRNA Splicing. PLoS Biol 4: e21. doi:10.1371/journal.pbio.0040021.

15. Nasim F-UH, Hutchison S, Cordeau M, Chabot B (2002) High-affinity hnRNP A1 binding sites and duplex-forming inverted repeats have similar effects on 5′splice site selection in support of a common looping out and repression mechanism. Rna 8: 1078–1089.

16. Howe KJ, Ares M (1997) Intron self-complementarity enforces exon inclusion in a yeast pre-mRNA. Proc Natl Acad Sci 94: 12467–12472.

17. Nakade K, Watanabe H, Sakamoto Y, Sato T (2011) Gene silencing of the Lentinula edodes lcc1 gene by expression of a homologous inverted repeat sequence. Microbiol Res 166: 484–493. doi:10.1016/j.micres.2010.09.004.

18. Hily J-M, Ravelonandro M, Damsteegt V, Bassett C, Petri C, et al. (2007) Plum pox virus coat protein gene Intron-hairpin-RNA (ihpRNA) constructs provide resistance to plum pox virus in Nicotiana benthamiana and Prunus domestica. J Am Soc Hortic Sci 132: 850–858.

19. Wesley SV, Helliwell CA, Smith NA, Wang M, Rouse DT, et al. (2001) Construct design for efficient, effective and high-throughput gene silencing in plants. Plant J 27: 581–590.

20. Yang G (2013) MITE Digger, an efficient and accurate algorithm for genome wide discovery of miniature inverted repeat transposable elements. BMC Bioinformatics 14: 186.

21. Lu C, Chen J, Zhang Y, Hu Q, Su W, et al. (2011) Miniature Inverted-Repeat Transposable Elements (MITEs) Have Been Accumulated through Amplification Bursts and Play Important Roles in Gene Expression and Species Diversity in Oryza sativa. Mol Biol Evol 29: 1005–1017. doi:10.1093/molbev/msr282.

22. Lu L, Jia H, Dröge P, Li J (2007) The human genome-wide distribution of DNA palindromes. Funct Integr Genomics 7: 221–227. doi:10.1007/s10142-007-0047-6.

23. Humphrey-Dixon EL, Sharp R, Schuckers M, Lock R, Gulick P (2011) Comparative genome analysis suggests characteristics of yeast inverted repeats that are important for transcriptional activity. Genome 54: 934–942. doi:10.1139/g11-058.

24. Sreeskandarajan S, Flowers MM, Karro JE, Liang C (2013) A MATLAB-based tool for accurate detection of perfect overlapping and nested inverted repeats in DNA sequences. Bioinformatics. doi:10.1093/bioinformatics/btt651.

25. Rice P, Longden I, Bleasby A (2000) EMBOSS: the European molecular biology open software suite. Trends Genet 16: 276–277.

26. Gupta R, Mittal A, Gupta S (2006) An efficient algorithm to detect palindromes in DNA sequences using periodicity transform. Signal Process 86: 2067–2073. doi:10.1016/j.sigpro.2005.10.008.

# Fine Mapping and Candidate Gene Search of Quantitative Trait Loci for Growth and Obesity Using Mouse Intersubspecific Subcongenic Intercrosses and Exome Sequencing

**Akira Ishikawa\*, Sin-ichiro Okuno**

Laboratory of Animal Genetics, Graduate School of Bioagricultural Sciences, Nagoya University, Nagoya, Aichi, Japan

## Abstract

Although growth and body composition traits are quantitative traits of medical and agricultural importance, the genetic and molecular basis of those traits remains elusive. Our previous genome-wide quantitative trait locus (QTL) analyses in an intersubspecific backcross population between C57BL/6JJcl (B6) and wild *Mus musculus castaneus* mice revealed a major growth QTL (named *Pbwg1*) on a proximal region of mouse chromosome 2. Using the B6.Cg-*Pbwg1* intersubspecific congenic strain created, we revealed 12 closely linked QTLs for body weight and body composition traits on an approximately 44.1-Mb wild-derived congenic region. In this study, we narrowed down genomic regions harboring three (*Pbwg1.12*, *Pbwg1.3* and *Pbwg1.5*) of the 12 linked QTLs and searched for possible candidate genes for the QTLs. By phenotypic analyses of F$_2$ intercross populations between B6 and each of four B6.Cg-*Pbwg1* subcongenic strains with overlapping and non-overlapping introgressed regions, we physically defined *Pbwg1.12* affecting body weight to a 3.8-Mb interval (61.5–65.3 Mb) on chromosome 2. We fine-mapped *Pbwg1.3* for body length to an 8.0-Mb interval (57.3–65.3) and *Pbwg1.5* for abdominal white fat weight to a 2.1-Mb interval (59.4–61.5). The wild-derived allele at *Pbwg1.12* and *Pbwg1.3* uniquely increased body weight and length despite the fact that the wild mouse has a smaller body size than that of B6, whereas it decreased fat weight at *Pbwg1.5*. Exome sequencing and candidate gene prioritization suggested that *Gcg* and *Grb14* are putative candidate genes for *Pbwg1.12* and that *Ly75* and *Itgb6* are putative candidate genes for *Pbwg1.5*. These genes had nonsynonymous SNPs, but the SNPs were predicted to be not harmful to protein functions. These results provide information helpful to identify wild-derived quantitative trait genes causing enhanced growth and resistance to obesity.

**Editor:** William Barendse, CSIRO, Australia

**Funding:** This work was supported by Grants-in-Aid for Scientific Research (B) from the Japan Society for the Promotion of Science to A Ishikawa. The funder had no role in study design, data collection and analysis, decision to publish, or preparation of the manuscript.

**Competing Interests:** The authors have declared that no competing interests exist.

\* Email: ishikawa@agr.nagoya-u.ac.jp

## Introduction

Body weight and body composition traits, including fat and organ weight, are quantitative in nature and are controlled by multiple genetic loci, referred to as QTLs (quantitative trait loci), environmental factors and their interactions. They are important economic traits in livestock [1]. For example, modern broiler chickens have been intensively selected for rapid growth rate, but they display excessive deposition of body fat. Since fat is a by-product of little economic value and often causes a decrease in feed efficiency, it is now an important selection criterion in chicken breeding programs [2]. In humans, obesity is characterized by excessive abdominal fat deposition and is now a main health concern worldwide because it is a predisposing factor of complex metabolic diseases such as type-2 diabetes and cardiovascular diseases [3]. The laboratory mouse has been long and widely used as the premier model animal for elucidating the genetic and molecular basis of these traits and other quantitative traits in livestock and humans because of its small body size, cost-effective

rearing, easy development of genetically engineered mice (e.g., knockouts and transgenics) and large amount of genomic information that is freely available [1,4]. Thousands of QTLs affecting various quantitative traits have been mapped to many chromosomal regions of the mouse and have been deposited in the Mouse Genome Database (MGD, release 5.19, August 2014) [5].

However, the genetic and molecular basis of quantitative traits remains elusive because it is not an easy task to pinpoint causative genes underlying QTLs, particularly for QTLs with small phenotypic effects on the traits. Most of the QTLs have small effects and only a few loci have moderate to large effects [4]. In mice, initial genome-wide QTL analysis is usually performed with a backcross or F$_2$ intercross population between two inbred strains and it provides a large confidence interval (10–50 cM) for a mapped QTL [6], where hundreds or thousands of genes are possibly located. Next, to reduce the confidence interval of the QTL to a level amenable to positional cloning, fine mapping is performed using a congenic mouse strain and subsequently

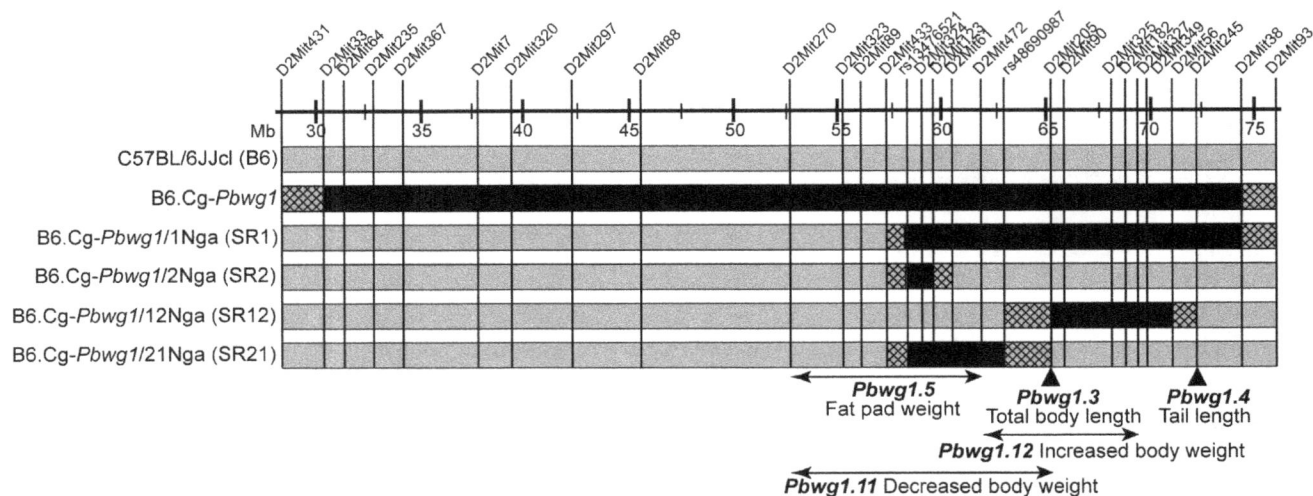

**Figure 1. Relative introgressed genomic intervals of four subcongenic strains (B6.Cg-*Pbwg1*/#Nga, abbreviation SR#) developed from the original B6.Cg-*Pbwg1* congenic strain carrying the *Pbwg1* growth QTL on mouse chromosome 2.** The black bar indicates the minimum introgressed interval derived from the wild *Mus musculus castaneus* mouse, and the gray bar indicates the interval from the background C57BL/6JJcl (B6) strain. The hatched bar shows a gray zone where recombination occurred. The physical map positions (Mb) of 27 microsatellite markers (*D2Mit#*) and two PCR-RFLP markers (*rs#*) developed in this study (Figure S1) are shown on the horizontal line. The triangle indicates the position of the peak LOD score for three body composition QTLs (*Pbwg1.3* to *Pbwg1.5*) previously identified [19,21]. The horizontal double-headed arrows indicate the maximum intervals of two body weight QTLs (*Pbwg1.11* and *Pbwg1.12*) previously defined [9].

developed subcongenic strains [7]. Then phenotypic values are compared between homozygous congenic/subcongenic strains and the background strain and/or among homozygous congenic/subcongenic strains with overlapping and non-overlapping introgressed regions. Often, the phenotypic effect of the QTL fails to be confirmed, illustrating the difficulty of identifying a causative gene for the QTL. If the QTL successfully is fine-mapped to a small region, the road from a QTL to a causative gene is still long [8]. In the present study, to overcome the problem of frequent failure in traditional congenic/subcongenic analyses, we used several $F_2$ populations obtained from intercrosses between each of the subcongenic strains and the background strain. In each of the $F_2$ populations, three possible diplotypes for the introgressed region are segregating: two are homozygous for either haplotype derived from the donor mouse or from the background mouse and the other is heterozygous for both haplotypes. Hence, using the $F_2$ mice can randomize environmental effects such as litter size and effects of contaminating donor and recipient alleles on unwanted small regions, both of which are produced by double recombination during recurrent backcrossing for development of subcongenic strains, as previously documented [9,10]. Moreover, the $F_2$ mice produced have genetically identical $F_1$ dams and $F_1$ sires. That is, the $F_1$ mice are heterozygotes for all loci on the introgressed region. Hence, using the $F_2$ mice can minimize genomic imprinting effects of alleles inherited from either $F_1$ dams or $F_1$ sires [11] and maternal genetic effects exerted from the $F_1$ dams [12] and epigenetic effects such as histone modification [13–15] exerted from either or both $F_1$ parents. Probably, some of these effects result in the missing QTL effect seen in traditional congenic/subcongenic analyses.

We previously discovered many QTLs for postnatal body weight and growth from an untapped resource of wild *M. m. castaneus* mice captured live in the Philippines, by genome-wide QTL analyses in an intersubspecific backcross population between C57BL/6JJcl (B6) inbred mice and the wild *castaneus* mice [16–18]. We further created the B6.Cg-*Pbwg1* congenic strain on the

B6 genetic background with an approximately 44.1-Mb wild-derived genomic region between *D2Mit33* and *D2Mit38* microsatellite markers, on which *Pbwg1*, a prominent growth QTL on a proximal region of mouse chromosome 2, is located. We developed more than 20 subcongenic strains derived from B6.Cg-*Pbwg1*. By phenotypic analysis of the $F_2$ intercross between B6.Cg-*Pbwg1* and B6 strains and by congenic/subcongenic analyses, we revealed 12 closely linked QTLs for body weight and body composition traits within the 44.1-Mb congenic region. These linked QTLs explained a small fraction of phenotypic variances [9,19–22]. Among the linked loci, several have unique QTL effects and are located on the distal half of the congenic region. For example, the wild-derived allele at *Pbwg1.12* and *Pbwg1.3* QTLs increases body weight and total body length, respectively, despite the fact that wild mice have a smaller body size than that of B6 [19,9]. In contrast, the allele at the *Pbwg1.5* QTL decreases abdominal white fat weight [19] and prevents obesity in mice fed both standard and high-fat diets [21].

In this study, we fine-mapped the three unique QTLs (*Pbwg1.12*, *Pbwg1.3* and *Pbwg1.5*) mentioned above by phenotypic analysis of $F_2$ mice obtained from intersubspecific subcongenic intercrosses. To search for possible candidate genes of the QTLs, we performed exome sequencing of genes on the congenic region and also prioritized candidate genes using bioinformatics analysis. Sequence data are not available for our wild *M. m. castaneus* mice captured in the Philippines, in contrast to the CAST/Eij inbred strain derived from wild *M. m. castaneus* mice trapped in Thailand, for which the whole genome has been already sequenced [23].

## Materials and Methods

### Ethics Statement

This study was carried out in accordance with the guidelines for the care and use of laboratory animals of the Graduate School of Bioagricultural Sciences, Nagoya University, Japan. The protocol was approved by the Animal Research Committee of Nagoya University.

## Animals

The B6.Cg-*Pbwg1* congenic strain was previously constructed [19]. Many subcongenic strains, named B6.Cg-*Pbwg1*/#Nga (old name: B6.Cg-*Pbwg1*/SR#, called SR# hereafter), were previously developed from descendants of B6.Cg-*Pbwg1* [9]. Previously developed SR1, SR3 and SR12 subcongenic strains and a newly developed SR21 subcongenic strain were used in this study (Figure 1). The background C57BL/6JJcl (B6) mice were purchased from Clea Japan (Tokyo, Japan). To develop four F$_2$ segregating populations, males of each subcongenic strain were crossed with B6 females to generate F$_1$ mice. The F$_1$ mice obtained were mated *inter se*. In total, the following F$_2$ individuals were produced: 273 (138 males and 135 females) for B6×SR1, 236 (113 males and 123 females) for B6×SR2, 132 (58 males and 74 females) for B6×SR12 and 291 (151 males and 140 females) for B6×SR21. Litter size was not standardized at birth to maximize the number of F$_2$ mice reared.

All mice were weaned at 3 weeks of age. Littermates of the same sex were housed in groups of up to four mice per cage. Standard chow (CA-1, Clea Japan), containing 27% crude protein, 5% crude fat, 3% crude fiber, 8% crude ash and 3.5 kcal/g, and tap water were provided *ad libitum*. The mice were reared in an environment with a temperature of 23±3°C, 55% relative humidity, and a light/dark cycle of 12:12.

## Genotyping

Genomic DNAs were extracted with a standard method from ear clips of the F$_2$ mice. Microsatellite markers located within each of the subcongenic intervals (Figure 1) were genotyped as described previously [19]. To reduce as short as possible the gray regions flanking the subcongenic intervals, where recombination occurred, we newly developed two PCR-RFLP markers based on two SNPs identified by exome sequencing in this study (Figure S1). Each of the F$_2$ mice had one of three diplotypes (B/B, B/C and C/C), where B is the haplotype derived from B6 mice and C is the haplotype derived from wild *castaneus* mice. Diplotype configuration was determined for each mouse of the four F$_2$ populations produced. F$_2$ mice having recombination within the subcongenic interval were excluded from this study.

## Phenotyping

Body weight of F$_2$ mice was measured to the nearest 0.01 g at 1, 3, 6, 10 and 14 weeks of age. Four body weight gains at 1–3 weeks, 3–6 weeks, 6–10 weeks and 10–14 weeks of age were calculated. After overnight fasting, mice were sacrificed under anesthesia. Total body length (from the tip of the nose to the end of the tail) and tail length (from the anus to the end of the tail) were immediately measured to the nearest 0.01 cm. Head-body length was obtained by subtracting tail length from total body length. After taking a blood sample by cardiopuncture, the lungs, spleen, liver, kidneys and testes were dissected and weighed to the nearest 0.001 g. In addition, the weights of two-sided inguinal and gonadal (epidydimal in males and parametrial in females) fat pads were recorded. In mice, the weight of white fat depots such as gonadal fat pads has been long and widely used as an indicator of fatness because the fat depots can be easily dissected out and are highly correlated with total body fat [1].

Body weight, weight gain and body composition data obtained for the F$_2$ populations were analyzed with a linear mixed model of the statistical discovery software JMP version 11.1.1 (SAS Institute, Cary, NC) in which diplotype, sex, parity, litter size and their

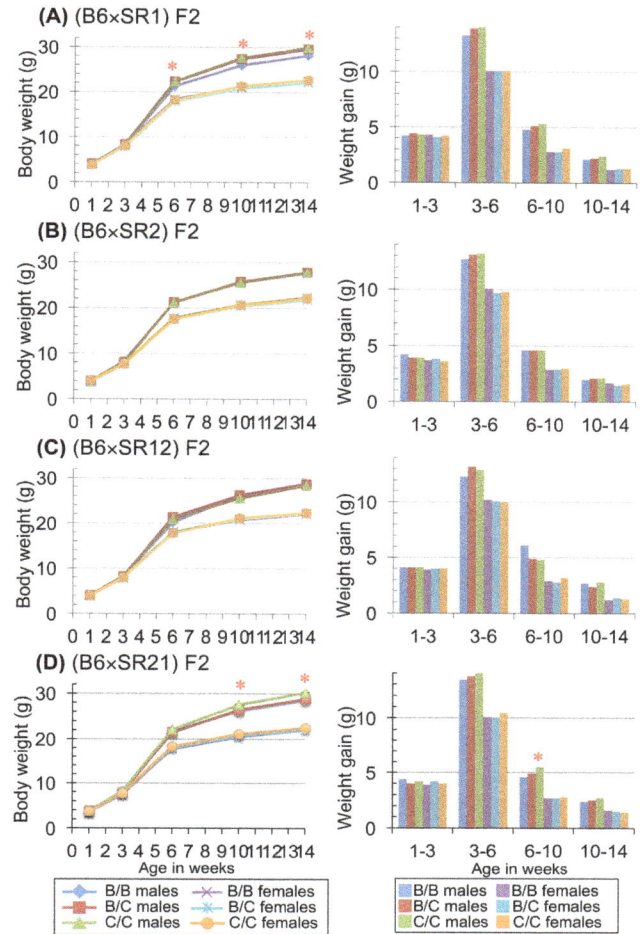

**Figure 2. Comparisons of body weight and body weight gain among three diplotypes (B/B, B/C and C/C) in four F$_2$ segregating populations obtained from (A) B6×SR1, (B) B6×SR2, (C) B6×SR12 and (D) B6×SR21 intercrosses.** B indicates the haplotype derived from B6 mice and C indicates the haplotype derived from wild *castaneus* mice. The asterisk shows significant differences in the trait among three diplotypes in each sex and the *P* values exceeded the Bonferroni-corrected 5% levels (see text for details).

possible two-way interactions were treated as fixed effects and dam was treated as a random effect. The fixed effects and interaction effects that were significant at the nominal 5% level were included in the final model. Phenotypic differences among diplotypes were determined by one-way analysis of variance followed by Tukey's HSD test. To adjust for multiple testing, Bonferroni-corrected 5% level was finally used as a significant threshold.

To estimate additive and dominance effects of diplotypes, a linear mixed model was fitted, with additive, dominance, sex, parity, litter size and their possible two-way interactions being included as fixed effects and dam being included as a random effect. The fixed effects and interaction effects that were significant at the nominal 5% level were finally included in the model. As previously defined for allelic effects of a QTL [24], additive diplotype effect is half of the difference between B/B and C/C homozygous diplotypes, and dominance diplotype effect is the difference between B/C heterozygous diplotypes and the average of B/B and C/C homozygous diplotypes. To estimate the additive

**Table 1.** Additive and dominance diplotype effects for body weight at # weeks of age and weight gain between # and # weeks of age in the $F_2$ populations obtained from B6×SR1 and B6×SR21 intercrosses.

| Sex | $F_2$ population | Trait | Diplotype effect[a] | | | | Degree of dominance | Inheritance[b] |
|---|---|---|---|---|---|---|---|---|
| | | | Additive | P value | Dominance | P value | | |
| Male | B6×SR1 | Wt1 | 0.039±0.046 | 0.40 | 0.049±0.069 | 0.47 | – | – |
| | | Wt3 | 0.041±0.111 | 0.71 | 0.193±0.170 | 0.26 | – | – |
| | | Wt6 | 0.525±0.173 | **0.0030** | 0.376±0.264 | 0.16 | 0.72 | **Add** |
| | | Wt10 | 0.788±0.167 | **0.0000076** | 0.582±0.256 | 0.025 | 0.73 | **Dom** |
| | | Wt14 | 0.874±0.178 | **0.0000034** | 0.586±0.271 | 0.033 | 0.67 | **Dom** |
| | | Gain1-3 | 0.023±0.095 | 0.81 | 0.151±0.142 | 0.29 | – | – |
| | | Gain3-6 | 0.395±0.128 | **0.0026** | 0.206±0.196 | 0.30 | 0.52 | Add |
| | | Gain6-10 | 0.267±0.123 | 0.031 | 0.040±0.185 | 0.83 | 0.15 | Add |
| | | Gain10-14 | 0.160±0.102 | 0.12 | -0.021±0.153 | 0.89 | – | – |
| | B6×SR21 | Wt1 | -0.001±0.043 | 0.98 | -0.034±0.062 | 0.59 | – | – |
| | | Wt3 | -0.066±0.097 | 0.49 | -0.319±0.140 | 0.024 | -4.83 | Overrec |
| | | Wt6 | 0.198±0.164 | 0.23 | -0.379±0.238 | 0.11 | – | – |
| | | Wt10 | 0.635±0.172 | **0.00032** | -0.483±0.249 | 0.055 | -0.76 | **Add** |
| | | Wt14 | 0.833±0.192 | **0.000030** | -0.552±0.279 | 0.050 | -0.66 | **Rec** |
| | | Gain1-3 | -0.074±0.070 | 0.29 | -0.282±0.102 | 0.0065 | 3.81 | Overrec |
| | | Gain3-6 | 0.277±0.144 | 0.016 | -0.045±0.166 | 0.78 | -0.16 | Add |
| | | Gain6-10 | 0.440±0.121 | **0.00039** | -0.099±0.176 | 0.57 | -0.23 | **Add** |
| | | Gain10-14 | 0.155±0.072 | 0.033 | -0.080±0.105 | 0.45 | -0.52 | Add |
| Female | B6×SR1 | Wt1 | -0.069±0.047 | 0.15 | -0.146±0.063 | 0.023 | -2.12 | Overrec |
| | | Wt3 | -0.156±0.091 | 0.089 | -0.278±0.120 | 0.023 | -1.78 | Overrec |
| | | Wt6 | 0.006±0.136 | 0.97 | -0.092±0.185 | 0.62 | – | – |
| | | Wt10 | 0.060±0.122 | 0.62 | -0.423±0.164 | 0.011 | -7.05 | Overrec |
| | | Wt14 | 0.099±0.136 | 0.47 | -0.379±0.182 | 0.040 | -3.83 | Overrec |
| | | Gain1-3 | -0.079±0.061 | 0.19 | -0.160±0.081 | 0.051 | -2.03 | Overrec |
| | | Gain3-6 | 0.010±0.111 | 0.93 | -0.030±0.152 | 0.84 | – | – |
| | | Gain6-10 | 0.177±0.096 | 0.069 | -0.143±0.130 | 0.27 | – | – |
| | | Gain10-14 | 0.033±0.079 | 0.68 | 0.058±0.107 | 0.59 | – | – |
| | B6×SR21 | Wt1 | 0.039±0.042 | 0.36 | 0.003±0.058 | 0.96 | – | – |
| | | Wt3 | 0.094±0.096 | 0.33 | 0.207±0.132 | 0.12 | – | – |
| | | Wt6 | 0.271±0.114 | 0.019 | -0.103±0.157 | 0.51 | -0.38 | Add |
| | | Wt10 | 0.301±0.119 | 0.013 | -0.201±0.165 | 0.22 | -0.67 | Add |
| | | Wt14 | 0.208±0.120 | 0.086 | -0.204±0.116 | 0.22 | – | – |
| | | Gain1-3 | 0.058±0.068 | 0.39 | 0.216±0.094 | 0.023 | 3.72 | Overdom |

**Table 1.** Cont.

| Sex | $F_2$ population | Trait | Diplotype effect[a] | | | | | Inheritance[b] |
|---|---|---|---|---|---|---|---|---|
| | | | Additive | P value | Dominance | P value | Degree of dominance | |
| | | Gain3-6 | 0.181±0.110 | 0.10 | −0.287±0.152 | 0.061 | - | - |
| | | Gain6-10 | 0.027±0.091 | 0.77 | −0.097±0.127 | 0.44 | - | - |
| | | Gain10-14 | −0.088±0.080 | 0.27 | −0.006±0.111 | 0.96 | - | - |

Data are means and standard errors. The P value in bold exceeded the Bonferroni-corrected 5% threshold level.
[a]Positive sign indicates that the haplotype derived form the wild castaneus mouse increased the trait value.
[b]Add, additive; Dom, dominant; Rec, recessive; Overrec, overrecessive; Overdom, overdominant; -, not applicable.

diplotype effect, diplotypes were assigned quantitatively as −1 for B/B homozygotes, zero for B/C heterozygotes and +1 for C/C homozygotes. To estimate the dominance diplotype effect, diplotypes were assigned as zero for two types of homozygotes and +1 for heterozygotes. The degree of dominance was calculated as the ratio of dominance diplotype effect to additive diplotype effect.

## Exome Sequencing

Genomic DNA was extracted with a standard method from the tail of a B6.Cg-*Pbwg1* congenic mouse. Enrichment of exon regions in the 44.1-Mb congenic interval on mouse chromosome 2 was performed using Roche NimbleGen sequence capture arrays that were custom-made on the basis of UCSC Mouse Genome Browser NCBI37/mm9 assembly (RefSeq mm9). Enrichment experiments and exome sequencing with the next-generation sequencer Roche GS FLX were outsourced to Hokkaido System Science Co., Ltd (Sapporo, Japan). Sequence reads obtained were mapped to RefSeq mm9, and then synonymous SNPs (sSNPs), nonsynonymous SNPs (nsSNPs), indels (insertions and deletions) and nonsense mutations were investigated.

## Candidate Gene Search

Endevour is a web-based computational software program that prioritizes candidate genes with respect to their biological processes or diseases of interest [25]. Genes on target regions carrying QTLs for body weight and obesity were prioritized on the basis of similarity to training genes that have already been shown to be involved in body weight and obesity regulation (Table S1). The training genes used were searched using Online Mendelian Inheritance in Man (OMIM) database (http://www.omim.org).

Effects of nsSNPs identified by exome sequencing on protein functions were investigated with two web-based software programs, SIFT [26] and PolyPhen-2 [27]. SIFT predicts tolerated and deleterious substitutions for nsSNPs based on the evolutionary conservation of amino acids within protein families [26]. PolyPhen-2 predicts possible impact of an amino acid substitution on the structure and function of a protein using straightforward physical and comparative considerations [27]. Since PolyPhen-2 was developed for human proteins, this software was implemented after converting the positions of amino acid substitutions in our mouse study to the corresponding positions of the human protein.

## Results

### Intersubspecific Subcongenic Intercross Analyses

Most growth and body composition traits examined in $F_2$ segregating populations between each of the four subcongenic strains (Figure 1) and the background B6 strain showed significant interactions between sex and trait (data not shown). We thus performed statistical comparisons of these traits among mice with three diplotypes in each sex separately.

Figure 2 shows measurements of body weight and body weight gain in the four $F_2$ segregating populations. In the B6×SR1 intercross, body weight of male mice with C/C diplotypes at 6 weeks of age was significantly higher than that of mice with the B/B diplotype ($P = 0.0041$, Tukey's HSD test) at the Bonferroni-corrected 5% significance level. However, it was not different from that of B/C males throughout ages examined. From 6 weeks onwards, the weight difference between C/C and B/B males remained significant ($P = 0.0000036$ at 10 weeks and 0.0000017 at 14 weeks of age). Additive diplotype effects for body weights at 6, 10 and 14 weeks of age exceeded the Bonferroni-corrected 5% level, whereas dominance diplotype effects were not significantly

**Table 2.** Body length and fat pad weight not adjusted for body weight at 14 weeks of age, and additive and dominance diplotype effects for body length and fat pad weight in the F$_2$ populations obtained from B6×SR1 and B6×SR21 intercrosses.

| Sex | F$_2$ population | Trait | Diplotype[c] | | | | Diplotype effect | | | | Degree of dominance | Inheritance[d] |
|---|---|---|---|---|---|---|---|---|---|---|---|---|
| | | | B/B | B/C | C/C | P value | Additive | P value | Dominance | P value | | |
| Male | B6×SR1 | No. of mice | 33 | 33–34 | 35–37 | | | | | | | |
| | | Tail length (cm) | 8.33±0.04 | 8.42±0.04 | 8.39±0.04 | 0.13 | 0.028±0.024 | 0.24 | 0.060±0.038 | 0.12 | - | - |
| | | Head-body length (cm) | 9.20±0.04[a] | 9.34±0.04[b] | 9.38±0.04[b] | **0.00079** | 0.088±0.025 | **0.00080** | 0.058±0.041 | 0.17 | 0.66 | **Add** |
| | | Total body length (cm) | 17.52±0.06[b] | 17.76±0.06[b] | 17.77±0.06[b] | **0.0014** | 0.118±0.038 | **0.0028** | 0.116±0.063 | 0.070 | 0.98 | **Add** |
| | | Inguinal fat pad weight (g) | 0.278±0.012[a] | 0.257±0.012[a] | 0.203±0.012[b] | **0.0000069** | −0.037±0.012 | **0.0000035** | 0.016±0.012 | 0.17 | 0.43 | **Add** |
| | | Gonadal fat pad weight (g) | 0.280±0.014[a] | 0.275±0.014[a] | 0.210±0.014[b] | **0.000038** | −0.034±0.008 | **0.000071** | 0.030±0.013 | 0.030 | 0.88 | **Rec** |
| | B6×SR21 | No. of mice | 32 | 38–39 | 38–39 | | | | | | | |
| | | Tail length (cm) | 8.38±0.04 | 8.38±0.03 | 8.45±0.03 | 0.13 | 0.034±0.021 | 0.11 | −0.039±0.033 | 0.24 | - | - |
| | | Head-body length (cm) | 9.26±0.04[a] | 9.27±0.03[a] | 9.41±0.03[b] | **0.00021** | 0.078±0.021 | **0.00035** | −0.070±0.033 | 0.038 | −0.90 | **Rec** |
| | | Total body length (cm) | 17.62±0.05[a] | 17.63±0.05[a] | 17.85±0.06[b] | **0.00022** | 0.115±0.032 | **0.00044** | −0.110±0.050 | 0.032 | −0.96 | **Rec** |
| | | Inguinal fat pad weight (g) | 0.384±0.016 | 0.382±0.015 | 0.404±0.015 | 0.35 | 0.010±0.009 | 0.25 | −0.012±0.014 | 0.39 | - | - |
| | | Gonadal fat pad weight (g) | 0.362±0.016 | 0.362±0.015 | 0.398±0.015 | 0.066 | 0.018±0.009 | 0.051 | −0.018±0.014 | 0.210 | - | - |
| Female | B6×SR1 | No. of mice | 24 | 34–36 | 32 | | | | | | | |
| | | Tail length (cm) | 8.14±0.03[ab] | 8.11±0.03[a] | 8.20±0.03[a] | 0.038 | 0.030±0.021 | 0.16 | −0.063±0.031 | 0.048 | −2.10 | Overrec |
| | | Head-body length (cm) | 8.88±0.05[ab] | 8.77±0.04[b] | 8.91±0.04[a] | 0.023 | 0.013±0.030 | 0.67 | −0.122±0.045 | 0.0081 | −9.38 | Overrec |
| | | Total body length (cm) | 17.02±0.06[ab] | 16.89±0.05[b] | 17.12±0.06[a] | **0.0043** | 0.049±0.039 | 0.22 | −0.178±0.059 | **0.0034** | −3.63 | **Overrec** |
| | | Inguinal fat pad weight (g) | 0.231±0.13[a] | 0.193±0.012[b] | 0.183±0.012[b] | **0.0037** | −0.024±0.007 | **0.0013** | −0.015±0.011 | 0.17 | −0.63 | **Add** |
| | | Gonadal fat pad weight (g) | 0.176±0.012[a] | 0.137±0.010[b] | 0.133±0.010[b] | **0.0019** | −0.022±0.006 | **0.0011** | −0.018±0.009 | 0.065 | −0.82 | **Add** |
| | B6×SR21 | No. of mice | 28 | 38–39 | 33 | | | | | | | |
| | | Tail length (cm) | 8.00±0.03[a] | 8.09±0.03[ab] | 8.16±0.03[b] | **0.0024** | 0.077±0.021 | **0.00052** | 0.011±0.032 | 0.73 | 0.14 | **Add** |
| | | Head-body length (cm) | 8.85±0.04[ab] | 8.88±0.03[b] | 8.98±0.03[a] | 0.012 | 0.065±0.023 | 0.0064 | −0.035±0.036 | 0.33 | −0.54 | Add |
| | | Total body length (cm) | 16.85±0.06[a] | 16.99±0.05[a] | 17.15±0.05[b] | **0.00051** | 0.150±0.004 | **0.00011** | −0.013±0.057 | 0.82 | −0.87 | **Rec** |
| | | Inguinal fat pad weight (g) | 0.382±0.14[ab] | 0.371±0.012[b] | 0.374±0.013[b] | 0.039 | −0.019±0.008 | 0.017 | 0.008±0.012 | 0.50 | 0.42 | Add |

**Table 2.** Cont.

| Sex | F$_2$ population | Trait | Diplotype[c] | | | | Diplotype effect | | | | Degree of dominance | |
| --- | --- | --- | --- | --- | --- | --- | --- | --- | --- | --- | --- | --- |
| | | | B/B | B/C | C/C | P value | Additive | P value | Dominance | P value | | Inheritance[d] |
| | | Gonadal fat pad weight (g) | 0.258±0.013 | 0.243±0.012 | 0.226±0.012 | 0.072 | −0.016±0.007 | 0.024 | 0.001±0.010 | 0.92 | 0.063 | Add |

Trait data are means and standard errors computed using a linear model including fixed and random effects (see Materials and Methods).
[a,b]Means with different superscript letters within a trait indicate significant differences among three deplotypes at $P \le 0.05$ (Tukey's HSD test). The $P$ value in bold exceeded the Bonferroni-corrected 5% threshold level.
[c]B denotes the haplotype where all alleles at marker loci on the congenic region are fixed for B6 alleles, and C indicates the haplotype on which all alleles are fixed for wild-derived *castaneus* alleles. Individuals with recombinant haplotypes were excluded from this analysis.
[d]Add, additive; Rec, recessive; Overrec, overrecessive; -, not applicable.

different from zero (Table 1). The C haplotype derived from wild mice increased body weight, despite the fact that the wild mice have approximately 60% of the body weight of B6 mice [16]. The mode of inheritance of this haplotype was additive or dominant (Table 1). In females, however, there were no significant differences in body weight at any age (Figure 2). For body weight gains at 1–3 weeks, 3–6 weeks, 6–10 weeks and 10–14 weeks, both sexes of mice with three diplotypes did not show significant differences at the Bonferroni-corrected 5% level. Although body weight gain in males at 3–6 weeks ($P = 0.0058$) was on the border of that level, its additive diplotype effect surpassed it (Table 1).

In B6×SR2 and B6×SR12 intercrosses, body weights at any age did not differ significantly among mice with three diplotypes in both sexes. Similarly, there were no significant diplotype differences in body weight gains at any age in both sexes (Figure 2 and Table S2).

On the other hand, in the B6×SR21 intercross, body weights of C/C males were significantly higher than those of B/C and B/B males at 10 weeks ($P = 0.00028$) and 14 weeks ($P = 0.000029$) of age (Figure 2). The wild-derived C haplotype was inherited in an additive or recessive fashion (Table 1). Furthermore, body weight gain at 6–10 weeks was significantly higher in the C/C males than in the B/C and B/B males ($P = 0.0016$) (Figure 2). The C haplotype was inherited in an additive fashion (Table 1). In contrast, neither body weight nor body weight gain was significantly different among females with three diplotypes at the Bonferroni-corrected 5% level (Figure 2 and Table 1).

Tables 2 and S3 show measurements of body composition traits not adjusted for final body weight at 14 weeks of age in two kinds of F$_2$ populations obtained from B6×SR1 and B6×SR21 intercrosses. In the B6×SR1 intercross, head-body length and total body length of C/C males were both significantly larger than those of B/B males at the Bonferroni-corrected 5% level (Table 2). The same tendency was observed for those traits in females, but the C haplotype was transmitted as different modes of inheritance in males (additive) and females (overrecessive) for an unknown reason. For inguinal fat pad weight and gonadal fat pad weight, C/C mice had the lowest values among mice with three diplotypes in both sexes. The C haplotype was inherited in an additive or recessive fashion depending on the sex. A significant difference in kidney weight was observed only in females (Table S3).

In the B6×SR21 intercross, unadjusted body length traits were significantly different in both sexes at the Bonferroni-corrected 5% level. C/C mice had the longest length in both sexes (Table 2). The C/C mice had significantly higher kidney weight in males but not in females (Table S3).

On the other hand, in both B6×SR2 and B6×SR12 intercrosses, no unadjusted traits were significantly different among mice with three diplotypes at the Bonferroni-corrected 5% level as well as the nominal 5% level (Table S4).

Tables 3 and S5 show measurements of body composition traits adjusted for final body weight at 14 weeks of age, which are body-size-free traits, in two F$_2$ populations obtained from B6×SR1 and B6×SR21 intercrosses. In the B6×SR1 intercross, significant diplotype differences were observed in inguinal and gonadal fat pad weights for both sexes and in testis weight for males at the Bonferroni-corrected 5% level. Both of the fat weights in C/C mice were lowest in each sex. Testis weight in C/C males was also lowest. The C haplotype for these three traits indicated an additive mode of inheritance. In contrast, in the B6×SR21 intercross, inguinal and gonadal fat pad weights were significantly lower in C/C females than in B/B females at the nominal 5% level, but those fat weights were not significantly different in males.

**Table 3.** Body length and fat pad weight adjusted for body weight at 14 weeks of age, and additive and dominance diplotype effects for body length and fat pad weight in the $F_2$ populations obtained from B6×SR1 and B6×SR21 intercrosses.

| Sex | $F_2$ population | Trait | Diplotype | | | | Diplotype effect | | | | Degree of dominance | Inheritance |
| --- | --- | --- | --- | --- | --- | --- | --- | --- | --- | --- | --- | --- |
| | | | B/B | B/C | C/C | P value | Additive | P value | Dominance | P value | | |
| Male | B6×SR1 | No. of mice | 33 | 33–34 | 35–37 | | | | | | | |
| | | Tail length (cm) | 8.37±0.04[a] | 8.40±0.04[a] | 8.37±0.04[a] | 0.70 | −0.003±0.024 | 0.91 | 0.032±0.037 | 0.39 | – | – |
| | | Head-body length (cm) | 9.27±0.03[a] | 9.30±0.03[a] | 9.32±0.03[a] | 0.52 | 0.022±0.021 | 0.31 | 0.010±0.033 | 0.76 | – | – |
| | | Total body length (cm) | 17.64±0.05[a] | 17.71±0.05[a] | 17.69±0.05[a] | 0.56 | 0.021±0.032 | 0.51 | 0.042±0.051 | 0.41 | – | – |
| | | Inguinal fat pad weight (g) | 0.287±0.013[a] | 0.249±0.012[b] | 0.198±0.012[c] | **0.00000049** | −0.042±0.008 | **0.00000097** | 0.013±0.012 | 0.28 | 0.31 | **Add** |
| | | Gonadal fat pad weight (g) | 0.290±0.014[a] | 0.269±0.014[a] | 0.203±0.014[b] | **0.0000028** | −0.042±0.009 | **0.0000036** | 0.023±0.013 | 0.088 | 0.55 | **Add** |
| | B6×SR21 | No. of mice | 32 | 38–39 | 38–39 | | | | | | | |
| | | Tail length (cm) | 8.40±0.03 | 8.39±0.03 | 8.41±0.03 | 0.83 | 0.008±0.002 | 0.69 | −0.016±0.032 | 0.63 | – | – |
| | | Head-body length (cm) | 9.31±0.02 | 9.32±0.02 | 9.35±0.02 | 0.49 | 0.017±0.016 | 0.28 | −0.013±0.024 | 0.58 | – | – |
| | | Total body length (cm) | 17.71±0.04 | 17.70±0.03 | 17.76±0.04 | 0.39 | 0.026±0.024 | 0.29 | −0.340±0.372 | 0.36 | – | – |
| | | Inguinal fat pad weight (g) | 0.386±0.016 | 0.384±0.015 | 0.398±0.016 | 0.71 | 0.006±0.009 | 0.53 | −0.008±0.014 | 0.57 | – | – |
| | | Gonadal fat pad weight (g) | 0.366±0.016 | 0.366±0.015 | 0.388±0.016 | 0.40 | 0.011±0.009 | 0.26 | −0.011±0.014 | 0.42 | – | – |
| Female | B6×SR1 | No. of mice | 24 | 34–36 | 32 | | | | | | | |
| | | Tail length (cm) | 8.10±0.03[a] | 8.14±0.03[a] | 8.18±0.03[a] | 0.096 | 0.027±0.018 | 0.14 | −0.018±0.028 | 0.51 | – | – |
| | | Head-body length (cm) | 8.86±0.03[a] | 8.82±0.03[a] | 8.87±0.03[a] | 0.43 | 0.006±0.022 | 0.78 | −0.043±0.035 | 0.21 | – | – |
| | | Total body length (cm) | 16.96±0.040[a] | 16.96±0.034[a] | 17.05±0.036[a] | 0.040 | 0.035±0.024 | 0.14 | −0.056±0.036 | 0.13 | – | – |
| | | Inguinal fat pad weight (g) | 0.227±0.012[a] | 0.200±0.010[ab] | 0.178±0.011[b] | **0.0030** | −0.025±0.007 | **0.00066** | −0.003±0.011 | 0.82 | −0.12 | **Add** |
| | | Gonadal fat pad weight (g) | 0.173±0.010[a] | 0.145±0.008[ab] | 0.127±0.009[b] | **0.0019** | −0.021±0.006 | **0.00071** | −0.003±0.009 | 0.71 | −0.14 | **Add** |
| | B6×SR21 | No. of mice | 28 | 38–39 | 33 | | | | | | | |
| | | Tail length (cm) | 8.03±0.03[a] | 8.10±0.03[ab] | 8.13±0.03[b] | 0.027 | 0.050±0.018 | 0.0076 | 0.014±0.027 | 0.60 | – | – |
| | | Head-body length (cm) | 8.88±0.03 | 8.89±0.03 | 8.95±0.03 | 0.093 | 0.039±0.020 | 0.054 | −0.026±0.030 | 0.38 | – | – |
| | | Total body length (cm) | 16.91±0.047[a] | 16.97±0.041[ab] | 17.08±0.044[b] | 0.011 | 0.086±0.028 | **0.0035** | −0.017±0.042 | 0.69 | −0.20 | Add |
| | | Inguinal fat pad weight (g) | 0.385±0.014[a] | 0.372±0.012[ab] | 0.341±0.013[b] | 0.016 | −0.022±0.008 | 0.0068 | 0.009±0.012 | 0.45 | – | – |

**Table 3.** Cont.

| Sex | F$_2$ population | Trait | Diplotype | | | | Diplotype effect | | | | Degree of dominance | Inheritance |
|---|---|---|---|---|---|---|---|---|---|---|---|---|
| | | | B/B | B/C | C/C | P value | Additive | P value | Dominance | P value | | |
| | | Gonadal fat pad weight (g) | 0.261±0.013[a] | 0.244±0.011[ab] | 0.222±0.012[b] | 0.021 | −0.020±0.007 | 0.0061 | 0.002±0.010 | 0.82 | - | - |

For abbreviations, see the footnotes to Tables 1 and 2.

On the other hand, in B6×SR2 and B6×SR12 intercrosses, neither males nor females showed significant diplotype differences in any adjusted body composition traits at the Bonferroni-corrected 5% level (Table S6).

Taking all results together, a growth QTL, the wild-derived allele of which increased body weight and body weight gain, was localized to an interval between *D2Mit433* (57.3 Mb) and *D2Mit205* (65.3), as summarized in Table 4. In a previous study using congenic strains [9], the growth QTL *Pbwg1.12* was physically localized to a maximum interval between *D2Mit472* (61.5) and *D2Mit327* (69.5) (see Figure 1), which overlapped with the interval of the growth QTL identified in this study. The wild-derived allele at *Pbwg1.12* increased body weight [9], which was exactly the same allelic effect of the growth QTL identified in this study. Therefore, in this study, we succeeded in confirming the presence of *Pbwg1.12* and in narrowing it down to a 3.8-Mb interval between *D2Mit472* (61.5) and *D2Mit205* (65.3).

Likewise, we were able to confirm the presence of the *Pbwg1.3* QTL affecting total body length that was previously revealed by interval mapping with an F$_2$ intercross population between the B6.Cg-*Pbwg1* original congenic and B6 strains [19]. *Pbwg1.3* was physically defined to an 8.0-Mb interval between *D2Mit433* (57.3) and *D2Mit205* (65.3) (Table 4).

In the B6×SR1 intercross, a QTL for which the wild-derived allele decreased inguinal and gonadal fat pad weights was clearly identified. In the B6×SR21 intercross, however, the presence of the obesity QTL was ambiguous, because *P* values for diplotype comparisons marginally exceeded the nominal 5% level but did not reach the Bonferroni-corrected 5% level. Previously, the *Pbwg1.5* QTL for resistance to obesity was physically mapped to an interval between *D2Mit270* (52.9) and *D2Mit472* (61.5) [21] (see Figure 1). Therefore, we were able to confirm the presence of *Pbwg1.5* in this study and localize it to a 2.1-Mb interval between *D2Mit123* (59.4) and *D2Mit472* (61.5) (Table 4).

## Exome Sequencing

Since no sequence data have so far been reported for the Philippine wild *castaneus* mice used in this study, we performed sequencing of 2,205 exons for 153 genes on the 44-Mb original congenic region of chromosome 2. According to RefSeq mm9, target bases for the exons were 767,440 bp. The NimbleGen sequence capture covered 97.1% of the target bases, i.e., 745,515 bp. Individual sequence coverage was 11.2 fold on average and ranged from 3 to 36 fold. As expected, some kinds of sequence variants, such as SNPs and indels, were observed in most genes derived from the wild mouse (Table S5). In total, 840 sSNPs and 334 nsSNPs were identified. Nine deletions and 10 insertions were detected in 13 genes. In addition, five nonsense mutations were identified within three genes. On the QTL regions narrowed by the above intersubspecific subcongenic intercross analyses, many SNPs and a few indels were identified, but no nonsense mutation was detected (Tables 5 and S7).

## Candidate Gene Search

As shown in Table 5, there were 11 genes on the 3.8-Mb region between *D2Mit472* and *D2Mit205*, where the growth QTL *Pbwg1.12* was located. Using training genes related to body weight (Table S1), Endevour prioritized *Gcg* (glucagon) and *Grb14* (growth factor receptor-bound protein 14) as the top two candidate genes for *Pbwg1.12*. *Gcg* had one nsSNP and *Grb14* had two nsSNPs. Both SIFT and PolyPhen-2 predicted that none of these nsSNPs inflicted possible damage on protein functions (Table 5).

On the 2.1-Mb region between *D2Mit123* and *D2Mit472* harboring the obesity QTL *Pbwg1.5*, 12 genes were located

**Table 4.** Summary of QTLs for growth and body composition confirmed in the four F$_2$ segregating populations.

| Trait | F$_2$ population[a] | | | | QTL | |
| | B6×SR1 | B6×SR2 | B6×SR12 | B6×SR21 | Symbol | Genomic interval (Mb) |
|---|---|---|---|---|---|---|
| Body weight & weight gain | Increased | ND | ND | Increased | Pbwg1.12[b] | D2Mit472-D2Mit205 (61.5–65.3) |
| Total body length (unadjusted) | Increased | ND | ND | Increased | Pbwg1.3 | D2Mit433-D2Mit205 (57.3–65.3) |
| Fat pad weight (unadjusted & adjusted) | Decreased | ND | ND | MG | Pbwg1.5[c] | D2Mit123-D2Mit472 (59.4–61.5) |

[a]The effect of the QTL allele derived from the wild mouse is shown: Increased, increased the trait value; Decreased, decreased the trait value; MG, decreased the trait value but the QTL effect was marginal; ND, QTL was not detected.
[b]Pbwg1.12 was previously defined to an interval between D2Mit472 and D2Mit327 [9] (see Figure 1 for their relative map positions).
[c]Pbwg1.5 was previously defined to an interval between D2Mit270 and D2Mit472 [21] (see Figure 1).

(Table 5). Endevour ranked *Ly75* (lymphocyte antigen 75) and *Itgb6* (integrin beta 6) as the top two candidate genes for *Pbwg1.5* using training genes related to obesity (Table S1). *Ly75* had nine nsSNPs and these were predicted to have no affect on protein function. In contrast, *Itgb6* had three nsSNPs. PolyPhen-2 predicted that none of the three nsSNPs caused possible damage to protein function, whereas SIFT predicted that one of them, i.e. A>C at the position of 2:60491216 leading to amino acid substitution of Sel302Ala, is harmful to protein function. This nsSNP has previously been reported as dbSNP rs28025203.

## Discussion

Previously we discovered the *Pbwg1.12* QTL for growth [9], the *Pbwg1.3* QTL for body length [19] and the *Pbwg1.5* QTL for obesity [21] from an untapped resource of wild *M. m. castaneus* mice caught in the Philippines. In this study, we were able to confirm the presence of these three QTLs by intersubspecific intercross analyses using four newly or previously constructed subcongenic strains with overlapping and/or non-overlapping genomic intervals. The unique effects of the wild-derived QTL allele at the QTLs revealed in previous studies [9,19,21] were duplicated in the present independent study. That is, this allele uniquely enhanced growth at *Pbwg1.12* and increased body length at *Pbwg1.3*, despite the fact that the wild mouse has approximately 60% of the body weight of B6 [16], whereas it decreased fat weight at *Pbwg1.5*. Furthermore, we were able to reduce the genomic interval harboring *Pbwg1.12* from 8.9 Mb [9] to 3.8 Mb in length and to narrow the interval of *Pbwg1.5* from 8.8 Mb [21] to 2.1 Mb. Although *Pbwg1.3* was previously localized to a 20-Mb confidence interval [19], it was physically mapped to an 8.0-Mb interval in this study.

Although *Pbwg1.3* and *Pbwg1.5* exerted phenotypic effects on both sexes, *Pbwg1.12* exhibited male-specific effect on body weight. In our previous study [9], the sex-specificity of *Pbwg1.12* on chromosome 2 was not tested because the sample size was small. Sex-specific QTLs for body weight are revealed on different chromosomal regions in our previous genome-wide QTL analysis [18] and in different mouse crosses [5]. In addition, sex-specific QTLs have commonly been observed in different quantitative traits of mice and other species, as mentioned previously [16,18]. However, the molecular mechanisms underlying sex-specific QTLs remain unclear. It is reported that androgen control of growth hormone secretion induces male-specific gene expression in the liver of mice [28]. We thus consider that the male-specific effect of *Pbwg1.12* may be mediated by sex hormones through male-specific expression of the causative gene of *Pbwg1.12*.

According to MGD [5], several QTLs affecting growth, body length and obesity were previously mapped to mouse chromosome 2 regions that are overlapped with our 8.0-Mb region between *D2Mit443* (57.3) and *D2Mit205* (65.3). Since the previous QTLs were mapped by genome-wide QTL analyses, confidence intervals of the QTLs are generally very large, spanning approximately 100 Mb or more. Thereafter, few map positions of QTLs have been determined physically, with a few exceptions. Four growth and obesity QTLs, named *Wg2a-Wg2d*, have been fine-mapped to the interval from *D2Ucd15* (74.7) to *D2Mit196* (160.5) by phenotypic analyses of subcongenic strains that possess the introgressed regions of the CAST/EiJ strain established from wild *M. m. castaneus* mice on the B6 genetic background [10]. However, this interval is outside of our QTL regions. The *Nidd5* QTL affecting adiposity has been fine-mapped by phenotypic analysis of congenic strains with donor regions derived from the BALB/cA strain on the genetic background of the obese/diabetic TSOD strain, and *Acvr1c* encoding activin receptor-like kinase 7 at the position of 58.1 Mb was very recently identified as a responsible gene for *Nidd5* [29]. The obese/diabetic TSOD strain unexpectedly has the wild-type allele at *Acvr1c*, whereas the normal BALB/cA strain has a nonsense mutation resulting in decreased fat mass phenotype [29]. In contrast, our exome sequencing analysis indicated that the *Acvr1c* gene derived from our wild *castaneus* mouse had neither nsSNPs nor nonsense mutations. Furthermore, *Acvr1c* lies outside the interval containing our *Pbwg1.12* for increased body weight and *Pbwg1.5* for decreased fat weight. *Acvr1c* is thus unlikely to be a candidate gene for our QTLs. In addition, as no nonsense mutation was identified for any of the genes located in the QTL interval, this kind of mutation could not become a sequence variant causing the differences in body weight and fat weight shown in this study.

In this study, exome sequencing and candidate gene prioritization strongly suggested that *Gcg* and *Grb14* are putative candidate genes for the *Pbwg1.12* QTL for enhanced growth. *Gcg* encodes proglucagon, a precursor of glucagon, glucagon-like peptide-1 (GLP-1) and several other components. Glucagon is generated in pancreatic α-cells and GLP-1 is yielded in intestinal L-cells, and these peptides paly key roles in glucose metabolism and homeostasis [30]. Mice lacking glucagon and GLP-1 are born normally without gross abnormalities and display α-cell hyperplasia and increased body weight [31]. It has very recently been revealed in rats fed a high-fat diet that hypothalamic glucagon signaling can suppress hepatic glucose production, suggesting that hypothalamic glucagon resistance may contribute to the hyperglycemia observed in obesity and diabetes [32]. *Grb14* encodes an adaptor protein belonging to the GRB7 family and it plays an important role in receptor-tyrosine kinase signaling pathways and

**Table 5.** Variants detected by exome sequencing of genes on the genomic regions harboring the growth QTL *Pbwg1.12* and the obesity QTL *Pbwg1.5*, prioritization of candidate genes and damage of protein functions caused by nsSNPs found in the candidate genes.

| Gene symbol | Position (bp) | | Number of variants[a] | | | | Candidate gene ranking[b] | | Damage to protein function[c] | |
| --- | --- | --- | --- | --- | --- | --- | --- | --- | --- | --- |
| | Start | End | sSNP | nsSNP | Deletion | Insertion | Body weight | Obesity | SIFT | PolyPhen-2 |
| Dapl1 | 59322709 | 59343075 | 1 | 0 | 0 | 0 | NA | | | |
| Tanc1 | 59450100 | 59684206 | 21 | 4 | 0 | 1 | NA | | | |
| Wdsub1 | 59690423 | 59720663 | 6 | 1 | 0 | 0 | NA | | | |
| Baz2b | 59737419 | 59963797 | 15 | 6 | 1 | 0 | NA | | | |
| March7 | 60047992 | 60086442 | 6 | 0 | 0 | 0 | NA | | | |
| Cd302 | 60090049 | 60122475 | 1 | 0 | 1 | 0 | NA | | | |
| Ly75 | 60131816 | 60221288 | 27 | 9 | 0 | 0 | NA | 1 | Tolerated | Benign |
| Pla2r1 | 60257095 | 60391318 | 18 | 8 | 0 | 0 | NA | | | |
| Itgb6 | 60436349 | 60511750 | 11 | 3 | 0 | 0 | NA | 2 | Affected | Benign |
| Rbms1 | 60590009 | 60801261 | 2 | 0 | 0 | 0 | NA | | | |
| Tank | 61416642 | 61492224 | 1 | 5 | 0 | 0 | NA | | | |
| Psmd14 | 61549750 | 61638433 | 1 | 0 | 0 | 0 | NA | | | |
| Tbr1 | 61642509 | 61652170 | 2 | 1 | 0 | 0 | | NA | | |
| Slc4a10 | 61884596 | 62164800 | 6 | 0 | 0 | 0 | | NA | | |
| Dpp4 | 62168131 | 62250288 | 6 | 0 | 0 | 0 | | NA | | |
| Gcg | 62312586 | 62321710 | 0 | 1 | 0 | 0 | 1 | NA | Tolerated | Benign |
| Fap | 62339001 | 62412078 | 2 | 2 | 0 | 0 | | NA | | |
| Ifih1 | 62433849 | 62484312 | 17 | 5 | 0 | 0 | | NA | | |
| Gca | 62502383 | 62532166 | 3 | 1 | 0 | 0 | | NA | | |
| Kcnh7 | 62541002 | 63022344 | 6 | 1 | 0 | 0 | | NA | | |
| Fign | 63815417 | 63936064 | 4 | 1 | 0 | 0 | | NA | | |
| Grb14 | 64750539 | 64860823 | 7 | 2 | 0 | 0 | 2 | NA | Tolerated | Benign |
| Cobll1 | 64926395 | 65076683 | 14 | 18 | 0 | 0 | | NA | | |

[a]SNP, synonymous SNP; nsSNP, nonsynonymous SNP; NA, not applicable because the QTL in question was not located on the region including the genes.

[b]The top two genes were prioritized as candidate genes for growth and obesity QTLs by the web-based software program Endevour [25].

[c]Damage caused by nsSNPs was investigated for the ranked genes by two software programs, SIFT [26] and PolyPhen-2 [27].

insulin signaling [33]. *Grb14* knockout mice are born normally, show a small reduction in body weight and exhibit improved glucose homeostasis and enhanced insulin signaling in the liver and skeletal muscle [34]. Judging from the phenotypic similarity between *Pbwg1.12* and knockout mice, *Gcg* is very likely to become a candidate gene for *Pbwg1.12*, although further studies such as pancreatic islet characterization and *Gcg* expression analysis will be needed in our subcongenic mice.

For the obesity *Pbwg1.5* QTL, *Ly75* and *Itgb6* were suggested to be putative candidate genes in this study. *Ly75* encodes DEC-205, a 205-kD integral membrane protein homologous to the macrophage mannose receptor, and DEC-205 is a novel endocytic receptor used by dendritic cells and thymic epithelial cells to direct captured antigens from the extracellular space to a specialized antigen-processing compartment [35]. *Ly75* knockout mice exhibit abnormalities in CD8-positive T cell morphology and cytotoxic T cell physiology [36]. *Itgb6* encodes the integrin β6 subunit, a member of the integrin family. This subunit heterodimerizes with the αv subunit to bind and/or activate latent transforming growth factor β. The expression of αvβ6 integrin is largely restricted to a subset of epithelial cells [37,38]. *Itgb6* knockout mice are born and grow normally but exhibit juvenile baldness associated with macrophage infiltration of the skin and accumulation of activated lymphocytes around conducting airways in the lungs, suggesting that alterations in this integrin may contribute to the development of inflammatory diseases of epithelial organs including the skin, lungs and kidney [39]. A previous microarray analysis revealed that 259 genes are differentially expressed in the liver between SM/J and LG/J mouse strains fed a high-fat diet, where SM/J is more responsive than LG/J for many obesity and diabetes traits. Most of these genes are associated with immune function, and 62 genes are located within intervals of QTLs previously mapped for obesity, diabetes and related traits [40]. High-fat diets are known to trigger an immune response through inflammation in many organs and tissues such as the liver and adipose tissue [41,42]. Hence, the genes associated with immune function can become candidate genes for obesity and related QTLs. Therefore, the *Ly75* and *Itgb6* genes with immune function may be good candidate genes for our *Pbwg1.5* that shows prevention of obesity when mice are fed both low-fat standard and high-fat diets [21].

The *Itgb6* gene on the *Pbwg1.5* region derived from a wild *castaneus* mouse caught in the Philippines had the nsSNP of g.2:60491216A>C, leading to amino acid substitution of Ser302Ala that was predicted to be harmful to protein function by SIFT but not by PolyPhen-2. In fact, the Ser residue is conserved among many mammals including humans, dogs, bovines, horses and rats [43]. In mice, both Ser and Ala residues are segregating among common inbred strains and also among wild-derived inbred strains [5]. It is noteworthy that the CAST/EiJ strain established from wild *castaneus* mice in Thailand has the same base substitution (C: Ala) as that of our wild *castaneus* mouse in the Philippines. In addition, QTLs for obesity and related traits identified from CAST/EiJ and other strains have so far not been fine-mapped to the *Pbwg1.5* region on mouse chromosome 2, as discussed earlier. These facts thus suggest that the A>C nsSNP might not act as a sequence variant causing our phenotypic variation.

Next-generation sequencing of 13 classical inbred mouse strains and four wild-derived inbred strains has recently revealed that QTLs with small effects on 100 phenotypes of disease and physiological traits, which were identified in more than 2,000 heterogeneous stock mice, are more likely to arise from intergenic sequence variants lying outside genes and are less likely to arise from nsSNPs and structural variants (indels, inversions, copy number gains and others) lying within genes. In contrast, it has

been shown that QTLs with large effects are more likely to arise from structural variants and are less likely to arise from intergenic variants [23,44]. We therefore consider that, since our QTLs have small effects on growth and obesity, their causative variants may be intergenic variants rather than nsSNPs and structural variants identified by exome sequencing in this study. As the next step, we will need to perform expression analysis of the four putative candidate genes searched in this study. Gene expression results will provide information helpful for identifying causative genes and further causative variants underlying our QTLs on chromosome 2.

In conclusion, by analysis using intersubspecific subcongenic intercrosses, we precisely fine-mapped three unique QTLs for enhanced growth, prevention of obesity and increased body length, which were discovered from a wild *M. m. castaneus* mouse, to small genomic intervals ranging from 2.1 to 8.0 Mb on mouse chromosome 2. By combined analysis of exome sequencing and bioinformatics, we identified four genes as putative candidate genes for the unique growth and obesity QTLs. We furthermore predicted that nsSNPs found in the candidate genes would not be harmful to protein functions.

## Supporting Information

**Figure S1  Two PCR-RFLP markers on mouse chromosome 2 developed in this study.** (A) The *rs13476521* PCR-RFLP marker was constructed on the basis of the *rs13476521* SNP located at 58,131,026 bp on the *Cytip* gene. B6 has the nucleotide base T and our exome sequencing revealed that our wild *castaneus* mouse has the base C being the same as that of CAST/EiJ. A pair of primers, 5′-CCTGGGGGAATGGA-TAAAGT-3′ and CCTGACTCGGACACTGGAAT, amplified a 364-bp fragment including this SNP. The restriction enzyme *EcoRV* cut the 364-bp fragment derived from B6 in two (195 and 169 bp), whereas it did not cut the 364-bp fragment derived from the wild mouse. (B) The *rs48690987* PCR-RFLP marker was developed on the basis of the *rs48690987* SNP at 62,606,356 bp on the *Ifih1* gene. B6 has the nucleotide base T, whereas our wild mouse has the base C being the same as that of CAST/EiJ. A pair of primers, AAATTCATCCGTTTCGTCCA and GGA-TAGTTTTCTGCCCTTTGC, amplified a 306-bp fragment. The enzyme *EcoT22I* generated two B6-derived fragments (160 and 146 bp), whereas it did not cut a wild-derived fragment. PCR was performed as described previously [19], and 2.0–2.5% agarose gels were used for electrophoresis.

**Table S1  Training genes used in prioritization of candidate genes.**

**Table S2  Additive and dominance diplotype effects for body weight and weight gain in the F₂ populations obtained from B6×SR2 and B6×SR12 intercrosses.**

**Table S3  Organ weight not adjusted for body weight at 14 weeks of age, and additive and dominance diplotype effects for organ weight in the F₂ populations obtained from B6×SR1 and B6×SR21 intercrosses.**

**Table S4  Measurements of body composition traits not adjusted for body weight at 14 weeks of age, and additive and dominance diplotype effects for body composition traits in the F₂ populations obtained from B6×SR2 and B6×SR12 intercrosses.**

**Table S5 Organ weight adjusted for body weight at 14 weeks of age, and additive and dominance diplotype effects for organ weight in the $F_2$ populations obtained from B6×SR1 and B6×SR21 intercrosses.**

**Table S6 Measurements of body composition traits adjusted for body weight at 14 weeks of age, and additive and dominance diplotype effects for body composition traits in the $F_2$ populations obtained from B6×SR2 and B6×SR12 intercrosses.**

**Table S7 Summary of variants identified by exome sequencing of 153 genes located on the original congenic region between *D2Mit33* and *D2Mit38* on mouse chromosome 2.**

## Author Contributions

Conceived and designed the experiments: AI. Performed the experiments: AI SO. Analyzed the data: AI. Contributed reagents/materials/analysis tools: AI. Wrote the paper: AI.

## References

1. Eisen EJ (2005) The mouse in animal genetics and breeding research. London: Imperial College Press. 364 p.
2. Baéza E, Bihan-Duval EL (2013) Chicken lines divergent for low or high abdominal fat deposition: a relevant model to study the regulation of energy metabolism. Animal 7: 965–973.
3. Fall T, Hägg S, Mägi R, Ploner A, Fischer K, et al. (2013) The role of adiposity in cardiometabolic traits: a Mendelian randomization analysis. PLoS Med 10(6): e1001474.
4. Flint J, Mackay TFC (2009) Genetic architecture of quantitative traits in mice, flies, and humans. Genom Res 19: 723–733.
5. Eppig JT, Blake JA, Bult CJ, Kadin JA, Richardson JE, the Mouse Genome Database Group (2012) The Mouse Genome Database (MGD): comprehensive resource for genetics and genomics of the laboratory mouse. Nucleic Acids Res 40: D881–886.
6. Darvasi A, Soller M (1997) A simple method to calculate resolving power and confidence interval of QTL map location. Behav Genet 27:125–132.
7. Darvasi A (1997) Interval-specific congenic strains (ISCS): an experimental design for mapping a QTL into a 1-centimorgan interval. Mamm Genome 8:163–167.
8. Drinkwater NR, Gould MN (2012) The long path from QTL to gene. PLoS Genet 8: e1002975.
9. Mollah MBR, Ishikawa A (2011) Intersubspecific subcongenic mouse strain analysis reveals closely linked QTLs with opposite effects on body weight. Mamm Genome 22: 282–289.
10. Farber CR, Medrano JF (2007) Dissection of a genetically complex cluster of growth and obesity QTLs on mouse chromosome 2 using subcongenic intercrosses. Mamm Genome 18: 635–645.
11. Cheverud JM, Hager R, Roseman C, Fawcett G, Wang B, Wolf JB (2008) Genomic imprinting effects on adult body composition in mice. PNAS 105: 4253–4258.
12. Jarvis JP, Kenney-Hunt J, Ehrich TH, Pletscher LS, Semenkovich CF, Cheverud JM (2005) Maternal genotype affects adult offspring lipid, obesity, and diabetes phenotypes in LG×SM recombinant inbred strains. J Lipid Res 46: 1692–1702.
13. Kilpinen H, Waszak SM, Gschwind AR, Raghav SK, Witwicki RM, et al. (2013) Coordinated effects of sequence variation on DNA binding, chromatin structure, and transcription. Science 342: 744–747.
14. McVicker G, van de Geijn B, Degner JF, Cain CE, Banovich NE, et al. (2013) Identification of genetic variants that affect histone modifications in human cells. Sicence 342: 747–749.
15. Kasowski M, Kyriazopoulou-Panagiotopoulou S, Grubert F, Zaugg JB, Kundaje A, et al. (2013) Extensive variation in chromatin states across humans. Science 342: 750–752.
16. Ishikawa A, Matsuda Y, Namikawa T (2000) Detection of quantitative trait loci for body weight at 10 weeks from Philippine wild mice. Mamm Genome 11: 824–830.
17. Ishikawa A, Namikawa T (2004) Mapping major quantitative trait loci for postnatal growth in an intersubspecific backcross between C57BL/6J and Philippine wild mice by using principal component analysis. Genes Genet Syst 79: 27–39.
18. Ishikawa A, Hatada S, Nagamine Y, Namikawa T (2005) Further mapping of quantitative trait loci for postnatal growth in an intersubspecific backcross of wild *Mus musculus castaneus* and C57BL/6J mice. Genet Res 85: 127–137.
19. Ishikawa A, Kim E-H, Bolor H, Mollah MBR, Namikawa T (2007) A growth QTL (*Pbwg1*) region of mouse chromosome 2 contains closely linked loci affecting growth and body composition. Mamm Genome 18: 229–239.
20. Ishikawa A (2009) Mapping an overdominant quantitative trait locus for heterosis of body weight in mice. J Hered 100: 501–504.
21. Mollah MBR, Ishikawa A (2010) A wild derived quantitative trait locus on mouse chromosome 2 prevents obesity. BMC Genet 11: 84.
22. Mollah MBR, Ishikawa A (2013) Fine mapping of quantitative trait loci affecting organ weights by mouse intersubspecific subcongenic strain analysis. Anim Sci J 84: 296–302.

23. Keane TM, Goodstadt L, Danecek P, White MA, Wong K, et al. (2011) Mouse genomic variation and its effect on phenotypes and gene regulation. Nature 477: 289–294.
24. Falconer DS, Mackay TF (1996) Introduction to quantitative genetics, fourth ed. Harlow, Essex, UK: Longmans Green. 464 p.
25. Tranchevent L-C, Barriot R, Yu S, Vooren SV, Loo PV, et al. (2008) Endeavour update: a web resource for gene prioritization in multiple speacies. Nucleic Acids Res 36: W377–W384.
26. Kumar P, Henikoff S, Ng PC (2009) Predicting the effects of coding non-synonymous variants on protein function using the SIFT algorithm. Nat Protocols 4: 1073–1082.
27. Adzhubei IA, Schmidt S, Peshkin L, Ramensky VE, Gerasimova A, et al. (2010) A method and server for predicting damaging missense mutations. Nat Methods 7: 248–249.
28. Robins DM (2005) Androgen receptor and molecular mechanisms of male-specific gene expression. Novartis Found Symp 268: 42–52.
29. Yogosawa S, Mizutani S, Ogawa Y, Izumi T (2013) Activin receptor-like kinase 7 suppresses lipolysis to accumulate fat in obesity through downregulation of peroxisome proliferator-activated receptor γ and C/EBPα. Diabetes 62: 115–123.
30. Hayashi Y (2011) Metabolic impact of glucagon deficiency. Diabetes Obes Metab 13 Suppl: 1151–157.
31. Hayashi Y, Yamamoto M, Mizoguchi H, Watanabe C, Ito R, et al. (2009) Mice deficient for glucagon gene-derived peptides display normoglycemia and hyperplasia of islet α-cells but not of intestinal L-cells. Mol Endocrinol 23: 1990–1999.
32. Mighiu PI, Yue JTY, Filippi BM, Abraham MA, Chari M, et al. (2013) Hypothalamic glucagon signaling inhibits hepatic glucose production. Nat Med 19: 766–772.
33. Cariou B, Bereziat V, Moncoq K, Kasus-Jacobi A, Perdereau D, et al. (2004) Regulation and functional roles of Grb14. Front Biosci 9: 1626–1636.
34. Cooney GJ, Lyons RJ, Crew AJ, Jensen TE, Molero JC, et al. (2004) Improved glucose homeostasis and enhanced insulin signalling in Grb14-deficient mice. EMBO J 23: 582–593.
35. Jiang W, Swiggard WJ, Heufler C, Peng M, Mirza A, et al. (1995) The receptor DEC-205 expressed by dendritic cells and thymic epithelial cells is involved in antigen processing. Nature 375: 151–155.
36. Guo M, Gong S, Maric S, Misulovin Z, Pack M, et al. (2000) A monoclonal antibody to the DEC-205 endocytosis receptor on human dendritic cells. Hum Immunol 61: 729–738.
37. Munger JS, Huang X, Kawakatsu H, Griffiths MJ, Dalton SL, et al. (1999) The integrin αvβ6 binds and activates latent TGFβ1: a mechanism for regulating pulmonary inflammation and fibrosis. Cell 96: 319–328.
38. Wang B, Dolinski BM, Kikuchi N, Leone DR, Peters MG, et al. (2007) Role of αvβ6 integrin in acute biliary fibrosis. Hepatology 46: 1404–1412.
39. Huang XZ, Wu JF, Cass D, Erle DJ, Corry D, et al. (1996) Inactivation of the integrin beta 6 subunit gene reveals a role of epithelial integrins in regulating inflammation in the lungs and skin. J Cell Biol 133(4): 921–928.
40. Partridge CG, Fawcett GL, Wang B, Semenkovich CF, Cheverud JM (2014) The effect of dietary fat intake on hepatic gene expression in LG/J and SM/J mice. BMC Genomics 15: 99.
41. Caspar-Bauguil S, Cousin B, Galinier A, Segafredo C, Nibbelink M, et al. (2005) Adipose tissue as an ancestral immune organ: site specific change in obesity. FEBS Lett 579: 3487–3492.
42. Radonjic M, de Haan JR, van Erk MJ, van Dijk KW, van den Berg SAA, et al. (2009) Genome-wide mRNA expression analysis of hepatic adaptation to high-fat diets reveals switch from an inflammatory to steatotic transcriptional program. PLoS One 4: e6646.
43. The UniProt Consortium (2014) Activities at the Universal Protein Resource (UniProt). Nucleic Acids Res 42: D191–D198.
44. Yalcin B, Wong K, Agam A, Goodson M, Keane TM, et al. (2011) Sequence-based characterization of structural variation in the mouse genome. Nature 477: 326–329.

# Comparative Transcriptomic Analysis of the Response to Cold Acclimation in *Eucalyptus dunnii*

Yiqing Liu[1,2]*, Yusong Jiang[2], Jianbin Lan[2], Yong Zou[2], Junping Gao[1]

**1** Department of Ornamental Horticulture, China Agricultural University, Beijing 100193, China, **2** College of Life Science & Forestry, Chongqing University of Art & Science, Yongchuan 402160, China

## Abstract

*Eucalyptus dunnii* is an important macrophanerophyte with high economic value. However, low temperature stress limits its productivity and distribution. To study the cold response mechanisms of *E. dunnii*, 5 cDNA libraries were constructed from mRNA extracted from leaves exposed to cold stress for varying lengths of time and were evaluated by RNA-Seq analysis. The assembly of the Illumina datasets was optimized using various assembly programs and parameters. The final optimized assembly generated 205,325 transcripts with an average length of 1,701 bp and N50 of 2,627 bp, representing 349.38 Mb of the *E. dunnii* transcriptome. Among these transcripts, 134,358 transcripts (65.4%) were annotated in the Nr database. According to the differential analysis results, most transcripts were up-regulated as the cold stress prolonging, suggesting that these transcripts may be involved in the response to cold stress. In addition, the cold-relevant GO categories, such as 'response to stress' and 'translational initiation', were the markedly enriched GO terms. The assembly of the *E. dunnii* gene index and the GO classification performed in this study will serve as useful genomic resources for the genetic improvement of *E. dunnii* and also provide insights into the molecular mechanisms of cold acclimation in *E. dunnii*.

**Editor:** Turgay Unver, Cankiri Karatekin University, Turkey

**Funding:** 1. Natural Science Foundation of Chongqing Province, China (Grant NO. cstc2013jcyjA80035), 2. Key Program for forestry of Chongqing University of Art and Science (Grant NO. 201302), 3. National Science Foundation for Young Scientists of China (Grant NO. 31340016). The funders had no role in study design, data collection and analysis, decision to publish, or preparation of the manuscript.

**Competing Interests:** The authors have declared that no competing interests exist.

* Email: liung906@163.com

## Introduction

Rapid population increase and the consequent increase in the requirement for different types of paper products, as well as the emphasis on paper as an environmentally friendly packaging material, have led to an increased demand for wood [1]. The imbalance between the supply and demand for forest products is growing. Eucalyptus is an economically important forest tree that grows in tropical and subtropical regions [2,3]. Eucalyptus trees can be highly productive over a short rotation period, tolerate a wide range of soils and commonly exhibit a straight stem form in those species utilized in production forestry. Furthermore, eucalypts, unlike many trees, do not have a true dormant period and retain their foliage, which enables growth during warm winter periods [4]. Nevertheless, in Eucalyptus plantations, low temperature stress limits their productivity and distribution. When the temperature drops to 8°C or below, Eucalyptus trees would exhibit various symptoms of cold injury due to their inability to adapt to the low temperature [5]. Cold stress also alters the physiological status, such as transient increases in hormone levels (e.g., ABA), changes in the membrane lipid composition, accumulates of compatible osmolytes (such as soluble sugars, betaine, and proline) and increases in antioxidant levels [6,7]. In contrast, temperate plants can withstand freezing temperatures following a period of low, but non-freezing temperatures, a process called cold acclimation. The mechanisms of cold acclimation have been extensively investigated in *Arabidopsis thaliana* [8] and other important crop species such as maize and barley [9,10]. Cold stress has been shown to induce changes in physiology and gene expression, and hundreds of cold-responsive genes have been identified so far [11]. However, in tropical and subtropical plants, especially *E. dunnii*, the molecular mechanisms of the cold response are not clear.

The physiological and biochemical changes that occur during plant cold acclimation result primarily from changes in the expression of cold-responsive genes. In general, Cold-responsive genes could be classified into two groups: 1) functional proteins, which directly protect plants against environmental stresses, and 2) regulatory proteins, which regulate the expression of downstream target genes in the stress response [4]. The first group mainly comprises enzymes involved in the biosynthesis of various osmoprotectants, such as late embryo genes is abundant (LEA) proteins, antifreeze proteins, chaperones, and detoxification enzymes[8,12]. The second group mainly includes transcription factors and protein kinases [12]. The best-characterized transcription factors (TFs) involved in the plant cold response are the class of AP2/ERF (APETALA2/ethylene-responsive element binding proteins), one kind of subfamily was known as CBF/DREB(C-repeat binding factor/dehydration resistance element binding protein), which regulate cold-responsive gene expression by binding to DRE/CRT cis-elements in the promoter region of cold-responsive genes [6,13]. Changes in the expression of cold-responsive contribute to the differences in plant cold tolerance. For example, *Solanum commersonii* and *S. tuberosum*, which are closely related species that differ in their cold acclimation abilities, exhibit considerable differences in the expression levels of cold-responsive genes [6,14]. Chen *et al*. found that the activities of some detoxification enzymes, such as catalase (CAT), superoxide dismutase (SOD), peroxidase (POD) and esterase (EST) are increased in response to cold stress, whereas the plant's metabolic activity is decreased [15–17]. Some cold-induced genes have been cloned from Eucalyptus plants. For example, four CBF paralogs were previously isolated from *E. gunnii*, and qRT-PCR analysis demonstrated that they exhibited complementary expression profiles in a range of natural standard and cold conditions [18]. Navarro *et al*.found overexpression of *EguCBF1a* or *EguCBF1b* in the cold-sensitive *E. urophylla·E. grandis* hybrid could enhance its freezing tolerance [19].

Given the importance of cold-responsive genes in plant cold tolerance, studying the cold response at the transcription level may be a key step in identifying specific tolerance mechanisms. Next generation sequencing (NGS) provides a high throughput approach for analyzing genes involved a particular process at transcription level. Compared to the traditional sequencing techniques, NGS is more robust and demonstrates greater resolution and inter-lab portability compared to several microarray platforms. NGS could detect millions of transcripts and is beneficial to explore new genes and their expression profiling independent of a reference genome [6,20,21]. For example, cDNA libraries for *E. gunnii* have been constructed to identify genes involved in cell protection (such as PCP, Lti6b and metallothionein), LEA/dehydrin accumulation, and cryoprotection [22,23]. Despite its obvious potential, these next generation sequencing methods have not been applied for *E. dunnii* yet.

The goal of this study was to construct a comprehensive transcriptome to investigate the molecular mechanism of cold tolerance in *E. dunnii*. The plants were exposed to low temperature (4°C) for 0, 3, 6, 12, and 24 h, and the first two expanded leaves below apical bud of *E. dunnii* were collected for high throughput RNA-Seq analysis. Paired-end (PE) reads from the RNA-Seq output were then assembled *de novo* to build an *E. dunnii* transcriptome, which was subjected to a comparative analysis. This analysis provides preliminary global insight into the molecular mechanism of cold tolerance and a good base for future basic research in *E. dunnii*.

## Results

### Physiological changes in *E. dunnii* in response to cold stress

Firstly, we detected the concentration of proline during the cold treatment (at 4°C) from 0 to 48 h. The concentration of proline decreased slightly from 0 to 3 h, but it increased rapidly as the cold stress prolonging (Fig. 1A). The decrease in proline content at 0 to 3 h might be caused by a transient stress response of *E. dunnii* to the low temperature shock. However, prolonged exposure to low temperature (24 h) resulted in proline accumulation.

Plant cells could accumulate amounts of reactive oxygen species under environmental stress, which result in severe damage of proteins, membrane lipid, DNA and other cellular components [15]. CAT could catalyze the decomposition of hydrogen peroxide to water and oxygen, and it is important in protecting the cell from oxidative damage by reactive oxygen species (ROS) [16–17]. The activity level of CAT changed during cold acclimation in *E. dunnii*. We observed an almost 25% increase in CAT activity after 3 h, and a nearly two-fold change after 24 h of cold stress

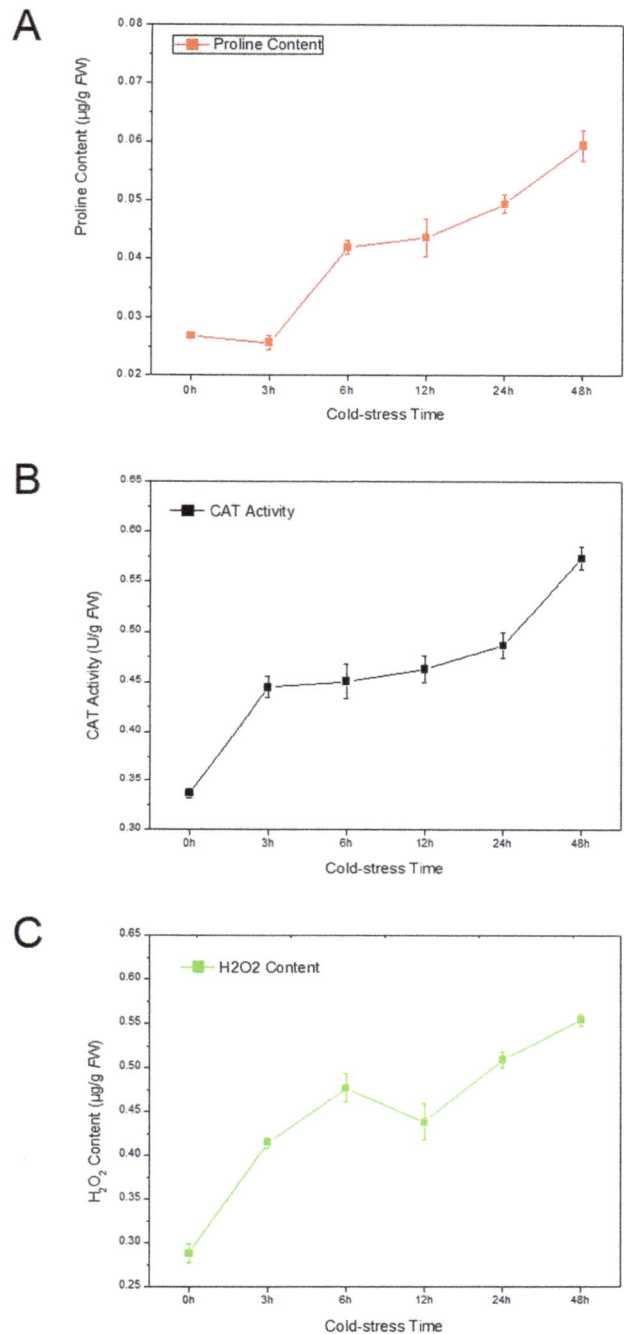

**Figure 1. Changes in proline content (A), CAT activity (B) and $H_2O_2$ content (C) under low temperature (4°C) treatment over time.**

(Fig. 1B). $H_2O_2$ is one kind of ROS molecule. The $H_2O_2$ concentration increased nearly 50% after 3 h, and then continued to increase at a more moderate rate, remaining at high levels until 24 h (Fig. 1C). These results indicated that *E. dunnii* plants are sensitive to the cold stress.

**Table 1.** Total number of reads for each treatment sample, as obtained by Illumina sequencing.

| Duration of low temperature (4°C) treatment | Paired-end reads | Total length | Total number of contigs | Average length | N50 of contigs | Alignment rate (%) |
|---|---|---|---|---|---|---|
| 0 h | 25,407,247 | 120,616,917 | 118,761 | 1,015 | 1,817 | 90.3 |
| 3 h | 24,817,373 | 179,471,852 | 149,467 | 1,200 | 2,136 | 91.7 |
| 6 h | 25,703,824 | 215,673,491 | 161,627 | 1,334 | 2,343 | 90.6 |
| 12 h | 34,870,702 | 302,614,952 | 195,733 | 1,546 | 2,630 | 92.4 |
| 24 h | 33,846,411 | 217,801,712 | 160,461 | 1,357 | 2,367 | 91.2 |
| TOTAL | – | 349,381,021 | 205,325 | 1,701 | 2,827 | 94.5 |

## RNA Sequencing, *de novo* assembly and functional annotation

To study the *E. dunnii* transcriptome in response to cold stress, we transferred plantlets with 10 leaves to a climate-chamber (4°C) and collected the first two expand leaves below apical bud at 0, 3, 6, 12, and 24 h time points, respectively. For the RNA-Seq analysis, we obtained 25,407,247, 24,817,373, 25,703,824, 34,870,702, and 33,846,411 clean paired-end reads, respectively (data not shown).

To obtain a more reliable and comprehensive transcriptome database, these five libraries were pooled together and then performed the *de novo* assembly. The pipeline for the bioinformatics analysis of the RNA-Seq data is shown in Fig. 2. The parameters of the contig databases assembled by each individual assembler, such as the alignment rate, sensitivity, accuracy and length distribution, were significantly different. Overall, the contig database produced by Trinity was significantly better than those from the other assemblers (Table S1). The optimal contig database contained 205,325 contigs ≥300 bp in length. The average length of these contigs was 1,701.6 bp, the N50 number was 2,827 bp, and the maximum length was 15,965 bp (Table 1). Additionally, there were 148,151 contigs with a length≥600 bp, 105,494 contigs with a length ≥1,200 bp, and 33,700 contigs with a length ≥ 3,000 bp (Fig. 3). The assembled contigs (≥300 bp) were deposited in the NCBI Transcriptome Shotgun Assembly (TSA) database under the accession number PRJNA208093.

Sequence similarity search against the NCBI non-redundant protein database (NR) was conducted using a locally installed BLAST program for functional annotation. Among all the assembled contigs (≥300 bp), 134,358 (65.4%) were annotated with BLASTx hits, matching 80,578 unique protein accessions (Table S2). For contigs longer than 600 bp, 80.9% had BLASTx hits, and for longer than 900 bp, the percentage increased to 88.8% (Fig. 3), indicating that most contigs, particularly the longer contigs, represent protein-encoding transcripts. As the completed genome information of *E. dunnii* was not available at this time, 70,967 contigs (34.6%) had no hits to any known proteins in the Nr database (Fig. 4A), suggesting that these contigs might be non-coding regions or potentially new genes [24].

In addition, among 134,358 contigs with BLASTx results, 52,265 (38.9%), 26,602 (19.8%), and 24,721 (18.4%) showed high sequence similarity to *Vitis vinifera*, *Populus* and *Glycine max*, respectively, but only 265 contigs (0.2%) shared homology with Eucalyptus (Fig. 4B). Alternatively, our results could indicate that *E. dunnii* is more closely to *V. vinifera* than *G. max* or *Arabidopsis* evolutionarily. Interestingly, some other plant transcriptomes, such as *Craterostigma plantagineum* [25] and *Fraxinus* spp. [26],

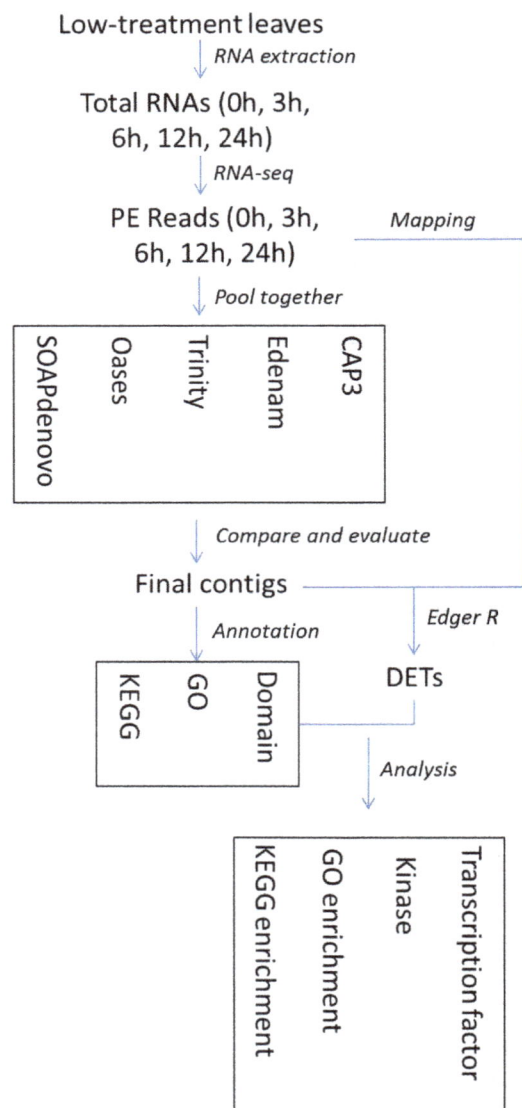

**Figure 2. The pipeline for the bioinformatics analysis of the deep sequencing data.**

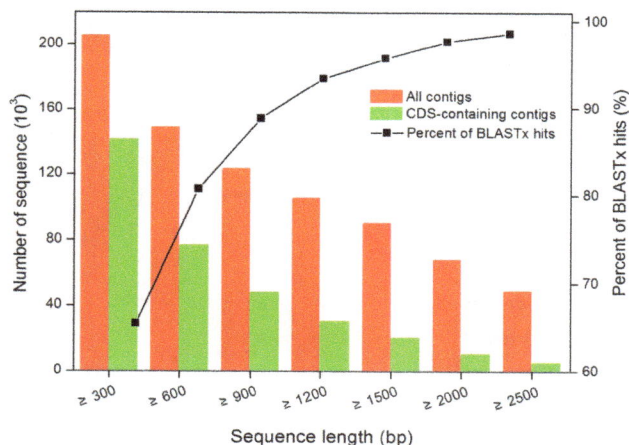

**Figure 3. Annotation rate and proportion of long-CDS-containing sequences.** A total of 205,325 contigs were used for the BLASTx search. The contig length is indicated on the X-axis. The size distributions of the final assembled contigs (red) and the number of long-CDS-containing contigs (green) are indicated on the left Y-axis. The percentage of BLASTx hits to size-grouped contigs is indicated by the diamond symbol.

display the same distribution pattern of top-hit species. These results could be simply explained by the number of genes deposited in Nr database. For example, by November 2013, the NCBI database contained 78,045 *V. vinifera* transcripts, 11,4590 *P. trichocarpa* transcripts, and 81,270 *G. max* transcripts, but only 7,146 Eucalyptus transcripts.

## Differential expression between the groups and qRT-PCR validation

To characterize the digital gene expression profiles of the *E. dunnii* in response to low temperature, we performed a short-read alignment of each library using Perl script provided by the Trinity software package. For samples treated at 4°C for 0, 3, 6, 12, and 24 h, a total of 90.3%, 91.6%, 92.1%, 91.2%, and 91.5% of the reads could be aligned back to the contigs, and 64.2%, 63.8%, 65.1%, 62.7% and 61.5% aligned concordantly exactly once. To eliminate the effect of library size, edgeR (empirical analysis of digital gene expression in R) was used to create an effective library size for each sample. The number of aligned reads per transcript was normalized to FPKM based on an RESM-based algorithm. Differentially expressed transcripts (DETs) with FDR $\leq 0.05$ and $\log_2$ fold-change ($\log_2$FC) $\geq 1$ between pairs of samples were identified by edgeR [27]. The edgeR analysis generated 10 DET sets (0 h vs. 3 h, 0 h vs. 6 h, 0 h vs. 12 h, 0 h vs. 24, 3 h vs. 6 h, 3 h vs. 12 h, 3 h vs. 24 h, 6 h vs. 12 h, 6 h vs. 24 h, and 12 h vs. 24 h) with 11,395, 11,908, 11,901, 12,671, 8,935, 10,641, 11,843, 9,531, 11,489, and 10,230 DETs, respectively (Fig. 5). We also found that 7,059, 7,348, 7,479, and 7,636 DETs were up-regulated in the 0 h vs. 3 h, 0 h vs. 6 h, 0 h vs. 12 h, and 0 h vs. 24 h comparison sets, respectively. These results demonstrated that the number of up-regulated DETs increased as the duration of cold stress prolonged.

To validate the expression patterns of each DET obtained from the comparative RNA-Seq studies, we randomly selected 31 transcripts from the annotated DETs for qRT-PCR analysis. Noteworthy, qRT-PCR results are often affected by the choice of reference genes. Previously, a report explored the expression stability of reference genes which are using in gene expression test in Eucalyptus in response to various abiotic stresses by qRT-PCR [28]. The authors found that expression of some genes, such as *PP2A-3/SAND*, *UPL7*, *UBC2* and *GAPDH*, are stable enough in all tested samples, while *ACT2* gene was not stable in response to environmental stimuli as expected. As mentioned in the paper of Cassan-Wang et al., GAPDH is a good choice as reference gene in qRT-PCR assay in Eucalyptus [28]. Therefore, we selected two most commonly used reference genes, beta-actin and GAPDH,- because these two genes could be mutual support, mutual correction, and minimize the experimental errors. The results showed that the expression patterns of 25 DETs were compatible with the RNA-Seq analysis (Table S3), suggesting that the differential expression analysis based on high-throughput RNA sequencing produced reliable expression data.

## Gene ontology (GO) and Kyoto Encyclopedia of Genes and Genomes (KEGG) enrichment analysis of DETs

GO (Gene Ontology) and KEGG (Kyoto Encyclopedia of Genes and Genomes) annotation were applied to the BLASTx results to provide comprehensive functional information for each transcript. In total, we obtained 198,528 GO annotations for 62,965 transcripts and 966 unique Enzyme Codes (ECs) for 28,295 transcripts (Table S2). Among the 62,965 transcripts with GO terms, 34,064 (54.1%) were assigned to the Biological process category, 19,959 (31.7%) to the Molecular function category, and 28,965 (46.0%) to the Cellular component category. In addition, 20,025 (31.8%) unique transcripts were assigned GO terms from all three categories (Fig. 6 and Table S4). To understand the mechanism of the cold stress response in *E. dunnii*, the DETs were subjected to GO and KEGG enrichment analysis. Under the GO category 'Biological process', the 'response to stress' and 'translational initiation' were the most highly enriched terms, with P.ad-values of 0 and 0.02, respectively. Under the category 'Molecular function', the 'quinolinate synthetase A activity' were the most highly enriched term, with a P.ad-values of 0.04. Under the category 'Cellular component', the 'cell part' was the most highly enriched term, with a P-value of 8.5E-11 (Table S4). KEGG analysis identified 27,688 contigs with pathway information were involved in 137 KEGG pathways. Among these 137 KEGG pathways, 'arginine and proline metabolism' and 'tropane, piperidine and pyridine alkaloid biosynthesis' were the two most significantly enriched KEGG pathways (Table S5).

## Cold-responsive transcription factors and protein kinases in *E. dunnii*

Transcription factors and protein kinases are crucial upstream regulators that respond to various biotic and abiotic stresses in plants [29,30]. In this study, we identified a total of 586 contigs involving in transcription factor activity, which were classified into 65 types of transcription factors, including AP2, bZIP, JmjC, and SRF-TF. In order to verify the expression pattern of these transcription factors, an additional 5 transcripts were selected to carry out qRT-PCR analysis. The results displayed that the expression trend of these transcripts agreed with the results of RNA-seq analysis (Table S3). The START and bzip domains transcription factor families were the largest groups represented in the cold-responsive transcription factors, containing 64 (39 up-regulated and 25 down-regulated) and 62 (43 up-regulated and 19 down-regulated) unique transcripts, respectively. The next largest groups were the UDF (31 up-regulated and 24 down-regulated), AP2 (27 up-regulated and 24 down-regulated) and Sigma70 (24 up-regulated and 9 down-regulated) families (Table S6). In

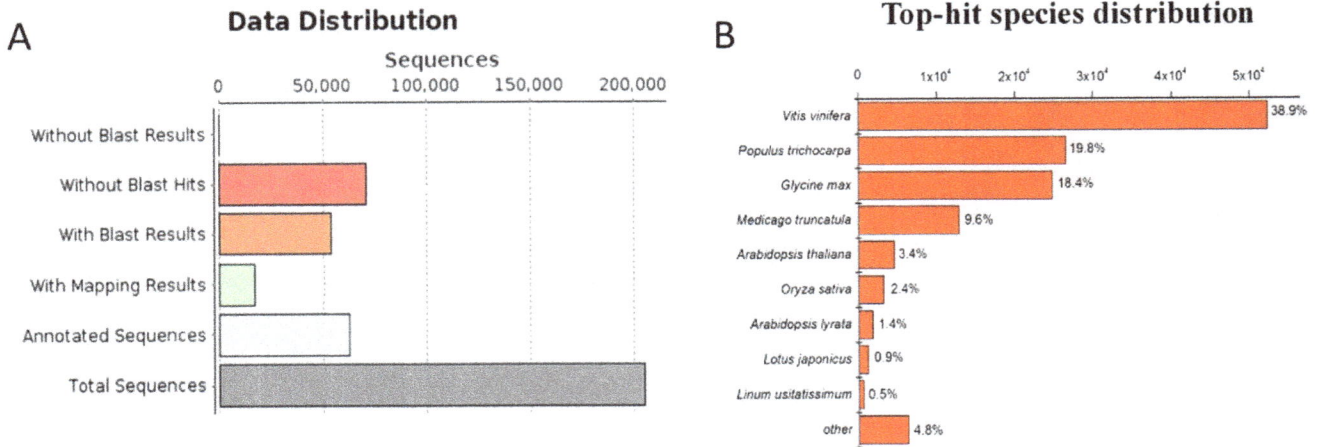

**Figure 4. Distribution of the BLASTx results (A) and the top-hit species distribution of the *E. dunnii* transcriptome (B).** A total of 205,325 contigs ≥300 bp in length were used for the sequence similarity searches, and 134,358 contigs produced BLASTx results. All of the contigs with BLASTx results were used for the species distribution analysis. Overall, 52,265 (38.9%), 26,602 (19.8%) and 24,721 (18.4%) contigs showed strong similarity to *Vitis vinifera*, *Populus* and *Glycine max*, but only 265 contigs (0.2%) shared the highest homology with Eucalyptus.

addition, we identified 169 contigs related to protein kinase activity, which were classified into 8 types of protein kinases based on their domains (Table S6).

## Discussion

### Improving *de novo* transcriptome assembly

The most critical step of an RNA-Seq study is the *de novo* assembly, especially for species without genome information [31–

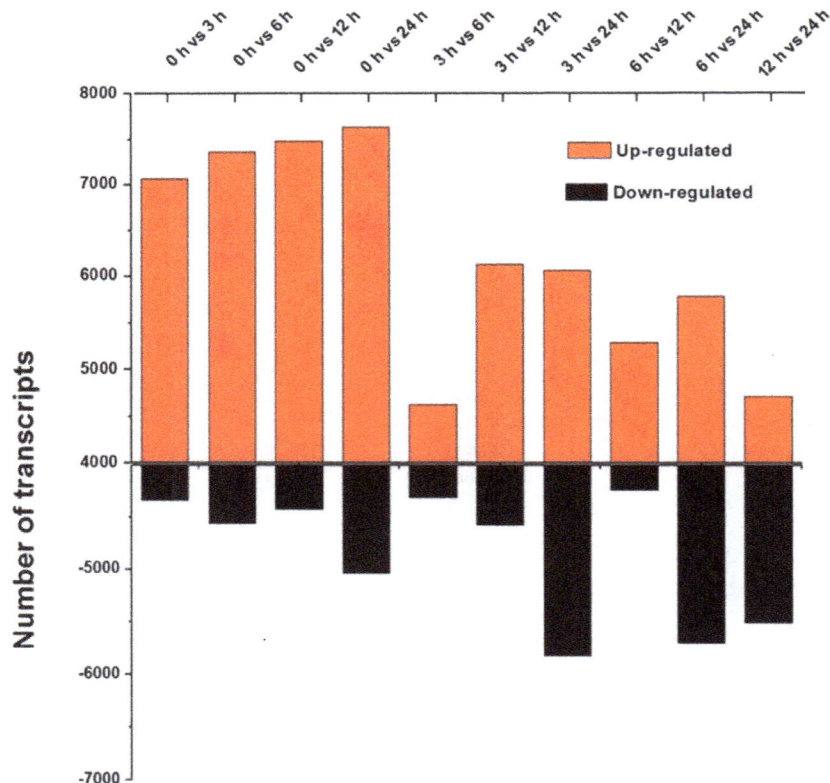

**Figure 5. Transcripts that exhibited differential expression pattern.** In total, 20,5325 contigs were used for the differential expression analysis, and the differential transcripts were identified by edgeR using the following parameters: FDR ≤0.05 and $\log_2$fold-change ≥1.

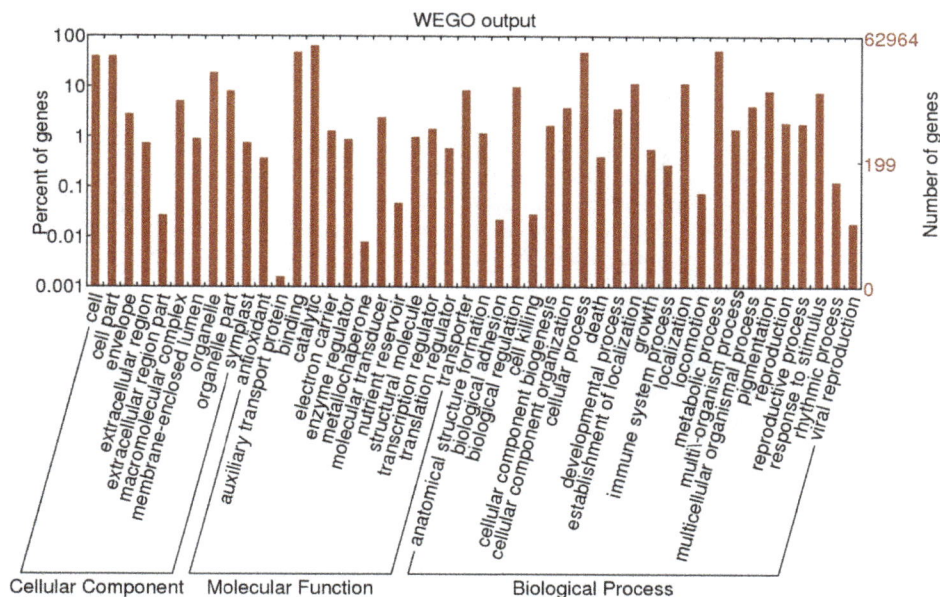

**Figure 6. GO assignment of all contigs in the *E. dunnii* transcriptome.** The contigs mapped to three main categories: Biological process, Cellular component and Molecular function. The right-hand y-axis indicates the number of annotated contigs.

34]. More and more genomes and/or transcriptomes sequences have been completed due to the development of high-throughput sequencing technologies. However, major published studies on transcriptome *de novo* assembly have typically used a single assembly program [35–37]. In this study, we compared the quality of 5 assemblers (Trinity, Osease, SOAP*denovo*, Edena, and Cap3) and then used the optimal combined strategy to construct the *E. dunnii* transcriptome database. When the reads were assembled using Trinity, Cap3, Edena, Oases, and SOAP *de novo*, the N50 (contig ≥300 bp) values were 2,827 bp, 2,551 bp, 1,368 bp, 1,838 bp, and 1,336 bp, respectively (Table S1). Although the accuracy and sensitivity of the contigs assembled by Trinity were the highest compared to the other assemblers, the assembly strategy still needs further optimization to obtain higher accuracy and sensitivity (Table S1). Different assembly software programs used different algorithms, such as the traditional OLC approach of the Edena assembler and the de Bruijn graph approach of the Oases and SOAP *de novo* assemblers [38]. For a particular species, these different algorithms have multiple advantages and disadvantages, which should be taken into account when selecting the most suitable assembler to complete the process of *de novo* assembly in different species. However, neither Trinity nor any other assembler is individually capable of assembling the results satisfactorily. When assembling the sweet potato transcriptome, Tao *et al.* [31] found that only 80% of the reads mapped back to contigs assembled by Trinity, implying that approximately 20% of the reads were not used effectively in the assembly process. In addition, sequencing quality, which is the foundation for obtaining an ideal assembly, should be improved. Xiao *et al.* [38] found that trimming all raw read sequences at the 3′-end and merging the assemblies from different assemblers significantly improved assembly outcome. Some researchers have also suggested that combining data produced by two or more sequencing methods, such as Illumina sequencing and 454 sequencing, could generate a more satisfactory assembly [39]. Combined assemblies use different assembling software and/or different assembling param-

eters, which means they benefit from the advantages of different software packages.

To date, there have been no standard criteria to evaluate the quality of transcriptome assemblies. Researchers appraise the quality of an assembly mainly by examining the data distribution of the assembly [40,41]. Besides the data distribution, we assessed the assembly quality using numerous metrics. Due to the lack of genomic resources for *E. dunnii*, we downloaded Eucalyptus genes with full-length from GenBank to use as reference sequences. The overlapping high-scoring segment pairs (HSPs) were only calculated once to determine the sensitivity. For each individual assembly, Trinity achieved higher sensitivity than Cap3, Oases and SOAP *de novo*. However, the final assembly generated by Edenam exhibited the highest sensitivity, which was slightly higher than that of Trinity (Table S1). To calculate the accuracy, we considered all unmatched components to be false positives, and Trinity exhibited a greatest accuracy. Taken together, the results from the above metrics indicate that our final assembly quality is optimal.

## DETs involving in proline metabolism and quinoline alkaloid biosynthesis

Free proline in plant cells can significantly improve cold resistance [42], as it acts as a type of osmotic adjuster that can reduce the cell freezing point and stabilize intracellular water. Furthermore, free proline can also protect the cell from excessive dehydration and lipid peroxidation [42]. The accumulation of proline is frequently associated with whole plant tolerance to chilling and other stresses [6]. In this study, we observed that the free proline content was increased more than two-fold after 48 h cold treatment (Fig. 1), which were consistent with that accumulation pattern in Arabidopsis and cassava [6,43].

KEGG analysis showed that the 'arginine and proline metabolism' pathway was significantly enriched (Table S5) in *E. dunnii* during cold-stress. A total of 576 transcripts were involved in the 'arginine and proline metabolism' pathway, with 79 transcripts being up-regulated in response to cold stress at 24 h (Tables S6, S2).

In higher plants, proline can be synthesized via the glutamate (Glu) pathway or the ornithine (Orn) pathway, depending on the initial substrate [44,45]. P5CS (delta 1-pyrroline-5-carboxylate synthetase), a key enzyme in the Glu pathway, functions as a bifunctional enzyme to transform Glu to GSA [46]. The accumulation of free proline could improve the ability of stress resistance in many plants, which regulated by the expression of *p5cs* [47,48]. In *E. dunnii*, 3 transcripts were annotated as *p5cs*, and all three transcripts were up-regulated, particularly contig_6788, whose expression increased more than 10-fold when the plants were exposure to low temperature (Fig. 7 and Table S2). This transcriptome result correlated well with the change in free proline content, suggesting that at 4°C, the Glu pathway was activated to increase the free proline content to protect the plant against cold stress. δ-OAT (ornithine-oxo-acid transaminase) is a key enzyme in the Orn pathway that catalyzes the transformation of L-Orn into GSA. Because δ-OAT can catalyze arginine to glutamate, it could be involved in proline synthesis and accumulation [49]. However, we only identified two contigs (contig_6006 and contig_60065) annotated as δ-OATin in the *E. dunnii* transcriptome, and neither was up- or down-regulated in response to cold stress (Fig. 7 and Table S2). Based on the expression profiles of these transcripts, we hypothesize that the Orn pathway may play a less important role than the Glu pathway during cold acclimation or that it may represent an alternative pathway for cold acclimation in *E. dunnii*.

Free proline accumulation is affected not only by the proline biosynthesis pathway but also by the proline degradation pathway. Under normal conditions, free proline functions as a feedback regulator to inhibit *p5cs* expression and concurrently induce ProDH (Proline dehydrogenase) gene expression. In contrast to the normal condition, *p5cs* expression is hyperactive during cold acclimation, whereas ProDH expression is inhibited, resulting in the accumulation of more and more free proline in plant cells. In *Arabidopsis* and other plants, proline levels are mainly determined by balance of biosynthetic and catabolic pathways, controlled by P5CS and ProDH genes, respectively [6]. Nanjo *et al.* found that proline degradation was inhibited in *Arabidopsis* transformed with *At*ProDH [43], suggesting that free proline levels increased in leaves.

Secondary metabolism and its products are also involved in the response to various stresses in plants, representing a process that formed over a long evolutionary period [50–52]. There is some evidence that secondary metabolic products and environmental factors (biotic and abiotic) are closely linked, as in the case of alkaloids, which play an important role in resisting insects and herbivores via chemical defense mechanism [53]. In addition to 'arginine and proline metabolism', the DETs were significantly enriched in 'quinoline alkaloid biosynthesis' pathway during cold acclimation, based on KEGG pathway analysis (Table S5). Early in the cold stress period (0–6 h), 40% of transcripts related to quinoline alkaloid metabolism were up-regulated more than 2-fold compared to the 0 h time point (Table S2), including contig_65006 and contig_65485. This suggests that the up-regulation of transcripts in response to low temperatures may play a crucial role in plant stress tolerance. However, when the duration of cold stress exceeded 6 h, the expression levels of these up-regulated transcripts decreased gradually, dropping to their initial levels(i.e., comparable to their expression at 0 h) by 24 h (Table S2). This suggests that there may be a relationship between quinoline alkaloid biosynthesis and abiotic factors, although this relationship may not be as simple and direct as the relationship between the biological environment and chemical defense [54,55]. Further research is needed to explore this relationship in depth. Many researchers believe that plants produce secondary metabolites such as alkaloids at the cost of slower growth [56,57]. However, when biotic and abiotic stresses become severe enough to affect their survival, the plants have no choice but to produce some secondary metabolites for protection against such rigorous stress conditions.

## 'Response to stress' and 'translational initiation' response to low temperature

Under the GO category 'Biological process', the terms 'response to stress' and 'translational initiation' accounted for1.76% of the total 198,528 GOs, but the DETs accounted for 15.4% of the transcripts involved in these GO terms. Additionally, both of two GO terms were significantly enriched in four comparison sets (0 h vs.3 h, 0 h vs. 6 h, 0 h vs. 12 h, and 0 h vs. 24 h) according to the GO enrichment analysis (Table S4). The largest proportion of the 'Biological process' terms included the 'metabolic process' (30.07%), 'cellular process' (27.99%), and 'biological regulation' (5.75%) (Table S4), indicating comprehensive changes in *E. dunnii* gene expression before and after the cold stress. However, although only a few transcripts were identified as belonging to 'response to cold stress', as up-term of 'response to stress', these transcripts represented the most important components that are directly involved in protecting plants from cold stress. A total of 50 transcripts were annotated under this term based on GO categorization, and most were up-regulated in response to low temperature treatment. In particular, 26 transcripts involved in the 'response to cold stress' were not expressed under normal conditions but were induced by exposure to low temperature (Table S7 and Fig. S1). ROS scavenging enzymes, including catalase (CAT), superoxide dismutase (SOD), and glutathione transferase (GST), have been demonstrated to play key roles in the removal of ROS [17,58–61]. During exposure to low temperature, the CAT activity was increased (from 0.34 to 0.56 U/g *F*w), which was in accordance with the expression level of the corresponding transcripts in the *E. dunnii* transcriptome (Figs. 1, and 7).

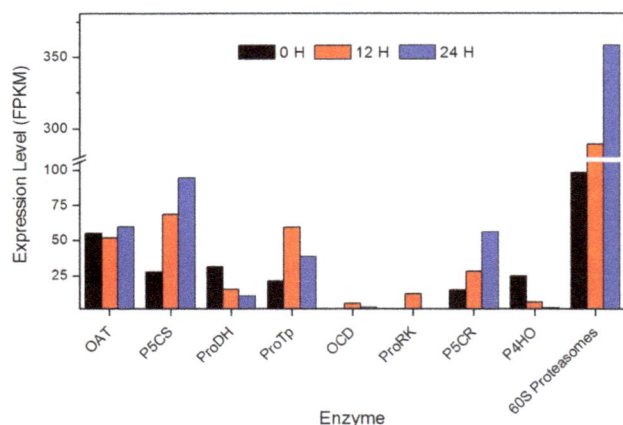

**Figure 7. The expression level of some key enzymes involved in the 'arginine and proline metabolism' pathway during cold acclimation.** Both the up-regulated expression of OAT (ornithine-oxo-acid transaminase), P5CS (pyrroline-5-carboxylatesynthase), ProTp (proline transporter), ProRK (proline-rich receptor protein kinase), P5CR (pyrroline-5-carboxylate reductase), and the down-regulated expression of ProDH (proline dehydrogenase), P4HO (prolyl 4-hydroxylase), OCD (ornithine cyclodeaminase) could result in proline accumulation.

Although the expression of peroxidases such as CAT and SOD increased significantly as the duration of cold exposure increased, these enzymes were still unable to completely clear the increased levels of $H_2O_2$, resulting in a significant increase in the amount of $H_2O_2$ during cold acclimation (Fig. 1). In this study, we found that some genes (e.g., MAP kinase and TCH2; Table S2) that are known to be involved in the response to other stresses (including salinity, heat and drought) in other plants are also involved in cold acclimation, which could support the hypothesis that the same gene have different functions in different plants.

The GO term 'translational initiation' was enriched in response to cold acclimation. A total of 254 transcripts were annotated under in this term, and 89 exhibited a greater than two-fold change in expression during the low temperature treatment (Table S8). Translation initiation in eukaryotes depends on many eukaryotic initiation factors (eIFs) that stimulate both the recruitment of the initiator tRNA, Met-tRNAiMet, and mRNA to the 40S ribosomal subunit and the subsequent scanning of the mRNA for the AUG start codon [62–64]. The largest of these initiation factors, the eIF-3 complex, organizes a web of interactions among several eIFs that assemble on the 40S subunit and participate in the different reactions involved in translation [62,65]. In plants, eIF-3plays the role of the central protein and interacts with many other translation initiation factors, such as eIF-4F, eIF-4G, eIF-4B, and eIF-1A [66]. Among the 89 contigs we identified that were annotated as 'translational initiation', 18 containedeIF-3 (eukaryotic translation initiation factor 3), and almost all were up-regulated during cold acclimation (Table S8). Daniel et al. [67] found that the expression level and phosphorylation state of these factors described above is subject to alteration during development, environmental stress (e.g., heat shock, and starvation), or viral infection. Tuteja [68] evaluated the roles of translation initiation, transcription factors, protein kinases, free proline, and reactive oxygen species in plant stress tolerance and found that these factors typically have synergistic effects in response to stress in plants. We also found that some transcripts encoding transcription factors, protein kinases (Table S6), translation initiation factors and antioxidant enzymes were up- or down-regulated in E. dunnii during cold acclimation, suggesting that the plant response to cold acclimation is a complex and global process.

## Cold-responsive transcription factor genes in E. dunnii

In Arabidopsis, at least 5 transcription factor families have been reported to be involved in the cold stress response process, including AP2-EREBP, MYB, NAC, bHLH and WRKY family [29]. Wang et al. found there were many families of transcription factor, such as bHLH family, MYB family, WRKY family, NAC family and so on, responding to cold acclimation in C. sinensis [16]. Meanwhile, An et al. identified 6 AP2-EREBP and 5 Myb transcription factors participated in the process of cold stress in treated cassava [6]. In our study, many transcripts were annotated as AP2 transcription factor based on the domain analysis. Among theses, 27 transcripts were up-regulated and 24 down-regulated under cold stress (Table S6). In present work, we tested 5 AP2 TF genes by qRT-PCR and found four were up-regulated, one was down-regulated during cold-stress, and the changing trend of the two methods was accordant (Table S3). The AP2-EREBP family plays a major role in the early stages of the cold response and is the major regulator that functions in activating cold-regulated effectors in Arabidopsis and other plants [69,6]. In Eucalyptus plant, the CBF proteins, belonging to A-1 subfamily of ERF/AP2 TF family has been reported involved in response to cold stress in E. gunnii and E. globules [19,70–72].

Besides the AP2 family, the bZIP family has also been demonstrated to be involved in the cold response in Arabidopsis and C. sinensis [69,16]. In this study, we found that bZIP family was the most enriched TF family, containing 62 genes (43 up-regulated and 9 down-regulated). Differential expression of bZIP TFs implies that other environmental or hormonal pathways may be involved in cold response in E. dunnii.

In addition, four novel transcription factor families (JmjC, SRF-TF, and Sigma70-like) were also identified. Although their homologous genes in other plant species have not yet been reported in response to cold stress, the expression level of these genes were markedly changed before and after cold stress, suggesting they might be specific to E. dunnii or attractive targets for further functional characterization in plant.

## Materials and Methods

### Plant materials

Eucalyptus dunnii was used in this study. The plantlets of E. dunnii with 10 leaves were grew in a climate-chamber, with a temperature of 25°C, 200 $\mu Em^{-2}s^{-1}$ illumination and a 14/10 h light/dark photoperiod. After eight weeks, the plants were moved into another climate-chamber with a temperature of 4°C and 200 $\mu Em^{-2}s^{-1}$ continuous illumination for cold-stress. For physiological measurement, we harvested the first two expanded leaves of these plantlets at 0, 3, 6, 12, 24 and 48 h after cold treatment, respectively. For RNA-seq, leaves from 6 plants treated by 0, 3, 6, 12, and 24 h were mixed for RNA isolation and sequencing. For test of physiological changes, leaves of plants treated by all time points were used. The harvested leaves were immediately frozen in liquid nitrogen for use.

### Analysis of physiological parameters

The proline content of the leaves was analyzed using a free proline ELISA kit (Omega, Georgia, USA) according to the manufacturer's instructions. The CAT activity and $H_2O_2$ content were measured using a CAT ELISA kit (Omega, Georgia, USA) and a $H_2O_2$ ELISA kit (Omega, Georgia, USA), respectively. All measurements were performed on the platform of Epoch-ELIASA (Shmadzu, Tokyo, Japan), and all analysis were repeated three times in this study.

### RNA extraction, library construction and RNA sequencing

Total RNA was isolated from the leaves by using Trizol reagent (Invitrogen, CA, USA) according to the manufacturer's instructions, and the RNAwas treated with RNase-free DNase I (TaKaRa, Dalian, China). The purity, concentration and RNA integrity number (RIN) were determined using a SMA3000 and/or Agilent 2100 Bioanalyzer. The total RNA was then sent to Beijing Genomics Institute (BGI) -Shenzhen (Shenzhen, China) for RNA sequencing.

More than 20 μg of total RNA extracted from each group of plants exposed to low temperatures (n>3) was used to construct the cDNA libraries. First, the polyadenylated RNAs (mRNAs) were purified and retrieved using magnetic beads coated with a poly-T oligo. These mRNAs were then mixed with fragmentation media and fragmented. The fragmented mRNAs were subjected to reverse transcription using reverse transcriptase and random primers. The second-strand cDNA synthesis was performed using DNA polymerase I and RNase H. Finally, the resulting dscDNAs were repaired by adding a single 'A' base, and specific Illumina adapters were ligated to the repaired ends. Fragments of approximately 200 bp in size were purified and retrieved from the gels. To construct the fragmented cDNA library, these fragments, which served as the template, were enriched by PCR using two primers that annealed to the ends of the adapters. The cDNA libraries constructed above were sequenced using an Illumina Hiseq2000. The PE read information and quality values were generated using the Illumina sequencing-by-synthesis, image analysis and base-calling procedures.

### Denovo assembly and functional annotation

Sequencing quality was assessed using fastQC software [http://www.bioinformatics.bbsrc.ac.uk/projects/fastqc/], and the PE reads were *de novo* assembled by five different assemblers: the Trinity software package (v2013-02-25) [73] with default parameters, the Oases software package (v0.1.21) [74] with a different K-value, the Edenam software package (v2013-07-15) [75] with a different M-value, the SOAP *de novo* software package (v2013-07-15) [76] with different K- and P-values, and the Cap3 software package (v12.07.21) [77] with default parameters. To evaluate the quality of the assemblies produced by the different assemblers, the PE reads were aligned back to the contigs assembled by a different assembler using Bowtie2 software (v2.0.0) [78], and the alignment rate was calculated. Subsequently, we analyzed the length distribution information of these contigs, such as the N50 number, average length, max length and total contig number, using common Perl scripts. Due to the lack of genomic information for Eucalyptus, 535 Eucalyptus sequences containing complete CDSs were downloaded from GenBank [http://www.ncbi.nlm.nih.gov/] and used as reference sequences to calculate the sensitivity and accuracy. Furthermore, we analyzed the best candidate coding sequence (CDS) for each contig from different assemblers and obtained the ratios of long CDS-containing transcripts to contigs with corresponding lengths.

All of the contigs (≥300 bp) produced by the Trinity software package were subjected to a similarity search against the NR database downloaded from GenBank utilizing local NCBI-BLAST software (v2.2.28+). The BLASTx searches were performed using a threshold E-value of $<10^{-3}$, max_target_seqs of 5, and an xml output file format. The BLASTx results were imported into Blast2GO software (v2.6.7) [79], and local functional annotation was performed. Enzyme codes, gene ontology (GO), and Kyoto Encyclopedia of Genes and Genomes (KEGG) pathways were retrieved from the KEGG web server (http://www.genome.jp/kegg/) [80]. GO classification [81] was performed using the WEGO program (http://wego.genomics.org.cn/cgibin/wego/index.pl) [82].

### Differential expression profiling and enrichment

To investigate the expression level of each transcript at the five treatment time points, the PE reads for each sample were aligned back to the optimal assembly result (assembled by the Trinity assembler) using Perl scripts provided by the Trinity software package. Using these scripts, we obtained the digital expression levels of each transcript and normalized these data with a RESM-based algorithm to obtain the FPKM (Fragments per Kilobase per Million Mapped Fragments) values of each transcript. Based on the normalized expression profiles, the effect and bias introduced by library size and/or RNA composition were eliminated using edgeR [83], and significant differentially expressed transcripts (DETs) were identified with a P.ad-value ≤0.05 and $\log_2$ fold-change ($\log_2$ FC) ≥1.

The DET enrichment analysis was performed using the common Perl and R scripts. We first counted the number of transcripts involved in each KEGG pathway from the Trinity assembled contigs and/or DETs. Based on the transcript numbers in the contig database and DETs, we determined the enriched

KEGG pathway. Then, the P-value was adjusted by the Bernoulli equation, and a P.ad-value<0.05 was the threshold value for significant enrichment results. We applied a similar approach for the GO enrichment analysis.

### Expression level verification

To verify the reliability and accuracy of the NGS-based expression level analysis, we randomly selected 31 transcripts from the contig database and evaluated the expression profiles among the five samples using quantitative real-time PCR. The primers for these transcripts are listed in Table S5. The first-strand cDNA was synthesized from 500 ng of total RNA using oligo (dT), random hexamers, and Moloney murine leukemia virus (M-MLV) reverse transcriptase (Invitrogen, CA, USA) according to the manufacturer's instructions. The real-time PCR was performed using the IQ5 Real-Time PCR System (Bio-Rad, CA, USA) in a total volume of 20 µL containing 100 ng of cDNA template, 1× SYBR *Premix Ex Taq*™II (Perfect Real Time, TaKaRa), and 400 nM of each primer. Serial dilutions of each cDNA were used to generate a quantitative PCR standard curve to calculate the corresponding PCR efficiencies. The PCR conditions were as follows: initial denaturation at 95°C for 30 s, followed by 40 cycles of denaturation at 95°C for 5 s, primer annealing at 60°C for 30 s, and DNA extension at 72°C for 30 s. Two most commonly used reference genes, beta-actin and GAPDH, were selected for internal controls. Three biological replicates were used, and melting curve analysis was performed to check the amplification specificity. The relative expression levels were calculated using the BIO-RAD IQ5 standard edition Optical System software (version 2.1) and a normalized expression (ddCt) model.

### Supporting Information

**Figure S1 Differences of 'response to stimulus' (A), 'response to cold' (B), 'transcription factor activity' (C) and 'kinase regulator activity' (D) between each pair of samples.** Overlap examinations were performed based on the resulting gene lists of four comparisons by VENNY. Overlap among four groups, D0 vs D3 (blue), D0 vs D6 (yellow), D0 vs D12 (yellow) and D0 vs D24 (red), were shown here.

**Table S1 The characteristics of contig databases assembled by different assembler.**

**Table S2 Sequence annotations of *E. dunnii* transcripts and gene expression profiling of five samples.**

**Table S3 Comparison of expression patterns between RNA-Seq expression and qRT-PCR.**

**Table S4 GO classification of *E. dunnii* teanscriptome and differentially.**

**Table S5 KEGG classification of *E. dunnii* teanscriptome and differentially expressed transcripts indentified between each pairs comparisons.**

**Table S6 Transcription factor and kinase of *E. dunnii* response to low-temperature stress.**

**Table S7 Expression patterns of some transcripts involved in 'Response to cold'.**

**Table S8 Expression patterns of some transcripts involved in 'translational initiation'.**

## Acknowledgments

The authors are extremely grateful to Professor Yizheng Zhang (Sichuan University) for technical advice and assistance in data processing.

## Author Contributions

Conceived and designed the experiments: YQL JPG. Performed the experiments: YQL YSJ JBL YZ. Analyzed the data: YSJ YQL JPG. Contributed reagents/materials/analysis tools: YQL YSJ. Wrote the paper: YSJ.

## References

1. Leslie AD, Mencuccini M, Perks M (2012) The potential for Eucalyptus as a wood fuel in the UK. Appl Energ 89: 176–182.
2. Munoz F, Valenzuela P, Gacitua W (2012) Eucalyptus nitens: nanomechanical properties of bark and wood fibers. Appl Phys A-Mater 108: 1007–1014.
3. Brawner JT, Lee DJ, Meder R, Almeida AC, Dieters MJ (2013) Classifying genotype by environment interactions for targeted germplasm deployment with a focus on Eucalyptus. Euphytica 191: 403–414.
4. Gomat HY, Deleporte P, Moukini R, Mialounguila G, Ognouabi N, et al. (2011) What factors influence the stem taper of Eucalyptus: growth, environmental conditions, or genetics? Ann For Sci 68: 109–120.
5. Sands PJ, Landsberg JJ (2002) Parameterisation of 3-PG for plantation grownEucalyptus globulus. Forest Ecol Manage 163: 273–292.
6. An D, Yang J, Zhang P (2012) Transcriptome profiling of low temperature-treated cassava apical shoots showed dynamic responses of tropical plant to cold stress. BMC Genomics 13: 64.
7. Pennycooke JC, Cox S, Stushnoff C (2005) Relationship of cold acclimation, total phenolic content and antioxidant capacity with chilling tolerance in petunia (Petuniax hybrida). Environ Exp Bot 53: 225–232.
8. Zhao ZG, Tan LL, Dang CY, Zhang H, Wu QB, et al. (2012) Deep-sequencing transcriptome analysis of chilling tolerance mechanisms of a subnival alpine plant, Chorispora bungeana. BMC Plant Biol 12: 222.
9. Fernandes J, Morrow DJ, Casati P, Walbot V (2008) Distinctive transcriptome responses to adverse environmental conditions in Zea mays L. Plant Biotechnol J 6: 782–798.
10. Ziemann M, Kamboj A, Hove RM, Loveridge S, El-Osta A, et al. (2013) Analysis of the barley leaf transcriptome under salinity stress using mRNA-Seq. Acta Physiol Plant 35: 1915–1924.
11. He XD, Li FG, Li M, Weng QJ, Shi JS, et al. (2012) Quantitative genetics of cold hardiness and growth in Eucalyptus as estimated from E. urophylla x E. tereticornis hybrids. New Forest 43: 383–394.
12. Yang QS, Wu JH, Li CY, Wei YR, Sheng O, et al. (2012) quantitative proteomic analysis reveals that antioxidation mechanisms contribute to cold tolerance in plantain (Musa paradisiaca L.; ABB Group) seedlings. Mol Cell Proteomics 11: 1853–1869.
13. Tatusov RL, Koonin EV, Lipman DJ (1997) A genomic perspective on protein families. Science 278: 631–637.
14. O'Rourke JA, Yang SS, Miller SS, Bucciarelli B, Liu JQ, et al. (2013) An RNA-Seq transcriptome analysis of orthophosphate-deficient white lupin reveals novel insights into phosphorus acclimation in plants. Plant Physiol 161: 705–724.
15. Wang XC, Yang YJ (2003) Research progress on resistance breeding of tea plant. J Tea Sci 23: 94–98.
16. Wang XC, Zhao QY, Ma CL, Zhang ZH, Cao HL, et al. (2013) Global transcriptome profiles of Camellia sinensis during cold acclimation. BMC Genomics 14: 415.
17. Torres R, Teixidó N, Usall J, Abadias M, Mir N, et al. (2011) Anti-oxidant activity of oranges after infection with the pathogen Penicillium digitatum or treatment with the biocontrol agent Pantoea agglomerans CPA-2. Biol Control 57: 103–109.
18. Fernandez M, Villarroel C, Balbontin C, Valenzuela S (2010) Validation of reference genes for real-time qRT-PCR normalization during cold acclimation in Eucalyptus globulus. Trees-Struct Funct 24: 1109–1116.
19. Navarro M, Ayax C, Martinez Y, Laur J, El Kayal W, et al. (2011) Two EguCBF1 genes overexpressed in Eucalyptus display a different impact on stress tolerance and plant development. Plant Biotechnol J 9: 50–63.
20. Ponciano G, McMahan CM, Xie WS, Lazo GR, Coffelt TA, et al. (2012) Transcriptome and gene expression analysis in cold-acclimated guayule (Parthenium argentatum) rubber-producing tissue. Phytochemistry 79: 57–66.
21. Liu SH, Wang NF, Zhang PY, Cong BL, Lin XZ, et al. (2013) Next-generation sequencing-based transcriptome profiling analysis of Pohlia nutans reveals insight into the stress-relevant genes in Antarctic moss. Extremophiles 17: 391–403.
22. Fernandez M, Aguila SV, Arora R, Chen KT (2012) Isolation and characterization of three cold acclimation-responsive dehydrin genes from Eucalyptus globulus. Tree Genet Genomes 8: 149–162.
23. Fernandez M, Valenzuela S, Barraza H, Latorre J, Neira V (2012) Photoperiod, temperature and water deficit differentially regulate the expression of four dehydrin genes from Eucalyptus globulus. Trees-Struct Funct 26: 1483–1493.
24. Li XY, Sun HY, Pei JB, Dong YY, Wang FW, et al. (2012) De novo sequencing and comparative analysis of the blueberry transcriptome to discover putative genes related to antioxidants. Gene 511: 54–61.
25. Rodriguez MC, Edsgard D, Hussain SS, Alquezar D, Rasmussen M, et al. (2010) Transcriptomes of the desiccation-tolerant resurrection plant Craterostigma plantagineum. Plant J 63: 212–228.
26. Bai X, Rivera-Vega L, Mamidala P, Bonello P, Herms D A, et al. (2011) Transcriptomic signatures of ash (Fraxinus spp.) phloem. PLoS One 6: e16368.
27. Dussert S, Guerin C, Andersson M, Joet T, Tranbarger TJ, et al. (2013) Comparative transcriptome analysis of three oil palm fruit and seed tissues that differ in oil content and fatty acid composition. Plant Physiol 162: 1337–1358.
28. Hua CW, Marcal S, Hong Y, Eduardo L, Victor C, et al. (2012) Reference Genes for High-Throughput Quantitative ReverseTranscription-PCR Analysis of Gene Expression in Organs andTissues of Eucalyptus Grown in Various EnvironmentalConditions. Plant Cell Physiology 53(12): 2101–2116.
29. Feng BM, Lu DH, Ma X, Peng YB, Sun J, et al. (2012) Regulation of the Arabidopsis anther transcriptome by DYT1 for pollen development. Plant J 72: 612–624.
30. Ragusa M, Statello L, Maugeri M, Majorana A, Barbagallo D, et al. (2012) Specific alterations of the microRNA transcriptome and global network structure in colorectal cancer after treatment with MAPK/ERK inhibitors. J Mol Med-JMM 90: 1421–1438.
31. Tao X, Gu YH, Wang HY, Zheng W, Li X, et al. (2012) Digital gene expression analysis based on integrated de novo transcriptome assembly of sweet potato [Ipomoea batatas (L.) Lam.]. PLoS One 7: e36234.
32. Birol I, Jackman SD, Nielsen CB, Qian JQ, Varhol R, et al. (2009) De novo transcriptome assembly with ABySS. Bioinformatics 25: 2872–2877.
33. Wang L, Li PH, Brutnell TP (2010) Exploring plant transcriptomes using ultra high-throughput sequencing. Brief Funct Genomics 9: 118–128.
34. Wu HL, Chen D, Li JX, Yu B, Xiao XY, et al. (2013) De Novo characterization of leaf transcriptome using 454 sequencing and development of EST-SSR markers in tea (Camellia sinensis). Plant Mol Biol Rep 31: 524–538.
35. Schafleitner R, Tincopa LR, Palomino O, Rossel G, Robles RF, et al. (2010) A sweet potato gene index established by de novo assembly of pyrosequencing and Sanger sequences and mining for gene-based microsatellite markers. BMC Genomics 11: 604.
36. Shi CY, Yang H, Wei CL, Yu O, Zhang ZZ, et al. (2011) Deep sequencing of the Camellia sinensis transcriptome revealed candidate genes for major metabolic pathways of tea-specific compounds. BMC Genomics 12: 131.
37. Duan J, Xia C, Zhao G, Jia J, Kong X (2012) Optimizing de novo common wheat transcriptome assembly using short-read RNA-Seq data. BMC Genomics 13: 392.
38. Xiao M, Zhang Y, Chen X, Lee EJ, Barber CJS, et al. (2013) Transcriptome analysis based on next-generation sequencing of non-model plants producing specialized metabolites of biotechnological interest. J Biotechnol 166: 122–134.
39. Ong WD, Voo LYC, Kumar VS (2012) De novo assembly, characterization and functional annotation of pineapple fruit transcriptome through massively parallel sequencing. PLoS One 7: e46937.
40. Paszkiewicz K, Studholme DJ (2010) De novo assembly of short sequence reads. Brief Bioinform 11: 457–472.
41. Verma P, Shah N, Bhatia S (2013) Development of an expressed gene catalogue and molecular markers from the de novo assembly of short sequence reads of the lentil (Lens culinaris Medik.) transcriptome. Plant Biotechnol J 11: 894–905.
42. Bates LS, Waldren RP, Teare ID (1973) Rapid determination of free proline for water-stress studies. Plant Soil 39: 205–207.
43. Nanjo T, Kobayashi M, Yoshiba Y, Kakubari Y, Yamaguchi-Shinozaki K, et al. (1999) Antisense suppression of proline degradation improves tolerance to freezing and salinity in Arabidopsis thaliana. Febs Letters 461: 205–210.
44. Kishor P, Hong Z, Miao GH, Hu C, Verma D (1995) Overexpression of [delta]-pyrroline-5-carboxylate synthetase increases proline production and confers osmotolerance in transgenic plants. Plant Physiol 108: 1387–1394.
45. Delauney AJ, Verma DPS (2002) Proline biosynthesis and osmoregulation in plants. Plant J 4: 215–223.
46. Forlani G, Scainelli D, Nielsen E (1997) [delta]1-pyrroline-5-carboxylate dehydrogenase from cultured cells of potato (purification and properties). Plant Physiol 113: 1413–1418.
47. Verslues PE, Sharma S (2010) Proline metabolism and its implications for plant-environment interaction. Arabidopsis Book 8: e0140.

48. Ábrahám E, Rigó G, Székely G, Nagy R, Koncz C, et al. (2003) Light-dependent induction of proline biosynthesis by abscisic acid and salt stress is inhibited by brassinosteroid in Arabidopsis. Plant Mol Biol 51: 363–372.

49. Delauney AJ, Hu CA, Kishor PB, Verma DP (1993) Cloning of ornithine delta-aminotransferase cDNA from Vigna aconitifolia by trans-complementation in *Escherichia coli* and regulation of proline biosynthesis. J Biol Chem 268: 18673–18678.

50. Tena G, Boudsocq M, Sheen J (2011) Protein kinase signaling networks in plant innate immunity. Curr Opin Plant Biol 14: 519–529.

51. Goossens A, Hakkinen ST, Laakso I, Seppanen-Laakso T, Biondi S, et al. (2003) A functional genomics approach toward the understanding of secondary metabolism in plant cells. Proc Natl Acad Sci U S A 100: 8595–8600.

52. Vom Endt D, Kijne JW, Memelink J (2002) Transcription factors controlling plant secondary metabolism: what regulates the regulators? Phytochemistry 61: 107–114.

53. Constabel CP, Lindroth RL (2010) The impact of genomics on advances in herbivore defense and secondary metabolism in Populus. In: Jansson S, Bhalerao R, Groover A, editors. Genetics and genomics of Populus, vol 8. London: Springer. pp. 279–305.

54. Mithofer A, Boland W (2012) Plant defense against herbivores: chemical aspects. Annu Rev Plant Biol 63: 431–450.

55. Osbourn A (2010) Secondary metabolic gene clusters: evolutionary toolkits for chemical innovation. Trends Genet 26: 449–457.

56. Dixon DP, Skipsey M, Edwards R (2010) Roles for glutathione transferases in plant secondary metabolism. Phytochemistry 71: 338–350.

57. Wu B, Li Y, Yan HX, Ma YM, Luo HM, et al. (2012) Comprehensive transcriptome analysis reveals novel genes involved in cardiac glycoside biosynthesis and mlncRNAs associated with secondary metabolism and stress response in *Digitalis purpurea*. BMC Genomics 13: 15.

58. Heller J, Tudzynski P (2011) Reactive oxygen species in phytopathogenic fungi: signaling, development, and disease. Annu Rev Phytopathol 49: 369–390.

59. Reape TJ, McCabe PF (2010) Apoptotic-like regulation of programmed cell death in plants. Apoptosis 15: 249–256.

60. Baxter A, Mittler R, Suzuki N (2013) ROS as key players in plant stress signalling. J Exp Bot 65: 1229–1240.

61. Dey S, Ghose K, Basu D (2010) Fusarium elicitor-dependent calcium influx and associated ros generation in tomato is independent of cell death. Eur J Plant Pathol 126: 217–228.

62. Hinnebusch AG (2006) eIF3: a versatile scaffold for translation initiation complexes. Trends Biochem Sci 31: 553–562.

63. Sizova DV, Kolupaeva VG, Pestova TV, Shatsky IN, Hellen CU (1998) Specific interaction of eukaryotic translation initiation factor 3 with the 5' nontranslated regions of hepatitis C virus and classical swine fever virus RNAs. J Virol 72: 4775–4782.

64. Dever TE (2002) Gene-specific regulation by general translation factors. Cell 108: 545–556.

65. Maquat LE, Tarn WY, Isken O (2010) The pioneer round of translation: features and functions. Cell 142: 368–374.

66. Preiss T, Hentze MW (2003) Starting the protein synthesis machine: eukaryotic translation initiation. Bioessays 25: 1201–1211.

67. Gallie DR, Le H, Caldwell C, Robert LT, Nam XH, et al. (1997) The phosphorylation state of translation initiation factors is regulated developmentally and following heat shock in wheat. J Biol Chem 272: 1046–1053.

68. Tuteja N (2007) Mechanisms of high salinity tolerance in plants. Methods Enzymol 428: 419–438.

69. Lee BH, Henderson DA, Zhu J-K (2005) The Arabidopsis Cold-Responsive Transcriptome and Its Regulation by ICE1. Plant Cell 17: 3155–3175.

70. Kayal WE, Navarro M, Marque G (2006) Expression profile of CBF-like transcriptional factor genes from Eucalyptus in response to cold. Journal Experiment Botany 57: 2455–2469.

71. Navarro M, Marque G, Ayax C (2009) Complementary regulation of four Eucalyptus CBF genes under various cold conditions. Journal Experiment Botany 60: 2713–2724.

72. Fernandez M, Aguila SV, Arora R (2012) Isolation and characterization of three cold acclimation-responsive dehydrin genes from Eucalyptus globulus. TREE GENETICS & GENOMES, 8: 149–155.

73. Grabherr MG, Haas BJ, Yassour M, Levin JZ, Thompson DA, et al. (2011) Full-length transcriptome assembly from RNA-Seq data without a reference genome. Nature Biotechnol 29: 644–652.

74. Schulz MH, Zerbino DR, Vingron M, Birney E (2012) Oases: robust de novo RNA-seq assembly across the dynamic range of expression levels. Bioinformatics 28: 1086–1092.

75. Hernandez D, Tewhey R, Veyrieras JB, Farinelli L, Østerås M, et al. (2014) *De novo* finished 2.8 Mbp Staphylococcus aureus genome assembly from 100 bp short and long range paired-end reads. Bioinformatics 30: 40–49.

76. Luo R, Liu B, Xie Y, Li Z, Huang W, et al. (2012) SOAP*denovo*2: an empirically improved memory-efficient short-read de novo assembler. Gigascience 1: 18.

77. Huang X, Madan A (1999) CAP3: a DNA sequence assembly program. Genome Res 9: 868–877.

78. Langmead B, Trapnell C, Pop M, Salzberg SL (2009) Ultrafast and memory-efficient alignment of short DNA sequences to the human genome. Genome Biol 10: R25.

79. Conesa A, Gotz S (2008) Blast2GO: a comprehensive suite for functional analysis in plant genomics. Int J Plant Genomics 2008. doi: 10.1155/2008/619832

80. Kanehisa M, Goto S (2000) KEGG: kyoto encyclopedia of genes and genomes. Nucleic Acids Res 28: 27–30.

81. Arasan SK, Park JI, Ahmed NU, Jung HJ, Lee IH, et al. (2013) Gene ontology based characterization of expressed sequence tags (ESTs) of *Brassica rapa cv. Osome*. Indian J Exp Biol 51: 522–530.

82. Ye J, Fang L, Zheng H, Zhang Y, Chen J, et al. (2006) WEGO: a web tool for plotting GO annotations. Nucleic Acids Res 34: W293–297.

83. Robinson MD, McCarthy DJ, Smyth GK (2010) edgeR: a Bioconductor package for differential expression analysis of digital gene expression data. Bioinformatics 26: 139–140.

# Complete Chloroplast Genome Sequence of Omani Lime (*Citrus aurantiifolia*) and Comparative Analysis within the Rosids

**Huei-Jiun Su[1], Saskia A. Hogenhout[2], Abdullah M. Al-Sadi[3], Chih-Horng Kuo[4,5,6]\***

**1** Institute of Ecology and Evolutionary Biology, National Taiwan University, Taipei, Taiwan, **2** Department of Cell and Developmental Biology, John Innes Centre, Norwich, United Kingdom, **3** Department of Crop Sciences, Sultan Qaboos University, Al Khoud, Oman, **4** Institute of Plant and Microbial Biology, Academia Sinica, Taipei, Taiwan, **5** Molecular and Biological Agricultural Sciences Program, Taiwan International Graduate Program, National Chung Hsing University and Academia Sinica, Taipei, Taiwan, **6** Biotechnology Center, National Chung Hsing University, Taichung, Taiwan

## Abstract

The genus *Citrus* contains many economically important fruits that are grown worldwide for their high nutritional and medicinal value. Due to frequent hybridizations among species and cultivars, the exact number of natural species and the taxonomic relationships within this genus are unclear. To compare the differences between the *Citrus* chloroplast genomes and to develop useful genetic markers, we used a reference-assisted approach to assemble the complete chloroplast genome of Omani lime (*C. aurantiifolia*). The complete *C. aurantiifolia* chloroplast genome is 159,893 bp in length; the organization and gene content are similar to most of the rosids lineages characterized to date. Through comparison with the sweet orange (*C. sinensis*) chloroplast genome, we identified three intergenic regions and 94 simple sequence repeats (SSRs) that are potentially informative markers with resolution for interspecific relationships. These markers can be utilized to better understand the origin of cultivated *Citrus*. A comparison among 72 species belonging to 10 families of representative rosids lineages also provides new insights into their chloroplast genome evolution.

**Editor:** Tongming Yin, Nanjing Forestry University, China

**Funding:** Funding for this work was provided by research grants from Biotechnology and Biological Sciences Research Council (BB/J004553/1) and the Gatsby Charitable Foundation to SAH, Sultan Qaboos University (SR/AGR/CROP/13/01) to AMA, and the Institute of Plant and Microbial Biology at Academia Sinica to CHK. The funders had no role in study design, data collection and analysis, decision to publish, or preparation of the manuscript.

\* Email: chk@gate.sinica.edu.tw

## Introduction

*Citrus* is in the family of Rutaceae, which is one of the largest families in order Sapindales. Flowers and leaves of *Citrus* are usually strong scented, the extracts from which contain many useful flavonoids and other compounds that are effective insecticides, fungicides and medicinal agents [1–3]. *Citrus* is of great economic importance and contains many fruit crops such as oranges, grapefruit, lemons, limes, and tangerines. However, due to a long cultivation history, wide dispersion, somatic bud mutation, and sexual compatibility among *Citrus* species and related genera, the taxonomy of *Citrus* remains controversial [4,5] and the origination of many *Citrus* species and hybrids is still unresolved [6,7].

The chloroplast (cp) genome sequence contains useful information in plant systematics because of its maternal inheritance in most angiosperms [8,9] and its highly conserved structures for developing promising genetic markers. The only complete cp genome available in *Citrus* is sweet orange (*Citrus sinensis*) [10], which has provided valuable information to the position of Sapindales in rosids. Although a genome sequencing project is in

progress for *C. clementine*, its complete chloroplast genome sequence is not available yet. To identify the cp genome regions that are polymorphic and may be used as molecular markers for resolving the evolutionary relationships among *Citrus* species, a second cp genome within the genus is necessary for comparative analysis. For this purpose, the major aim of this study is to determine the complete cp genome sequence of *C. aurantiifolia*.

*C. aurantiifolia*, which is commonly known as Key lime, Mexican lime, Omani lime, Indian lime, or acid lime, is native to Southeast Asia and widely cultivated in tropics and subtropics. Oman is known to be a transit country for lime, from which lime spread to Africa and the New World [11]. In Oman, Omani lime is considered the fourth most important fruit crop in terms of cultivated area and production. The products of Omani lime can be used for beverage, food additives and cosmetic industries [12]. Omani lime is sensitive to several biotic agents, the most serious of which is '*Candidatus* Phytoplasma aurantifolia', the cause of witches' broom disease of lime (WBDL). Recent studies on WBDL focused on effect of genetic diversity of Omani limes on the disease [13], transcriptome and proteomic analysis of lime response to infection by phytoplasma [14–16] and effect of phytoplasma on seed germination, growth and metabolite content in lime [17,18].

Here, we present the complete chloroplast genome sequence of Omani lime (*C. aurantiifolia*). To identify loci of potential utility for the molecular identification and phylogenetic analyses of *Citrus* cultivars and species, we compared the intergenic regions and SSRs in the cp genomes of *C. aurantiifolia* and *C. sinensis*. Furthermore, we performed phylogenetic analyses to infer the history of gene losses in the cp genome evolution among representative rosids lineages.

## Materials and Methods

### Sample Preparation and Sequencing

The Omani lime leaves were collected from a 5-year-old lime tree at a private farm located in the Omani territory of Madha (GPS coordinates: 25.276318, 56.318909). This farm is owned by one of the co-authors of this work, Dr. Abdullah M. Al-Sadi, whom should be contacted for future permissions. This study does not involve endangered or protected species and does not require specific permission from regulatory authority concerned with protection of wildlife. The sample was stored in a cool box and transported to the Plant Pathology Research Laboratory at Sultan Qaboos University (Al Khoud, Oman) for DNA extraction

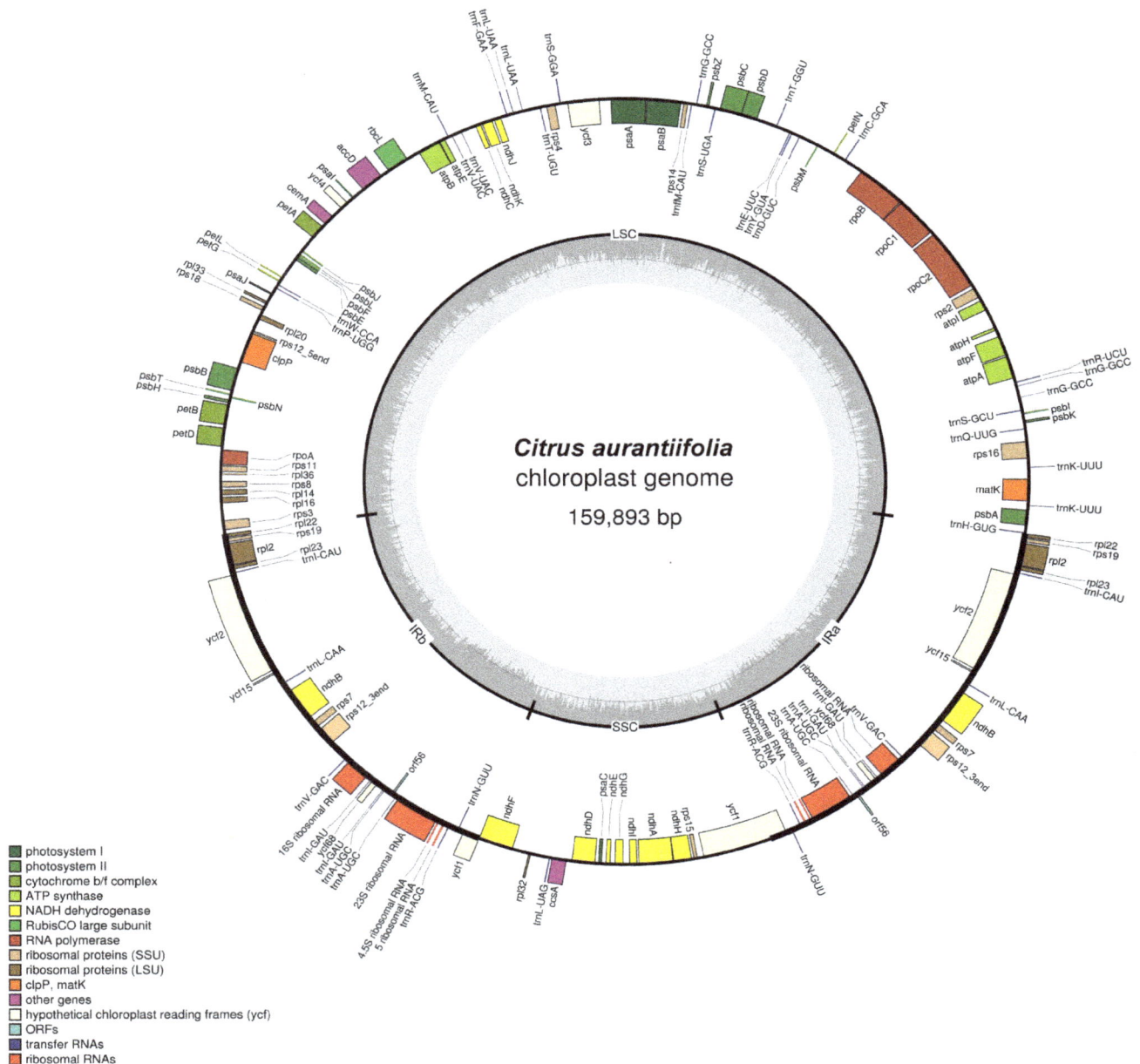

**Figure 1. Chloroplast genome map of *Citrus aurantiifolia*.** Gene drawn inside the circle are transcribed clockwise, whereas those outside are counterclockwise. The within-genome GC content variation is indicated in the middle circles.

following a protocol of Maixner et al. [19]. The leaves were washed with clear water before the isolation procedure. 1 g of leaves were used and crushed in 3 ml CTAB extraction buffer (2% CTAB, 1.4 M NaCl, 500 mM EDTA pH8, 1 M Tris-HCl pH8 and 0.2% beta-mercaptol). 1.5 ml of the leave extract was transferred to a 2 ml tube and incubated in a water-bath at 65°C for 15 min. The tube was turned up and down twice during incubation, centrifuged at 960 g for 5 min, and the supernatant was subsequently transferred to a clean eppendorf tube. An equal volume of chloroform-isoamyl alcohol mix (24:1) was added and the tube was centrifuged at 21000 g for 20 min. The supernatant was transferred to a new tube and then 0.6 volume of isopropanol was added to the supernatant and incubated at −20°C for 30 min. The DNA pellet was collected by centrifugation at 21000 g for 20 min and then washed with 1 ml of 70% ethanol. The final DNA was resuspended in 100 µl TE (Tris 10 mM, EDTA 1 mM pH8) and was stored at −80°C until used.

The library construction and sequencing were done at the Genome Analysis Centre (Norwich, UK). The Illumina TruSeq DNA Sample Preparation v2 Kit was used to prepare an indexed library. The DNA sample was sheared to a fragment size of 500–600 bp using a sonicator, followed by end-repair and the addition of a single A base for binding of the indexed adapter. The appropriate sized library (500 bp) was selected by gel electrophoresis, followed by PCR enrichment. The 251 bp paired-end sequencing run was performed on an Illumina MiSeq instrument using the SBS chemistry and Illumina software MCS v2.3.0.3 and RTA v1.18.42. The raw reads were deposited at the NCBI Sequence Read Archive under the accession number SRR1611615.

## Genome Assembly and Analyses

The procedures for genome assembly and annotation were based on our previous studies of cp genomes [20,21]. In addition to the standard de novo assembly approach by using Velvet v1.2.10 [22] with the k-mer size set to 243, a reference-based approach for assembly as described below was used in parallel. All of the raw reads were initially mapped onto the published cp genome of C. sinensis [10] using BWA v0.6.2 [23]. The sequence variations were identified with SAMtools v0.1.19 [24] and visually inspected using IGV v2.3.25 [25]. The variants were corrected with the raw reads and the regions without sufficient coverage were converted into gaps. This corrected sequence was then used as the new draft reference for the next iteration of verification. Gaps were filled using the reads overhang at margins and the process was repeated until the reference was fully supported by all mapped raw reads. The final assembly, which was supported by our de novo and reference-based approaches, resulted in an average of 1,441-fold coverage of paired-end reads with a mapping quality of 60 and the region with the lowest coverage is 506-fold.

The preliminary annotations of the C. aurantiifolia cp genome were performed online using the automatic annotator DOGMA [26] and verified using BLASTN [27,28] searches (e-value cutoff = 1e-10) against other land plant cp genomes. Each annotated gene was manually compared with C. sinensis cp genome for start and stop codons or intron junctions to ensure accurate annotation. The codon usage was analyzed by using the seqinr R-cran package [29]. A circular map of genome was produced using OGDRAW [30].

To identify the differences between C. aurantiifolia and C. sinensis, the two sequences were aligned using Mauve v2.3.1 [31] and the result was analyzed using custom Perl scripts. Intergenic gene regions were parsed out from the two Citrus cp genomes and aligned using MUSCLE v3.8.31 [32] with the default settings. The pairwise distances were calculated using the DNADIST program in the PHYLIP package v3.695 [33].

The positions and types of simple sequence repeats (SSRs) in the two Citrus cp genomes were detected using MISA (http://pgrc. ipk-gatersleben.de/misa/). The minimum number of repeats were set to 10, 5, 4, 3, 3, and 3 for mono-, di-, tri-, tetra-, penta-, and hexanucleotides, respectively. For long repeats, the program REPuter [34] was used to identify the number and location of direct and inverted (i.e., palindromic) repeats. A minimum repeat

**Table 1.** Summary of the Citrus chloroplast genome characteristics.

| Attribute | C. aurantiifolia (KJ865401) | C. sinensis (NC_008334) |
|---|---|---|
| Size (bp) | 159,893 | 160,129 |
| overall GC content (%) | 38.4 | 38.5 |
| LSC size in bp (% total) | 87,148 (54.5%) | 87,744 (54.8%) |
| SSC size in bp (% total) | 18,763 (11.7%) | 18,393 (11.5%) |
| IR size in bp (% total)[a] | 26,991 (16.9%) | 26,996 (16.9%) |
| Protein-coding regions size in bp (% total) | 81,468 (51.0%) | 79,773 (49.8%) |
| rRNA and tRNA size in bp (% total) | 11,850 (7.5%) | 11,850 (7.4%) |
| Introns size in bp (% total) | 17,129 (10.7%) | 18,252 (11.4%) |
| Intergenic spacer size in bp (% total) | 49,446 (30.9%) | 50,254 (31.4%) |
| Number of different genes | 115 | 113[b] |
| Number of different protein-coding genes | 81 | 79[b] |
| Number of different rRNA genes | 4 | 4 |
| Number of different tRNA genes | 30 | 30 |
| Number of different genes duplicated by IR | 22 | 20 |
| Number of different genes with introns | 17 | 17 |

[a]Each cp genome contains two copies of inverted repeats (IRs).
[b]According to the original annotation, not including orf56.

**Table 2.** Differences between the *C. aurantiifolia* and *C. sinensis* cp genomes.

| Indel | | Length (bp) | | Count | |
|---|---|---|---|---|---|
| | | 1 | | 43 | |
| | | 2–10 | | 20 | |
| | | 11–100 | | 18 | |
| | | 101–1,000 | | 3 | |
| | Sum | 1,780 | | 116 | Percentage[a]: 1.11% |
| **Substitution** | | | | | |
| | | Type | | Count | |
| | | A <-> T | | 34 | |
| | | C <-> G | | 15 | |
| | | A <-> C | | 81 | |
| | | T <-> C | | 64 | |
| | | A <-> G | | 51 | |
| | | T <-> G | | 85 | |
| | Sum | | | 330 | Percentage[a]: 0.21% |
| **10 most divergent intergenic regions** | | | | | |
| | | Region | Length[b] (bp) | Pairwise distance | |
| | | *rps3 - rpl22* (LSC) | 234 | 0.027 | |
| | | *ndhE - ndhG* (SSC) | 276 | 0.018 | |
| | | *psaC - ndhE* (SSC) | 231 | 0.017 | |
| | | *psbH - petB* (LSC) | 118 | 0.017 | |
| | | *trnY-GUA-trnE-UCC* (LSC) | 59 | 0.017 | |
| | | *trnH-GUG - psbA* (LSC) | 449 | 0.016 | |
| | | *rpl32 - trnL-UAG* (SSC) | 1,141 | 0.015 | |
| | | *psbT-psbN* (LSC) | 66 | 0.015 | |
| | | *trnG-GCC-trnR-UCU* (LSC) | 204 | 0.015 | |
| | | *trnD-GUC-trnY-GUA* (LSC) | 469 | 0.013 | |

[a]Relative to the length of *C. aurantiifolia*.
[b]Length in *C. aurantiifolia*.

size of 30 bp and sequence identity greater than 90% setting were used according to the study of *C. sinensis* cp genome [10]. The redundant or overlapping repeats were identified and filtered manually.

## Phylogenetic Inference

Phylogenetic analysis of the representative rosids lineages with complete cp genomes available was performed using PhyML v20120412 [35] with the GTR+I+G model. A total of 72 rosids species were chosen as the ingroups and *Vitis venifera* was included as the outgroup, the accession numbers were provided in Table S1. The protein-coding and rRNA genes were parsed from the selected cp genomes and clustered into ortholog groups using OrthoMCL [36]. The presence/absence of orthologous genes in each genome was examined and further verified using TBLASTN [27,28] searches (e-value cutoff = 1e-10). The nucleotide sequences of the conserved genes were aligned individually by using MUSCLE with the default settings. The concatenated alignment was used to infer a maximum likelihood phylogeny as described above. The bootstrap supports were estimated from 1,000 resampled alignments generated by the SEQBOOT program in the PHYLIP package.

## Investigations of *orf56* and *ycf68*

To investigate the presence/absence of *orf56* and *ycf68* in the selected cp genomes, the gene sequences from *C. aurantiifolia* was used as the queries to perform BLASTN [27,28] searches (e-value cutoff = 1e-10). The significant hits were examined to investigate the presence of intact open reading frames (ORFs). Phylogenetic analysis of the cp *orf56* genes and the homologous mitochondrial sequences was performed as described above. The final alignment contains 190 aligned nucleotide sites and a total of 70 sequences, including two sequences of *Amborella* as the outgroup.

## Results and Discussion

### General Features of the Omani Lime Chloroplast Genome

The complete cp genome of *C. aurantiifolia* (Christm.) Swingle (GenBank accession number KJ865401.1) is 159,893 bp in length, including a large single copy (LSC) region of 87,148 bp, a small single copy (SSC) region of 18,763 bp, and a pair of inverted repeats (IRa and IRb) of 26,991 bp each (Figure 1 and Table 1). A total of 137 different genes, including 93 protein-coding genes, 30 tRNA genes, and four rRNA genes, were annotated (Table S2). Among these, 12 protein-coding genes and 7 tRNA genes are duplicated in the IR regions. Most of the protein-coding genes are

**Table 3.** List of simple sequence repeats.

| Repeat unit | Length (bp) | Number of SSRs | Start position[a] |
|---|---|---|---|
| A | 10 | 6 | 4512; 47812; 53871; **72614**; 121748; **159288** |
| | 11 | 6 | **6866**; 10130; 69481; 71892; **117725**; **134802** |
| | 12 | 9 | **8332**; 31399; 47307; 63928; **111804**; **113977** (*ycf1*); 118367; 140302 (*ycf68*); 144255 |
| | 13 | 2 | 10107; 84557 |
| | 14 | 1 | **385** |
| | 15 | 1 | 32360 |
| | 16 | 2 | 69965; 118302 |
| | 17 | 3 | 7620; 39139; 74176 |
| | 19 | 1 | **12023** |
| | 22 | 1 | 70289 |
| T | 10 | 10 | **2424** (*matK*); 19786; **26964** (*rpoB*); 37622; 46938; **63632**; **87731**; 117742; 117871; 118851 |
| | 11 | 11 | 9401; 10416; **17001**; 30912; 46021; 63530; 112216; 117988; 118224; **121703**; **131189** (*ycf1*) |
| | 12 | 6 | **14722**; 29024; 102773; 106715 (*ycf68*); **133040** (*ycf1*); **135213** |
| | 13 | 2 | **73946**; 80423 |
| | 14 | 2 | 1776; 85274 |
| | 15 | 2 | 54209; 57817 |
| | 17 | 1 | 45965 |
| | 18 | 3 | 52748; 68339; 81409 |
| | 20 | 1 | 49202 |
| | 23 | 2 | 23694; **33282** |
| C | 10 | 2 | **28769; 104247** |
| G | 10 | 1 | **142772** |
| AT | 10 | 4 | **20631** (*rpoC2*); **33636**; 11817; **121517** (*ndhD*) |
| AAG | 12 | 1 | **97331** |
| AAT | 12 | 2 | **38604; 122629** |
| ATA | 12 | 1 | 70220 |
| ATT | 12 | 3 | **10283; 53810; 54088** |
| | 18 | 1 | **1760;** |
| CTT | 12 | 2 | **37353** (*psbC*); **149686** |
| TAA | 12 | 2 | **30250**; 61945 |
| TAT | 12 | 1 | **83297** |
| TTC | 12 | 1 | **73084** |
| TAAA | 12 | 2 | 4866; **45088** |
| AAAT | 12 | 3 | **30423; 32502; 71394** |
| ATAC | 12 | 1 | **51167** |
| ATTT | 12 | 1 | 49193 |
| | 20 | 1 | **117168** |
| TTAA | 12 | 2 | 39175; 39188 |
| TTAG | 12 | 1 | **61483** |
| TTTC | 12 | 1 | **14352** |
| TCTT | 12 | 1 | **46961** |
| AATAA | 20 | 1 | 144226 |
| TTTTA | 20 | 1 | 102781 |
| TTCAAA | 18 | 1 | 63817 |

[a]The SSR-containing coding regions are indicated in parentheses. SSRs that are identical in the *C. sinensis* chloroplast genome are highlighted in bold; SSRs that are conserved but with different lengths are highlighted by underline.

**Table 4.** List of long repeat sequences.

| Repeat size | Type[a] | Start position of 1st repeat | Start position the repeat found in other region | Location[b] | Region |
|---|---|---|---|---|---|
| 30 | D | 1759 | 1762 | IGS (psbA-trnK-UUU) | LSC |
| 30 | P | 1771 | 12015 | IGS (psbA-trnK-UUU, atpA-atpF) | LSC |
| 30 | P | 8231 | 37726, 47606 | **IGS (trnS-GCU, trnS-UGA, trnS-GGA),** | LSC |
| 30 | D | 23226 | 85067 | **intron (rpoC1), IGS (rpl16-rps3)** | LSC |
| 30 | D | 23686 | 52733 | intron (rpoC1), IGS (ndhK-ndhC) | LSC |
| 30 | P | 23687 | 70291 | intron (rpoC1), IGS (trnP-UGG-psaJ) | LSC |
| 30 | D | 23692 | 33280 | intron (rpoC1), IGS (trnE-UUC-trnT-GGU) | LSC |
| 30 | D | 49192 | 117171 | IGS (psbA-trnK-UUU), IGS (atpA-atpF) | LSC, IR |
| 30 | D, P | 49197 | 102764, 144233 | IGS (trnT-UGU-trnL-UAA), IGS (rps12-trnV-GAC) | LSC, IR |
| 30 | D, P | 51215 | 102768, 144229 | IGS (trnF-GAA-ndhJ), IGS (rps12-trnV-GAC) | LSC, IR |
| 30 | P | 71344 | 71344 | **IGS (rpl33-rps18)** | LSC |
| 30 | D, P | 102768 | 102773, 144224 | IGS (rps12-trnV-GAC) | IR |
| 30 | D | 144225 | 144230 | IGS (trnV-GAC-rps12) | IR |
| 31 | P | 4492 | 117868 | IGS (trnK-UUU-rps16), IGS (rpl32-trnL-UAG) | LSC, IR |
| 31 | P | 10106 | 49188 | IGS (trnG-GCC-trnR-UCU, trnT-UGU-trnL-UAA) | LSC |
| 31 | P | 29811 | 29811 | **IGS (petN-psbM)** | LSC |
| 31 | P | 33281 | 70282 | **IGS (trnE-UUC-trnT-GGU, trnP-UGG-psaJ)** | LSC |
| 31 | P | 119977 | 119977 | intron (ccsA) | IR |
| 32 | D | 7615 | 74171 | IGS (psbK-psbI), intron (clpP) | LSC |
| 32 | P | 39166 | 39166 | IGS (trnG-GCC-trnfM-CAU) | LSC |
| 34 | P | 38774 | 38782 | **IGS (psbZ-trnG-GCC)** | LSC |
| 34 | P | 49186 | 70288 | rps4, IGS (trnP-UGG-psaJ) | LSC |
| 34 | D, P | 111432 | 111464, 135529, 135561 | **IGS (rrn4.5-rrn5)** | IR |
| 35 | P | 10097 | 49193 | IGS trnG-GCC-trnR-UCU, trnT-UGU-trnL-UAA) | LSC |
| 36 | P | 27648 | 27648 | IGS (rpoB-trnC-GCA) | LSC |
| 40 | P | 77776 | 77776 | IGS (psbT-psbN) | LSC |
| 41 | D | 41294 | 43518 | **psaB, psaA** | LSC |
| 41 | D, P | 102353 | 124945, 144633 | **IGS (rps12-trnV-GAC), intron (ndhA)** | IR |
| 48 | P | 30626 | 30626 | **IGS (petN-psbM)** | LSC |
| 50 | D | 39020 | 39044 | **IGS (trnG-GCC-trnfM-CAU)** | LSC |
| 51 | P | 9984 | 9984 | IGS (trnG-GCC-trnR-UCU) | LSC |
| 53 | P | 8869 | 31095 | **IGS (trnS-GCU-trnG-GCC, psbM-trnD-GUC)** | LSC |
| 54 | P | 441 | 441 | IGS (trnH-GUG-psbA) | LSC |

[a]D: direct repeat; P: palindrome inverted repeat.
[b]IGS: intergenic spacer region. Sequences conserved in the *C. sinensis* chloroplast genome are highlighted in bold.

composed of a single exon, while 14 contain one intron and three contain two introns. The gene *rps12* was predicted to undergo trans-splicing, with the 5′ exon located in the LSC region and the other two exons located in the IR regions.

The protein-coding regions contain a total of 27,159 codons (Table S3). Isoleucine and cysteine are the most and least frequent amino acids and have 2,892 (10.7%) and 359 (1.2%) codons, respectively. The codon usage is biased towards a high ratio of A/T at the third position, which is also observed in many land plant cp genomes [37].

## Sequence Comparisons with Sweet Orange

The general characteristics of the two *Citrus* cp genomes are summarized in Table 1, overall the compositions are quite similar. The GC content of these *Citrus* cp genomes is approximately 38.5%, which is slightly higher than the average of the 72 representative rosids lineages (36.7%). In these two *Citrus* cp genomes, the genic regions, introns, and intergenic regions account for ca. 58%, 11%, and 31%, respectively (Table 1).

The pairwise sequence alignment between the two *Citrus* cp genomes revealed approximately 1.3% sequence divergence (Table 2), including 1,780 indels (1.11%) and 330 substitutions (0.21%). The LSC region contains more sequence polymorphisms than expected by its size, including 1,360 (76.4%) indels and 235 (71.2%) substitutions. In contrast, the two IR regions account for ca. 34% of the cp genome yet contain only 16 (0.9%) indels and 12 (3.6%) substitutions. The size differences in the LSC and SSC regions between these two cp genomes are mostly explained by one large indel in each region. The LSC sizes differ by 596 bp and a 523-bp indel was found in the spacer between *rps16* and *trnQ-*

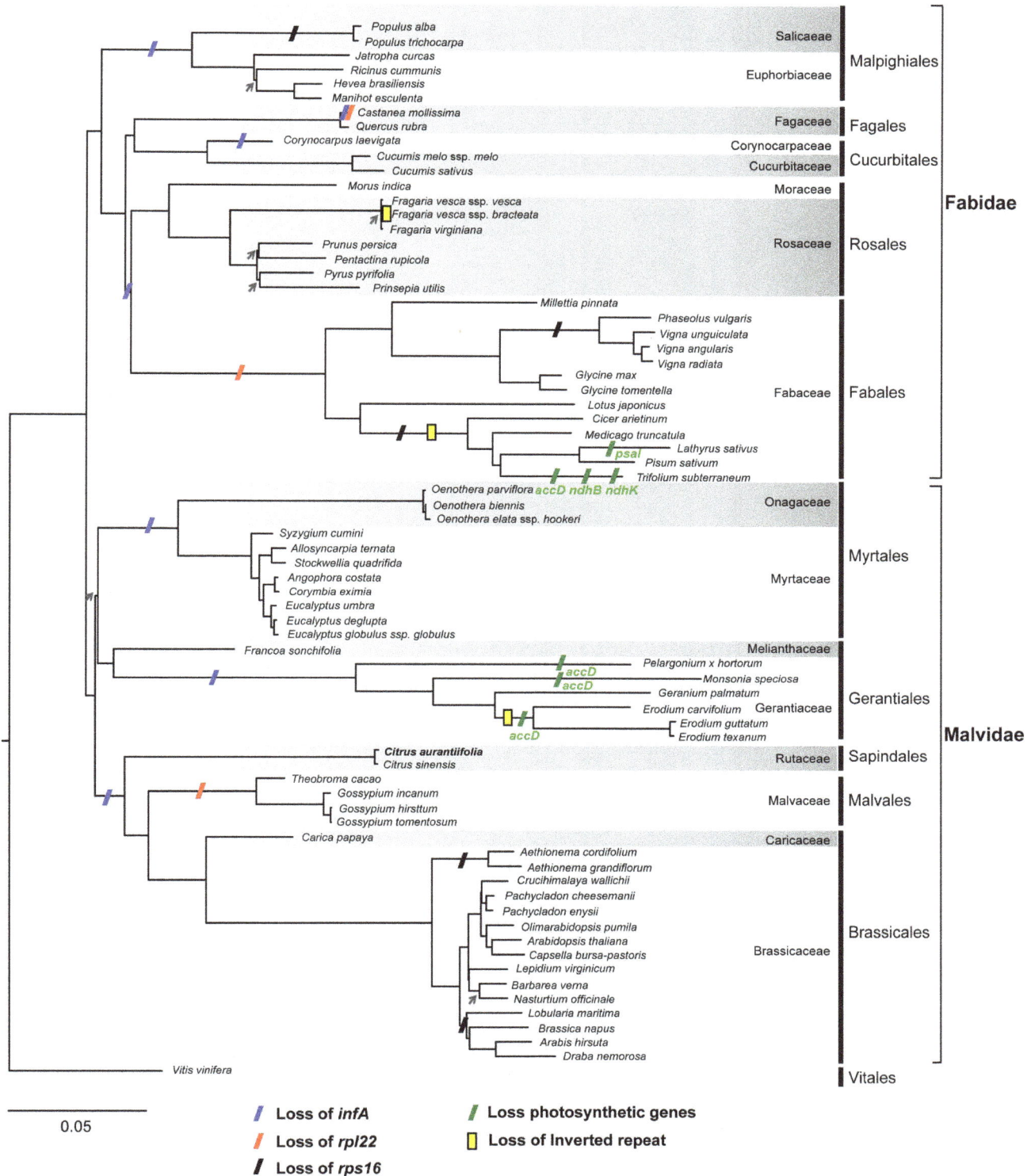

**Figure 2. Maximum likelihood phylogeny of the representative rosids lineages.** The common grape vine (*Vitis vinifera*) is included as the outgroup to root the tree. The concatenated alignment includes 62 conserved chloroplast genome genes and 54,689 aligned nucleotide sites. Nodes received <70% bootstrap support are indicated by gray arrows. The putative events of gene losses are inferred based on the most parsimonious scenario.

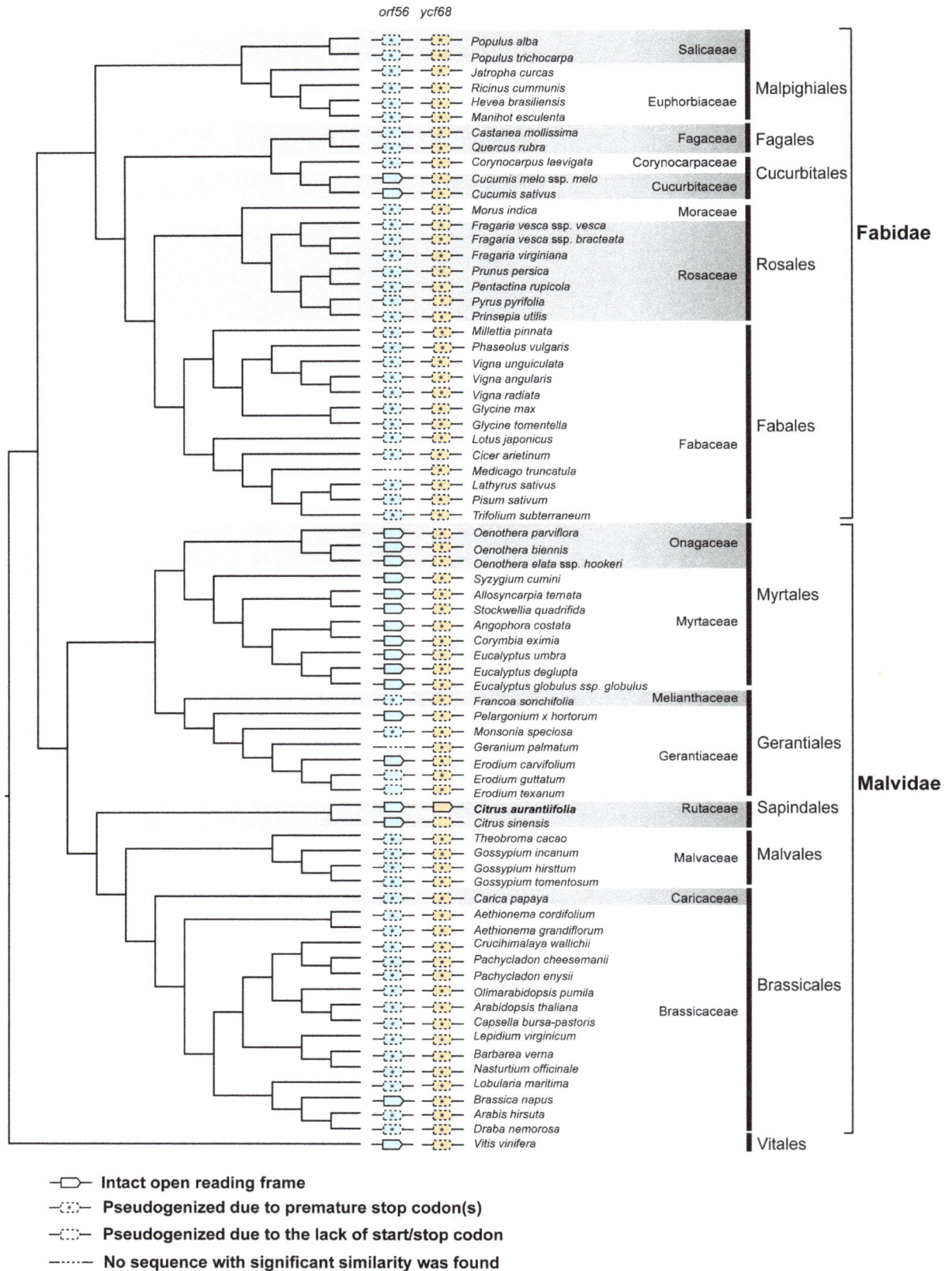

**Figure 3. The phylogenetic distribution patterns of** *orf56* **and** *ycf68*.

UUG. The SSC sizes differ by 370 bp and a 354-bp indel was found in the spacer between *rpl32* and *trnL-UAA*.

To identify the intergenic regions that may be useful for phylogenic analysis or molecular identification, we searched for the spacers that are >400 bp in length and exhibit above-average sequence divergence between the two *Citrus* species (i.e., >1.3%). A total of three regions satisfied these criteria, including the spacer between *trnH-GUG* and *psbA* (449 bp, 1.6% divergence), the spacer between *rpl32* and *trnL-UAG* (1141 bp, 1.5% divergence), and the spacer between *trnD-GUC* and *trnY-GUA* (469 bp, 1.3% divergence).

The junctions between the IR, LSC, and SSC regions in *C. aurantiifolia* are similar to that of *C. sinensis* except for the LSC-IRb boundary. A total of 23 indels and five substitutions were found at this region, resulting in one copy of *rpl22* spanning across the LSC-IRb junction in *C. aurantiifolia*. Comparing the IR junctions of *Citrus* with *Theobroma* and *Gossypium* in Malvaceae [38], it was found that the IRs in *Citrus* have expanded to include *rps19* and 252 nt of *rpl22*, whereas in Malvaceae, *rps19* is located in LSC and *rpl22* was missing [38–40].

## Analyses of Repetitive Sequences

A total of 109 SSR loci were found in the cp genome of *C. aurantiifoliaa*, accounting for 1,352 bp of the total sequence (ca. 0.8%). Among these, 94 were also found in *C. sinensis* and 42 exhibit length polymorphism (Table 3). Most SSRs are located in intergenic regions, but some were found in coding genes such as *matK* and *ycf1*. Concerning the controversial status of *Citrus* taxonomy, the SSRs identified in this study may provide new perspective to refine the phylogeny and elucidate the origin of the cultivars. Furthermore, these SSRs may be used as molecular markers for population studies.

In addition, 62 large repeats that are longer than 30 bp were found in the *C. aurantiifolia* cp genome (Table 4). Most of these repeats are located in intergenic spacers, except for three that are located in the coding regions of *rps4*, *psaA* and *psaB*. Twelve of these long repeats were also found in *C. sinensis,* indicating that these repeats might be widespread in the genus.

## Gene Content Analyses within the Rosids

A maximum likelihood phylogenetic analysis of 72 representative rosids lineages was conducted based on a concatenated alignment of four rRNA and 58 protein-coding genes with 54,689 sites (Figure 2). *Citrus* represents Sapindales and is sister to the clade containing Malvales and Brassicales. These relationships are congruent with the previous reports [10,41–43]. Based on this phylogeny and the gene content, we inferred the gene loss events during the cp genome evolution in rosids.

The translation initiation factor gene *infA* in cp has been lost independently at least 24 times in angiosperms and evidence provided from some cases suggested functional replacement by a nucleus copy [44]. Although the majority of *infA* in our selected cp genomes were found to be pseudogenized or completely lost, an intact *infA* was found in *Quercus*, *Francoa*, and two *Cuscumis* species.

The *rpl22* were found to be lost in Fabaceae [45] and *Castanea* of Fagaceae [46] following independent transfers to nucleus. Furthermore, another putative loss of *rpl22* was detected in *Passiflora* [46]. The *rpl22* in Malvaceae, including *Theobroma* and three *Gossypium* species, were found to be pseudogenized in our analysis. In *Citrus*, the ORF of *rpl22* was shortened to 252–264 nt compared to the typical length of 399–489 nt in other rosids [10,46]. However, compared with the pseudogenized *rpl22* found in Malvalvace, the *rpl22* homologs in *Citrus* still show high

sequence conservation. Additionally, the *rpl22* transcripts can be identified in the EST database for various *Citrus* species (data not shown). Taking account into the above consideration, we did not annotate *rpl22* as a pseudogene in *Citrus*.

The parallel losses of *rps16* were found in several rosids lineages (Figure 2), including one time in Salicaceae, two times in Fabaceae and another two times in Brassicaceae. The loss of *rps16* in *Medicago* and *Populus* was found to be substituted by a nuclear-encoded copy that transferred from the mitochondrion (mt) [47]. Because the nuclear-encoded RPS16 was found to target both mt and cp in *Arabidopsis*, *Lycopersicon*, and *Oryza* [47], it is possible that the cp genome-encoded *rps16* would not be maintained by selection and will eventually become lost in these lineages.

There are only a few gene loss events of photosynthetic genes found in rosids. In addition to the loss of *psaI* in *Lathyrus sativus* [48], the *accD* seems to be lost independently in *Trifolium subterraneum* and several Gerantiaceae species except for *Geranium palmatum*. In *Trifolium,* a nuclear-encoded *accD* copy has been reported [48], which presented another example of horizontal gene transfer from cp to nucleus. Successful gene transfers from cp to the nucleus in angiosperms are rare and have been only documented for four genes in rosids. Other than the three genes described above (i.e., *infA*, *rpl22*, and *accD*), the *rpl32* in *Populus* (Salicaceae) is the fourth example [49–51].

The IR has been reported to be independently lost at least five times among seed plants, two of which are within rosids [51]. In addition to the inverted repeat lacking clade (IRLC) of papilionoid Fabaceae [52] and *Erodium* of Gerantiaceae [53,54], the IR was found to be lost in two lineages of *Fragaria* (Rosaceae), which are *F. vesca* ssp. *bracteatea* and *F. mandschurica* (accession: NC_018767, not shown in Figure 2). Based on the *Fragaria* phylogeny shown in a previous study [55], it seems that IR loss was not a single event in *Fragaria*.

## Molecular Evolution of *orf56* and *ycf68* within the Rosids

In the comparison of gene content between the two *Citrus* cp genomes, *C. aurantiifolia* was found to contain two additional protein-coding genes. The first gene, *orf56*, is located in the *trnA-UGC* intron that contains one sequence homologous to previously recognized mitochondrial *ACRS* (ACR-toxin sensitivity gene) in *Citrus* [56]. In addition to the 171-bp identical sequences between cp *orf56* and the ORF sequences of *ACRS* in mt, the full length of 355-bp region of *ACRS* that conferred sensitivity to ACR-toxin in *E. coil* are also identical. Furthermore, the whole *trnA-UGC* among two *Citrus* cp regions and *C. jambhiri* mitochondrial *ACRS* shared more than 96% identity (Figure S1), which highlight the conservation of this region between cp and mt.

The gene *orf56* has also been included in the annotation of complete cp genomes of *Calycanthus* [57] and *Pelargonium* [58]. Our BLAST search against the rosids genome database revealed that in addition to *Citrus* and *Pelargonium*, all of the species examined in Cucurbitaceae and Myrtales also contain an intact *orf56* (Figure 3). Moreover, an intact *ACRS* ORF is also present in the mt genomes of *Liriodendron* [59] and *Silene* [60] and the ORF sequences between cp and mt are identical. Goremykin et al. [57] suggested that the *ACRS* gene was relative recently transferred from cp to mt. Based on the phylogeny containing the cp *orf56* and the mt *ACRS* (Figure S2), it appears that *orf56* has been independently transferred from cp to mt in different lineages.

The second gene, *ycf68*, is located in the *trnI-GAU* intron. A nearly identical sequence was found in *C. sinensis* but an additional T insertion near the C-terminus abolished the stop codon at the corresponding position. The intact *ycf68* can be detected in several monocots and Nymphaeaceae [61,62]. However, in the majority of other rosids (Figure 3) and the rest of the eudicots [61], the *ycf68* homologs all contain premature stop codons. Although Raubeson et al. [61] argued that *ycf68* is not a protein-coding gene based on the lack of intron-folding pattern, the high levels of sequence conservation among the ORFs of identified homologs suggest that the true identity and functionality of this putative gene remains to be further investigated.

## Conclusions

We reported the complete cp genome sequence of *Citrus aurantiifolia* (Rutaceae) in this study. The genome organization and gene content is typical of most angiosperms and highly similar to that of *C. sinensis* (i.e., 98.7% identical at the nucleotide level). The only difference in the gene content between the two *Citrus* cp genomes is the *C. aurantiifolia*-specific presence of a protein-coding gene (*ycf68*) in the *trnI-GAU* intron. Notably, three long intergenic spacers with high sequence divergence and 94 shared SSR regions were identified in the *C. aurantiifolia*-*C. sinensis* comparison. These regions may provide phylogenetic utility at low taxonomic levels and could be applied to the molecular identification of *Citrus* cultivars. Finally, our comparative analysis of gene content among 72 representative rosids lineages highlighted multiple events of gene losses within this group.

## Supporting Information

**Figure S1   Alignment of the *orf56*-containing sequences of two *Citrus* cp genomes and *C. jambhiri* mitochondrial *ACRS* sequences.**

**Figure S2   The maximum likelihood phylogeny of the cp *orf56* and mt *ACRS* ORF sequences.**

**Table S1   List of the complete chloroplast genome sequences included in the phylogenetic analysis.**

**Table S2   List of the genes found in the *C. aurantiifolia* cp genome.**

**Table S3   Codon usage of the *C. aurantiifolia* cp genome.**

## Acknowledgments

We thank Sam T. Mugford and Allyson M. MacLean for help with purification and quality controls of DNA samples.

## Author Contributions

Conceived and designed the experiments: HJS CHK. Performed the experiments: SAH AMA. Analyzed the data: HJS CHK. Contributed reagents/materials/analysis tools: HJS AMA CHK. Contributed to the writing of the manuscript: HJS SAH AMA CHK.

## References

1. Mabberley DJ (2004) *Citrus* (Rutaceae): a review of recent advances in etymology, systematics and medical applications. Blumea 49: 481–198.
2. Tripoli E, Guardia ML, Giammanco S, Majo DD, Giammanco M (2007) *Citrus* flavonoids: molecular structure, biological activity and nutritional properties: a review. Food Chem 104: 466–479.
3. Ezeabara CA, Okeke CU, Aziagba BO, Ilodibia CV, Emeka AN (2014) Determination of saponin content of various parts of six *Citrus* species. Int Res J Pure Appl Chem 4: 137–143.
4. Nicolosi E, Deng ZN, Gentile A, La Malfa S, Continella G, et al. (2000) *Citrus* phylogeny and genetic origin of important species as investigated by molecular markers. Theor Appl Genet 100: 1155–66.
5. Hynniewta M, Malik SK, Rao SR (2014) Genetic diversity and phylogenetic analysis of Citrus (L.) from north-east India as revealed by meiosis, and molecular analysis of internal transcribed spacer region of rDNA. Meta Gene 2: 237–251.
6. Liu Y, Heying E, Tanumihardjo SA (2012) History, global distribution, and nutritional importance of citrus fruits. Compr Rev Food Sci Food Saf 11: 530–545.
7. Penjor T, Yamamoto M, Uehara M, Ide M, Matsumoto N, et al. (2013) Phylogenetic relationships of *Citrus* and its relatives based on *matK* gene sequences. PLoS ONE 8: e62574.
8. Corriveau JL, Coleman AW (1988) Rapid screening method to detect potential biparental inheritance of plastid DNA and results for over 200 angiosperm species. Am J Bot 75: 1443–1458.
9. Zhang Q, Liu Y, Sodmergen (2003) Examination of the cytoplasmic DNA in male reproductive cells to determine the potential for cytoplasmic inheritance in 295 angiosperm species. Plant Cell Physiol 44: 941–951.
10. Bausher MG, Singh ND, Lee SB, Jansen RK, Daniell H (2006) The complete chloroplast genome sequence of *Citrus sinensis* (L.) Osbeck var 'Ridge Pineapple': organization and phylogenetic relationships to other angiosperms. BMC Plant Biol 6: 21.
11. Davies FS, Albrigo LG (1994). Citrus. CABI International, Wiltshire, UK. 1–2.
12. Vand SH, Abdullah TL (2012) Identification and introduction of Thornless Lime (*Citrus aurantiifolia*) in Hormozgan, Iran. Indian J Sci Technol 5: 3670–3673.
13. Al-Sadi AM, Al-Moqbali HS, Al-Yahyai RA, Al-Said FA (2012) AFLP data suggest a potential role for the low genetic diversity of acid lime (*Citrus aurantifolia* Swingle) in Oman in the outbreak of witches' broom disease of lime. Euphytica 188: 285–297.
14. Taheri F, Nematzadeh G, Zamharir MG, Nekouei MK, Naghavi M, et al. (2011) Proteomic analysis of the Mexican lime tree response to "*Candidatus* Phytoplasma aurantifolia" infection. Mol Biosyst 7: 3028–3035.
15. Zamharir MG, Mardi M, Alavi SM, Hasanzadeh N, Nekouei MK, et al. (2011) Identification of genes differentially expressed during interaction of Mexican lime tree infected with "*Candidatus* Phytoplasma aurantifolia". BMC Microbiol 11: 1.
16. Monavarfeshani A, Mirzaei M, Sarhadi E, Amirkhani A, Khayam Nekouei M, et al. (2013) Shotgun proteomic analysis of the Mexican lime tree infected with "*Candidatus* Phytoplasma aurantifolia." J Proteome Res 12: 785–795.
17. Faghihi MM, Bagheri AN, Bahrami HR, Hasanzadeh H, Rezazadeh R, et al. (2011) Witches'-broom disease of lime affects seed germination and seedling growth but is not seed transmissible. Plant Disease 95: 419–422.
18. Zafari S, Niknam V, Musetti R, Noorbakhsh SN (2012) Effect of phytoplasma infection on metabolite content and antioxidant enzyme activity in lime (*Citrus aurantifolia*). Acta Physiol Plant 34: 561–568.
19. Maixner M, Ahrens U, Seemüller E (1995) Detection of the German grapevine yellows (Vergilbungskrankheit) MLO in grapevine, alternative hosts and a vector by a specific PCR procedure. Eur J Plant Pathol 101: 241–250.
20. Ku C, Hu J-M, Kuo C-H (2013) Complete plastid genome sequence of the basal asterid *Ardisia polysticta* Miq. and comparative analyses of asterid plastid genomes. PLoS ONE 8: e62548.
21. Ku C, Chung W-C, Chen L-L, Kuo C-H (2013) The complete plastid genome sequence of Madagascar periwinkle *Catharanthus roseus* (L.) G. Don: plastid genome evolution, molecular marker identification, and phylogenetic implications in asterids. PLoS ONE 8: e68518.
22. Zerbino DR, Birney E (2008) Velvet: algorithms for *de novo* short read assembly using de Bruijn graphs. Genome Res 18: 821–829.
23. Li H, Durbin R (2009) Fast and accurate short read alignment with Burrows-Wheeler transform. Bioinformatics 25: 1754–1760.
24. Li H, Handsaker B, Wysoker A, Fennell T, Ruan J, et al. (2009) The Sequence Alignment/Map format and SAMtools. Bioinformatics 25: 2078–2079.
25. Robinson JT, Thorvaldsdottir H, Winckler W, Guttman M, Lander ES, et al. (2011) Integrative genomics viewer. Nat Biotechnol 29: 24–26.
26. Wyman SK, Jansen RK, Boore JL (2004) Automatic annotation of organellar genomes with DOGMA. Bioinformatics 20: 3252–3255.
27. Altschul SF, Gish W, Miller W, Myers EW, Lipman DJ (1990) Basic local alignment search tool. J Mol Biol 215: 403–410.
28. Camacho C, Coulouris G, Avagyan V, Ma N, Papadopoulos J, et al. (2009) BLAST+: architecture and applications. BMC Bioinformatics 10: 421.
29. Charif D, Lobry JR (2007) SeqinR 1.0–2: a contributed package to the R project for statistical computing devoted to biological sequences retrieval and analysis. In: Bastolla DU, Porto PDM, Roman DHE, Vendruscolo DM, editors.

Structural Approaches to Sequence Evolution. Biological and Medical Physics, Biomedical Engineering. Springer Berlin Heidelberg. 207–232.

30. Lohse M, Drechsel O, Bock R (2007) OrganellarGenomeDRAW (OGDRAW): a tool for the easy generation of high-quality custom graphical maps of plastid and mitochondrial genomes. Curr Genet 52: 267–274.

31. Darling ACE, Mau B, Blattner FR, Perna NT (2004) Mauve: multiple alignment of conserved genomic sequence with rearrangements. Genome Res 14: 1394–1403.

32. Edgar RC (2004) MUSCLE: multiple sequence alignment with high accuracy and high throughput. Nucl Acids Res 32: 1792–1797.

33. Felsenstein J (1989) PHYLIP - Phylogeny Inference Package (version 3.2). Cladistics 5: 164–166.

34. Kurtz S, Schleiermacher C (1999) REPuter: fast computation of maximal repeats in complete genomes. Bioinformatics 15: 426–427.

35. Guindon S, Gascuel O (2003) A simple, fast, and accurate algorithm to estimate large phylogenies by maximum likelihood. Syst Biol 52: 696–704.

36. Li L, Stoeckert CJ, Roos DS (2003) OrthoMCL: Identification of ortholog groups for eukaryotic genomes. Genome Res 13: 2178–2189.

37. Clegg MT, Gaut BS, Learn GH, Morton BR (1994) Rates and patterns of chloroplast DNA evolution. Proc Natl Acad Sci U S A 91: 6795–6801.

38. Kane N, Sveinsson S, Dempewolf H, Yang JY, Zhang D, et al. (2012) Ultrabarcoding in cacao (*Theobroma* spp.; Malvaceae) using whole chloroplast genomes and nuclear ribosomal DNA. Am J Bot 99: 320–329.

39. Lee SB, Kaittanis C, Jansen RK, Hostetler JB, Tallon LJ, et al. (2006) The complete chloroplast genome sequence of *Gossypium hirsutum*: organization and phylogenetic relationships to other angiosperms. BMC Genomics 7: 61.

40. Xu Q, Xiong G, Li P, He F, Huang Y, et al. (2012) Analysis of complete nucleotide sequences of 12 *Gossypium* chloroplast genomes: origin and evolution of allotetraploids. PLoS ONE 7: e37128.

41. Bremer B, Bremer K, Chase MW, Fay MF, Reveal JL, et al. (2009) An update of the Angiosperm Phylogeny Group classification for the orders and families of flowering plants: APG III. Bot J Linn Soc 161: 105–121.

42. Worberg A, Alford MH, Quandt D, Borsch T (2009) Huerteales sister to Brassicales plus Malvales, and newly circumscribed to include *Dipentodon*, *Gerrardina*, *Huertea*, *Perrottetia*, and *Tapiscia*. Taxon 58: 468–478.

43. Ruhfel BR, Gitzendanner MA, Soltis PS, Soltis DE, Burleigh JG (2014) From algae to angiosperms-inferring the phylogeny of green plants (*Viridiplantae*) from 360 plastid genomes. BMC Evol Biol 14: 23.

44. Millen RS, Olmstead RG, Adams KL, Palmer JD, Lao NT, et al. (2001) Many parallel losses of *infA* from chloroplast DNA during angiosperm evolution with multiple independent transfers to the nucleus. Plant Cell 13: 645–658.

45. Gantt JS, Baldauf SL, Calie PJ, Weeden NF, Palmer JD (1991) Transfer of *rpl22* to the nucleus greatly preceded its loss from the chloroplast and involved the gain of an intron. EMBO J 10: 3073–3078.

46. Jansen RK, Saski C, Lee S-B, Hansen AK, Daniell H (2011) Complete plastid genome sequences of three rosids (*Castanea*, *Prunus*, *Theobroma*): evidence for at least two independent transfers of *rpl22* to the nucleus. Mol Biol Evol 28: 835–847.

47. Ueda M, Nishikawa T, Fujimoto M, Takanashi H, Arimura S, et al. (2008) Substitution of the gene for chloroplast RPS16 was assisted by generation of a dual targeting signal. Mol Biol Evol 25: 1566–1575.

48. Magee AM, Aspinall S, Rice DW, Cusack BP, Sémon M, et al. (2010) Localized hypermutation and associated gene losses in legume chloroplast genomes. Genome Res 20: 1700–1710.

49. Cusack BP, Wolfe KH (2007) When gene marriages don't work: divorce by subfunctionalization. Trends Genet 23: 270–272.

50. Ueda M, Fujimoto M, Arimura S, Murata J, Tsutsumi N, et al. (2007) Loss of the *rpl32* gene from the chloroplast genome and subsequent acquisition of a preexisting transit peptide within the nuclear gene in *Populus*. Gene 402: 51–56.

51. Jansen RK, Ruhlman TA (2012) Plastid genomes of seed plants. In: Bock R, Knoop V, editors. Genomics of chloroplasts and mitochondria. Advances in photosynthesis and respiration. Springer Netherlands. 103–126.

52. Wojciechowski MF, Lavin M, Sanderson MJ (2004) A phylogeny of legumes (Leguminosae) based on analysis of the plastid *matK* gene resolves many well-supported subclades within the family. Am J Bot 91: 1846–1862.

53. Blazier JC, Guisinger MM, Jansen RK (2011) Recent loss of plastid-encoded *ndh* genes within *Erodium* (Geraniaceae). Plant Mol Biol 76: 263–272.

54. Guisinger MM, Kuehl JV, Boore JL, Jansen RK (2011) Extreme reconfiguration of plastid genomes in the angiosperm family Geraniaceae: rearrangements, repeats, and codon usage. Mol Biol Evol 28: 583–600.

55. Njuguna W, Liston A, Cronn R, Ashman TL, Bassil N (2013) Insights into phylogeny, sex function and age of *Fragaria* based on whole chloroplast genome sequencing. Mol Phylogenet Evol 66: 17–29.

56. Ohtani K, Yamamoto H, Akimitsu K (2002) Sensitivity to *Alternaria alternata* toxin in citrus because of altered mitochondrial RNA processing. Proc Natl Acad Sci U S A 99: 2439–2444.

57. Goremykin V, Hirsch-Ernst KI, Wölfl S, Hellwig FH (2003) The chloroplast genome of the "basal" angiosperm *Calycanthus fertilis* – structural and phylogenetic analyses. Plant Syst Evol 242: 119–135.

58. Chumley TW, Palmer JD, Mower JP, Fourcade HM, Calie PJ, et al. (2006) The complete chloroplast genome sequence of *Pelargonium* × *hortorum*: organization and evolution of the largest and most highly rearranged chloroplast genome of land plants. Mol Biol Evol 23: 2175–2190.

59. Richardson AO, Rice DW, Young GJ, Alverson AJ, Palmer JD (2013) The "fossilized" mitochondrial genome of *Liriodendron tulipifera*: ancestral gene content and order, ancestral editing sites, and extraordinarily low mutation rate. BMC Biol 11: 29.

60. Sloan DB, Müller K, McCauley DE, Taylor DR, Šorchová H (2012) Intraspecific variation in mitochondrial genome sequence, structure, and gene content in *Silene vulgaris*, an angiosperm with pervasive cytoplasmic male sterility. New Phytol 196: 1228–1239.

61. Raubeson LA, Peery R, Chumley TW, Dziubek C, Fourcade HM, et al. (2007) Comparative chloroplast genomics: analyses including new sequences from the angiosperms *Nuphar advena* and *Ranunculus macranthus*. BMC Genomics 8: 174.

62. Ahmed I, Biggs PJ, Matthews PJ, Collins LJ, Hendy MD, et al. (2012) Mutational dynamics of aroid chloroplast genomes. Genome Biol Evol 4: 1316–1323.

# Characterization of the *Runx* Gene Family in a Jawless Vertebrate, the Japanese Lamprey (*Lethenteron japonicum*)

**Giselle Sek Suan Nah[1], Boon-Hui Tay[1], Sydney Brenner[1,2], Motomi Osato[1,3,4]\*, Byrappa Venkatesh[1,5]\***

1 Institute of Molecular and Cell Biology, Agency for Science, Technology and Research (A\*STAR), Singapore, Singapore, 2 Okinawa Institute of Science and Technology Graduate University, Onna-son, Okinawa 904-0495, Japan, 3 Cancer Science Institute of Singapore, National University of Singapore, Singapore, Singapore, 4 Institute of Bioengineering and Nanotechnology, Agency for Science, Technology and Research, Singapore, Singapore, 5 Department of Paediatrics, Yong Loo Lin School of Medicine, National University of Singapore, Singapore, Singapore

## Abstract

The cyclostomes (jawless vertebrates), comprising lampreys and hagfishes, are the sister group of jawed vertebrates (gnathostomes) and are hence an important group for the study of vertebrate evolution. In mammals, three *Runx* genes, *Runx1*, *Runx2* and *Runx3*, encode transcription factors that are essential for cell proliferation and differentiation in major developmental pathways such as haematopoiesis, skeletogenesis and neurogenesis and are frequently associated with diseases. We describe here the characterization of *Runx* gene family members from a cyclostome, the Japanese lamprey (*Lethenteron japonicum*). The Japanese lamprey contains three *Runx* genes, *RunxA*, *RunxB*, and *RunxC*. However, phylogenetic and synteny analyses suggest that they are not one-to-one orthologs of gnathostome *Runx1*, *Runx2* and *Runx3*. The major protein domains and motifs found in gnathostome Runx proteins are highly conserved in the lamprey Runx proteins. Although all gnathostome *Runx* genes each contain two alternative promoters, P1 (distal) and P2 (proximal), only lamprey *RunxB* possesses the alternative promoters; lamprey *RunxA* and *RunxC* contain only P2 and P1 promoter, respectively. Furthermore, the three lamprey *Runx* genes give rise to fewer alternative isoforms than the three gnathostome *Runx* genes. The promoters of the lamprey *Runx* genes lack the tandem Runx-binding motifs that are highly conserved among the P1 promoters of gnathostome *Runx1*, *Runx2* and *Runx3* genes; instead these promoters contain dispersed single Runx-binding motifs. The 3′UTR of lamprey *RunxB* contains binding sites for miR-27 and miR-130b/301ab, which are conserved in mammalian *Runx1* and *Runx3*, respectively. Overall, the *Runx* genes in lamprey seem to have experienced a different evolutionary trajectory from that of gnathostome *Runx* genes which are highly conserved all the way from cartilaginous fishes to mammals.

**Editor:** Nikolas Nikolaidis, California State University Fullerton, United States of America

**Funding:** Biomedical Research Council of A\*STAR, Singapore. The funders had no role in study design, data collection and analysis, decision to publish, or preparation of the manuscript.

**Competing Interests:** The authors have declared that no competing interests exist.

\* Email: csimo@nus.edu.sg (MO); mcbbv@imcb.a-star.edu.sg (BV)

## Introduction

The polyomavirus enhancer-binding protein 2 (PEBP2) or core-binding factor (CBF) is an ancient Runt domain heterodimeric transcription factor of α and β subunits. In human, the α-subunit is encoded by three *Runt* gene family members, *RUNX1*, *RUNX2* and *RUNX3* that contain an evolutionarily conserved 128 amino acid Runt domain responsible for DNA binding and heterodimerization with the β-subunit. The β-subunit comprises a single protein, RUNXβ (also known as PEBP2β or CBFβ) that does not directly interact with DNA but serves to allosterically enhance the DNA-binding activity of the α-subunit and regulate its turnover by protecting it from ubiquitin-proteasome-mediated degradation [1,2]. Basally branching chordates such as amphioxus (*Branchiostoma floridae*) and tunicates (*Ciona* and *Oikopleura*) possess a single *Runx* gene. By contrast, vertebrates such as cartilaginous fishes (e.g. elephant shark, dogfish) and tetrapods contain three *Runx* genes (*Runx1*, *Runx2* and *Runx3*) [3,4,5] owing to the two rounds of whole genome duplication at the base of vertebrates [6,7]. Teleost fishes such as zebrafish contain a duplicated *Runx2* gene (*Runx2b*) that resulted from an additional teleost-specific genome duplication event [8]. On the other hand, pufferfishes (fugu and *Tetraodon*) possess three *Runx* genes like tetrapods and an additional fourth *Runx* gene called *frRunt* [9,10], which is hypothesized to represent an ancestral vertebrate *Runx* gene that was lost in tetrapods and zebrafish [9]. In contrast to multiple copies of α-subunit encoding *Runx* genes, *Runxb* is present as a single copy in vertebrates as well as in invertebrates analysed, with the exception of the fruit fly (*Drosophila melanogaster*) that possesses two orthologs of *Runxβ* genes, *Brother* and *Big brother* [11,12,13].

In mammals, the three *Runx* genes play pivotal roles in several developmental processes including haematopoiesis, neurogenesis, and skeletogenesis. *Runx1* is required for the emergence and maintenance of hematopoietic stem cells [14,15] and is frequently mutated in human leukaemia [16,17,18]. *Runx2* is indispensable for osteogenesis as evidenced by the complete lack of ossified skeleton in *Runx2*-deficient mice [19]. Haploinsufficiency of *RUNX2* results in the autosomal dominant bone disease, cleidocranial dysplasia [20]. *Runx3* is involved in diverse biological pathways including the regulation of epithelial homeostasis in the gastrointestinal tract [21,22], development of T cells during thymopoiesis [23] and the differentiation of various cell types of the immune system [24,25,26]. *Runx3* is also essential for the differentiation of proprioceptive neurons [27] and chondrocyte maturation during skeletogenesis [28].

The living vertebrates are divided into two broad groups: the jawless vertebrates (cyclostomes) and jawed vertebrates (gnathostomes). While the gnathostomes are represented by cartilaginous fishes, ray-finned fishes, lobe-finned fishes and tetrapods, the cyclostomes are represented by only lampreys and hagfish which constitute a monophyletic group. Cyclostomes diverged from gnathostomes ~500 Mya and are morphologically and physiologically quite distinct from gnathostomes. They possess a single median dorsal "nostril" as opposed to ventrally located nostrils in gnathostomes. In addition, they lack mineralized tissues, hinged jaws, paired appendages, pancreas and spleen [29]. Although cyclostomes lack an immunoglobulin-based adaptive immune system, a thymus-like organ called the 'thymoid', has been identified at the tips of gill filaments in the gill basket [30] and is the site of antigen receptor gene assembly of T-cell like lymphocytes [31]. The adaptive immune system of cyclostomes makes use of an alternative immune receptor system different from that of gnathostomes. Instead of T-cell and B-cell receptors generated from the immunoglobulin superfamily, cyclostomes utilize variable lymphocyte receptors (VLR) that are assembled from leucine-rich repeat modules (reviewed in [32]). These distinctive morphological and physiological traits along with the unique phylogenetic position of cyclostomes make them an important group for understanding the evolution and function of vertebrate *Runx* genes.

Full-length coding sequences of two *Runx* genes, designated *MgRunxA* and *MgRunxB* have been previously cloned in a cyclostome, the Atlantic hagfish (*Myxine glutinosa*). These *Runx* genes were found to be expressed in cartilaginous tissues, suggestive of their involvement in early vertebrate skeletogenesis [4]. Mining of the genome assembly of the sea lamprey (*Petromyzon marinus*) led to the identification of partial sequences for only two *Runx* genes, *PmRunxA* and *PmRunxB* and the conclusion that this is likely to be the full complement of *Runx* genes in lamprey [33]. However, it should be noted that the sea lamprey genome assembly was generated from DNA extracted from a somatic tissue (liver) that has been shown to lose ~20% of genomic DNA due to developmentally programmed rearrangement during early embryonic development [34] and hence does not contain the full complement of genes present in the germline genome. Recently, the whole genome sequence of another lamprey, the Japanese lamprey (*Lethenteron japonicum*) was generated using DNA from the testis [35] (http://jlampreygenome.imcb.a-star.edu.sg/). The Japanese lamprey and sea lamprey are northern hemisphere lampreys that diverged from each other 30–10 Mya ago [36]. Compared to the 2.3 Gb genome of the sea lamprey, the Japanese lamprey has a relatively smaller genome of ~1.6 Gb [37]. To improve our understanding of the evolution, function and regulation of *Runx* genes, we mined the Japanese lamprey genome assembly for *Runx* genes and completed the coding sequences for *Runx* genes by RT-PCR and RACE. Our analyses show that the Japanese lamprey genome codes for three *Runx* genes like gnathostomes but they are not the exact one-to-one orthologs of gnathostome *Runx1*, *Runx2* and *Runx3*.

## Materials and Methods

### Ethics statement

The Japanese lamprey specimens were collected from commercial fishermen who routinely catch them like other commercial fishes in Ishikari River near Ebetsu in Hokkaido, Japan. The procedure for extraction of DNA and RNA from lamprey tissues was approved by the Institutional Animal Care and Use Committee of the Biological Resource Centre, Agency for Science, Technology and Research (A*STAR), Singapore.

### Identification of *Runx* genes in the Japanese lamprey genome assembly

The genome assembly of the Japanese lamprey (http://jlampreygenome.imcb.a-star.edu.sg/) was searched with human RUNX1, RUNX2, RUNX3 and RUNXβ protein sequences using the 'TBLASTN' algorithm. Scaffolds #47, #165, #769 and #850 containing sequences homologous to human and elephant shark Runx1, Runx2, Runx3 and Runxβ protein sequences were searched against the non-redundant protein database at NCBI using BlastX to identify the coding exons of lamprey *Runx* genes. Missing exons in the scaffolds (presumably due to gaps in the scaffolds) were identified by sequencing RACE (Rapid Amplification of cDNA Ends) products as described below.

### Full-length cDNA cloning by RACE

Total RNA was extracted from various tissues of adult Japanese lamprey using Trizol reagent (Gibco BRL, Grand Island, NY) according to manufacturer's protocol. One μg of total RNA was reverse transcribed into 5′RACE-ready or 3′RACE-ready single-strand cDNA by using the SMART RACE cDNA Amplification kit (Clontech, Palo Alto, CA). Primers were designed for representative exons identified in the Japanese lamprey scaffolds and full-length *Runx* transcripts were obtained by 5′ RACE and/or 3′ RACE (primer sequences available upon request). 5′ and 3′ RACE were each performed in a nested PCR. The first round of PCR was performed with the Universal Primer and a gene-specific primer. The resulting PCR product was diluted 20× and 1 μl used for the nested PCR with the Nested Universal Primer and a second gene-specific primer. All RACE products were cloned into the pGEM-T Easy Vector (Promega, USA), and sequenced completely using the BigDye Terminator Cycle Sequencing Kit (Applied Biosystems, USA) on an ABI 3730×l capillary sequencer (Applied Biosystems, USA). The genome structures of the Japanese lamprey *Runx* genes, including the exon-intron structures, UTRs and transcription start sites were deciphered by mapping the cloned full length cDNA sequences to the Japanese lamprey genomic sequence. Sequences for various isoforms of the Japanese lamprey *Runx* genes have been deposited in GenBank (KJ787775–KJ787788).

### Amino acid alignment and phylogenetic analysis

The full-length protein sequences for human and other chordate *Runx* genes, including the two Runx sequences (RunxA and RunxB) from the Atlantic hagfish were retrieved from the National Center for Biotechnology Information (NCBI) database. RunxA and RunxB from the sea lamprey were not used as they are partial sequences. Multiple sequence alignments were generated using MUSCLE (www.ebi.ac.uk/Tools/msa/muscle/). For phylogenet-

**A  LjRunxA (29 kb)**

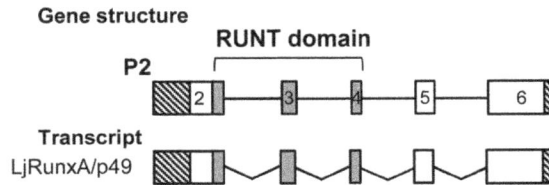

**B  LjRunxB (134 kb)**

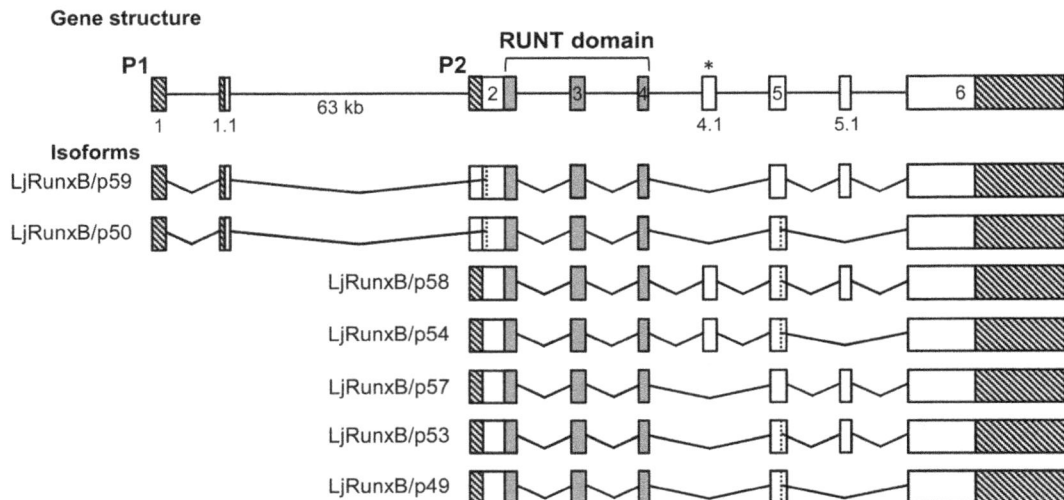

**C  LjRunxC (35 kb)**

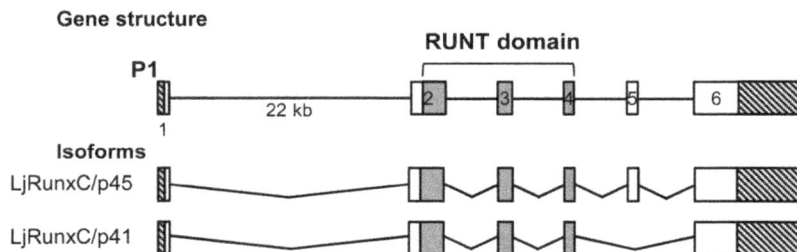

**Figure 1. Exon-intron organization of lamprey (*Lj*) *Runx* genes.** Schematic representation of the gene structures and transcript isoforms of (A) *LjRunxA*, (B) *LjRunxB* and (C) *LjRunxC*. Exons are indicated by boxes. The vertical dashed lines indicate internal splice sites located within the coding exon. Exons constituting the Runt domain are indicated in grey. The two alternative promoters are denoted as P1 and P2. Crosshatched boxes indicate 5'- and 3'-UTRs. The asterisk (*) indicates an exon in *LjRunxB* that is absent in gnathostome *Runx*. Not drawn to scale.

ic analysis, the alignments were trimmed using the Gblocks Server (ver. 0.91b) with less stringent selection parameters [38]. MEGA 6.06 (http://www.megasoftware.net/) was used to determine the most appropriate amino acid substitution model for each dataset. Maximum Likelihood (ML) and Bayesian (BI) methods were used for phylogenetic analyses employing the JTT+G+I or JTT+G model. MEGA 6.06 was used for ML analyses and 100 bootstrap replicates were used for node support. BI analyses were carried out using MrBayes 3.2 (http://mrbayes.csit.fsu.edu/). Two independent runs starting from different random trees were run for 1 million generations with sampling done every 100 generations. A consensus tree was built from all sampled trees excluding the first 2,500 trees (burn-in).

## Synteny analysis

The order of genes in the *Runx* loci of human, chicken, zebrafish, sea anemone (*Nematostella vectensis*) and sponge (*Amphimedon queenslandica*) were obtained from Ensembl (www.ensembl.org); amphioxus (*Branchiostoma floridae*) from the UCSC Genome Browser (http://genome.ucsc.edu/); elephant shark from the elephant shark genome assembly [39] (http://esharkgenome.imcb.astar.edu.sg/) and Japanese lamprey from the Japanese lamprey genome assembly (http://jlampreygenome.imcb.a-star.edu.sg/).

## Expression profiling by qRT-PCR

Purified total RNA was reverse-transcribed into cDNA with Superscript II (Invitrogen, Carlsbad, CA). The single strand cDNA

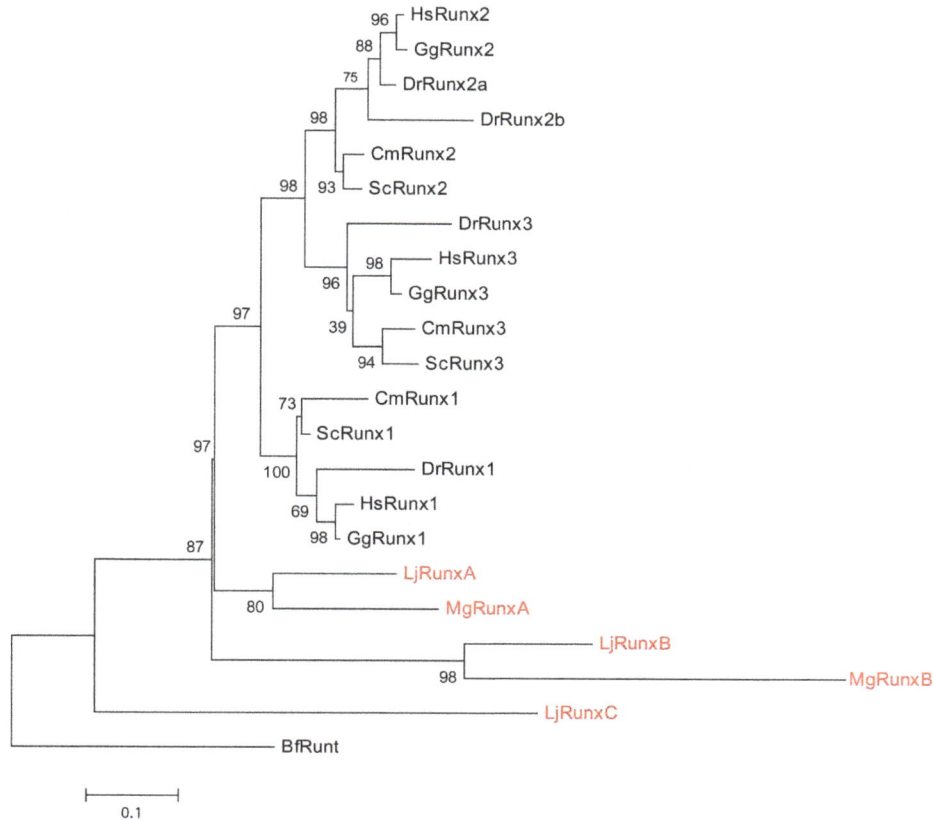

**Figure 2. Phylogenetic analysis of chordate Runx sequences.** Protein sequences of Japanese lamprey *Runx* genes were aligned with homologous sequences from selected chordates. The gaps in the alignments were trimmed using the Gblocks Server. The resulting protein alignment used for phylogenetic analyses is provided in Fig. S3. Maximum Likelihood (ML) trees were generated for the alignments. Statistical support values for the nodes are shown as ML bootstrap percentages. Hagfish and Japanese lamprey Runx proteins are highlighted in red. Lancelet (*Branchiostoma floridae*) Runt (BfRunt) was used as the outgroup. Hs, *Homo sapiens*; Gg, *Gallus gallus*; Dr, *Danio rerio*; Cm, *Callorhinchus milii*; Sc, *Scyliorhinus canicula*; Mg, *Myxine glutinosa*; Lj, *Lethenteron japonicum*.

was used as a template in qRT-PCR reactions with KAPA SYBR FAST qPCR Kit reagents (KAPA Biosystems, Boston, MA). Sequences of primers used in qRT-PCR are given in Table S1. All primer pairs span at least one intron which helps to distinguish cDNA sequences from genomic DNA products. Expression levels of *Runx* genes were normalized using *beta-actin* gene as the reference. Quantification of gene expression levels was performed using the comparative $C_T$ method [40]. The relative expression levels of each *Runx* gene in different tissues were estimated in relation to a reference tissue (one that showed the lowest level of expression among the tissues analysed). Note, however, that these relative expression levels are not comparable between different *Runx* genes.

## Results and Discussion

### Cloning and orthology of Japanese lamprey α-subunit *Runx* family members

To identify *Runx* genes in the Japanese lamprey, we BLAST-searched the Japanese lamprey genome assembly [35] using human RUNX protein sequences. We identified three scaffolds (Scaffold_850, 47 and 769) that contained fragments of *Runx* genes. The missing exons and full-length coding sequences of these genes were identified by RT-PCR and/or RACE using cDNA from gill, intestine or skin (Fig. 1 and S1).

To determine the relationships of the three lamprey *Runx* genes to *Runx1*, *Runx2* and *Runx3* of gnathostomes, we carried out phylogenetic analysis using two different algorithms, ML and BI. The *Runt* gene from amphioxus was used as the outgroup. Both ML and BI analyses showed that the Japanese lamprey *RunxA* and *RunxB* genes are orthologs of the hagfish *RunxA* and *RunxB* genes, respectively (Fig. 2 and S2). However, the three lamprey and two hagfish genes were found to cluster outside the three gnathostome genes. In the ML tree, cyclostome *RunxA* formed a sister clade to all three gnathostome *Runx* genes while *RunxB* and *RunxC* genes were represented on separate branches outside the *RunxA*+gnathostome *Runx* clade (Fig. 2). In BI analysis, the combined clade of cyclostome *RunxA* and *RunxB* genes constituted the sister group to all three gnathostome *Runx* genes with *RunxC* being an outgroup (Fig. S2). This pattern of clustering of gnathostome and cyclostome *Runx* genes suggests that three gnathostome *Runx* genes are the result of duplications in the gnathostome ancestor after it diverged from the cyclostome lineage. This implies that the three lamprey *Runx* genes are not the exact one-to-one orthologs of the three gnathostome *Runx* genes.

In addition, we generated an ML tree that included single *Runx* sequences from two more well characterized invertebrate species, *Ciona intestinalis* an urochordate and *Nematostella vectensis* a cnidarian (Fig. S4 and S5). In this tree, the relationships of

**Runx1 locus in gnathostomes**

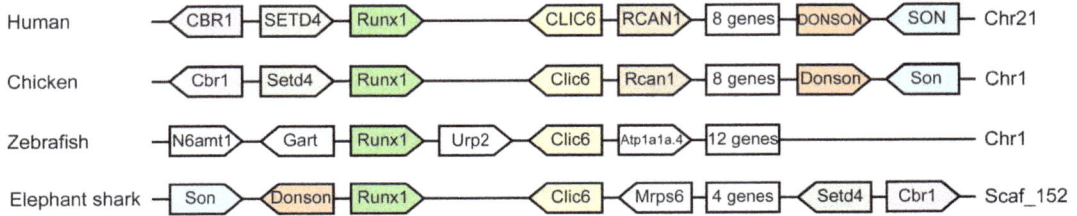

**Runx2 locus in gnathostomes**

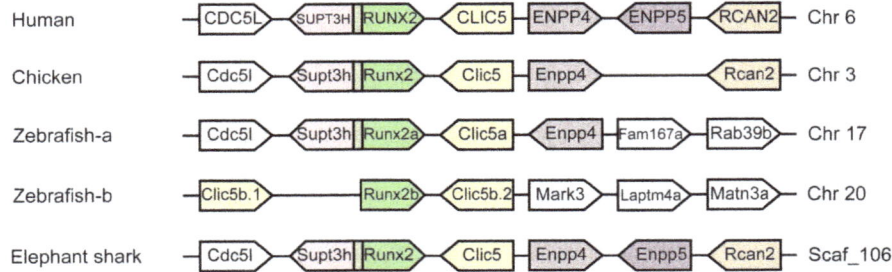

**Runx3 locus in gnathostomes**

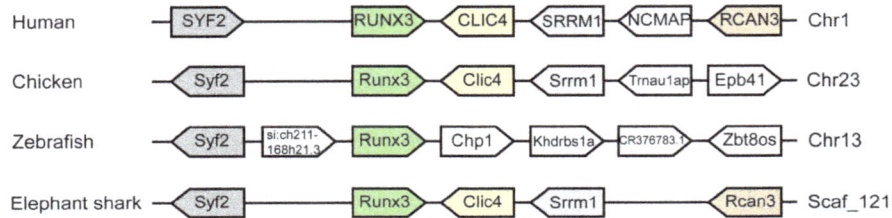

**Japanese Lamprey Runx loci**

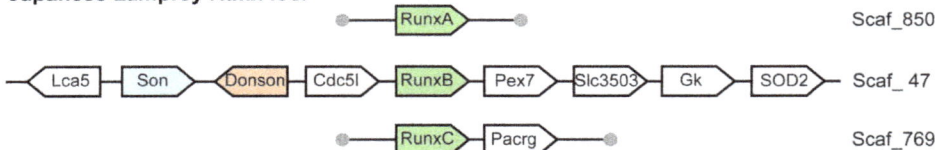

**Amphioxus Runt locus**

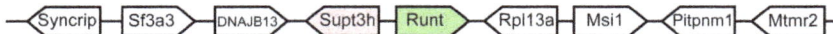

**Sea anemone Runt locus**

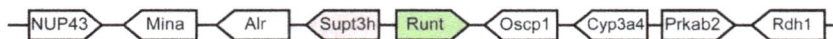

**Sponge Runt locus**

**Figure 3. Synteny of genes in the *Runx* loci of Japanese lamprey and selected metazoans.** Genes are represented by block arrows. Genes with conserved synteny are coloured. Clusters of some non-syntenic genes are represented as white boxes. Grey circles indicate the end of scaffolds.

gnathostome *Runx* genes are essentially similar to that depicted in Fig. 2, and the cyclostome genes cluster outside the gnathostome clade. However, some differences in topology are observed among the cyclostome branches (Fig. S5). These branches, though, are not well supported (bootstrap values of 40 and 37), suggesting that these relationships are not reliable. We believe that such changes

to the topology are due to the highly divergent sequences of *Ciona* and *Nematostella* (see alignment in Fig. S4).

The three *Runx* genes in the gnathostomes are likely to be the result of two rounds of whole genome duplication (2R) that occurred early during the evolution of vertebrates [6,41,42]. Consistent with this hypothesis, synteny of sets of paralogous genes are conserved in the gnathostome *Runx1*, *Runx2* and *Runx3* gene loci. For example, synteny of paralogs of *Runx*, *Clic* and *Rcan* are conserved in the *Runx1*, *Runx2* and *Runx3* loci of all gnathostomes analysed so far (Fig. 3). Conserved synteny of blocks of genes between genomes provides additional support for the orthologous relationship of genes in the syntenic block. For instance, orthologs of several genes in the human *RUNX2* locus (e.g. *CDC5L*, *SUPT3H*, *CLIC5*) are conserved in the *Runx2* locus of chicken, zebrafish and elephant shark indicating that the *Runx* gene in this locus of these gnathostomes is indeed an ortholog of human *RUNX2* (Fig. 3). An interesting pair of syntenic genes in the gnathostome *Runx2* locus is the tightly linked *RUNX2* and *SUPT3H* genes whose first exons overlap. The intertwined organization of these two genes is conserved in all gnathostomes examined, with the exception of the duplicate locus of zebrafish *Runx2b* in which the duplicate *Supt3hb* gene has been lost (Fig. 3). The close linkage between *Runx* and *Supt3h* genes seems to be an ancestral state, because in invertebrates such as amphioxus, sea anemone (*Nematostella vectensis*) and demosponge (*Amphimedon queenslandica*), the *Supt3h* gene is located next to the single *Runt* gene (Fig. 3). It can therefore be inferred that after the duplication of the ancestral vertebrate *Runx* locus in the gnathostome ancestor, the paralogs of *Supt3h* linked to *Runx1* and *Runx3* loci were lost whereas the paralog in *Runx2* locus was retained.

In order to verify the orthology relationships of the three lamprey *Runx* genes to gnathostome *Runx* genes, we analysed the synteny of genes at the lamprey *Runx* gene loci. Interestingly, none of the lamprey *Runx* genes is linked to a *Supt3h* gene. A TBLASTN search of the Japanese lamprey genome assembly using human and elephant shark Supt3h protein sequences identified a single *Supt3h* gene located on a scaffold (Scaf_1) distinct from those containing *Runx* genes. Either this gene previously located next to a *Runx* gene was translocated to a new location in the lamprey genome or the *Runx* gene linked to it has been lost. Of the three Japanese lamprey *Runx* genes, the most extensive synteny information is available for *RunxB*. The syntenic genes of the lamprey *RunxB* include three genes (*Son*, *Donson* and *Cdc5l*) whose orthologs are conserved in the *Runx* loci of gnathostomes. However, while the ortholog of *Son* and *Donson* are found in the gnathostome *Runx1* locus, the ortholog of *Cdc5l* is found linked to gnathostome *Runx2*. The synteny of genes in the lamprey *RunxB* locus therefore suggests that *RunxB* is not an exact ortholog of either *Runx1* or *Runx2* of gnathostomes. This provides further support to the inference of the phylogenetic

analysis that the three lamprey *Runx* genes are not one-to-one orthologs of the three gnathostome *Runx* genes.

## Genomic organization of Japanese lamprey α-subunit *Runx* genes

The exon-intron organization of the three gnathostome *Runx* genes is largely similar except for *Runx3* which characteristically lacks an exon (exon 5.1) at the C-terminal end, that is present in both *Runx1* and *Runx2* (Fig. S1). The three lamprey α-subunit *Runx* genes share several structural similarities with those of gnathostomes, including three exons that encode the highly conserved Runt domain (Fig. 1 and S1). Interestingly, of the three lamprey *Runx* genes, only *RunxB* contains the exon equivalent of exon 5.1 of gnathostome *Runx1* and *Runx2* genes and is therefore structurally more similar to these two gnathostome *Runx* genes. In addition, *RunxB* contains an extra exon (exon 4.1) that is not equivalent to any of the gnathostome exons including exon 4.1 in elephant shark *Runx1* (Fig. 1 and S1). This feature distinguishes the lamprey *RunxB* gene from gnathostome *Runx1* and *Runx2*. Lamprey *RunxA* and *RunxC* both lack the exon equivalent to exon 5.1 and are thus more similar to the gnathostome *Runx3* gene (Fig. 1A, C and S1). Since phylogenetic and synteny analyses do not support an orthology relationship between these two lamprey genes and the gnathostome *Runx3* gene, their similar structure appears to be the result of convergent evolution.

All gnathostome *Runx* genes are transcribed from two alternative promoters, P1 (distal) and P2 (proximal) that are separated by a characteristically large intron (Fig. S1). Of the three lamprey *Runx* genes, only *RunxB* contains both P1 and P2 promoters (Fig. 1). Lamprey *RunxA* contains only the proximal (P2) promoter. Our efforts to identify a distal promoter by 5′RACE using cDNA from several tissues failed to yield transcripts originating from a distal promoter and led us to conclude that the P1 promoter is absent in *RunxA* (Fig. 1A). By contrast, lamprey *RunxC* contains the P1 promoter but lacks the P2 promoter (Fig. 1C). The translation initiation codon (ATG) associated with P2 promoter has been mutated to another codon in this gene and the open reading frame which is devoid of an ATG codon runs all the way to the 5′end of exon 2. Thus, *RunxC* has lost the potential to generate transcripts from the proximal promoter. The presence of two alternative promoters in all gnathostome *Runx* genes and the lamprey *RunxB* gene suggests that the two alternative promoters had already emerged in the single ancestral *Runx* gene in the common ancestor of gnathostomes and cyclostomes. Following the duplication of the ancestral *Runx* gene, lamprey *RunxA* and *RunxC* genes secondarily lost P1 and P2 promoters, respectively.

Each of the gnathostome *Runx* genes express a diverse repertoire of isoforms arising from two alternative promoters (P1 and P2) as well as by alternative splicing of exons (Fig. 1). These isoforms are differentially expressed during development and exert a variety of biological functions [43,44]. Of the three lamprey *Runx* genes, only *RunxB* has the potential to generate isoforms originating from two alternative promoters. In addition, it gives rise to multiple isoforms resulting from alternative splicing of exons similar to gnathostome *Runx* genes (Fig. 1B). By contrast, we could identify only one transcript for lamprey *RunxA* and two isoforms for *RunxC* (Fig. 1A and C). Thus, these lamprey genes give rise to fewer isoforms than the lamprey *RunxB* gene and the three gnathostome *Runx* genes.

The P1 and P2 promoter regions of gnathostome *Runx* genes harbour binding sites for transcription factors that mediate transcriptional regulation of *Runx* genes. Notably, the P1

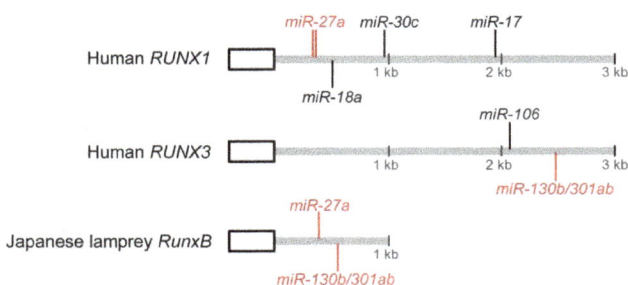

**Figure 4. miRNA binding sites in the 3′UTR of human *Runx1* and *Runx3* and Japanese lamprey *RunxB* genes.** The last coding region of *Runx* gene is represented by a rectangle and 3′UTR by a grey line. Positions of miRNA binding sites are indicated by vertical lines. Binding sites conserved in human and Japanese lamprey are shown in red.

**P1**

```
1  MASDSIFESFPS-YPQCFMRECILGMNPSRDVHD          HsRUNX1
1  MASNSLFSTVTP-CQQNFFWD                        HsRUNX2
1  MASNSIFDSFPT-YSPTFIRD                        HsRUNX3
1  MASNSLFESFPS-YQQCFMRD                        CmRunx1
1  MASNSLFSSVTP-CQQNFFWD                        CmRunx2
1  MASNSIFDSFPT-YPQPYLRD                        CmRunx3
1  --------------------                         LjRunxA
1  MASNSIFETF-SPYQQPFLRE                        LjRunxB
1  MASNSIFESFPSPYAQSFSRG                        LjRunxC
```

**P2**

```
1  MRIPVD  AS--TSRRFTPPSTALSP--------------GKMSEALPLGAPD-----               HsRUNX1
1  MRIPVD  PS--TSRRFSPPSSSLQP------------GKMSDVSPVVAAQQQQQQQQQQQQQQQQ       HsRUNX2
1  MRIPVD  PS--TSRRFTPPSPAFPCGG--------GGGKMGENSGALSAQ-                      HsRUNX3
1  MRVPVD  PS--TSRRFTPPPCTSLGS-------------GKMTDSLPLPPPP-                    CmRunx1
1  MRIPVD  PS--TSRRFTPPSTTLPAS-----------GKMSDVSGMAPHQESGP-                  CmRunx2
1  MRIPVD  PN--TSRRFTPPSTTLSSSG-----------SG-KMSEP-VSLSQ-                    CmRunx3
1  MHIPVD  AG--VSRRFTPPSTALLHAAHHHHHHHHGSSSSTSKMSSDAPPLSHQDG-               LjRunxA
1  MRTLLE  TGGGAGRRFAAPPAAAKLAAEG----------LGGSSDAALLHHHHHHLQQQQQQLGTG----   LjRunxB
1  ------  KPVEISSPLLHGSQVHAQQQQQHHQHQDHHHHQQQQQPSAPCSPPLSLGKLQQQQQQ-       LjRunxC
```

```
                                                                    Ⓟ      *         *
35  ------AGAALAG----------KLRSGD-RSMVEVLADHPGELVRTDSPNFLCSVLPTHWRCNKTL    HsRUNX1
53  QQQQQEAAAAAAAAAAAAAAAAAAVPRLRPPHDNRTMVEIIADHPAELVRTDSPNFLCSVLPSHWRCNKTL HsRUNX2
39  ------AAVGPGG----------RARPEV-RSMVDVLADHAGELVRTDSPNFLCSVLPSHWRCNKTL    HsRUNX3
35  ------PAPAASHPPHPDSALTKLRPAD-RTMVDVLADHPGELVRTDSPNFLCSVLPTHWRCNKTL     CmRunx1
40  -----GVAVAGAAAAAAAAALGRSLIRPHENRSMVDIIADHPAELVRTDSPNFLCSVLPSHWRCNKTL   CmRunx2
36  ------PSPGVLS----------RVRPET-RTVVDVLADHPGELVRTDSPNFLCSVLPSHWRCNKTL    CmRunx3
53  -----ALMGVKMRGAGGGGGGGGIGGAHDRPMGDVLADHPGELVRTDSPNFLCSVLPSHWRCNKTL     LjRunxA
55  ----PAGAMGVPAANCQDAAAGGAGAAGSLLHSKRATADVLADHPGELVRTDSPNFLCSVLPSHWRCNKTL LjRunxB
54  --------TPSHAQAGHPGPGSQVGGAPPEEAGLALGLGEQSSEMLRTDSPNFLCSALPPHWRCNKTL   LjRunxC
```

```
86   PIAFKVVALGDVPDGTLVTVMAGNDENYSAELRNATAAMKNQVARFNDLRFVGRSGRGKSFTLTITVFTN  HsRUNX1
123  PVAFKVVALGEVPDGTVVTVMAGNDENYSAELRNASAVMKNQVARFNDLRFVGRSGRGKSFTLTITVFTN  HsRUNX2
90   PIAYKVVSLGDVPDGTLVTVLAGNDENYSAELRNATAVMKNQVARFNDLRFVGRSGRGKSFTLTITVFTN  HsRUNX3
95   PVAFKVVSLGDVPDGTLVTVMAGNDENYSAELRNATAVMKNQVARFNDLRFVGRSGRGKSFTLTITVFTN  CmRunx1
105  PVAFKVVALGDVPDGTLVTVMAGNDENYSAELRNASAVMKNAVARFNDLRFVGRSGRGKSFTLTITVFTS  CmRunx2
87   PVAFKVVALGDVPDGTLVTVGVMAGNDENYSAELRNASAVMKQQVARFNDLRFVGRSGRGKSFTLTITVFTS CmRunx3
117  PVAFKVVALGDVPDGTLVTVLAGNDENYAAELRNATAVMKQQVARFNDLRFVGRSGRGKSFTLTITVFTS  LjRunxA
122  PVAFKVMAMADVPDGTPVAVMAGNDENYSAELRNASAVMKGHVARFNDLRFVGRSGRGKSLNLTITVFTS  LjRunxB
118                                                                           LjRunxC
```

```
156  PPQVATYHRAIKITVDGPREPRR-----------------------             HsRUNX1
193  PPQVATYHRAIKVTVDGPREPRR-----------------------             HsRUNX2
160  PTQVATYHRAIKVTVDGPREPRR-----------------------             HsRUNX3
165  PPQVATYHRAIKITVDGPREPRRSVLQTKLTYQWFGFCLEG----              CmRunx1
175  PPQVATYHRAIKVTVDGPREPRR-----------------------             CmRunx2
157  PPQVATYHRAIKVTVDGPREPRR-----------------------             CmRunx3
187  PPQVATYHRAIKVTVDGPREPRR-----------------------             LjRunxA
192  PPQVATYHRAIKVTVDGPREPRIIGDWAETIHSLHVTTRFNRGKLPGLRPGAQRCQLGALTSELPGHPLG  LjRunxB
188  PPQVATYQRAIKVTVDGPREPRR-----------------------             LjRunxC
```

```
179  HRQKLDDQTKPGSL--SFSERLSELEQLRRTAMRVSPHHP-APTPNPRASLN-HSTAFNPQPQSQMQDT   HsRUNX1
216  HRQKLDD-SKPS---LFSDRLSDLGRIPHPSMRVG----VPPQNPRPSLNSAPSPFNPQGQSQITDP     HsRUNX2
183  HRQKLEDQTKP----FPDRFGDLERLR--MRVTP-----STPSPRGSLS-TTSHFSSQPQTPIQ-       HsRUNX3
188  HRQKLEEQSKSAGM--AFSERLSELEQIRRSAMRGSPHHP-TATPNPRHTLNPTTTAFNPQAHSQIQDT   CmRunx1
198  HRQKPEDQPKVG---LFSERLSELERLRQTAMRVG----ASTQSPRHSLNPTPVPFSSQGQAQITDP     CmRunx2
180  HRQKLEDQAKPNT--LFSDRLSELERYRQTAMRVGP----STPSPRPQLN-APSHFSSSAQTQMP-      CmRunx3
210  HRQRLEDPVKAGAA--MFPERLCELEQFRRSGMRVPPHAGGGAHPAAPRASLG-AHTAFNAQPQPQMQ-   LjRunxA
262  HRQRLEDAGKALS---FSERLSELEHLRRTALRVGAHHVP-PPTPALRPPLN-PPPHYNAQT-----DG   LjRunxB
211  HRQRQEDMLKSNSMGGLFPERLLQLGPVS---------------PCYVTPTH-                   LjRunxC
```

```
                  Ⓟ                           Ⓟ
244  RQIQPSPPWSYDQSYQ-YLGSIASPSVHPATPISPGRASGMTTLSAELSSRLS----------         HsRUNX1
275  RQAQSSPPWSYDQSYPSYLSQMTSPSIHSTTPLSSTRGTGLPAIT-DVPRRISDDDTATSDFCLWPSTLS  HsRUNX2
234  -------------------------------------------------                      HsRUNX3
254  RQVQPSPPWTYDQSYP-YIGGISTPSVHPATPISPARASGMPTLTAEISSRLS----              CmRunx1
258  RQTQSSPPWSYDQSYPSYLSQMPSPSIHSTAPLSPARATGLPAITADMPRRLS----              CmRunx2
237  -------------------------------------------------                      CmRunx3
274  -------------------------------------------------                      LjRunxA
323  ---QPSPPWSFDQPYGPYLSQIAAPALHTNAGLSPSRAVGIA-                            LjRunxB
256  -------------------------------------------------                      LjRunxC
```

```
295  ----TAPDLTAFSDP-----RQFP----ALPSIS-----DPRMHYP------                 HsRUNX1
344  KKSQAGASELGPFSDP----RQFP----SISSLTES-RFSNPRMHYP----                 HsRUNX2
234  ----GTSELNPFSDP-----RQFDRSFPTLPTLTES-RFPDPRMHYPG---------AMS        HsRUNX3
305  ----VASDLTAFGDPR-MSIDRQFS----ALPPLSDS-RFPDPRMHYP----                CmRunx1
310  ----GASDLGAFSDH-----RQFERQF-GLSSLTDS-RFSNPRMHYP----                 CmRunx2
237  ----GSSDLSPFSDP-----RQFERQFPGLSSLADS-RFADPRVHYPG---------TMP        CmRunx3
274  ----ASSELGAFSAPPRVSLESQFAA---GLSSLPEG-RFQDPRVHYP----                LjRunxA
357  ----GGSELGAFSDP-----RQFA----SLSYLPDPSRNTDPRMHYSGPLSVYPTQQVGAGSLGLPVP  LjRunxB
256  ----DLQGVFNPSRMGLHHDG-QFA----LPDPRLHYP----                          LjRunxC
```

```
324  GAFTYSPTPVTSGIG-IGMSAMG--------------------SATRYH----TYLPPPYPGS       HsRUNX1
382  ATFTYTP-PVTSGMS-LGMSAT---------------------THYH------TYLPPPYPGS       HsRUNX2
276  AAFPYSATPSGTSISSLSVAGMP---------------------ATSRFHH---TYLPPPYPGA      HsRUNX3
344  GAFAYTPTPVTNGIG-LGMSAMT--------------------TAARYH----TYLPPPYPGS       CmRunx1
347  ATFTYTPTPVTSGMS-LGMSAT---------------------THYH------TYLPPPYPGS       CmRunx2
279  AAFTYSATPPTTGIGGISMSSMA--------------------STARYH----TYLPPPYPGS       CmRunx3
315  GAFAYTPGGAASSALGLGMSPMGGG---------PAASAAVAAAAARYHAAAASYHHLPPPYPGS     LjRunxA
415  ARYAYLPPPYPAHTPPHSHQSSPFTGSSPFTTAGSSPFNAASAPFTSGSSHFSAGPAPFQTSSASFQPGP LjRunxB
277  GAFAYPHLTPAPPPS-----------------------PRYH-----AYLPPPYPGS            LjRunxC
```

```
                                                           Ⓟ
361  --SQAQGGPFQASS----PSYH-LYYGAS---AGSYQFSMV-G---GERSPP-RILPPCTNAST---    HsRUNX1
415  -SQSQSGPFQTSS----TPY-LYYGTS---SGSYQFPMVPG---GDRSPP-RMLPPCTTTSN---     HsRUNX2
315  --PQNQSGPFQANP----APYH-LYYGTS---SGSYQFSMVAGSSSGGDRSPT-RMLASCTSSAASVA  HsRUNX3
381  -GQGQGGPFQTSS----APYH-LYYGTS---AGSYQFSMMSG---GDRSPP-RILPCTNAST---     CmRunx1
381  -PTNQSGPFQTST----APY-LYYGTS---SGSYQFSMVAG---GERSPS-RMLP-CTSAST---    CmRunx2
317  --AQNQGGTFQAGA----SPYH-LYYGTS---AGSYQFSMVAG----ERSPT-RMLPSCTSASS---   CmRunx3
374  QQGQGQGGSAFQAAAS--APFQHLYYGAAP-STSSYQFPPPVMQ----QQQQP-GAVGEERPLAT---  LjRunxA
485  AHFQASTAPFQSSSASFQSGFNQLYYGHS---GGPAYSLSVMAGD-ERQSAVGGRMVATGVGVCVGAG   LjRunxB
306  PPRHAPSSAYQNGGP--SGPYALFYGAPPSGAAGGYQLSLVAGAS-DQRSPE--LPGSVGACA-      LjRunxC
```

```
411  GSALLNPSLPNQSD---VVEAEGSH-SNSPTNMAPSARLEEA-----VWRPY----            HsRUNX1
465  GSTLLNPNLPNQND---GVDADGSH-SNSPTVLNSSGRMDES-----VWRPY----            HsRUNX2
373  AGNLMNPSLGGQSD---GVEADGSH-SNSPTALSTPGRMDEA-----VWRPY----            HsRUNX3
432  GSSLLNPSLPNQNE---VGETNGSH-SNSPTNMSTTARLEEG-----VWRPY----            CmRunx1
430  GSTLLNPNLPNQND---GVDADGSH-SSSPTVLSTGTGRIDES-----VWRPY----           CmRunx2
367  GSSLMNPNLANQND---GVDADGSH-SNSPT-LTTTGRLDES-----VWRPY----            CmRunx3
430  APLINNPALPSAHQDAGGHDPEGSQ-GSSPTSLSS-GRLEEA-----VWRPY----            LjRunxA
549  DSVLLTPGLSSPDGD-TADGGDGSS-SSSPSNANSLADAVPSRLEPES-VWRPY----          LjRunxB
364  -----------------EGTESDGSCPSVSPGALGAVRAMEEA------VWRPYQRPV          LjRunxC
```

**Figure 5. Runx α-subunit proteins in human, elephant shark and Japanese lamprey. Alignment of human, elephant shark and Japanese lamprey Runx α-subunit proteins.** The first block shows the amino-terminal part of the protein derived from the P1 promoter that differs from that derived from the P2 promoter. The highly conserved Runt domain is highlighted in pink. Within the Runt domain, surfaces involved in DNA contact and interaction with the β-subunit are denoted by black and blue lines, respectively. Cysteine residues involved in the redox regulation of DNA-binding activity are indicated with asterisks. Nuclear localization signal (NLS) is demarcated by a green dashed box. The PY and VWRPY motifs are indicated in red and blue, respectively. The transactivation domain (TAD) is highlighted in blue and the inhibitory domain (ID) is boxed in blue. The nuclear matrix targeting signal (NMTS) is boxed in black. Minimal consensus sequences for phosphorylation by Erk are boxed by dashed black lines. The consensus phosphorylation site for Cdc2 is indicated in green. The residue targeted for phosophorylation is indicated by Ⓟ. Hs, *Homo sapiens*; Cm, *Callorhinchus milii*; Lj, *Lethenteron japonicum*.

promoters of gnathostome *Runx1, 2* and *3* contain two tandem binding sites for Runx (Pu/TACCPuCA) at similar locations in the 5′UTRs [3]. These highly conserved binding sites have been implicated in the auto and cross-regulation of the three gnathostome *Runx* genes [45,46]. The presence of these sites in the three gnathostome *Runx* genes suggests that they were present in their single ancestral *Runx* gene. In contrast to gnathostome *Runx* genes, the P1 promoters of the lamprey *RunxB* and *RunxC* genes lack these characteristic tandem Runx binding sites. Instead, the P1 promoter of *RunxB* contains dispersed single Runx binding sites at positions −506, −282 and +131. In addition, its P2 promoter contains one Runx binding site at position −328 (data not shown). However, no such single Runx binding site is found in the P1 promoter of *RunxC* or in the P2 promoter of *RunxA*. It remains to be seen if these dispersed Runx motifs in the lamprey *RunxB* promoters are involved in its auto-regulation and/or in its cross-regulation by RunxA and RunxC.

The expression of gnathostome *Runx1–3* genes is post-transcriptionally regulated by miRNAs that target highly conserved miRNA binding sites in the 3′UTR of their transcripts [47,48,49]. To verify if the lamprey *Runx* genes are also regulated by miRNAs, we searched the 3′UTRs of the lamprey *Runx* genes for seed miRNA binding sites that are well characterized in the mammalian *Runx* genes. While no such binding sites were found in *RunxA* and *RunxC*, *RunxB* was found to contain binding sites for miR-27 and miR-130b/301ab (Fig. 4). In mammals, the binding site for miR-27 is present in *Runx1* whereas that for miR-130b/301ab is present in *Runx3* (Fig. 4). It is therefore possible that the single ancestral vertebrate *Runx* gene contained sites for both miRNAs and that they were differentially lost in mammalian *Runx1* and *Runx3* genes. Murine miR-27 has been shown to be

involved in the regulation of *Runx1* in megakaryocytic and granulocytic differentiation [48] whereas human miR-130b and miR-301a have been implicated in the down-regulation of *RUNX3* in gastric cancer [50,51]. These three miRNAs have been previously shown to be expressed in cyclostomes such as the sea lamprey and the Atlantic hagfish [52]. It is therefore likely that the predicted binding sites for these miRNAs in Japanese lamprey *RunxB* are functional and mediate regulation of *RunxB*.

## Comparison of Japanese lamprey Runx α-subunit protein sequences

The domain structures and motifs of Runx proteins are highly conserved among gnathostome Runx1–3. These include the 128 amino acid Runt domain that is required for their DNA binding and heterodimerization with the β-subunit, as well as the transactivation domain (TAD) and inhibitory domain (ID) located at their carboxy termini. In addition, gnathostome Runx proteins contain a nuclear localization signal (NLS) located contiguous to the Runt domain that mediates their nuclear import as well as a nuclear matrix targeting signal (NMTS) in the TAD that directs Runx transcription factors to specific nuclear-matrix associated sites involved in the regulation of gene expression. Other distinctive features of gnathostome Runx proteins include the PY and VWPRY motifs that mediate the transcriptional activity of Runx proteins by recruiting different interacting partners. The proline-rich PY motif in the TAD mediates the binding of Runx proteins to WW domain-containing proteins, such as Yes-associated protein (YAP), transcriptional co-activator with PDZ-binding motif (TAZ), and Smurf [53,54,55]; while the C-terminal VWRPY motif recruits the Groucho/Transducin-like enhancer

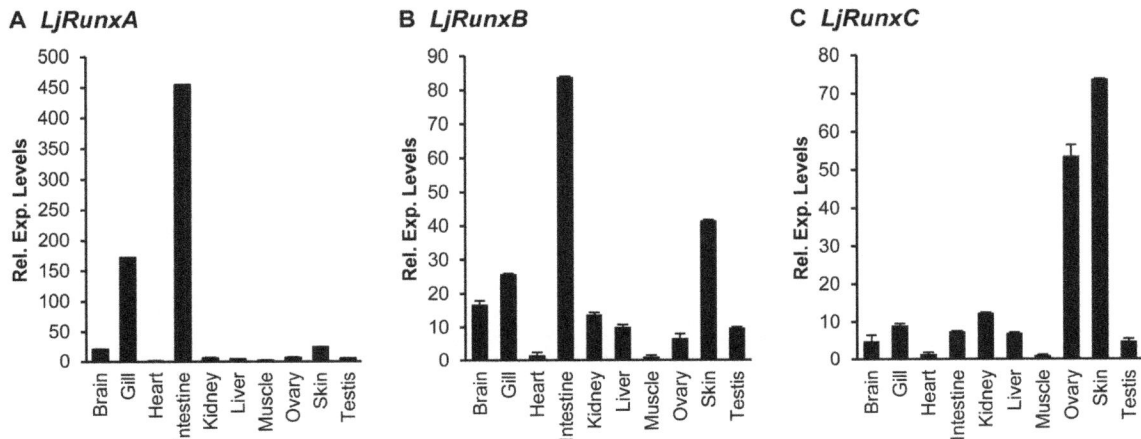

**Figure 6. Expression patterns of Japanese lamprey α-subunit *Runx* genes.** Relative expression levels of (A) *LjRunxA*, (B) *LjRunxB* and (C) *LjRunxC* in various tissues of the Japanese lamprey determined by qRT-PCR. Note that the relative expression levels of each *Runx* gene between different tissues are estimated in relation to the expression level of the tissue showing the lowest expression, and hence the expression levels are not comparable between different *Runx* genes.

(TLE) transcriptional co-repressor for transcriptional repression [56].

All of these domains and motifs are conserved in the Japanese lamprey RunxA, B and C proteins (Fig. 5). All three proteins contain the characteristic Runt domain, within which, residues for DNA-binding and interaction with the β-subunit [57,58] are remarkably well conserved with those of gnathostome Runx proteins. The NLS, NMTS, PY and VWPRY motifs are also conserved in lamprey Runx proteins albeit with slight variations: while gnathostome Runx proteins terminate at the VWPRY motif, the open reading frame of lamprey RunxC continues beyond the VWPRY motif to include four additional amino acids (Fig. 5). Among all known Runx proteins, this peculiar feature is seen only in the single Runt protein of *Ciona*. The biological significance of these extra residues is unknown. Additionally, lamprey RunxB contains a unique stretch of 47 amino acids within the NLS domain of RunxB (Fig. 5, boxed in green dotted lines). This sequence is encoded by the lamprey-specific extra exon 4.1 that is alternatively spliced in two of the isoforms of *RunxB* (Fig. 1B). Whether these intervening amino acids affect the nuclear translocation of these RunxB isoforms remains to be verified. The N-termini of the lamprey Runx P1 isoforms beginning with MAS(N/D)S are highly conserved with gnathostome Runx proteins. However, the N-termini of the P2 isoforms of the lamprey proteins vary from those of gnathostomes which comprise MR(I/V)PV. The N-terminal sequences of the lamprey RunxA and RunxB P2 isoforms contain MHIPV and MRTLL, respectively. Similar divergent N-terminal sequences have also been noted in the P2 promoters of *Runx1* (MVFLW), *Runx2* (MRPIV) and *Runx3* (MHIPV) genes of teleost fishes such as fugu and zebrafish [10]. The implications of such divergent N-terminal sequences on the functions of Runx proteins are not known.

The stability and activity of gnathostome Runx proteins are affected by several post-translational modifications including phosphorylation. In human RUNX1, the serine and threonine residues S249, S266, S276 and T273 followed by a proline residue act as phosphorylation sites for ERK [59,60]. Among these sites, S249 and S276 are conserved in lamprey RunxB but not in the other two lamprey Runx proteins (Fig. 5). Phosphorylation at serine residue S104 within the Runt domain of human RUNX2 has been shown to negatively regulate RUNX2 activity by inhibiting its heterodimerization with RUNXβ [61]. This serine residue is invariably conserved among all gnathostome Runx. It is also conserved in all three lamprey Runx proteins. Additionally, the consensus phosphorylation site for CDC2, (S/T)PX(R/K), at

which serine residue S451 of human RUNX2 was reported to be phosphorylated [62], is highly conserved in gnathostome Runx1–3 proteins. However, it is conserved only in lamprey RunxC and not in the other two lamprey Runx proteins (Fig. 5).

## Expression profile of Japanese lamprey α-subunit *Runx* genes

We investigated the expression patterns of *Runx* genes in various tissues of adult Japanese lamprey by quantitative RT-PCR. *RunxA* and *RunxB* are highly expressed in the gills and intestine- tissues that comprise the primary lymphoid organs of the lamprey (Fig. 6A and B). In the lamprey, lymphocytes develop in the typhlosole, an invaginated spiral valve spanning the length of the intestine as well in the "thymoid", the lamprey thymus-equivalent that is located at the tips of gill filaments [30]. *Runx* expression in the gills is reminiscent of that seen in the hagfish and amphioxus, where the gills and associated blood vessels are enriched with lymphocyte-like cells [4,63]. Expression of *Runx* in the lymphoid compartments of these phylogenetically ancient chordates is consistent with its integral function in the specification of immune-related cells in gnathostomes, and may indicate a possible role for *Runx* in the primordial immune system of the chordate ancestor. Furthermore, significant *Runx* expression in the intestine of the adult lamprey appears to support the prevailing view of an ancestral function of *Runx* genes in the gut, since sea urchin [64], nematode (*C.elegans*) [65], amphioxus [4] and gnathostomes all express *Runx* in the developing gut.

Particularly striking is the significant expression of *RunxB* and *RunxC* in the skin of the lamprey (Fig. 6B and C). Apart from the primary lymphoid organs such as the gill and intestine, the lamprey epidermis is also home to abundant numbers of lymphocytes, in particular, the VLRC+ lymphocytes that display dendritic morphology and which have been posited to phenotyp-ically resemble mammalian dendritic epidermal T cells (DETC) [31]. Though yet to be verified, based on the significant role *Runx3* plays in the development of DETC [66], it may be postulated that the lamprey *RunxC* similarly contributes to the specification of these immune cell types in the lamprey. In addition, the lamprey brain is another domain of significant *RunxB* expression (Fig. 6B). *Runx* expression in the brain and cranial ganglia has been previously demonstrated in the early larvae of the sea lamprey [33]. These expression patterns are consistent with those observed in the neuronal tissues of gnathostomes [27,67,68] as well as a basal metazoan, the sea

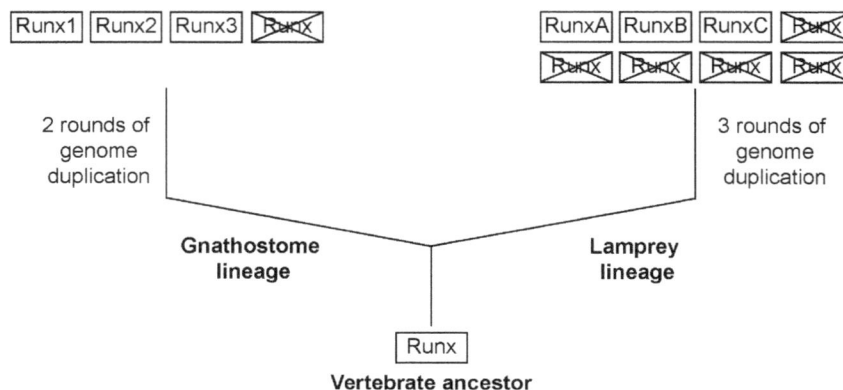

**Figure 7. A model depicting the evolution of *Runx* genes in vertebrates.** The phylogenetic analysis and synteny maps suggest that the three *Runx* genes in lamprey are not one-to-one orthologs of the three *Runx* genes in gnathostomes.

## A  LjRunxβ (72 kb)

**Gene structure**

1 [2] [3] [4] [4.1] [5] 6

**Isoforms**

LjRunxβ Type 1

LjRunxβ Type 3

LjRunxβ Type 4

LjRunxβ Type 5

**B**

```
  1  MPRVVPDQRSKFENEEFFRKLSRECEIKYTGFRDRPHEERQARFQNACRDGRSEIAFVATGTNLSLQFFP     HsRunxβ Type 1
  1  MPRVVPDQRSKFENEEFFRKLSRECEIKYTGFRDRPHEERQMRFQNASRDGRSEIAFVATGTNLSLQFFP     CmRunxβ Type 1
  1  MPRVVPDQRSKFDNEEFFRKLSRECEVKYTGYRDRPLEERQMRFQSACREGRSDLAFVATGTNLSLQFLP     LjRunxβ Type 1

 71  ASWQGE-QRQTPSREYVDLEREAGKVYLKAPMILNGVCVIWKGWIDLQRLDGMGCLEFDEERAQ---QED     HsRunxβ Type 1
 71  ANWHGE-QRQTPTREYVDFDRESGKVYLKAPMILNGVCVIWKGWIDLQRLDGMGFLEFDEERAQ---QED     CmRunxβ Type 1
 71  PTFHTEGQRPAPTRDYVDFEREQGKVHLKAPMILNGVCVIWRGWVDLQRLDGMACLEFDEERAR---QED     LjRunxβ Type 1
                                                                    ---QED     HsRunxβ Type 2
                                                                    ---QED     CmRunxβ Type 2
                                                                    ------     HsRunxβ Type 3
                                                                    ------     CmRunxβ Type 3
                                                                    ------     LjRunxβ Type 3
                                                                    LVA---     LjRunxβ Type 4
                                                                    ---QED     LjRunxβ Type 5

137  ALAQQAFEEARRRTREFEDRDRSHREEME----------ARRQQDPSPGSN-LGGGDDLKLR            HsRunxβ Type 1
137  TLAQQAYEDVRRRARDFEDRDRSHRDDME----------ARRQQDPSPGSN-LGSGDELKLR            CmRunxβ Type 1
138  LLAQQAFEEARRRARDFEDRERVHREDLE----------GRRPDTSPNSGGMNSADELKLR            LjRunxβ Type 1
137  ALAQQAFEEARRRTREFEDRDRSHREEMEVRVSQLLAVTGKKTTRP------                    HsRunxβ Type 2
137  TLAQQAYEDVRRRARDFEDRDRSHRDDMEVR--------GTQAAGPQSRLQPGERR------            CmRunxβ Type 2
133  -----------------------------------ARRQQDPSPGSN-LGGGDDLKLR            HsRunxβ Type 3
133  -----------------------------------ARRQQDPSPGSN-LGSGDELKLR            CmRunxβ Type 3
134  -----------------------------------GRRHPDTSPNSGGMNSADELKLR            LjRunxβ Type 3
137  --------------------------------------------------------             LjRunxβ Type 4
138  LLAQQAFEEARRRARDFEDRERVHREDLEVR-------------------------             LjRunxβ Type 5
```

**Figure 8. Exon-intron organization and protein sequence encoded by the Japanese lamprey *Runxβ*.** (A) Schematic representation of the genomic structure and the four transcripts cloned (*LjRunxb* types 1, 3, 4 and 5). Exons are indicated by boxes. The 5'- and 3'-UTRs are represented as crosshatched boxes. (B) Alignment of Japanese lamprey, elephant shark and human RUNXβ amino acid sequences using ClustalW. Conserved residues are shaded grey. Hs, *Homo sapiens*; Cm, *Callorhinchus milii*; Lj, *Lethenteron japonicum*.

anemone, *Nematostella* [11], suggestive of an evolutionarily ancient function in neural development.

Prominent expression of *RunxC* was also observed in the Japanese lamprey ovary (Fig. 6C). This expression pattern is concordant with *Runx* function in the female reproductive system of mammals [69,70] and *Drosophila* [71] and may therefore hint at a possible role of *Runx* in conserved pathways of ovarian function. In gnathostomes, Runx proteins are central regulators in the development of cellular cartilage and ossification of endochondral bone [28,72]. Expression of *Runx* has been described in the cartilage of vertebrates that lack mineralized bone, such as the hagfish and cartilaginous fishes [4], suggesting that *Runx* genes have a conserved functional role in cartilage formation in stem vertebrates. In the sea lamprey, Cattell *et. al* [33] detected *RunxA* and *RunxB* expression in the mucocartilage of larvae but noted

the apparent absence of *Runx* expression in the branchial basket cartilage, skeletal elements that are believed to be homologous to the cellular cartilage of gnathostomes. This led to the speculation that *Runx* genes may have lost their ancestral function in cartilage development in the lamprey lineage or are not engaged in gene regulatory networks of the ancestral vertebrate cartilage [33]. Here, we have reported the presence of three *Runx* genes in the Japanese lamprey, as compared to only two previously identified in the sea lamprey. The characterization of the expression of all three *Runx* genes in the lamprey may shed new light into the involvement of *Runx* in early vertebrate skeletal development.

### Evolution of *Runx* family genes in vertebrates

Invertebrate chordates such as the amphioxus, *Branchiostoma floridae* (cephalochordate) and *Ciona intestinalis* (urochordate) contain a single *Runx* gene, known as *Runt* genes. The functions of these genes are not well characterized, although reports documenting *Runx* expression in the developing gut, pharyngeal endoderm and regenerating oral cirral skeletal cells of the amphioxus [4,73] reflect functions in endodermal specification and roles in the rudimentary skeletogenic programs of these phylogenetically ancient chordates.

Based on the identification of the three *Runx* genes in the lamprey and their comparative analysis with *Runx* genes known in gnathostomes, we propose that the stem vertebrate ancestor contained a single *Runx* gene. As part of the two rounds of whole genome duplications, this gene locus underwent duplications (Fig. 7) in the lineage that gave rise to gnathostomes such as cartilaginous fishes, lobe-finned fishes and tetrapods giving rise to four *Runx* paralogs [6,7]. Since these vertebrates contain only three *Runx* paralogs each (*Runx1*, *Runx2* and *Runx3*), we infer that the fourth paralog was lost in the common ancestor of gnathostomes. The remaining three *Runx* genes in the gnathostome ancestor acquired specialized and distinct functions such that *Runx1* is involved mainly in hematopoiesis [18], *Runx2* in skeletogenesis [19] and *Runx3* in neurodevelopment and immunity [24,25,27].

The whole-genome duplication history in the lamprey lineage is not well resolved. Although previous studies suggested that lampreys shared the two rounds of duplication with gnathostomes, a more recent study suggested that the two rounds of duplications may have occurred independently in the lamprey lineage followed by an additional round of whole genome duplication [35]. In any case, the presence of six *Hox* clusters in the Japanese lamprey and the relative age of paralogous genes in the lamprey genome as measured by the transversion rate of four-fold degenerate sites [35] suggests that its lineage has experienced three rounds of genome duplication. According to this hypothesis, the genome duplications would have given rise to eight paralogs of *Runx* genes in the lamprey lineage. Since only three *Runx* genes are present in the Japanese lamprey, we infer that five of the eight *Runx* genes have been lost (Fig. 7). Although the exact functions of the lamprey *Runx* genes have not been investigated, the differential expression patterns of *Runx* genes in various tissues of the Japanese lamprey (Fig. 6), sea lamprey [33] and the Atlantic hagfish [4] are suggestive of distinct functions of each of the cyclostome *Runx* genes. However, further investigations will be required to understand the specific functions of the *Runx* gene family in cyclostomes and how they differ from those in gnathostomes. Since lampreys lack mineralized tissues and an Ig-based adaptive immune system, some differences in the functions of gnathostome and lamprey *Runx* genes are expected.

## Characterization of a Japanese lamprey β-subunit *Runx* gene

Gnathostome α-subunit Runx proteins form heterodimeric complexes with their β-subunit partner. This association not only enhances the DNA binding affinity of the complex by stabilizing the interaction of the α-subunit Runt domain with DNA, it also protects against the ubiquitin-mediated degradation of the α-subunit [74]. In addition to α-subunit *Runx* genes in the Japanese lamprey, we also cloned full-length coding sequence and various isoforms of the β-unit encoding *Runxb*. The genomic organization of the Japanese lamprey *Runxb* gene is largely similar to that of its gnathostome ortholog except for the presence of an additional exon 4.1 (Fig. 8A). Through alternative splicing, the gnathostome *Runxb* gene gives rise to three isoforms, *Runxb* Type 1, 2 and 3. Type 1 and Type 2 differ in several amino acids at their carboxyl termini while

Type 1 and Type 3 isoforms are almost identical, except for the absence of exon 5 in Type 3 owing to exon skipping [1]. The Japanese lamprey *Runxb* is also transcribed into these three isoforms (Fig. 8A). In addition, we identified two isoforms (Type 4 and 5) that are unique to the Japanese lamprey (Fig. 8A). *Runxb* Type 4 terminates at the alternative exon 4.1, while *Runxb* Type 5 appears to be a truncated form of Type 1, terminating prematurely in exon 5 (Fig. 6A). The *Runxb* type 1, 3, 4 and 5 isoforms encode proteins of 189, 157, 137 and 168 amino acids, respectively (Fig. 6B). At the protein level, lamprey and gnathostome Runxβ show high conservation of amino acid residues 1–165 (Fig. 8B). Of these, the N-terminal 135 amino acids required for its heterodimerization with the α-subunit and stimulation of DNA-binding activity [2] as well as essential residues 68–93 for its interaction with filamin A in the cytoplasm [75] are almost perfectly conserved Functional significance of the various isoforms of gnathostomes Runxβ are not well documented, though one might speculate that Runxβ Type 3, which is missing the domain required for the heterodimerization with the α-subunit, may perform functions that are independent of its interaction with α-subunit Runx proteins. Such a function is supported by the recent observation that the retention of Runxβ in the cellular midbody during cytokinesis was not seen to be compromised in the absence of α-subunit Runx [76]. It remains to be demonstrated whether the various isoforms of lamprey Runxβ exhibit different functions.

## Supporting Information

**Table S1 Primers used for qRT-PCR of Japanese lamprey *Runx* genes.**

**Figure S1 Exon-intron organization of elephant shark and lamprey *Runx* genes.** Schematic representation of the gene structures of elephant shark *Runx1*, *Runx2* and *Runx3* and lamprey *RunxA*, *RunxB* and *RunxC*. Exons are indicated by boxes. Exons constituting the Runt domain are indicated in grey. The two alternative promoters are denoted as P1 and P2. Crosshatched boxes indicate 5′- and 3′-UTRs. The asterisk (*) indicates an exon in *LjRunxB* that is absent in mammals and different from exon 4.1 in elephant shark. Not drawn to scale.

**Figure S2 Phylogenetic analysis of chordate Runx sequences (Bayesian Inference).** Protein sequences of Japanese lamprey *Runx* genes were aligned with homologous sequences from selected chordates. A Bayesian inference (BI) tree was generated for the alignment. Statistical support values for the nodes are shown as Bayesian posterior probability values. Hagfish and Japanese lamprey Runx proteins are highlighted in red. Lancelet (*Branchiostoma floridae*) Runt (BfRunt) was used as the outgroup. Hs, *Homo sapiens*; Gg, *Gallus gallus*; Dr, *Danio rerio*; Cm, *Callorhinchus milii*; Sc, *Scyliorhinus canicula*; Mg, *Myxine glutinosa*; Lj, *Lethenteron japonicum*.

**Figure S3 Runx protein sequence alignment used for phylogenetic tree in Fig. 2 and Fig. S2.** Alignment obtained after trimming the gaps using the Gblocks Server (ver. 0.91b). Hs, *Homo sapiens*; Gg, *Gallus gallus*; Dr, *Danio rerio*; Cm, *Callorhinchus milii*; Sc, *Scyliorhinus canicula*; Mg, *Myxine glutinosa*; Lj, *Lethenteron japonicum*; Bf, *Branchiostoma floridae*.

**Figure S4 Runx protein sequence alignment used for phylogenetic tree in Fig. S5.** Alignment obtained after trimming the gaps using the Gblocks Server (ver. 0.91b). Hs, *Homo sapiens*; Gg, *Gallus gallus*; Dr, *Danio rerio*; Cm, *Callorhinchus milii*; Sc, *Scyliorhinus canicula*; Mg, *Myxine glutinosa*; Lj, *Lethenteron japonicum*; Ci, *Ciona intestinalis*; Bf, *Branchiostoma floridae*; Nv, *Nematostella vectensis*.

**Figure S5 Phylogenetic analysis of chordate Runx sequences (including CiRunt and NvRunx).** Protein sequences of Japanese lamprey *Runx* genes were aligned with homologous sequences from selected chordates. A Maximum Likelihood (ML) tree employing the JTT+G model was generated for the alignment. Sea anemone (*Nematostella vectensis*) Runx (NvRunx) was used as the outgroup. Hs, *Homo sapiens*; Gg, *Gallus gallus*; Dr, *Danio rerio*; Cm, *Callorhinchus milii*; Sc, *Scyliorhinus*

*canicula*; Mg, *Myxine glutinosa*; Lj, *Lethenteron japonicum*; Bf, *Branchiostoma floridae*; Ci, *Ciona intestinalis*.

## Acknowledgments

We thank Seiji Yanai for helping in the collection of Japanese lamprey specimens, and Vydianathan Ravi and Alison P. Lee for critical reading the manuscript.

## Author Contributions

Conceived and designed the experiments: SB MO BV. Performed the experiments: GSN BT. Analyzed the data: GSN BV. Wrote the paper: GSN BV.

## References

1. Ogawa E, Inuzuka M, Maruyama M, Satake M, Naito-Fujimoto M, et al. (1993) Molecular cloning and characterization of PEBP2 beta, the heterodimeric partner of a novel Drosophila runt-related DNA binding protein PEBP2 alpha. Virology 194: 314–331.
2. Kagoshima H, Akamatsu Y, Ito Y, Shigesada K (1996) Functional dissection of the alpha and beta subunits of transcription factor PEBP2 and the redox susceptibility of its DNA binding activity. J Biol Chem 271: 33074–33082.
3. Nah GS, Lim ZW, Tay BH, Osato M, Venkatesh B (2014) Runx Family Genes in a Cartilaginous Fish, the Elephant Shark (Callorhinchus milii). PloS one 9: e93816.
4. Hecht J, Stricker S, Wiecha U, Stiege A, Panopoulou G, et al. (2008) Evolution of a core gene network for skeletogenesis in chordates. PLoS Genet 4: e1000025.
5. Levanon D, Groner Y (2004) Structure and regulated expression of mammalian RUNX genes. Oncogene 23: 4211–4219.
6. Dehal P, Boore JL (2005) Two rounds of whole genome duplication in the ancestral vertebrate. PLoS Biol 3: e314.
7. Putnam NH, Butts T, Ferrier DE, Furlong RF, Hellsten U, et al. (2008) The amphioxus genome and the evolution of the chordate karyotype. Nature 453: 1064–1071.
8. Christoffels A, Koh EG, Chia JM, Brenner S, Aparicio S, et al. (2004) Fugu genome analysis provides evidence for a whole-genome duplication early during the evolution of ray-finned fishes. Mol Biol Evol 21: 1146–1151.
9. Glusman G, Kaur A, Hood L, Rowen L (2004) An enigmatic fourth runt domain gene in the fugu genome: ancestral gene loss versus accelerated evolution. BMC Evol Biol 4: 43.
10. Ng CE, Osato M, Tay BH, Venkatesh B, Ito Y (2007) cDNA cloning of Runx family genes from the pufferfish (Fugu rubripes). Gene 399: 162–173.
11. Sullivan JC, Sher D, Eisenstein M, Shigesada K, Reitzel AM, et al. (2008) The evolutionary origin of the Runx/CBFbeta transcription factors–studies of the most basal metazoans. BMC Evol Biol 8: 228.
12. Fujioka M, Yusibova GL, Sackerson CM, Tillib S, Mazo A, et al. (1996) Runt domain partner proteins enhance DNA binding and transcriptional repression in cultured Drosophila cells. Genes Cells 1: 741–754.
13. Golling G, Li L, Pepling M, Stebbins M, Gergen JP (1996) Drosophila homologs of the proto-oncogene product PEBP2/CBF beta regulate the DNA-binding properties of Runt. Mol Cell Biol 16: 932–942.
14. Okuda T, van Deursen J, Hiebert SW, Grosveld G, Downing JR (1996) AML1, the target of multiple chromosomal translocations in human leukemia, is essential for normal fetal liver hematopoiesis. Cell 84: 321–330.
15. Jacob B, Osato M, Yamashita N, Wang CQ, Taniuchi I, et al. (2010) Stem cell exhaustion due to Runx1 deficiency is prevented by Evi5 activation in leukemogenesis. Blood 115: 1610–1620.
16. Osato M (2004) Point mutations in the RUNX1/AML1 gene: another actor in RUNX leukemia. Oncogene 23: 4284–4296.
17. Osato M, Ito Y (2005) Increased dosage of the RUNX1/AML1 gene: a third mode of RUNX leukemia? Crit Rev Eukaryot Gene Expression 15: 217–228.
18. Speck NA, Gilliland DG (2002) Core-binding factors in haematopoiesis and leukaemia. Nat Rev Cancer 2: 502–513.
19. Komori T, Yagi H, Nomura S, Yamaguchi A, Sasaki K, et al. (1997) Targeted disruption of Cbfa1 results in a complete lack of bone formation owing to maturational arrest of osteoblasts. Cell 89: 755–764.
20. Otto F, Kanegane H, Mundlos S (2002) Mutations in the RUNX2 gene in patients with cleidocranial dysplasia. Hum Mutat 19: 209–216.
21. Li QL, Ito K, Sakakura C, Fukamachi H, Inoue K, et al. (2002) Causal relationship between the loss of RUNX3 expression and gastric cancer. Cell 109: 113–124.
22. Brenner O, Levanon D, Negreanu V, Golubkov O, Fainaru O, et al. (2004) Loss of Runx3 function in leukocytes is associated with spontaneously developed colitis and gastric mucosal hyperplasia. Proc Natl Acad Sci U S A 101: 16016–16021.
23. Taniuchi I, Osato M, Egawa T, Sunshine MJ, Bae SC, et al. (2002) Differential requirements for Runx proteins in CD4 repression and epigenetic silencing during T lymphocyte development. Cell 111: 621–633.
24. Ohno S, Sato T, Kohu K, Takeda K, Okumura K, et al. (2008) Runx proteins are involved in regulation of CD122, Ly49 family and IFN-gamma expression during NK cell differentiation. Int Immunol 20: 71–79.
25. Fainaru O, Woolf E, Lotem J, Yarmus M, Brenner O, et al. (2004) Runx3 regulates mouse TGF-beta-mediated dendritic cell function and its absence results in airway inflammation. EMBO J 23: 969–979.
26. Watanabe K, Sugai M, Nambu Y, Osato M, Hayashi T, et al. (2010) Requirement for Runx proteins in IgA class switching acting downstream of TGF-beta 1 and retinoic acid signaling. J Immunol 184: 2785–2792.
27. Inoue K, Shiga T, Ito Y (2008) Runx transcription factors in neuronal development. Neural Dev 3: 20.
28. Yoshida CA, Yamamoto H, Fujita T, Furuichi T, Ito K, et al. (2004) Runx2 and Runx3 are essential for chondrocyte maturation, and Runx2 regulates limb growth through induction of Indian hedgehog. Genes Dev 18: 952–963.
29. Osorio J, Retaux S (2008) The lamprey in evolutionary studies. Dev Genes Evol 218: 221–235.
30. Bajoghli B, Guo P, Aghaallaei N, Hirano M, Strohmeier C, et al. (2011) A thymus candidate in lampreys. Nature 470: 90–94.
31. Hirano M, Das S, Guo P, Cooper MD (2011) The evolution of adaptive immunity in vertebrates. Adv Immunol 109: 125–157.
32. Kasahara M, Sutoh Y (2014) Two forms of adaptive immunity in vertebrates: similarities and differences. Adv Immunol 122: 59–90.
33. Cattell M, Lai S, Cerny R, Medeiros DM (2011) A new mechanistic scenario for the origin and evolution of vertebrate cartilage. PloS one 6: e22474.
34. Smith JJ, Antonacci F, Eichler EE, Amemiya CT (2009) Programmed loss of millions of base pairs from a vertebrate genome. Proceedings of the National Academy of Sciences of the United States of America 106: 11212–11217.
35. Mehta TK, Ravi V, Yamasaki S, Lee AP, Lian MM, et al. (2013) Evidence for at least six Hox clusters in the Japanese lamprey (Lethenteron japonicum). Proc Natl Acad Sci U S A 110: 16044–16049.
36. Kuraku S, Kuratani S (2006) Time scale for cyclostome evolution inferred with a phylogenetic diagnosis of hagfish and lamprey cDNA sequences. Zool Sci 23: 1053–1064.
37. Smith JJ, Kuraku S, Holt C, Sauka-Spengler T, Jiang N, et al. (2013) Sequencing of the sea lamprey (Petromyzon marinus) genome provides insights into vertebrate evolution. Nat Genet 45: 415–421, 421e411–412.
38. Castresana J (2000) Selection of conserved blocks from multiple alignments for their use in phylogenetic analysis. Mol Biol Evol 17: 540–552.
39. Venkatesh B, Lee AP, Ravi V, Maurya AK, Lian MM, et al. (2014) Elephant shark genome provides unique insights into gnathostome evolution. Nature 505: 174–179.
40. Livak KJ, Schmittgen TD (2001) Analysis of relative gene expression data using real-time quantitative PCR and the 2(-Delta Delta C(T)) Method. Methods 25: 402–408.
41. Van de Peer Y, Maere S, Meyer A (2010) 2R or not 2R is not the question anymore. Nature reviews Genetics 11: 166.
42. Van de Peer Y, Maere S, Meyer A (2009) The evolutionary significance of ancient genome duplications. Nature reviews Genetics 10: 725–732.
43. Bee T, Swiers G, Muroi S, Pozner A, Nottingham W, et al. (2010) Nonredundant roles for Runx1 alternative promoters reflect their activity at discrete stages of developmental hematopoiesis. Blood 115: 3042–3050.
44. Liu JC, Lengner CJ, Gaur T, Lou Y, Hussain S, et al. (2011) Runx2 protein expression utilizes the Runx2 P1 promoter to establish osteoprogenitor cell number for normal bone formation. J Biol Chem 286: 30057–30070.
45. Drissi H, Luc Q, Shakoori R, Chuva De Sousa Lopes S, Choi JY, et al. (2000) Transcriptional autoregulation of the bone related CBFA1/RUNX2 gene. J Cell Physiol 184: 341–350.

46. Spender LC, Whiteman HJ, Karstegl CE, Farrell PJ (2005) Transcriptional cross-regulation of RUNX1 by RUNX3 in human B cells. Oncogene 24: 1873–1881.

47. Xu Y, Wang K, Gao W, Zhang C, Huang F, et al. (2013) MicroRNA-106b regulates the tumor suppressor RUNX3 in laryngeal carcinoma cells. FEBS Lett 587: 3166–3174.

48. Rossetti S, Sacchi N (2013) RUNX1: A MicroRNA Hub in Normal and Malignant Hematopoiesis. Int J Mol Sci 14: 1566–1588.

49. Lian JB, Stein GS, van Wijnen AJ, Stein JL, Hassan MQ, et al. (2012) MicroRNA control of bone formation and homeostasis. Nat Rev Endocrinol 8: 212–227.

50. Lai KW, Koh KX, Loh M, Tada K, Subramaniam MM, et al. (2010) MicroRNA-130b regulates the tumour suppressor RUNX3 in gastric cancer. Eur J Cancer 46: 1456–1463.

51. Wang M, Li C, Yu B, Su L, Li J, et al. (2013) Overexpressed miR-301a promotes cell proliferation and invasion by targeting RUNX3 in gastric cancer. J Gastroenterol 48: 1023–1033.

52. Heimberg AM, Cowper-Sal-lari R, Semon M, Donoghue PC, Peterson KJ (2010) microRNAs reveal the interrelationships of hagfish, lampreys, and gnathostomes and the nature of the ancestral vertebrate. Proc Natl Acad Sci U S A 107: 19379–19383.

53. Kanai F, Marignani PA, Sarbassova D, Yagi R, Hall RA, et al. (2000) TAZ: a novel transcriptional co-activator regulated by interactions with 14-3-3 and PDZ domain proteins. EMBO J 19: 6778–6791.

54. Yagi R, Chen LF, Shigesada K, Murakami Y, Ito Y (1999) A WW domain-containing yes-associated protein (YAP) is a novel transcriptional co-activator. EMBO J 18: 2551–2562.

55. Jin YH, Jeon EJ, Li QL, Lee YH, Choi JK, et al. (2004) Transforming growth factor-beta stimulates p300-dependent RUNX3 acetylation, which inhibits ubiquitination-mediated degradation. J Biol Chem 279: 29409–29417.

56. Javed A, Guo B, Hiebert S, Choi JY, Green J, et al. (2000) Groucho/TLE/R-esp proteins associate with the nuclear matrix and repress RUNX (CBF(alpha)/AML/PEBP2(alpha)) dependent activation of tissue-specific gene transcription. J Cell Sci 113 (Pt 12): 2221–2231.

57. Akamatsu Y, Ohno T, Hirota K, Kagoshima H, Yodoi J, et al. (1997) Redox regulation of the DNA binding activity in transcription factor PEBP2. The roles of two conserved cysteine residues. J Biol Chem 272: 14497–14500.

58. Tahirov TH, Inoue-Bungo T, Morii H, Fujikawa A, Sasaki M, et al. (2001) Structural analyses of DNA recognition by the AML1/Runx-1 Runt domain and its allosteric control by CBFbeta. Cell 104: 755–767.

59. Tanaka T, Kurokawa M, Ueki K, Tanaka K, Imai Y, et al. (1996) The extracellular signal-regulated kinase pathway phosphorylates AML1, an acute myeloid leukemia gene product, and potentially regulates its transactivation ability. Mol Cell Biol 16: 3967–3979.

60. Zhang Y, Biggs JR, Kraft AS (2004) Phorbol ester treatment of K562 cells regulates the transcriptional activity of AML1c through phosphorylation. J Biol Chem 279: 53116–53125.

61. Wee HJ, Huang G, Shigesada K, Ito Y (2002) Serine phosphorylation of RUNX2 with novel potential functions as negative regulatory mechanisms. EMBO reports 3: 967–974.

62. Qiao M, Shapiro P, Fosbrink M, Rus H, Kumar R, et al. (2006) Cell cycle-dependent phosphorylation of the RUNX2 transcription factor by cdc2 regulates endothelial cell proliferation. J Biol Chem 281: 7118–7128.

63. Huang G, Xie X, Han Y, Fan L, Chen J, et al. (2007) The identification of lymphocyte-like cells and lymphoid-related genes in amphioxus indicates the twilight for the emergence of adaptive immune system. PloS one 2: e206.

64. Robertson AJ, Dickey CE, McCarthy JJ, Coffman JA (2002) The expression of SpRunt during sea urchin embryogenesis. Mech Dev 117: 327–330.

65. Nam S, Jin YH, Li QL, Lee KY, Jeong GB, et al. (2002) Expression pattern, regulation, and biological role of runt domain transcription factor, run, in Caenorhabditis elegans. Mol Cell Biol 22: 547–554.

66. Woolf E, Brenner O, Goldenberg D, Levanon D, Groner Y (2007) Runx3 regulates dendritic epidermal T cell development. Dev Biol 303: 703–714.

67. Park BY, Saint-Jeannet JP (2010) Expression analysis of Runx3 and other Runx family members during Xenopus development. Gene expression patterns: GEP 10: 159–166.

68. Kalev-Zylinska ML, Horsfield JA, Flores MV, Postlethwait JH, Chau JY, et al. (2003) Runx3 is required for hematopoietic development in zebrafish. Dev Dyn 228: 323–336.

69. Jeong JH, Jin JS, Kim HN, Kang SM, Liu JC, et al. (2008) Expression of Runx2 transcription factor in non-skeletal tissues, sperm and brain. J Cell Physiol 217: 511–517.

70. Park ES, Park J, Franceschi RT, Jo M (2012) The role for runt related transcription factor 2 (RUNX2) as a transcriptional repressor in luteinizing granulosa cells. Mol Cell Endocrinol 362: 165–175.

71. Sun J, Spradling AC (2012) NR5A nuclear receptor Hr39 controls three-cell secretory unit formation in Drosophila female reproductive glands. Current biology: CB 22: 862–871.

72. Otto F, Thornell AP, Crompton T, Denzel A, Gilmour KC, et al. (1997) Cbfa1, a candidate gene for cleidocranial dysplasia syndrome, is essential for osteoblast differentiation and bone development. Cell 89: 765–771.

73. Kaneto S, Wada H (2011) Regeneration of amphioxus oral cirri and its skeletal rods: implications for the origin of the vertebrate skeleton. Journal of experimental zoology Part B, Molecular and developmental evolution 316: 409–417.

74. Ito Y (2004) Oncogenic potential of the RUNX gene family: 'overview'. Oncogene 23: 4198–4208.

75. Yoshida N, Ogata T, Tanabe K, Li S, Nakazato M, et al. (2005) Filamin A-bound PEBP2beta/CBFbeta is retained in the cytoplasm and prevented from functioning as a partner of the Runx1 transcription factor. Mol Cell Biol 25: 1003–1012.

76. Lopez-Camacho C, van Wijnen AJ, Lian JB, Stein JL, Stein GS (2014) Core Binding Factor beta (CBFbeta) Is Retained in the Midbody During Cytokinesis. J Cell Physiol.

# Transcriptome and Allele Specificity Associated with a 3BL Locus for Fusarium Crown Rot Resistance in Bread Wheat

Jian Ma[1,2], Jiri Stiller[1], Qiang Zhao[3], Qi Feng[3], Colin Cavanagh[4], Penghao Wang[4], Donald Gardiner[1], Frédéric Choulet[5], Catherine Feuillet[5], You-Liang Zheng[2], Yuming Wei[2], Guijun Yan[6], Bin Han[3], John M. Manners[1], Chunji Liu[1,6]*

1 CSIRO Agriculture Flagship, St Lucia, QLD, Australia, 2 Triticeae Research Institute, Sichuan Agricultural University, Wenjiang, Chengdu, China, 3 National Center for Gene Research, Shanghai Institutes for Biological Sciences, Chinese Academy of Sciences, Shanghai, China, 4 CSIRO Agriculture Flagship, Black Mountain, ACT, Australia, 5 INRA-UBP Joint Research Unit 1095, Genetics, Diversity and Ecophysiology of Cereals, Clermont-Ferrand, France, 6 School of Plant Biology, Faculty of Science and The UWA Institute of Agriculture, The University of Western Australia, Perth, WA, Australia

## Abstract

Fusarium pathogens cause two major diseases in cereals, Fusarium crown rot (FCR) and head blight (FHB). A large-effect locus conferring resistance to FCR disease was previously located to chromosome arm 3BL (designated as Qcrs-3B) and several independent sets of near isogenic lines (NILs) have been developed for this locus. In this study, five sets of the NILs were used to examine transcriptional changes associated with the Qcrs-3B locus and to identify genes linked to the resistance locus as a step towards the isolation of the causative gene(s). Of the differentially expressed genes (DEGs) detected between the NILs, 12.7% was located on the single chromosome 3B. Of the expressed genes containing SNP (SNP-EGs) detected, 23.5% was mapped to this chromosome. Several of the DEGs and SNP-EGs are known to be involved in host-pathogen interactions, and a large number of the DEGs were among those detected for FHB in previous studies. Of the DEGs detected, 22 were mapped in the Qcrs-3B interval and they included eight which were detected in the resistant isolines only. The enrichment of DEG, and not necessarily those containing SNPs between the resistant and susceptible isolines, around the Qcrs-3B locus is suggestive of local regulation of this region by the resistance allele. Functions for 13 of these DEGs are known. Of the SNP-EGs, 28 were mapped in the Qcrs-3B interval and biological functions for 16 of them are known. These results provide insights into responses regulated by the 3BL locus and identify a tractable number of target genes for fine mapping and functional testing to identify the causative gene(s) at this QTL.

**Editor:** Meixue Zhou, University of Tasmania, Australia

**Funding:** This publication is based upon work supported by a joint CAS/CSIRO project (Project No R-1910-1) and an ARC Linkage project (Project No. LP120200830). JM is grateful to the Sichuan Agricultural University and the China Scholarship Council for funding his visit to CSIRO Agriculture Flagship. The funders had no role in study design, data collection and analysis, decision to publish, or preparation of the manuscript.

**Competing Interests:** The authors have declared that no competing interests exist.

* Email: chunji.liu@csiro.au

## Introduction

Fusarium pathogens cause two serious diseases in cereals, Fusarium crown rot (FCR) and Fusarium head blight (FHB). FCR is a chronic problem in many parts of the semi-arid cereal producing regions worldwide including Australia [1]. In contrast, FHB favours environments with high humidity and temperature. It is a sporadic problem in Australia but causes massive annual losses worldwide [2]. Both diseases can produce mycotoxins which can be harmful if present in foods or feeds [3,4].

FHB is one of the most intensively studied diseases. Sources of resistance have been intensively searched [5,6] and numerous quantitative trait loci (QTL) conferring resistance have been reported [3]. The best known source of FHB resistance is from the genotype Sumai 3 and the QTL on chromosome arm 3BS from this genotype is the most potent locus conferring resistance to this

disease [7]. The 3BS locus contains a glycosyltransferase gene that has the potential to detoxify the mycotoxin deoxynivalenol which is also a virulence factor and this may explain the resistance mechanism [7]. This 3BS QTL does not confer any significant level of resistance to FCR in wheat [8]. In addition to the effort of map-based cloning of the 3BS resistance locus [9], transcriptome analysis has also been conducted in recent years to identify genes differentially expressed between FHB resistant and susceptible genotypes [10–15]. Efforts have also been made in transforming defence-related genes into susceptible or moderately susceptible wheat varieties to obtain transgenic plants with improved FHB resistance [16–19].

Compared with the studies on FHB, our knowledge of FCR and its possible resistance mechanisms is limited. There have been some studies on host transcriptional responses during the infection of susceptible genotypes following application of defence inducing compounds that can reduce FCR symptom development [20,21]. Several QTL have been reported [22–26]. Among these, the one located on chromosome arm 3BL from the *Triticum spelta* accession 'CSCR6' (designated as *Qcrs-3B*) seems to be highly effective. This QTL accounted for up to 49% of phenotypic variance and provided significant effects in multiple hexaploid genetic backgrounds [24]. Recently, several independent sets of resistant and susceptible near-isogenic lines (NILs) that differ in the 3BL locus for FCR have been developed [27]. These genetic resources provide an ideal tool for studying the host responses to infection associated with resistance to this disease and for identifying genes that co-locate with the *Qcrs-3B* locus.

RNA sequencing (RNA-seq) has become a powerful tool for transcriptome analysis. The approach is not only highly sensitive and efficient for identifying differentially expressed genes (DEGs) [28] but, when combined with genomic and genetic analysis can also be used for detecting SNPs in transcribed genes that co-locate with a target locus [29]. These features of RNA-seq analysis are particularly attractive for applications in hexaploid wheat where multiple homoeologous alleles exist for most genes and transcripts. We thus conducted an RNA-seq analysis against five sets of the NILs developed for the *Qcrs-3B* locus [27], examined DEGs and SNPs between the NIL lines. This analysis provides candidate genes for both the response determined by the 3BL locus as well as those co-located with the 3BL locus that may be genetically causative for resistance. In addition to identifying genes underlying the FCR resistance locus, we were also interested in finding out if expressed genes associated with resistance to FCR were related to those observed by others for FHB.

## Materials and Methods

### Plant materials

Five sets of NILs generated using the heterogeneous inbred family method for the FCR QTL on chromosome arm 3BL reported by Ma et al. [27] were used in this study. Four of these NIL sets, including '1R/1S', '2R/2S', '3R/3S', and '4R/4S', were derived from the population of 'Janz'*2/'CSCR6'. The other set, '9R/9S', was derived from the population of 'Lang'/CSCR6'. 'R' isolines are those carrying the resistant allele and 'S' isolines are those carrying the susceptible allele at the *Qcrs-3B* locus. The FCR donor 'CSCR6' is a genotype belonging to the taxon *T. spelta* [24]. The NIL set '1R/1S' (designated as Family A) was used for the primary analysis. The other four NIL sets (designated as Family B) were used for validating results obtained from Family A.

### Determination of the QTL *Qcrs-3B* chromosomal interval

The wheat 3B pseudomolecule 'traes3bPseudomoleculeV1'of Chinese Spring was downloaded from Generic Genome Browser version 2.3 (https://urgi.versailles.inra.fr/gb2/gbrowse/wheat_annot_3B/) hosted by Unité de Recherche Génomique Info (URGI) in February, 2014 [30]. The DArT marker wPt-7301, which locates proximally to the QTL, was placed at about 736 Mb in the 3B pseudomolecule. Another marker, wPt-7514, which locates near the centre of the QTL was placed at about 765 Mb on the 3B pseudomolecule (Fig. 1). Sequences for the two DArT markers located distally to the QTL, however, were not available. Considering the terminal location of the QTL on this chromosome [24], the most distally located gene on the genomic sequence for this chromosome arm was used as the distal border of this QTL. Thus the size of the QTL interval used in this study could be over-estimated.

### Fusarium crown rot inoculation and experimental design

A single isolate of *F. pseudograminearum* (*Fp*, CS3427) was used in this study. This is one of the most aggressive isolates collected from northern New South Wales, Australia and maintained in the CSIRO collection [31]. Inoculum was prepared based on the method described before [32].

Surface-sterilized seeds were germinated in Petri dishes on three layers of filter paper saturated with water. Ten seedlings were used in each of the biological replications for all of the experiments conducted. Two-day-old seedlings were inoculated following the method described before [33] with the *F. pseudograminearum* isolate (*Fp*-infection) or water (mock). Samples were taken by cutting the shoot base (0–4 cm) at 3 or 5 days post inoculation (dpi) and frozen in liquid nitrogen immediately and kept at −80°C until processed.

Two datasets of RNA sequences were obtained in this study. The first dataset was obtained from Family A. The experimental design for Family A contained two treatments (mock and *Fp*-infection), two time points (3 and 5 dpi) and six biological replicates (Table S1). The second one was generated from Family B, which were used only to validate those expressed genes with SNPs (SNP-EGs) detected from Family A. A single trial with three biological replications was conducted for these four sets of NILs. Mocks were not used and samples were collected at 5 dpi only. Before RNA isolation, samples from the three biological replications for each of the eight isolines were pooled. Thus, a total of 8 samples (each for a different line of the four sets of NILs) were used for RNA extraction.

### RNA extraction, library construction and Illumina sequencing

Total RNA was isolated using a QIAGEN RNeasy plant mini kit (Qiagen, Hilden, Germany) according to the manufacturer's instructions, using RLT buffer and including the optional on-column DNase I digestion. The yield and purity of each RNA sample was determined by the absorbance (Abs) at 260 and 280 nm and the integrity of all RNA samples was assessed on 1% agarose gels. Each sample of 10 μg of total RNA was sent to Australian Genome Research Facility Ltd (Parkville, Victoria, Australia) for further processing before Illumina HiSeq sequencing. Two technical replications were run for each of the 56 (48 for '1R/1S' and 8 for the other 4 sets of NILs) cDNA libraries with 6 lanes of 100 bp paired-end sequencing. Raw reads were trimmed using a SolexaQA package 2.2 with minimum Phred quality value of 30 and minimum length of 70 bp. The RNA sequences were available at the National Center for Biotechnology Information (NCBI) with the accession number of SRP048912.

### Data quality control

For Family A, RNA-seq Illumina fastq sequence datasets were pooled by replicate (6 replicates per time point) at time points 3 dpi and 5 dpi. For Family B RNA sequences were pooled by response to FCR inoculation only (+/−) due to the fact that there was no biological replication used. FastQC (version 0.10.1) was used as a preliminary check that the Phred scores were acceptable. BioKanga (version 2.76.2; developed by bioinformatics team, unpublished) filtering was then employed and reads containing polymorphic variation but not supported by at least two other overlapping reads were removed from the datasets. Additionally, all but one instance of any duplicated reads were removed to reduce the effect of PCR artefacts. All retained reads for alignment were unique.

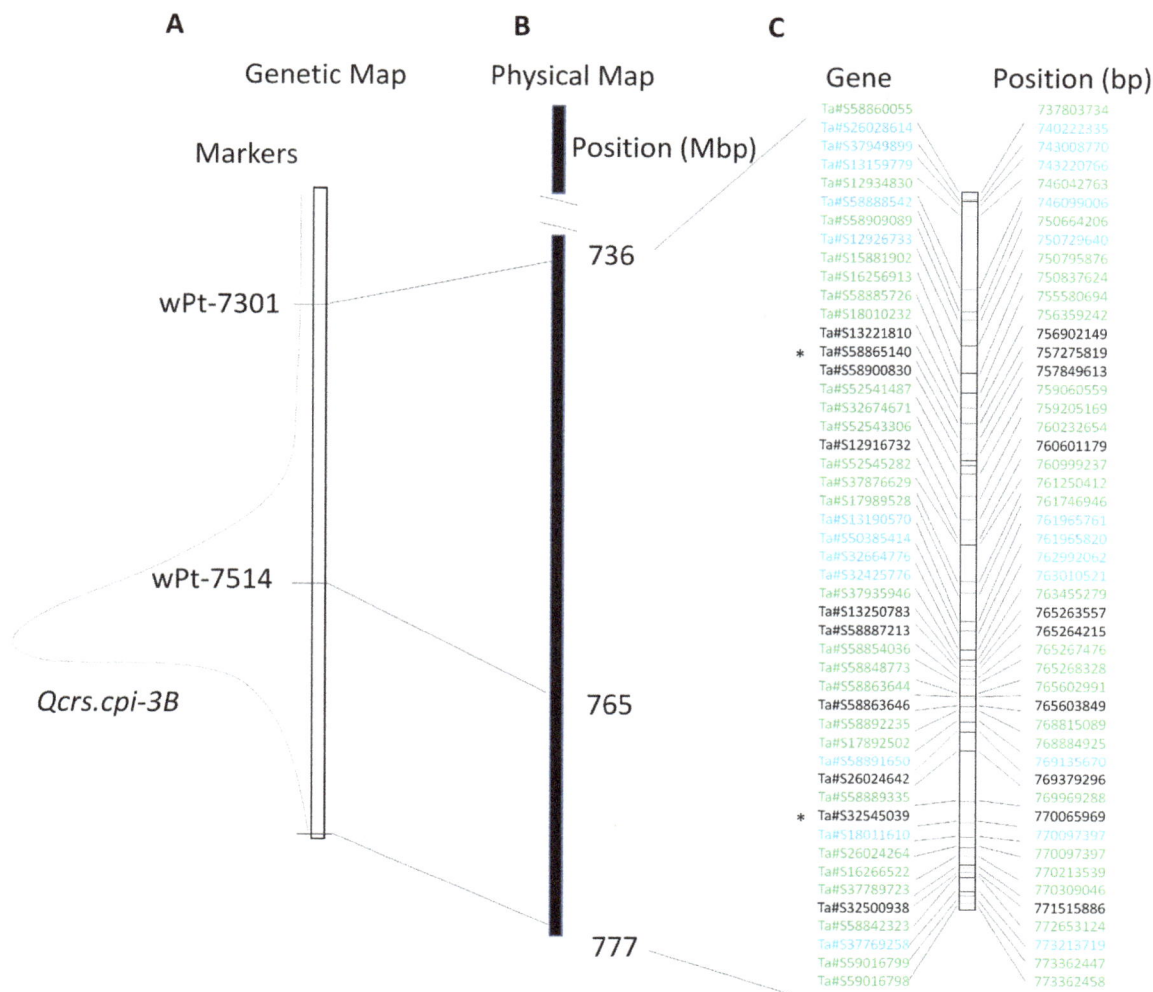

**Figure 1. The *Qcrs-3B* interval and 48 unique genes identified in the interval.** (A) Genetic map showing the *Qcrs-3B* interval with two DArT markers which were successfully placed on the 3B pseudomolecule of Chinese Spring. (B) The *Qcrs-3B* interval on the physical map of the 3B pseudomolecule; and (C) Genes located in the*Qcrs-3B* interval. Differentially expressed genes (DEGs) between the resistant and susceptible isolines were indicated as black for the up- and blue for the down-regulated genes. The expressed genes containing SNP were indicated as green. '*' indicates that they were also DEGs.

## RNA-seq analyses

Two methods were used in this study to analyse the trimmed RNA reads. One was based on the complete IWGSC (International Wheat Genome Sequencing Consortium) chromosome shotgun sequence contigs (CSS-contigs) (www.wheatgenome.org) due to their increased genome coverage for estimating distributions of DEGs and SNPs between the 'R' and 'S' isolines in the wheat genome. The other was based on UniGene (NCBI, www.ncbi.nlm.nih.gov/) aiming at identifying specific gene(s) underlying FCR resistance.

**Distributions of DEGs and SNPs between chromosomes and on 3B.** Following BioKanga (version 2.76.2) filtering, the retained reads were independently aligned against the complete CSS-contigs and the 3B pseudomolecule using BioKanga. Alignment parameters were set such that a maximum of two substitutions were allowed and no multi-aligned reads were accepted (except for reads mapping to the CSS once and the pseudomolecule). Two additional mismatches were allowed for the first 12 bp of the reads to account for primer artefact. Reads were aligned utilising the paired-end reads with insert sizes from 100 to

2 kbp. PCR differential amplification artefacts were reduced within the sequence alignment processing using BioKanga's sliding window mechanism. A counts matrix was generated and loaded into R (version 3.0.1) for downstream statistical analysis.

BioKanga SNP calling was run with raw, non-filtered reads aligned against the CSS contigs and 3B pseudomolecule allowing at most two mismatches and at most two additional mismatches for the first 12 bp of the reads accounting for primer artefact. Only unique alignments were accepted. A custom R function was written to identify candidate SNPs underlying the QTL. This method applied multiple criteria to identify trait linked SNPs which identified polymorphisms between the two isolines for a given set of NILs across replicates and was able to detect presence or absence polymorphism. The criteria used for identification of SNPs included: (1) The loci present in all replicates must be homozygous (the dominating nucleotide must have a ratio higher than 0.9; (2) any candidate SNP must have had a minimum of 3 reads coverage. Each candidate SNP was visually checked using the Integrative Genomics Viewer (version 2.3.12).

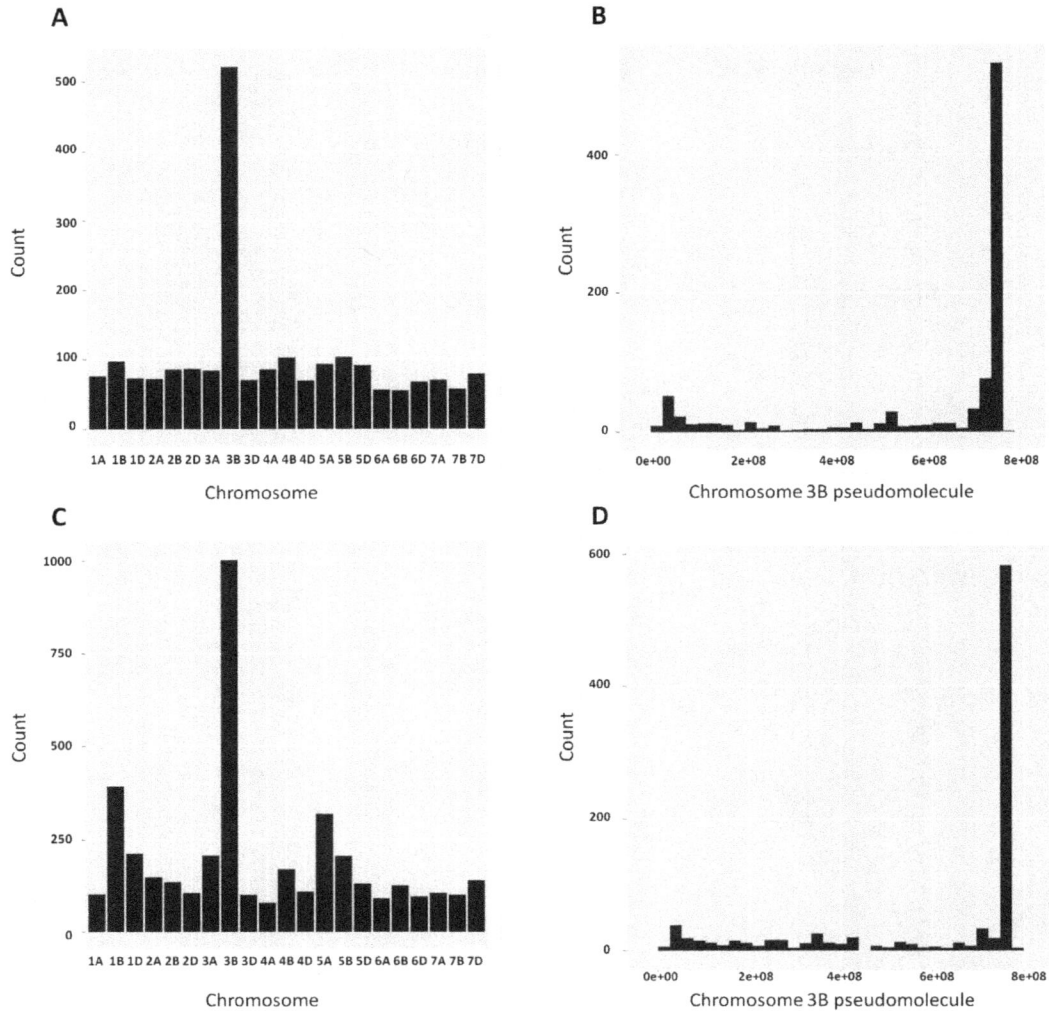

**Figure 2. Distribution of DEGs (A, B) and SNPs (C, D) across the 21 chromosomes (left) and along the 3B pseudomolecule (right).** Note: 3B pseudomolecule is from short (left) to long (right) arm in base pair (bp).

**UniGene-based analysis of DEGs and SNPs.** Analysis of gene expression was performed using BioKanga. Reference sequences used in this analysis included 58,596 of the wheat UniGenes [downloaded from ftp://ftp.ncbi.nih.gov/repository/UniGene/Triticum_aestivum/of NCBI in Jun of 2012] and the genome sequences of the diploid wheat A-genome progenitor *T. urartu* and the diploid D-genome progenitor *Aegilops tauschii* (downloaded from NCBI) [34,35].

The obtained reads were aligned against the 58 K UniGenes and the diploid A and B genome reference sequences with no more than two mismatches allowed. Only those reads matching best with the wheat UniGenes were selected for expression analysis. BioKanga 'maploci' was used to normalize counts based on RPKM (Reads per kilobase per million reads). Prior to the differential expression analysis, the 12 replicates (6 biological and 2 technical) for each genotype-treatment-timepoint sample were merged together for a given pair of comparison by BioKanga genDEseq. In total, four pairwise comparisons between genotypes were conducted. These are summarised throughout the paper in the following way: $S^M3\_v\_R^M3$; $S^M5\_v\_R^M5$; $S^I3\_v\_R^I3$ and $S^I5\_v\_R^I5$. Symbols are 'M' for mock; 'I' for *Fp-infection*; '3' for 3 dpi; '5' for 5 dpi; 'R' for the resistant isoline R; 'S' for the

susceptible isoline S, and 'a_v_b' for comparing object 'a' with 'b', in which 'a' is the control and 'b' is the treatment. DEGs were determined with the threshold of FDR≤0.01 and the absolute value of $\log_2$FoldChange $\geq 1$ or $\leq -1$ or 'inf' (the value of one comparative object is zero and the other is not). The R software (version 3.1.0) was used to generate heat maps.

Genes responsive to FCR infection were identified by four pairwise comparisons between the treatments: $R^M3\_v\_R^I3$; $R^M5\_v\_R^I5$; $S^M3\_v\_S^I3$; $S^M5\_v\_S^I3$. The responsive genes after *Fp*-infection compared with mock were identified with the same method as DEGs: threshold of FDR≤0.01 and the absolute value of $\log_2$FoldChange $\geq 1$ or $\leq -1$ or 'inf'.

The trimmed sequences for each line-treatment-timepoint sample were pooled together. SNPs between 4 pairwise comparisons were identified: $S^M3\_v\_R^M3$, $S^M5\_v\_R^M5$, $S^I3\_v\_R^I3$, and $S^I5\_v\_R^I5$. The alignment of reads to the reference sequences was performed with a maximum of 2 mismatches per read. Minimum coverage for the declaration of an SNP was 4 nucleotides. SNPs between the resistant and susceptible isolines were identified using the Biokanga snpmarkers sub-process with a minimum 90% score, i.e., the percentage of a given nucleotide at a SNP position is at least 90% in the resistant or susceptible isolines.

**A**

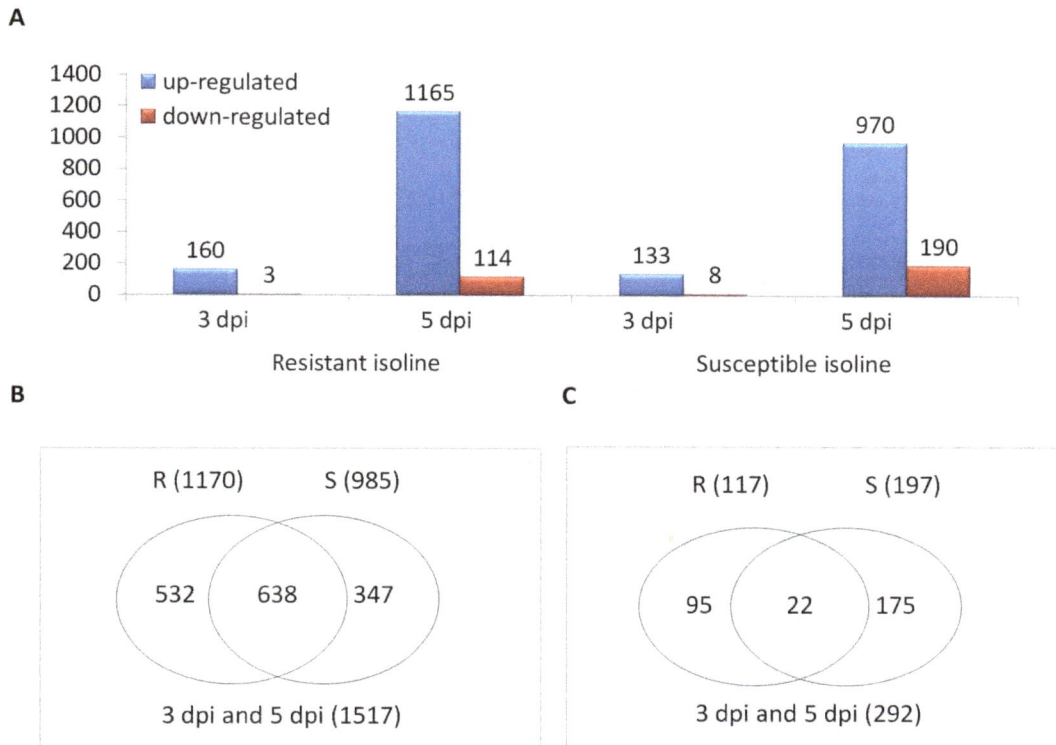

**Figure 3. Differentially expressed genes (DEGs) between the resistant and susceptible isolines in Family A following** *Fusarium pseudograminearum* *(Fp)* **-infection (compared to those in the mock).** (A) Overview of the number of DEGs. (B) Venn diagrams showing the number of up- or (C) down-regulated genes in the resistant isoline compared with those in the susceptible isoline. DEGs were determined with the threshold of FDR ≤0.01 and the absolute value of log$_2$FoldChange ≥1 or ≤−1 or 'inf' (the value of one comparative object is zero and the other one is not). Symbols are 'M' for mock; 'I' for *Fp*-infection; 'R' for resistant isoline; and 'S' for susceptible isoline.

DEGs and expressed genes containing SNP (SNP-EGs) was mapped in the targeted QTL interval by blasting them against the 3B pseudomolecue. All sequence comparisons were performed using the BLASTN 2.2.26+ algorithm with e-value <10$^{-5}$ and length >100 bp.

## Functional annotation

Functions of UniGenes were annotated using the Blast2GO program (version 2.6.6) with default parameters except that e-value threshold of 10$^{-10}$ was used when executing steps of 'Blast' and 'Annotation'. Alignments with a higher score were visually inspected and annotated if a reasonable degree of homology was observed.

## Validation of DEGs by real-time qualitative PCR

Among the identified DEGs between the 'R' and 'S' isolines from Family A, a total of 4 genes were randomly selected and assessed using real-time quantitative PCR (RT-qPCR) given that six biological replicates were used for RNA-seq analysis. *Ta.27922.1.S1_x_at*, encoding a cyclin family protein, was used as the internal reference gene [36]. Primers were designed based on the tool of Primer-BLAST (http://www.ncbi.nlm.nih.gov/tools/primer-blast/) and listed in Table S2. The validations were conducted using Family A. FCR inoculation, tissue sampling and RNA extraction were based on the methods as described earlier and three biological replications were used. RNA extraction, cDNA synthesis, and expression analyses were carried out as described by Ma et al. [37]. Each biological replication was

analysed in two separate wells (technical replication). The average values from the two replications were used for each biological replication. Calculations of the relative fold change were conducted using the method of 2−ΔΔCT. Transcripts with Ct values >40 cycles were regarded as having no expression value.

## Validation of SNP-EGs by re-sequencing

Three genes with SNPs, *Ta#S37789723*, *Ta#S52545282*, and *Ta#S58887817*, were randomly selected for validation by re-sequencing. Primers were designed based on alignments between wheat UniGene and sequences of *T. urartu* and *Ae. tauschii* (Table S2). Genomic DNA from the isolines '1R' and '1S' were extracted from 20-day old seedlings using the hexadecyltrimethy-lammonium bromide (CTAB) method [38]. PCR amplification and sequencing were conducted based on the methods described by Ma et al. [39] with annealing temperatures ranging from 62°C to 65°C depending on the primers (Table S2).

## Results

### Exploratory analysis of variance factors in gene expression patterns

A total of 152 Gb sequences were obtained from Family A and 77 Gb sequences from Family B. Prior to fitting models, some basic exploratory analysis was conducted looking at the variance components for both Family A and B based on sequence reads. All samples were consistent and no outliers were observed. For Family A, the largest proportion of the variance was driven by time point. Principal component analysis (PCA) demonstrated that the

**A**

RᴹM3 RᴵI3 RᴹM5 RᴵI5 SᴹM3 SᴵI3 SᴹM5 SᴵI5

Cell wall-related proteins (6)
WIR1 proteins (4)

Glutathione-transferases (17)

Germin and Germin-like proteins (14)

Receptor-like kinases (42)

Disease resistance-related proteins (14)

Calcium-binding related proteins (13)

Bowman-birk inhibitors (6)
Ascorbate peroxidases (3)
Phenylalanine ammonia lyases (6)
Salicylic acid-related protein (1)
Ethylene-related protein (1)
WRKY transcription factors (7)
Detoxifying-related proteins (10)

Pathogenesis-related proteins (46)

Cytochrome p450 (21)

Jasmonic acid –related  protein (1)

1400 1200 1000 800 600 400 200   0

**B**

RᴹM3 RᴵI3 RᴹM5 RᴵI5 SᴹM3 SᴵI3 SᴹM5 SᴵI5

MYB-related protein mybas2

Methyl chloride transferase

Acid phosphatase 1

Disease resistance protein RGA1

Alpha-L-fucosidase 2-like

Bidirectional sugar transporter sweet15-like

Hippocampus abundant transcript 1 protein

Histidine decarboxylase

Cytochrome P45

C-4 methylsterol oxidase

Premnaspirodiene oxygenase-like

Subtilisin-like protease

Lysine-specific demethylase 8-like

Calcium binding ef-hand protein

Unknown

O-methyltransferase 2-like

Monoglyceride lipase-like

Polyamine oxidase precursor

NAC domain protein NAC5

Membrane protein

Membrane protein

Senescence-associated protein

250   200   150   100   50   0

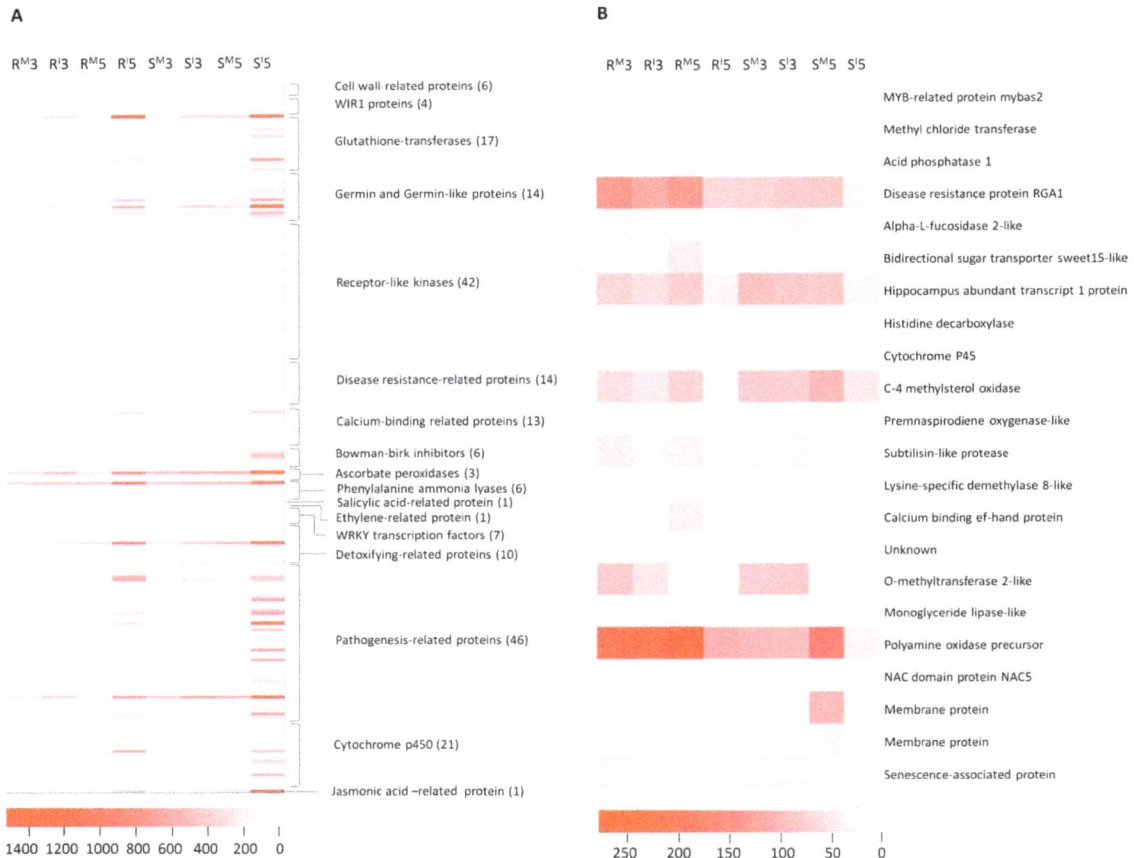

**Figure 4. Heat maps showing up- (A) or down - regulated genes (B) belonging to various classes in Family A detected between** *pseudograminearum* **infection (***Fp***-infection) and water treatment (mocks) at 3 dpi and 5 dpi, respectively.** Up- or down-regulated genes were determined with the threshold of FDR≤0.01 and the absolute value of $\log_2$ FoldChange ≥1 or ≤ −1 or 'inf' (the value of one comparative object is zero and the other one is not). Symbols are 'M' for mock; 'I' for *Fp*-infection; '3' for 3 dpi; '5' for 5 dpi; 'R' for resistant isoline; 'S' for susceptible isoline. The color key represents the RPKM normalized value. Lighter color indicates greater transcript accumulation. Each column represents a sample and each row a gene.

samples may be separated by the first two components according to the time point and the genotype (Figs. S1A and S1B). The hierarchical clustering (Fig. S1C) was consistent with the PCA. For Family B, the experimental run effect was quite strong. However, a large proportion of the variance still contributed to the genotypic effect (data not shown).

### Distribution of DEGs and SNPs in the wheat genome

For increased genome coverage, the exploratory analysis on distribution of the detected DEGs and SNP-EGs in the wheat genome was conducted using the CSS contigs. This analysis detected 2,020 DEGs between the two isolines of Family A. Chromosome 3B had significantly more DEGs compared with other chromosomes families (Fig. 2A) and that a large proportion of those mapped on chromosome 3B was concentrated on the distal end where the targeted QTL resides (Fig. 2B).

Based on the use of the CSS contigs, 955 SNPs were detected between the 'R' and 'S' isolines of Family A and chromosome 3B had significantly more SNPs compared with other chromosomes (Fig. 2C) and that a large proportion of those on chromosome 3B co-located with the targeted QTL at the distal end of this chromosome arm (Fig. 2D).

### Genes induced by *Fusarium* infection

Only data from Family A were suitable for this analysis as 'mock' controls were not used in assessing the NILs of Family B. Following *Fp*-inoculation, the numbers of up-regulated genes detected from the 'R' lines were 160 at 3 dpi and 1,165 at 5 dpi; and from the 'S' lines were 133 and 970, respectively, at the two different time points. The numbers of down-regulated genes detected at the two different time points following *Fp*-inoculation were 3 and 114, respectively, from the 'R' line, and 8 and 190, respectively, from the 'S' line (Fig. 3A).

In total, 1,809 induced genes (1,517 up- and 292 down-regulated) were detected between the two isolines following *Fp*-infection (Fig. 3B and 3C). Of them, 638 were up-regulated and 22 down-regulated in both isolines (Table S3). The 638 up-regulated genes contain 46 encoding pathogenesis-related proteins, 42 encoding receptor-like kinases, 21 encoding cytochrome P450 s, 17 encoding glutathione transferases, and 10 encoding detoxifying-related proteins (Fig. 4A). They also contain genes encoding proteins involved in host-pathogen interactions: 14 for disease resistance-related proteins, 6 for cell wall-related proteins, 4 for WIR1 (wheat induced resistance 1) proteins, 7 for WRKY transcription factors, 3 for ascorbate peroxidises, 6 for phenylalanine ammonia lyases, and 14 for germin and germin-like proteins (Fig. 4A). They also contain genes for the biosynthesis of plant

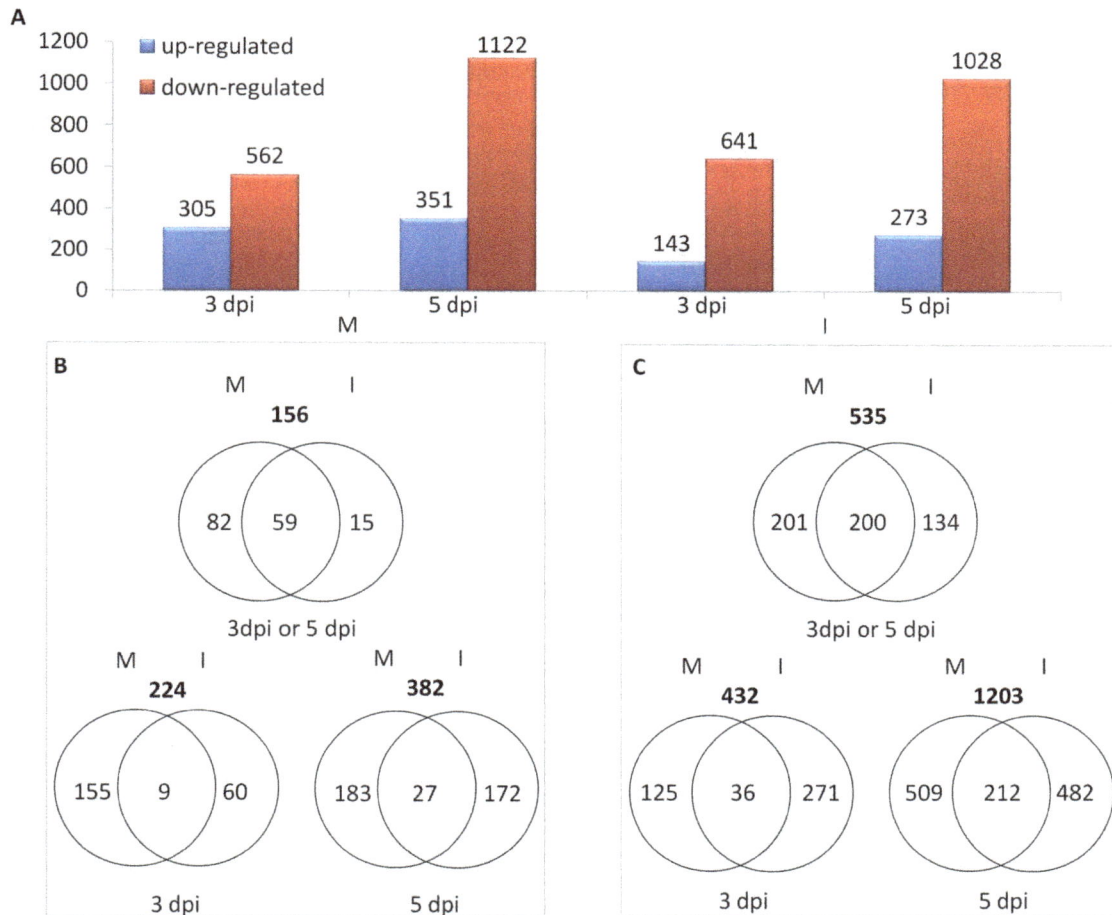

**Figure 5. Genes responsive to** *Fusarium pseudograminearum* **(Fp) infection.** (A) Numbers of genes exhibiting differential accumulation between *Fp*-infected and mock plants at either 3 dpi or 5 dpi. (B) Venn diagrams showing the number of up-regulated or (C) down-regulated genes. These genes were determined with the threshold of FDR≤0.01 and the absolute value of log2FoldChange ≥1 or ≤−1 or 'inf' (the value of one comparative object is zero and the other one is not). Symbols are 'M' for mock; 'I' for *Fp*-infection; 'R' for resistant isoline; 'S' for susceptible isoline.

hormones which are known to be involved in wheat-*Fusarium* interactions: jasmonic acid (JA), ethylene (ET) and salicylic acid (SA) (Fig. 4A). Three of the 22 down-regulated genes seem to be highly relevant to disease resistance as they encode disease resistance protein RGA1, an EF-hand calcium binding protein, and a senescence-associated protein (Fig. 4B). Fifty-two of the up-regulated genes and three of the down-regulated ones co-located with the targeted QTL interval (Table S3), and five of them were among those showing differences between the 'R' and 'S' lines (below).

## Transcriptome differences between resistant and susceptible isolines

An important part of this investigation was to identify transcripts that are differentially expressed between the 'R' isoline and the 'S 'isoline to provide an indication of what molecular mechanisms may be associated with resistance. The numbers of DEGs from the 'mock' treatment in Family A were: 562 down-regulated and 305 up-regulated at 3 dpi and 1,122 down-regulated and 351 up-regulated at 5 dpi. The numbers of DEGs detected from the *Fp* treatment between the NILs were: 641 down-regulated and 143 up-regulated at 3 dpi; and 1,028 down-regulated and 273 up-regulated at 5 dpi (Fig. 5). The DEGs

obtained between the two treatments at the two time points assessed between the two isolines represent a total of 691 (156 up- and 535 down-regulated in the 'R' isoline) unique transcripts (Fig. 5 and Table S4).

Of these 691 genes differentially expressed between the 'R' and 'S' lines, 88 (12.7%) were derived from chromosome 3B. Twenty-two (10 up- and 12 down-regulated) of them on chromosome 3B were mapped in the *Qcrs-3B* interval (Table S5). Functions for 6 of the 10 up-regulated genes and 7 of the down-regulated genes are known (Table S5).

The 691 DEGs also contained 8 genes which were detected from the resistant isolines only and they were absent in the 'S' isoline of Family A assessed (Table 1). Biological functions are known for only one of these 8 genes. That is *Ta#S32500938*, encoding the gibberellin 2-beta-dioxygenase 8-like. Two of the 8 DEGs (*Ta#S12916732* and *Ta#S32500938*) were mapped in the *Qcrs-3B* interval with *Ta#S32500938* being the only one annotated with known biological function.

## Expressed genes containing SNPs (SNP-EGs) between the NILs

A total of 255 unique SNP-EGs between the NIL set '1R/1S' of Family A were detected. Among them, 203 were detected from the

**Table 1.** Genes expressed in the resistant isoline only[a].

| Unigene name | Annotation[b] | Expression value[c] | | | | | | | |
|---|---|---|---|---|---|---|---|---|---|
| | | $S^{M}3$ | $R^{M}3$ | $S^{M}5$ | $R^{M}5$ | $S^{I}3$ | $R^{I}3$ | $S^{I}5$ | $R^{I}5$ |
| Ta#S12916732 | Unknown | 0 | 3.3 | 0 | 3.3 | 0 | 3.1 | 0 | 2.9 |
| Ta#SS856178 | Unknown | 0 | 11.8 | 0 | 14.6 | 0 | 9.8 | 0 | 9.7 |
| Ta#SS856166 | Unknown | 0 | 14 | 0 | 14.7 | 0 | 11.8 | 0 | 14.5 |
| Ta#SS850454 | Unknown | 0 | 5.5 | 0 | 5.5 | 0 | 4.8 | 0 | 4.4 |
| Ta#S32500938 | Gibberellin 2-beta-dioxygenase 8-like | 0 | 2.4 | 0 | 1.4 | 0 | 3.3 | 0 | 2.7 |
| Ta#S1626095 | Unknown | 0 | 2.1 | 0 | 3.5 | 0 | 3.2 | 0 | 2.8 |
| Ta#SS857100 | Unknown | 0 | 4.6 | 0 | 6.1 | 0 | 3.5 | 0 | 2.6 |
| Ta#SS856177 | Unknown | 0 | 3.7 | 0 | 3.2 | 0 | 3 | 0 | 4.7 |

[a]Genes were identified with the threshold of FDR≤0.01.
[b]Annotations were performed using the software Blast2Go with BLASTX E-score of less than $10^{-10}$.
[c]RPKM normalized values are shown. The symbols are: 'M' for mock; 'I' for Fp-infection; '3' for 3 dpi; '5' for 5 dpi; 'R' for resistant isoline; 'S' for susceptible isoline.

Fp-infected samples, 116 from the mock samples, and 64 were detected from both of the samples (Fig. 6). Of these 255 unique SNP-EGs, 60 (or 23.5%) were mapped on chromosome 3B, and 28 genes with a total of 71 SNPs were mapped in the Qcrs-3B interval (Table S6). Biological functions for 16 of these 28 genes are known (Table S6).

Of the 28 SNP-EGs mapped in the targeted QTL interval in the Family A, 18 (64.3%) were also detected among the four sets of NILs in Family B. These 18 SNP-EGs contain a total of 56 SNPs. Forty-one of these 56 SNPs (73.2%) were among those detected in Family A (Tables 2 and S6).

## Validation of DEGs and SNP-EGs detected from the RNA sequence analysis

To verify the RNA-seq results obtained, RT-qPCR analysis was conducted against 4 genes that were randomly selected from the DEGs between the 'R' and 'S' isolines detected based on UniGene analysis. The expression patterns of these 4 genes assessed by RT-qPCR (Fig. S2) were consistent with those obtained from the RNA-seq analysis.

Three of the genes with a total of 10 SNPs identified between the resistant and susceptible isolines were also randomly selected for validation based on re-sequencing the NIL set '1R/1S'. All of the SNPs were identified correctly in the re-sequencing experiment (Fig. S3).

## Discussion

RNA-seq analysis was conducted against five sets of NILs for a large-effect locus conferring FCR resistance on chromosome arm 3BL in wheat. Not unexpectedly, the numbers of detected SNP-EGs from these NILs on the targeted chromosome 3B were significantly higher compared with those on any other chromosome and a large proportion of them were concentrated on the distal end of chromosome arm 3BL where the targeted Qcrs-3B locus locates. Interestingly, this was also observed for DEGs between 'R' and 'S' isolines suggesting that there may be a regulation of proximally located genes by the Qcrs-3B locus. The use of the multiple sets of NILs allowed the identification of better defined sets of candidate genes underlying the targeted locus. Functions of these genes provide insights into responses regulated by the 3BL locus and the 48 genes mapped in the targeted QTL interval are tractable in further efforts to functionally test gene(s) underlying this QTL for causal effects. These genes are being used as markers in fine mapping the Qcrs-3B locus based on a NIL-derived population.

## Candidate genes underlying the Qcrs-3B QTL for FCR resistance

Of the large numbers of DEGs and SNP-EGs detected in this study, targeting those located in the Qcrs-3B interval could be productive in further efforts of characterizing the FCR locus and cloning genes underlying the QTL. The 22 DEGs and the 28 SNP-EGs represent a total of 48 unique genes as two of the DEGs also contain SNPs (Fig. 1).

Five of the 22 DEGs mapped in the targeted QTL interval do not only differ between the 'R' and 'S' lines but were also induced by Fp-infection (Table S6). Several of these DEGs have been previously associated with host-pathogen interactions. They include those encoding the pathogenesis-related protein 4, the disease resistance response protein 206-like, and the disease resistance protein RPM1. One of the up-regulated genes mapped in the Qcrs-3B interval encodes a homologue of resistance protein

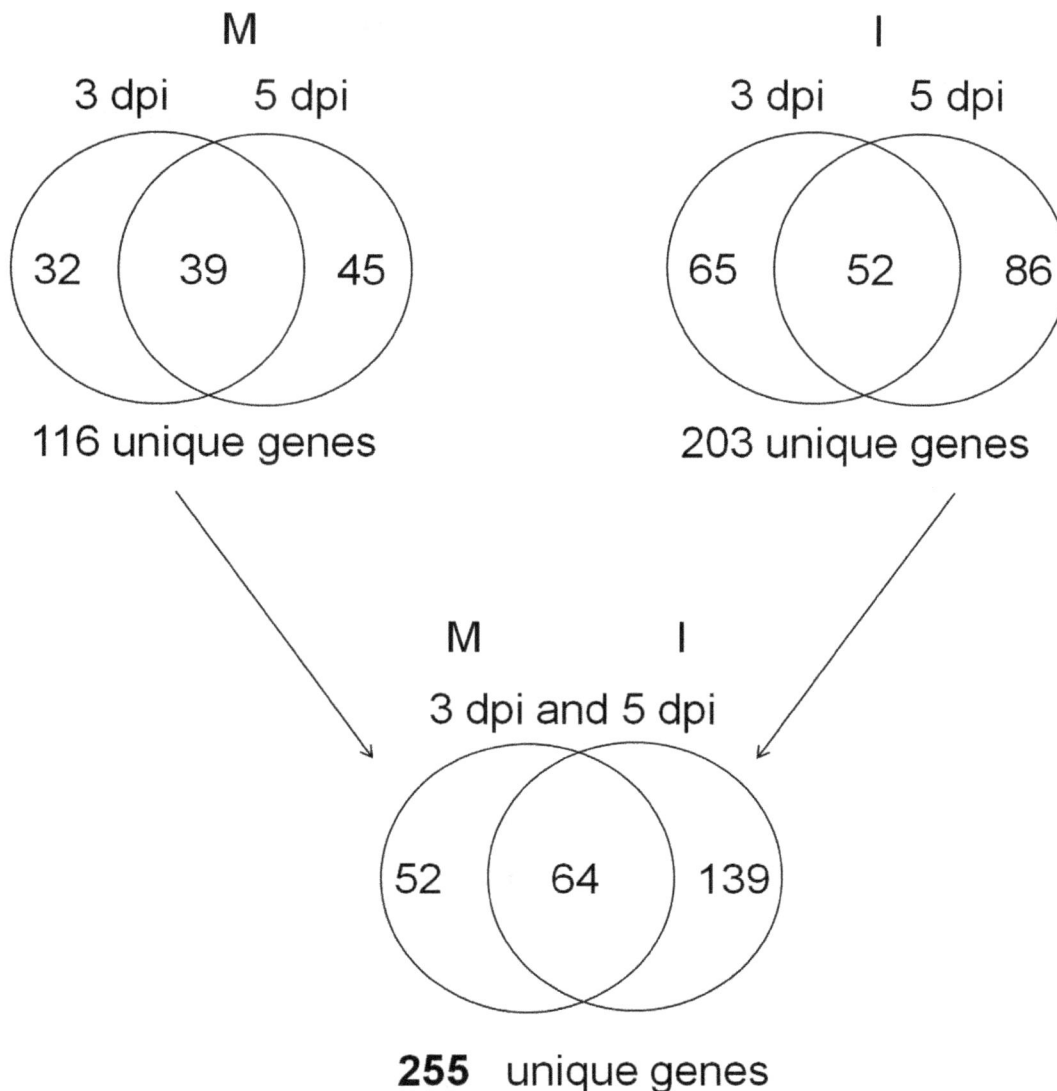

**Figure 6. Venn diagrams showing the number of the expressed genes with SNPs between the resistant and susceptible isolines of Family A (1R/1 S).** Symbols are 'M' for mock or water-treatment; 'I' for *Fp*-infection; '3' for 3 dpi and '5' for 5 dpi.

RGA2, which is known to confer broad spectrum resistance to potato late blight [40].

The 22 DEGs mapped in the targeted QTL interval also include two of those expressed in the resistant isolines only. Of these two genes, only *Ta#S32500938* was annotated with known

biological function. This gene encodes a gibberellin 2-beta-dioxygenase 8-like enzyme. It degrades active gibberellin and is involved in cold response [41]. Clearly, together with the other one gene expressed in the resistant isolines only and located in the

**Table 2.** Validation of SNP-EGs and SNPs identified from Family A in Family B.

| NILs | SNP-EGs | SNPs | Novel SNPs |
|------|---------|------|------------|
| 2R/2S | 14 | 32 | 9 |
| 3R/3S | 13 | 28 | 7 |
| 4R/4S | 11 | 25 | 8 |
| 9R/9S | 14 | 29 | 7 |
| Unique | 18 | 41 | 15 |

targeted QTL interval (*Ta#S12916732*), possible roles of *Ta#S32500938* in FCR resistance need to be clarified.

Of the 28 SNP-EGs located in the targeted interval, only two were significantly differentially expressed between the 'R' and 'S' isolines assessed. Considering that a difference in a gene sequence does not necessarily result in significantly changed expression but could affect the expression of down-stream genes [37], the gene(s) underlying the *Qcrs-3B* could also be among the 18 non-DEGs located in the QTL interval. Out of these SNP-EGs, three encode MYB transcription factors which have been previously implicated in host-pathogen interactions [42]. Another one of the 28 SNP-EGs with known functions is particularly interesting as it encodes an NADH-quinone oxidoreductase subunit which is a well-known defence protein involved in detoxification reaction [43].

### DEGs induced by FCR and FHB infections

There is a plethora of transcriptomic studies on FHB in recent years [15,44,45] and it is known that all *Fusarium* pathogens which cause FHB can cause FCR [1]. Considering the shared aetiology, pathogen biology and epidemiology between FHB and FCR [1], a comparison of host genes induced between these two diseases could be interesting.

As a transcriptomic analysis routinely detects hundreds of DEGs, it is not unexpected that such a comparison identified many genes which were induced by both FHB and FCR. Genes responsive to both diseases include those encoding some well-known defence proteins [e.g. pathogenesis-related (PR) proteins, receptor-like kinases (RLKs), glutathione-S-transferases, and cytochrome P450s] [20,21,32,46,47], those encoding responsive proteins during disease infection, those encoding xylanase inhibitors that can hinder the xylanase released by *Fusarium* pathogens from degrading the primary component of cell walls [48], and those of WRKY transcription factors which are known to function in biotic and abiotic stress responses [49]. The shared genes induced by these two diseases also include those encoding detoxifying-related proteins involved in detoxifying and transporting mycotoxins produced by *Fusarium* species [50,51], and those encoding germ and germin-like proteins that are related to metabolism of reactive oxygen species (ROS) [52,53]. Such ROS as superoxide and hydrogen peroxide can induce programmed cell death (PCD) [54]. Another gene induced by both FCR and FHB was one of those encoding PCD-related proteins [12]. Similarly, two of the DEGs detected between the 'R' and 'S' lines of FCR in this study, *Ta#S22386683* and *Ta#S37823230,* were among those detected between the NILs for a FHB locus [12].

When focused on the targeted QTL interval, however, overlap of genes detected between FHB and FCR was not obvious. For example, none of the DEGs located near the *Fhb1* locus using the deletion line [15] was among those located in the *Qcrs-3B* region in this study. This is not difficult to comprehend in considering that different DEGs were not only detected between NILs for different loci of FHB [44,45] but also detected for the same *Fhb1* locus between the use of NILs [44] and a deletion line [15].

### Supporting Information

**Figure S1 Principal component analysis (PCA) of RNA sequences from Family A by time point (A), genotype (B)** and hierarchical clustering with Genotype×Treatment× Timepoint (C).** IDs for each database were listed in Table S1.

**Figure S2 RT-qPCR validation of 4 genes showing differential expression between the resistant and susceptible isolines of the NIL set '1R/1S'.** The columns represented the average expression ratios calculated from all three biological replications. For *TA#S58850454*, no expression was detected in the susceptible isoline. The values were presented as the averages of $2^{\wedge}$ ($Ct_{TA\#S58850454}-Ct_{reference\ gene}$). Error bar shows standard deviation. Symbols are 'M' for mock; 'I' for *Fp*-infection, '3' for 3 dpi and '5' for 5 dpi.

**Figure S3 Multiple alignments of *Ta#S37789723* (A), *Ta#S52545282* (B) and *Ta#S58887817* (C) with sequences from the resistant and susceptible isolines.** The SNPs in the red box were validated. Those in black box were identified by re-sequencing.

**Table S1 Sample summary of the first set of the near isogenic lines (Family A).**

**Table S2 Primers used for real-time qualitative PCR analysis and for the validation of single nucleotide polymorphism detected from the RNA sequence analysis.**

**Table S3 Transcripts that exhibited differential accumulation in both the resistant and the susceptible isolines following *Fusarium pseudograminearum* (*Fp*) treatment compared to those in the mock at 3 dpi and 5 dpi.**

**Table S4 Differential expressed genes (DEGs) in the resistant isoline compared with the susceptible isoline of Family A following *Fusarium pseudograminearum* (*Fp*) or water (mock) treatment at 3 dpi or 5 dpi.**

**Table S5 Differentially expressed genes (DEGs) between the resistant and susceptible isolines that mapped in the *Qcrs-3B* interval.**

**Table S6 Genes with homozygous SNPs between the resistant and susceptible isolines mapped in the *Qcrs-3B* interval.**

### Author Contributions

Conceived and designed the experiments: JMM CL BH GY Y-LZ YW. Performed the experiments: JM JS QZ QF CC PW DG. Analyzed the data: JM JS QZ QF CC PW CL. Contributed reagents/materials/analysis tools: FC CF. Contributed to the writing of the manuscript: JM JS PW DG JMM CL.

### References

1. Chakraborty S, Liu C, Mitter V, Scott J, Akinsanmi O, et al. (2006) Pathogen population structure and epidemiology are keys to wheat crown rot and Fusarium head blight management. Australasian Plant Pathology 35: 643–655.
2. Goswami RS, Kistler HC (2004) Heading for disaster: Fusarium graminearum on cereal crops. Molecular Plant Pathology 5: 515–525.
3. Buerstmayr H, Ban T, Anderson JA (2009) QTL mapping and marker-assisted selection for Fusarium head blight resistance in wheat: a review. Plant breeding 128: 1–26.
4. Mudge AM, Dill-Macky R, Dong Y, Gardiner DM, White RG, et al. (2006) A role for the mycotoxin deoxynivalenol in stem colonisation during crown rot

disease of wheat caused by *Fusarium graminearum* and *Fusarium pseudogra-minearum*. Physiological and Molecular Plant Pathology 69: 73–85.

5. Dubin HJ (1997) Fusarium Head Scab: Global Status and Future Prospects: In: Dubin HJ, Gilchrist L, Reeves J, McNab A (eds) CIMMYT, Mexico, DF.

6. Lu W, Cheng S, Wang Y (2001) Wheat scab research in China. Scientific Publication Ltd, Beijing.

7. Lemmens M, Scholz U, Berthiller F, Dall'Asta C, Koutnik A, et al. (2005) The ability to detoxify the mycotoxin deoxynivalenol colocalizes with a major quantitative trait locus for Fusarium head blight resistance in wheat. Molecular Plant-Microbe Interactions 18: 1318–1324.

8. Li HB, Xie GQ, Ma J, Liu GR, Wen SM, et al. (2010) Genetic relationships between resistances to Fusarium head blight and crown rot in bread wheat (Triticum aestivum L.). Theoretical and applied genetics 121: 941–950.

9. Liu S, Pumphrey MO, Gill BS, Trick HN, Zhang JX, et al. (2008) Toward positional cloning of Fhb1, a major QTL for Fusarium head blight resistance in wheat. Cereal Research Communications 36: 195–201.

10. Cho S-H, Lee J, Jung K-H, Lee Y-W, Park J-C, et al. (2012) Genome-wide analysis of genes induced by Fusarium graminearum infection in resistant and susceptible wheat cultivars. Journal of Plant Biology 55: 64–72.

11. Gottwald S, Samans B, Lück S, Friedt W (2012) Jasmonate and ethylene dependent defence gene expression and suppression of fungal virulence factors: two essential mechanisms of Fusarium head blight resistance in wheat? BMC genomics 13: 369.

12. Jia H, Cho S, Muehlbauer GJ (2009) Transcriptome analysis of a wheat near-isogenic line pair carrying Fusarium head blight-resistant and-susceptible alleles. Molecular plant-microbe interactions 22: 1366–1378.

13. Li G, Yen Y (2008) Jasmonate and ethylene signaling pathway may mediate Fusarium head blight resistance in wheat. Crop Science 48: 1888–1896.

14. Blume B, Nürnberger T, Nass N, Scheel D (2000) Receptor-mediated increase in cytoplasmic free calcium required for activation of pathogen defense in parsley. The Plant Cell Online 12: 1425–1440.

15. Xiao J, Jin X, Jia X, Wang H, Cao A, et al. (2013) Transcriptome-based discovery of pathways and genes related to resistance against *Fusarium* head blight in wheat landrace Wangshuibai. BMC genomics 14: 197.

16. Anand A, Zhou T, Trick HN, Gill BS, Bockus WW, et al. (2003) Greenhouse and field testing of transgenic wheat plants stably expressing genes for thaumatin-like protein, chitinase and glucanase against *Fusarium graminearum*. Journal of Experimental Botany 54: 1101–1111.

17. Li W, Faris J, Muthukrishnan S, Liu D, Chen P, et al. (2001) Isolation and characterization of novel cDNA clones of acidic chitinases and β-1, 3-glucanases from wheat spikes infected by Fusarium graminearum. Theoretical and Applied Genetics 102: 353–362.

18. Mackintosh CA, Lewis J, Radmer LE, Shin S, Heinen SJ, et al. (2007) Overexpression of defense response genes in transgenic wheat enhances resistance to Fusarium head blight. Plant cell reports 26: 479–488.

19. Zhu X, Li Z, Xu H, Zhou M, Du L, et al. (2012) Overexpression of wheat lipid transfer protein gene TaLTP5 increases resistances to Cochliobolus sativus and Fusarium graminearum in transgenic wheat. Functional & integrative genomics 12: 481–488.

20. Desmond OJ, Edgar CI, Manners JM, Maclean DJ, Schenk PM, et al. (2006) Methyl jasmonate induced gene expression in wheat delays symptom development by the crown rot pathogen *Fusarium pseudograminearum*. Physiological and Molecular Plant Pathology 67: 171–179.

21. Desmond OJ, Manners JM, Schenk PM, Maclean DJ, Kazan K (2008) Gene expression analysis of the wheat response to infection by *Fusarium pseudo-graminearum*. Physiological and Molecular Plant Pathology 73: 40–47.

22. Bovill WD, Ma W, Ritter K, Collard B, Davis M, et al. (2006) Identification of novel QTL for resistance to crown rot in the doubled haploid wheat population 'W21MMT70'×'Mendos'. Plant Breeding 125: 538–543.

23. Collard B, Grams R, Bovill WD, Percy C, Jolley R, et al. (2005) Development of molecular markers for crown rot resistance in wheat: mapping of QTLs for seedling resistance in a '2–49'×'Janz'population. Plant Breeding 124: 532–537.

24. Ma J, Li H, Zhang C, Yang X, Liu Y, et al. (2010) Identification and validation of a major QTL conferring crown rot resistance in hexaploid wheat. Theoretical and applied genetics 120: 1119–1128.

25. Wallwork H, Butt M, Cheong J, Williams K (2004) Resistance to crown rot in wheat identified through an improved method for screening adult plants. Australasian Plant Pathology 33: 1–7.

26. Zheng Z, Kilian A, Yan G, Liu CJ (2014) QTL Conferring Fusarium Crown Rot Resistance in the Elite Bread Wheat Variety EGA Wylie. PloS one 9: e96011.

27. Ma J, Yan G, Liu C (2012) Development of near-isogenic lines for a major QTL on 3BL conferring Fusarium crown rot resistance in hexaploid wheat. Euphytica 183: 147–152.

28. Wang Z, Gerstein M, Snyder M (2009) RNA-Seq: a revolutionary tool for transcriptomics. Nature Reviews Genetics 10: 57–63.

29. Cavanagh CR, Chao S, Wang S, Huang BE, Stephen S, et al. (2013) Genome-wide comparative diversity uncovers multiple targets of selection for improve-ment in hexaploid wheat landraces and cultivars. Proc Natl Acad Sci USA 110: 8057–8062.

30. Choulet F, Alberti A, Theil S, Glover N, Barbe V, et al. (2014) Structural and functional partitioning of bread wheat chromosome 3B. Science 345: 1249721.

31. Akinsanmi O, Mitter V, Simpfendorfer S, Backhouse D, Chakraborty S (2004) Identity and pathogenicity of Fusarium spp. isolated from wheat fields in Queensland and northern New South Wales. Crop Pasture Sci 55: 97–107.

32. Li X, Zhang J, Song B, Li H, Xu H, et al. (2010) Resistance to Fusarium head blight and seedling blight in wheat is associated with activation of a cytochrome P450 gene. Phytopathology 100: 183–191.

33. Yang X, Ma J, Li H, Ma H, Yao J, et al. (2010) Different genes can be responsible for crown rot resistance at different developmental stages of wheat and barley. European journal of plant pathology 128: 495–502.

34. Jia J, Zhao S, Kong X, Li Y, Zhao G, et al. (2013) *Aegilops tauschii* draft genome sequence reveals a gene repertoire for wheat adaptation. Nature 496: 91–95.

35. Ling H-Q, Zhao S, Liu D, Wang J, Sun H, et al. (2013) Draft genome of the wheat A-genome progenitor *Triticum urartu*. Nature 496: 87–90.

36. Long XY, Wang JR, Ouellet T, Rocheleau H, Wei YM, et al. (2010) Genome-wide identification and evaluation of novel internal control genes for Q-PCR based transcript normalization in wheat. Plant molecular biology 74: 307–311.

37. Ma J, Jiang QT, Zhao QZ, Zhao S, Lan XJ, et al. (2013) Characterization and expression analysis of waxy alleles in barley accessions. Genetica 141: 227–238.

38. Murray M, Thompson WF (1980) Rapid isolation of high molecular weight plant DNA. Nucleic acids research 8: 4321–4326.

39. Ma J, Jiang Q-T, Zhang X-W, Lan X-J, Pu Z-E, et al. (2013) Structure and expression of barley starch phosphorylase genes. Planta 238: 1081–1093.

40. Song J, Bradeen JM, Naess SK, Raasch JA, Wielgus SM, et al. (2003) Gene RB cloned from Solanum bulbocastanum confers broad spectrum resistance to potato late blight. Proc Natl Acad Sci USA 100: 9128–9133.

41. Thomas SG, Phillips AL, Hedden P (1999) Molecular cloning and functional expression of gibberellin 2-oxidases, multifunctional enzymes involved in gibberellin deactivation. Proc Natl Acad Sci USA 96: 4698–4703.

42. Liu Y, Schiff M, Dinesh-Kumar S (2004) Involvement of MEK1 MAPKK, NTF6 MAPK, WRKY/MYB transcription factors, COI1 and CTR1 in N-mediated resistance to tobacco mosaic virus. The Plant Journal 38: 800–809.

43. Ross D, Kepa JK, Winski SL, Beall HD, Anwar A, et al. (2000) NAD (P) H: quinone oxidoreductase 1 (NQO1): chemoprotection, bioactivation, gene regulation and genetic polymorphisms. Chemico-biological interactions 129: 77–97.

44. Schweiger W, Steiner B, Ametz C, Siegwart G, Wiesenberger G, et al. (2013) Transcriptomic characterization of two major *Fusarium* resistance quantitative trait loci (QTLs), *Fhb1* and *Qfhs. ifa-5A*, identifies novel candidate genes. Molecular plant pathology 14: 772–785.

45. Kugler KG, Siegwart G, Nussbaumer T, Ametz C, Spannagl M, et al. (2013) Quantitative trait loci-dependent analysis of a gene co-expression network associated with *Fusarium* head blight resistance in bread wheat (*Triticum aestivum* L.). BMC genomics 14: 728.

46. Mauch F, Dudler R (1993) Differential induction of distinct glutathione-S-transferases of wheat by xenobiotics and by pathogen attack. Plant Physiology 102: 1193–1201.

47. Song W-Y, Wang G-L, Chen L-L, Kim H-S, Pi L-Y, et al. (1995) A receptor kinase-like protein encoded by the rice disease resistance gene, Xa21. Science 270: 1804–1806.

48. Carpita NC, Gibeaut DM (1993) Structural models of primary cell walls in flowering plants: consistency of molecular structure with the physical properties of the walls during growth. The Plant Journal 3: 1–30.

49. Rushton PJ, Somssich IE, Ringler P, Shen QJ (2010) WRKY transcription factors. Trends in plant science 15: 247–258.

50. Poppenberger B, Berthiller F, Lucyshyn D, Sieberer T, Schuhmacher R, et al. (2003) Detoxification of the Fusarium mycotoxin deoxynivalenol by a UDP-glucosyltransferase from *Arabidopsis thaliana*. Journal of Biological Chemistry 278: 47905–47914.

51. Muhitch MJ, McCormick SP, Alexander NJ, Hohn TM (2000) Transgenic expression of the TRI 101 or PDR 5 gene increases resistance of tobacco to the phytotoxic effects of the trichothecene 4, 15-diacetoxyscirpenol. Plant Science 157: 201–207.

52. Hancock JT, Desikan R, Clarke A, Hurst RD, Neill SJ (2002) Cell signalling following plant/pathogen interactions involves the generation of reactive oxygen and reactive nitrogen species. Plant Physiology and Biochemistry 40: 611–617.

53. Dunwell JM, Gibbings JG, Mahmood T, Saqlan Naqvi S (2008) Germin and germin-like proteins: evolution, structure, and function. Critical Reviews in Plant Sciences 27: 342–375.

54. Desmond OJ, Manners JM, Stephens AE, Maclean DJ, Schenk PM, et al. (2008) The Fusarium mycotoxin deoxynivalenol elicits hydrogen peroxide production, programmed cell death and defence responses in wheat. Molecular Plant Pathology 9: 435–445.

# Variability among the Most Rapidly Evolving Plastid Genomic Regions is Lineage-Specific: Implications of Pairwise Genome Comparisons in *Pyrus* (Rosaceae) and Other Angiosperms for Marker Choice

**Nadja Korotkova**[1,2,3☌], **Lars Nauheimer**[1,2☌], **Hasmik Ter-Voskanyan**[3,4], **Martin Allgaier**[5], **Thomas Borsch**[1,2,3]*

1 Institut für Biologie/Botanik, Systematische Botanik und Pflanzengeographie, Freie Universität Berlin, Berlin, Germany, 2 Dahlem Centre of Plant Sciences (DCPS), Berlin, Germany, 3 Botanischer Garten und Botanisches Museum Berlin-Dahlem, Berlin, Germany, 4 Institute of Botany, National Academy of Sciences of Republic Armenia, Yerevan, Armenia, 5 The Berlin Center for Genomics in Biodiversity Research (BeGenDiv), Berlin, Germany

## Abstract

Plastid genomes exhibit different levels of variability in their sequences, depending on the respective kinds of genomic regions. Genes are usually more conserved while noncoding introns and spacers evolve at a faster pace. While a set of about thirty maximum variable noncoding genomic regions has been suggested to provide universally promising phylogenetic markers throughout angiosperms, applications often require several regions to be sequenced for many individuals. Our project aims to illuminate evolutionary relationships and species-limits in the genus *Pyrus* (Rosaceae)—a typical case with very low genetic distances between taxa. In this study, we have sequenced the plastid genome of *Pyrus spinosa* and aligned it to the already available *P. pyrifolia* sequence. The overall *p*-distance of the two *Pyrus* genomes was 0.00145. The intergenic spacers between *ndhC–trnV, trnR–atpA, ndhF–rpl32, psbM–trnD,* and *trnQ–rps16* were the most variable regions, also comprising the highest total numbers of substitutions, indels and inversions (potentially informative characters). Our comparative analysis of further plastid genome pairs with similar low *p*-distances from *Oenothera* (representing another rosid), *Olea* (asterids) and *Cymbidium* (monocots) showed in each case a different ranking of genomic regions in terms of variability and potentially informative characters. Only two intergenic spacers (*ndhF–rpl32* and *trnK–rps16*) were consistently found among the 30 top-ranked regions. We have mapped the occurrence of substitutions and microstructural mutations in the four genome pairs. High AT content in specific sequence elements seems to foster frequent mutations. We conclude that the variability among the fastest evolving plastid genomic regions is lineage-specific and thus cannot be precisely predicted across angiosperms. The often lineage-specific occurrence of stem-loop elements in the sequences of introns and spacers also governs lineage-specific mutations. Sequencing whole plastid genomes to find markers for evolutionary analyses is therefore particularly useful when overall genetic distances are low.

**Editor:** Damon P. Little, The New York Botanical Garden, United States of America

**Funding:** The study was carried out as part of the project "Developing tools for conserving the plant diversity of the Transcaucasus" funded by VolkswagenStiftung (http://www.volkswagenstiftung.de/nc/en.html), grant number AZ85021. The funders had no role in study design, data collection and analysis, decision to publish, or preparation of the manuscript.

**Competing Interests:** The authors have declared that no competing interests exist.

* Email: t.borsch@bgbm.org

☌ These authors contributed equally to this work.

## Introduction

Clarifying species limits and reconstructing phylogenetic relationships in clades with recently diverged species is challenging. Levels of genetic divergence are often low while at the same time large numbers of samples need to be analysed. The same applies to analysing phylogeographic patterns, where many individuals from different populations need to be included. Due to the often complex modes of speciation in angiosperms, evidence from uniparentally inherited organellar genomes and the recombined nuclear genome is needed to unravel evolutionary histories [1–3]. This is also the case in the genus *Pyrus* where — like in many Rosaceae — polyploidy, hybridization, and reticulate evolution

occur. Estimates of *Pyrus* diversity vary between 50 and 80 species [4,5] and 20 taxa alone have been described from the southern Caucasus [6,7]. Similarly, the numbers of accepted species differ between treatments as a consequence of poorly understood species limits. *Pyrus* is a typical case for evolutionary and taxonomic analyses of diverse species groups in flowering plants that require the inclusion of hundreds of individuals. Before entering into large-scale sampling, we were interested to find the genomic regions with the best information potential for generating haplotype networks and inferring phylogenetic relationships. In this study, we focus on the plastid genome.

Along the same line of argumentation, Shaw et al. [8,9] inspired to employ a broader spectrum of noncoding and rapidly evolving plastid markers in phylogenetic analyses of closely related species. Shaw et al. [8] sequenced a wide range of plastid markers for three species across angiosperms and later compared plastid genome pairs of three lineages of angiosperms (*Atropa* and *Nicotiana* for the asterids, *Lotus* and *Medicago* for the rosids, and *Oryza* and *Saccharum* for the monocots) [9]. Their studies resulted in a set of 32 regions that ranked highest in their number of potentially informative characters (defined as sum of substitutions, indels and inversions following [8] and abbreviated as "PICs"). This set was consequently suggested to generally contain the most variable and phylogenetically most informative genomic regions in angiosperm plastid genomes. However, the question remains how to best select four or five of the total top 32 regions, as many species-level evolutionary studies require.

Noncoding genomic regions such as introns and spacers often contain stem-loops and other specific structural elements that can be highly dynamic and are AT-rich. This results in a mosaic-like pattern of conserved and variable elements [10]. Considering that certain stem-loop elements within given introns and spacers are often unique to restricted lineages [11,12], lineage specificity in the overall variability of genomic regions is to be expected. In several recent comparative analyses of angiosperm plastid genomes [13,14] different genomic regions were depicted as the most variable. Nonetheless, these results need to be considered with care because some of the respective authors worked with pairs of hardly differentiated genomes while others had pairs of genomes with high *p*-distances. We expect that taxon-specific differences caused by certain sequence elements will be less prominent when more distant genomes are studied.

Next-generation sequencing techniques greatly facilitate the analysis of whole plastid genomes [15–17]. To date, phylogenomic studies of plastid genomes in land plants often just relied on concatenated sequences of the conserved genes, neglecting the information from the noncoding regions. In other cases, the authors included rather few taxa for which plastid genome sequences were automatically assembled from the respective 454 or Illumina runs, without completing parts of low coverage or areas with difficulties to obtain correct sequences. However, especially those might be informative at and below the species level (e.g., AT-rich stretches of DNA including microsatellites) [18–20]. On the other hand, there are recent studies which used completely annotated plastid genomes to detect infraspecific variability in species of *Olea* [21], *Colocasia* [22], or *Phalaenopsis* [23], or to find genomic regions with the highest number of potentially informative characters in more distant genome pairs of angiosperm genera [9,24–26].

We have sequenced the plastid genome of *Pyrus spinosa* using 454 pyrosequencing in order to compare it with the published plastid genome sequence of *P. pyrifolia* [27]. In our *Pyrus* genome pair, the proportion of sites at which the two sequences are different (*p*-distances) is almost 10-fold lower than in the genome pairs studied by Shaw et al. [9]. For further comparison, we selected three fully annotated plastid genome pairs using the criterion of low *p*-distances (≤0.005) similar to *Pyrus*. Here we wanted to represent another rosid pair (*Oenothera parviflora* and *O. argillicola*; Onagraceae), an asterid pair (*Olea europaea* and *O. woodiana*; Oleaceae) and a monocot pair (*Cymbidium tortisepalum* and *C. sinense* (Orchidaceae).

The goals of this study were (1) to find the most variable regions of the *Pyrus* plastid genome and to propose plastid markers for species-level evolutionary studies in *Pyrus*, (2) to assess the variability of plastid genome regions based on comparable

genome-pairs with overall low *p*-distances (0.0005 to 0.005) in major lineages of angiosperms, (3) to clarify if there are universal or lineage-specific rankings of variability within the group of about 35 top variable genomic regions, and (4) to evaluate if there are lineage specific differences in molecular evolutionary patterns that could cause the variability of genomic regions.

## Material and Methods

### DNA extraction, 454 pyrosequencing, genome assembly and annotation

*Pyrus spinosa* was sampled from the living collection of the Botanical Garden Berlin-Dahlem (Acc. No. 248458110, IPEN-Nr. TR-0-B-2484581, origin: Turkey: Kastamonu, Pontic Mountains around Küre, leg.: Ern, Krone 7145, 9/1981, voucher at B). The leaf tissue was silica-dried and total genomic DNA was extracted using the NucleoSpin Plant II kit (Macherey Nagel) according to the manufacturer's instructions.

Shotgun sequencing from total genomic DNA was performed on a Roche 454 GS-FLX Titanium sequencer (Roche Applied Science, Indianapolis, Indiana, USA). The 454 run (1/4 plate) resulted in 120,255 reads with an average of 400 bp after removing the adaptor sequences.

An initial mapping assembly with MIRA 4 [31] using *Pyrus pyrifolia* as reference resulted in 4191 reads mapped to a single contig with an average coverage of 13.44. However, reads with larger indels, not occurring in the reference, were not incorporated into the contigs what lead to an incorrect genome sequence. To remove the bias of the reference sequence, the reads were *de novo* assembled to contigs using the Roche GS *De Novo* Assembler (Newbler) v.2.6 which resulted in 836 large contigs (N50 = 829), and with Mira 4 [28], which resulted in 1125 large contigs (N50 = 1072, N90 = 538, N95 = 519). All these contigs were mapped on the *Pyrus pyrifolia* plastid genome (GenBank acc. no. NC015996; Terakami et al. [27]) using Geneious 7 to produce a consensus sequence. The combined method of mapping *de novo* contigs recovered nine indels (maximum length 71 bp), which were not found with mapping alone. Finally the second inverted repeat was manually inserted into the consensus sequence.

The positions of protein coding genes, rRNAs, tRNAs and the inverted repeats were annotated with the help of DOGMA [29] and Geneious 7. All coordinates of exons, reading frames and the positions of tRNAs were manually checked by aligning the respective genes of *Nicotiana tabacum* L. (NC001879) to the *Pyrus spinosa* sequence in PhyDe [30] because DOGMA tends to incorrectly place the start and stop codons and often does not annotate small exons. In case of more deviating gene sequences (e.g. *matK* or *ycf1*), the *Pyrus* gene sequences were translated to amino acid sequences to correctly annotate the reading frame.

**Verification by Sanger sequencing.** Pyrosequencing is limited in that the exact number of nucleotides within longer homonucleotide stretches (polyAs or polyTs) cannot be reliably determined [16,31]. Our initial assembly contained several homonucleotide stretches and AT-rich sequence motifs. In our data, ambiguously called bases were frequent in homonucleotide stretches with more than six of the same nucleotides. To validate the sequence in such parts, we applied the Sanger method (electrophoresis was done at Macrogen Europe, Amsterdam, The Netherlands). Primers for amplification and sequencing were taken from the literature or designed in this study (see Table S1). Pherograms were checked by eye for peaks and corresponding quality scores to ensure that the polyA/T stretch was correctly read. All Sanger sequencing reads were unambiguous with no overlapping peaks after the polyA/T stretches. The respective

**Table 1.** GenBank accession numbers and references for the plastid genomes used in this study.

| Species | GenBank accession number | Reference |
| --- | --- | --- |
| Pyrus spinosa | HG737342 | this study |
| Pyrus pyrifolia | NC015996 | Terakami et al. [27] |
| Cymbidium tortisepalum | NC021431 | Yang et al. [24] |
| Cymbidium sinense | NC021430 | Yang et al. [24] |
| Oenothera parviflora | NC010362 | Greiner et al. [66] |
| Oenothera argillicola | EU262887 | Greiner et al. [67] |
| Olea woodiana | NC015608 | Besnard et al. [68] |
| Olea europaea | NC015401 | Besnard et al. [68] |

reads were aligned with the previously assembled genome sequence in Geneious 7 and the consensus sequence was corrected accordingly. The *Pyrus spinosa* plastid genome sequence is available in EMBL under accession HG737342.

**Pairwise genome comparisons and calculation of sequence divergence.** In addition to *Pyrus*, we took three other plastid genome pairs from published sources to represent closely related species, a further rosid genus, an asterid and a monocot genus. Genome sequences had to be complete and fully annotated. The aligned genome pairs had to show an overall distance of $p < 0.005$ (Table 1). All genome sequences were aligned in PhyDe using a motif alignment approach [32,33]. The pairwise alignments are provided as File S1, S2, S3, and S4.

Sequences of all introns and intergenic spacers larger than 100 bp were extracted from the alignments. The number of single nucleotide polymorphisms (SNPs) and indels for each sequence pair were counted with a script in R (v. 3.0.2). PICs were then determined in the sense of Shaw et al. [8] as the sum of all substitutions and indels. *P*-distances (proportion of differing nucleotide sites in the two sequences compared) of the regions were calculated by dividing the number of SNPs by the length of the regions without counting indel positions. The two parts of the *trnK* intron were analysed separately.

To assess the *p*-distances of the genome pairs used by Shaw et al. [8], we have aligned the genomes of *Lotus japonicus* (NC002694) and *Medicago truncatula* (AC093544); *Nicotiana tabacum* (NC001879) and *Atropa belladonna* (NC004561.1); *Saccharum* hybrid (NC005878) and *Oryza sativa* (NC008155) using MAFFT v. 7 [34], and calculated the *p*-distances of these genomes using PAUP* v. 4.0b10 [35].

To compare the whole genome variability apart from specific regions, a sliding window approach was performed counting the number of SNPs and indels and calculating the AT-content for 500 bp slots of the consensus sequences. The genome comparisons were visualized using Circos v. 0.64 [36].

## Molecular evolution within genomic regions

In order to assess the role of the base composition in variable sequence parts, i.e., indels and nucleotides around SNPs, we calculated their AT contents and compared them with the overall AT content of the whole genomes (consensus of pairwise aligned genomes). Three groups of indels were distinguished: (1) length variable poly-n loci that consist of a single nucleotide that is repeated at least sevenfold, (2) simple sequence repeats (SSRs) that show one repetition of a motif of multiple nucleotides, inverted repeats, or inversions, and (3) indels that do not fall in the former categories.

Further, AT contents of nucleotides adjacent to SNPs were calculated in intervals of increasing size (1–10, 20, 50, and 100 bp in each direction). A script was written in R v.3.0.2, which distinguishes the indels and regions around SNPs, calculates the AT contents, and displays their distributions.

The lineage-specific occurrence of substitutions and microstructural mutations was examined in more detail on the example of group II introns (*atpF*, *rpl16*) that strongly deviated in variability among our four genome pairs. These introns possess a mosaic-like structure of conserved and variable sequence elements. The variable parts usually correspond to the structurally and functionally least constrained terminal stem-loops, which appear in the respective RNA secondary structure. We first annotated the domains of the *atpF* and *rpl16* introns by comparing our sequences with the consensus alignment of Michel et al. [37]. The RNA secondary structures of individual domains were then predicted using RNAstructure 5.6 (available at http://rna.urmc.rochester.edu/RNAstructure.html) using the algorithm of Mathews et al. [38]. The "fold as RNA" option was implemented to allow for U–G pairings.

**Selecting genomic regions as markers for evolutionary studies in *Pyrus*.** Our aim was not only to find the most variable plastid regions in *Pyrus* but also to select several regions to be best used in evolutionary studies of *Pyrus*. Thus, efficient

**Table 2.** Sequence statistics for the four genome pairs compared.

| Genome pair | p-distance | Aligned length [bp] | Length difference | SNPs | Indels |
| --- | --- | --- | --- | --- | --- |
| Pyrus spinosa/P. pyrifolia | 0.00145 | 160607 bp | 227 bp | 230 | 173 |
| Olea europaea/O. woodiana | 0.00294 | 156091 bp | 30 bp | 458 | 112 |
| Oenothera parviflora/O. argillicola | 0.00122 | 165952 bp | 1690 bp | 199 | 173 |
| Cymbidium tortisepalum/C. sinense | 0.0008 | 155833 bp | 79 bp | 124 | 62 |

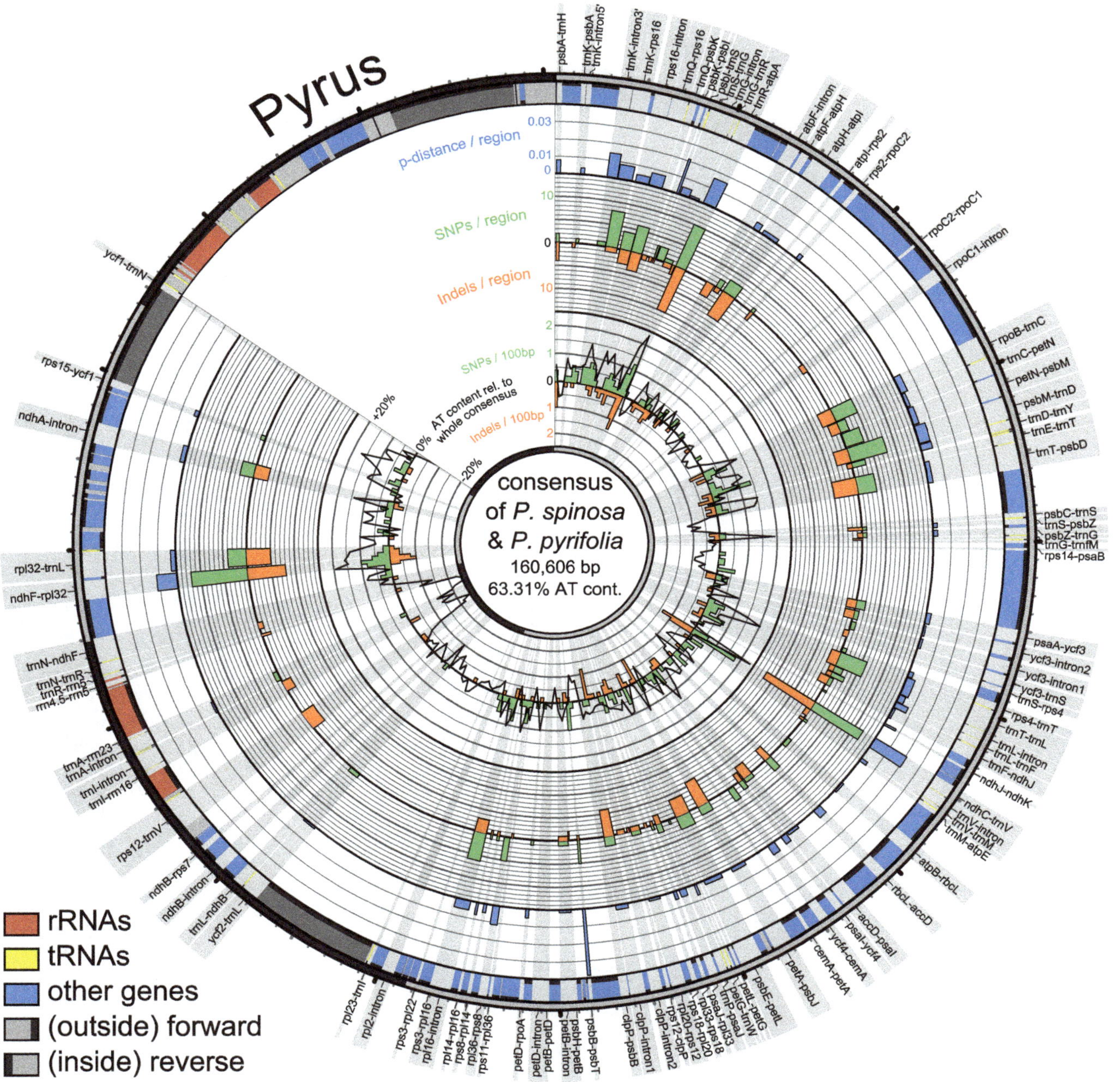

**Figure 1. Circular representation of plastid genome pair in *Pyrus*.** Shown are consensus sequences of compared species pairs of *Pyrus spinosa* and *P. pyrifolia* with their differing *p*-distances, numbers of SNPs and indels across the consensus. Radial grey highlights show the regions in focus of study with their names. Circular graphs from outside to inside: outermost circle with ticks for every 1,000 bp (small) and 10,000 bp (big) indicates part of genome, single copy regions in light grey and inverted repeats in dark grey; bands show locations of genes (blue), tRNAs (yellow) and rRNAs (red); the three outermost histograms display *p*-distances (blue), number of SNPs (green) and indels (orange) per spacer region; innermost graph shows number of SNPs (green histogram), indels (orange histogram), and AT content relative to the whole consensus (black line graph) of 500 bp long parts of the whole consensus.

amplification and sequencing strategies including primer binding sites, region size and the information content per primer read had to be considered in addition to a high rank in terms of variability. Furthermore, polyA/T stretches larger than seven nucleotides (microsatellites) had to be considered. Their presence usually require two primer reads for sequencing that start from both ends of the amplicon because slippage is likely to occur after the polyA/T stretch. Since a region >1000 bp usually requires two primers to sequence, one microsatellite was not considered a problem, while several microsatellites within the same region led to dismiss it. Considering that current technology generates reliable read lengths of 800–1000 bases, we selected fragments of 900–1300 bp in size _ a size range that can be easily amplified and then sequenced with a maximum of two primers.

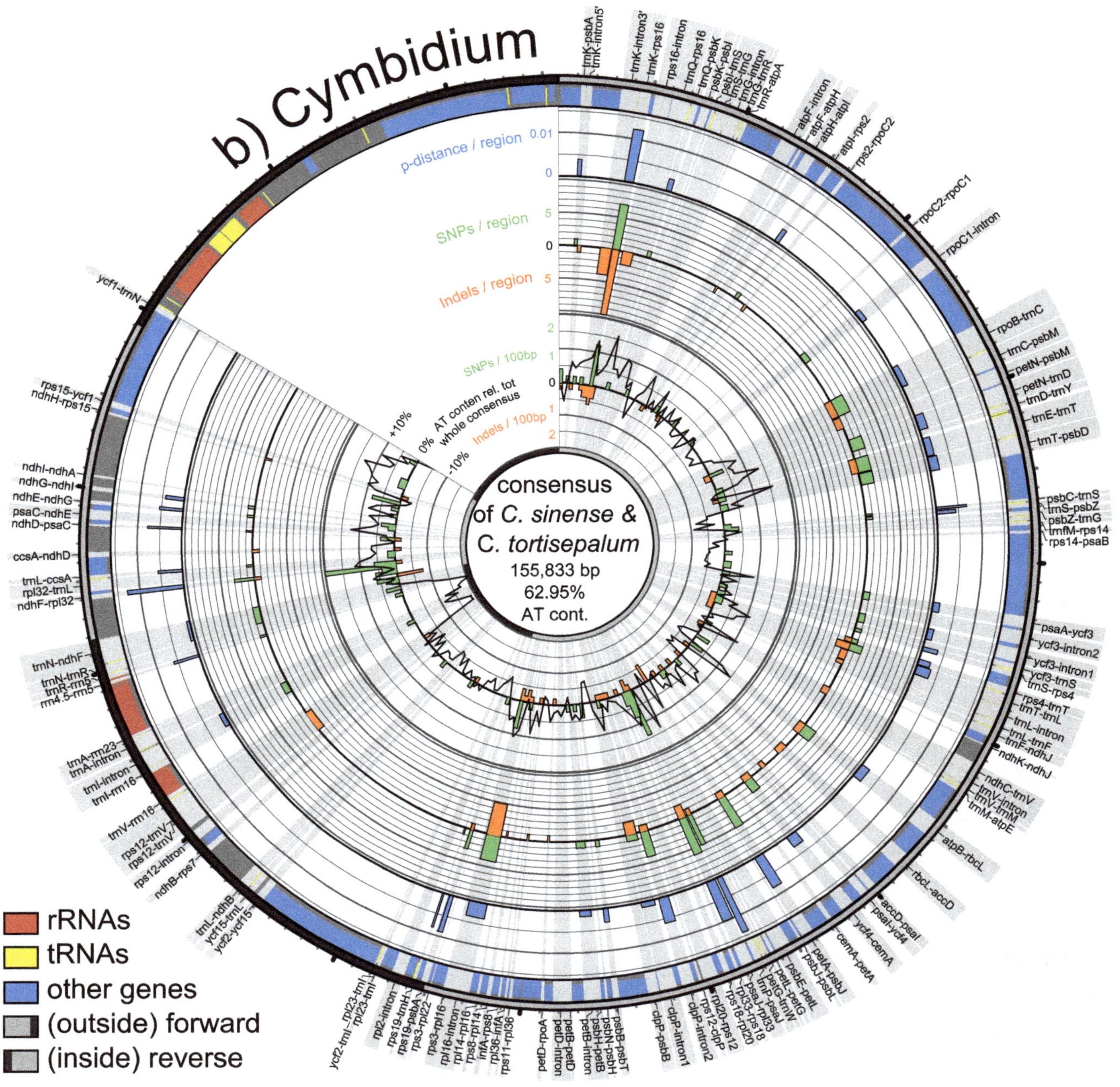

**Figure 2. Circular representation of plastid genome pair in *Cymbidium*.** Shown are consensus sequences of compared species pairs of *Cymbidium tortisepalum* and *C. sinense* with their differing *p*-distances, numbers of SNPs and indels across the consensus. Radial grey highlights show the regions in focus of study with their names. Circular graphs from outside to inside: outermost circle with ticks for every 1,000 bp (small) and 10,000 bp (big) indicates part of genome, single copy regions in light grey and inverted repeats in dark grey; bands show locations of genes (blue), tRNAs (yellow) and rRNAs (red); the three outermost histograms display *p*-distances (blue), number of SNPs (green) and indels (orange) per spacer region; innermost graph shows number of SNPs (green histogram), indels (orange histogram), and AT content relative to the whole consensus (black line graph) of 500 bp long parts of the whole consensus.

## Results and Discussion

### Size and structure of the *Pyrus* plastid genome

The plastid genome of *Pyrus spinosa* is 159,694 bp in length, and the inverted repeats (IRs) account for 26,396 bp. The large single-copy region (LSC) is 87,694 bp in length and the small single-copy region (SSC) 19,205 bp. The genome has a GC content of 36.6%. Gene content and order are identical to *Pyrus*

*pyrifolia*, with 113 unique genes and 17 duplicates in the IR [30]. The extension of IRs is identical to *P. pyrifolia*, while a 137 bp gap in the LSC of *P. spinosa* directly adjacent to IRa leads to a different IR boundary. The *p*-distance between the two genomes is 0.00145 (Table 2). The consensus structure of the two *Pyrus* genomes and the variability between them is illustrated in Fig. 1. Most of the variation occurs in the noncoding parts, especially in intergenic spacers of the LSC region. The SSC is less variable and

**Table 3.** Ranking and comparison of p-distances and differences in the four plastid genome pairs.

| Rank | Pyrus | | | | Cymbidium | | | | Oenothera | | | | Olea | | | |
|---|---|---|---|---|---|---|---|---|---|---|---|---|---|---|---|---|
| | Region | Aligned length [bp] | PICs (SNPs/Indels) | p-distance [*10^-3] | Region | Aligned length [bp] | PICs (SNPs/Indels) | p-distance [*10^-3] | Region | Aligned length [bp] | PICs (SNPs/Indels) | p-distance [*10^-3] | Region | Aligned length [bp] | PICs (SNPs/Indels) | p-distance [*10^-3] |
| 1 | psbB-psbT | 184 | 6 (5/1) | 37.88 | trnP-psaJ | 366 | 6 (5/1) | 14.04 | ycf1-ndhF | 381 | 15 (9/6) | 36.73 | trnG-trnR | 170 | 5 (4/1) | 23.81 |
| 2 | psbI-trnS | 149 | 4 (3/1) | 22.06 | ndhF-rpl32 | 259 | 4 (3/1) | 13.04 | psbI-psbL | 134 | 2 (2/0) | 14.93 | psbC-trnS | 243 | 6 (5/1) | 21.1 |
| 3 | ndhC-trnV | 760 | 24 (12/12) | 20.34 | trnK-rps16 | 613 | 17 (7/10) | 12.15 | rps4-trnT | 332 | 5 (4/1) | 12.16 | trnR-atpA | 112 | 3 (2/1) | 18.02 |
| 4 | trnR-atpA | 909 | 20 (10/10) | 13.61 | psaJ-rpl33 | 629 | 8 (6/2) | 9.93 | trnG-trnfM | 172 | 3 (2/1) | 11.9 | trnS-trnG | 715 | 12 (10/2) | 14.33 |
| 5 | ndhF-rpl32 | 1078 | 20 (12/8) | 11.41 | rps19-psbA | 345 | 4 (3/1) | 8.7 | ndhG-ndhI | 408 | 5 (4/1) | 9.9 | accD-psaI | 702 | 11 (10/1) | 14.27 |
| 6 | rpl36-rps8 | 459 | 5 (5/0) | 10.89 | rps19-trnH | 122 | 1 (1/0) | 8.2 | accD-psaI | 577 | 6 (5/1) | 8.68 | ycf15-trnL | 354 | 5 (5/0) | 14.12 |
| 7 | trnK-rps16 | 974 | 9 (8/1) | 8.38 | petA-psbJ | 635 | 6 (5/1) | 7.9 | trnQ-psbK | 355 | 6 (3/3) | 8.52 | psbA-trnH | 447 | 8 (6/2) | 13.51 |
| 8 | trnQ-rps16 | 905 | 10 (6/4) | 8.3 | ndhD-psaC | 129 | 1 (1/0) | 7.75 | ndhF-rpl32 | 932 | 7 (7/0) | 8.26 | rps4-trnT | 326 | 5 (4/1) | 12.31 |
| 9 | psbA-trnH | 268 | 6 (2/4) | 7.81 | psbC-trnS | 146 | 1 (1/0) | 6.85 | trnQ-accD | 2615 | 23 (12/11) | 5.59 | trnG-trnM | 174 | 2 (2/0) | 11.49 |
| 10 | trnL-trnF | 403 | 4 (3/1) | 7.59 | rrn4.5-rrn5 | 168 | 1 (1/0) | 5.95 | rps12-clpP | 397 | 4 (2/2) | 5.13 | psbB-psbT | 186 | 2 (2/0) | 10.75 |
| 11 | ndhJ-ndhK | 137 | 1 (1/0) | 7.3 | clpP intron 2 | 676 | 5 (4/1) | 5.93 | rps16-tbcL | 976 | 8 (4/4) | 5.06 | trnK-psbD | 1322 | 15 (14/1) | 10.6 |
| 12 | rpl14-rpl16 | 145 | 2 (1/1) | 6.99 | trnL-ccsA | 180 | 1 (1/0) | 5.56 | atpI-rps2 | 216 | 1 (1/0) | 4.63 | trnK-rps16 | 899 | 12 (9/3) | 10.06 |
| 13 | trnD-trnY | 448 | 4 (3/1) | 6.79 | ndhE-ndhG | 185 | 1 (1/0) | 5.41 | rps2-rpoC2 | 219 | 2 (1/1) | 4.59 | rps2-rpoC2 | 209 | 4 (2/2) | 9.66 |
| 14 | psbM-trnD | 1235 | 11 (8/3) | 6.51 | trnS-psbZ | 230 | 1 (1/0) | 4.35 | trnP-psaI | 515 | 4 (2/2) | 4.37 | trnS-rps4 | 314 | 3 (3/0) | 9.55 |
| 15 | trnW-trnP | 156 | 1 (1/0) | 6.41 | ndhG-ndhI | 233 | 1 (1/0) | 4.29 | trnL-ycf2 | 462 | 2 (2/0) | 4.33 | psbN-psbH | 105 | 1 (1/0) | 9.52 |
| 16 | rpl16 intron | 1003 | 9 (6/3) | 6.01 | trnK-psbA | 257 | 1 (1/0) | 3.89 | atpH-atpI | 939 | 5 (4/1) | 4.26 | trnD-trnY | 107 | 2 (1/1) | 9.43 |
| 17 | ycf4-cemA | 526 | 3 (3/0) | 5.7 | trnS-rps4 | 287 | 1 (1/0) | 3.48 | trnK intron 5' | 249 | 2 (1/1) | 4.03 | psbM-trnD | 657 | 7 (6/1) | 9.15 |
| 18 | rbcL-accD | 569 | 6 (3/3) | 5.33 | rpl16 intron | 1191 | 9 (4/5) | 3.46 | petN-psbM | 926 | 3 (3/0) | 3.24 | trnQ-psbK | 335 | 5 (3/2) | 9.01 |
| 19 | trnT-trnL | 1241 | 8 (6/2) | 4.94 | trnT-trnL | 610 | 4 (2/2) | 3.36 | psaA-ycf3 | 669 | 3 (2/1) | 3.01 | atpI-rps2 | 222 | 2 (2/0) | 9.01 |
| 20 | psaI-ycf4 | 413 | 4 (2/2) | 4.89 | atpI-rps2 | 300 | 1 (1/0) | 3.33 | trnS-psbZ | 348 | 1 (1/0) | 2.87 | atpF intron | 697 | 7 (6/1) | 8.62 |
| 21 | rps8-rpl14 | 207 | 2 (1/1) | 4.83 | psbB-psbT | 323 | 1 (1/0) | 3.1 | trnK-rps16 | 758 | 4 (2/2) | 2.67 | petN-psbM | 1171 | 12 (10/2) | 8.62 |
| 22 | rpl33-rps18 | 218 | 2 (1/1) | 4.67 | trnQ-psbK | 348 | 1 (1/0) | 2.87 | petD intron | 761 | 3 (2/1) | 2.65 | trnK-psbA | 236 | 2 (2/0) | 8.47 |
| 23 | trnS-trnG | 651 | 3 (3/0) | 4.62 | ycf4-cemA | 728 | 3 (2/1) | 2.75 | trnS-trnG | 788 | 2 (2/0) | 2.54 | ndhF-rpl32 | 479 | 4 (4/0) | 8.37 |
| 24 | rps16 intron | 909 | 7 (4/3) | 4.48 | rps4-trnT | 367 | 2 (1/1) | 2.74 | trnG intron | 804 | 4 (2/2) | 2.5 | psbI-trnS | 120 | 1 (1/0) | 8.33 |
| 25 | petD-rpoA | 225 | 1 (1/0) | 4.48 | trnT-psbD | 947 | 2 (2/0) | 2.11 | psaI-ycf4 | 412 | 5 (1/4) | 2.45 | petG-trnW | 121 | 1 (1/0) | 8.26 |
| 26 | atpF-atpH | 451 | 3 (2/1) | 4.44 | atpB-rbcL | 960 | 3 (2/1) | 2.11 | psbE-petL | 984 | 4 (2/2) | 2.05 | rps14-psaB | 122 | 1 (1/0) | 8.2 |
| 27 | trnM-atpE | 242 | 2 (1/1) | 4.29 | petN-trnD | 1020 | 2 (2/0) | 1.96 | trnT-trnL | 1114 | 4 (2/2) | 1.89 | psaA-ycf3 | 741 | 6 (6/0) | 8.1 |
| 28 | psaJ-rpl33 | 472 | 4 (2/2) | 4.29 | trnE-trnT | 1216 | 3 (2/1) | 1.68 | trnT-psbD | 1441 | 6 (2/4) | 1.43 | trnK intron 5' | 268 | 3 (2/1) | 7.49 |
| 29 | rpoB-trnC | 1216 | 8 (5/3) | 4.13 | psaA-ycf3 | 638 | 1 (1/0) | 1.57 | ycf3 intron 2 | 720 | 1 (1/0) | 1.39 | rpl32-trnL | 835 | 10 (6/4) | 7.26 |
| 30 | trnL intron | 514 | 3 (2/1) | 3.9 | rpoB-trnC | 1461 | 3 (2/1) | 1.38 | petB intron | 775 | 4 (1/3) | 1.34 | ndhC-trnV | 1119 | 12 (8/4) | 7.24 |

The regions are sorted according to p-distances.

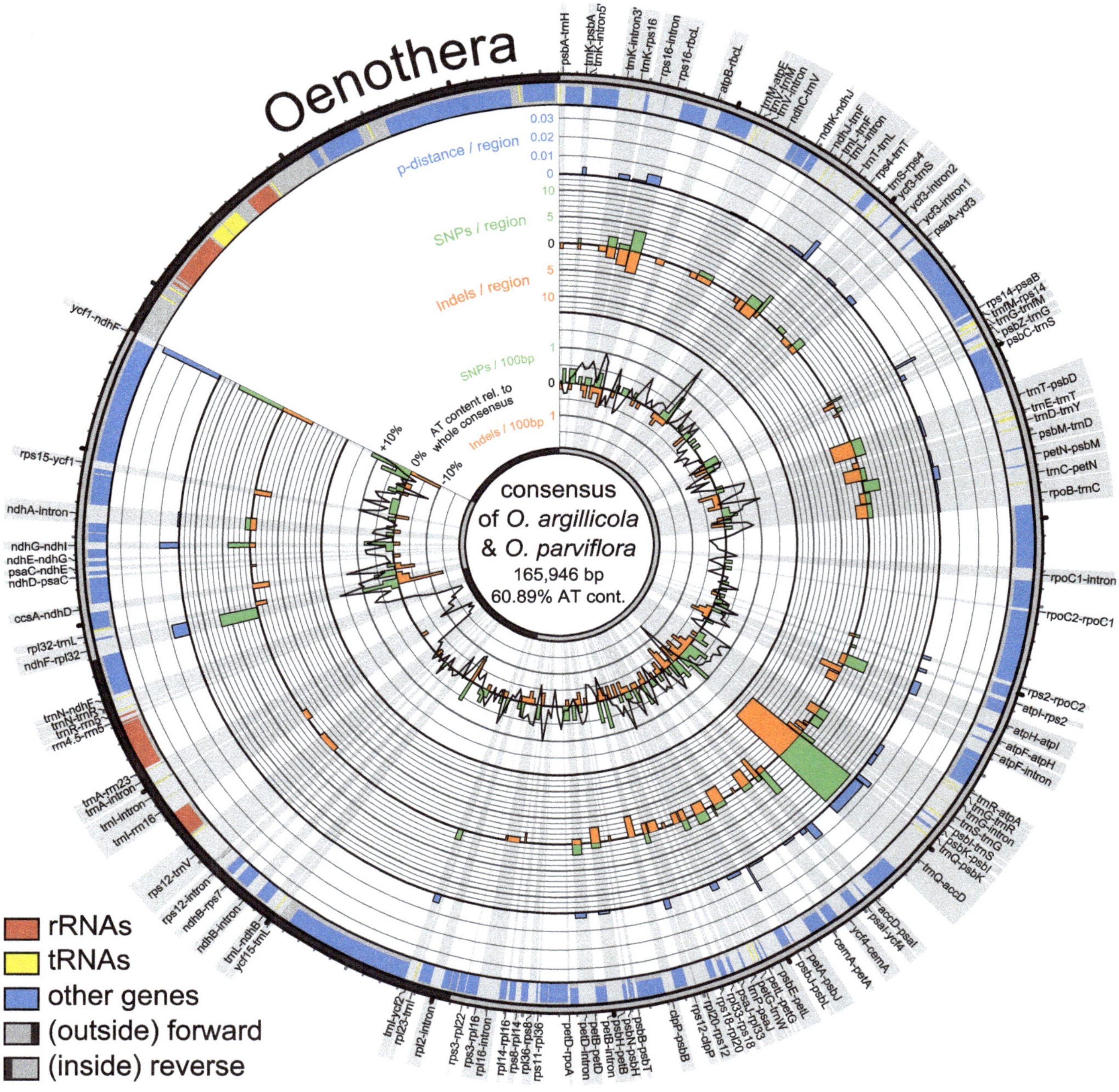

**Figure 3. Circular representation of plastid genome pairs in *Oenothera*.** Shown are consensus sequences of compared species pairs of *Oenonthera parviflora* and *O. argillicola* with their differing *p*-distances, numbers of SNPs and indels across the consensus. Radial grey highlights show the regions in focus of study with their names. Circular graphs from outside to inside: outermost circle with ticks for every 1,000 bp (small) and 10,000 bp (big) indicates part of genome, single copy regions in light grey and inverted repeats in dark grey; bands show locations of genes (blue), tRNAs (yellow) and rRNAs (red); the three outermost histograms display *p*-distances (blue), number of SNPs (green) and indels (orange) per spacer region; innermost graph shows number of SNPs (green histogram), indels (orange histogram), and AT content relative to the whole consensus (black line graph) of 500 bp long parts of the whole consensus.

almost no variation is found in the IRs. There are some genome parts with intergenic spacers alternating tRNA genes where variation appears to accumulate. This is especially the case in the region from *trnK* to *trnA* and from *rpoB* to *psbD* (Figs. 1, 2).

**Finding the most variable regions of the *Pyrus* plastid genome.** The five regions with the highest *p*-distances are the intergenic spacers *psbB–psbT*, *psbI–trnS*, *ndhC–trnV*, *trnR–atpA*,

and *ndhF–rpl32*. Taking the PICs as a basis, the five top-ranked regions are *ndhC–trnV*, *trnR–atpA*, *ndhF–rpl32*, *psbM–trnD*, and *trnQ–rps16* (Table 3, Fig. 1–4).

Comparing our results with the ranking of Shaw et al. [9] it appears that 17 of our 30 top-ranked regions in *Pyrus* are also among the 32 top-ranked in their study. However, their ranks are different. For example, in Shaw et al. [8], the *rpl32–trnL* spacer

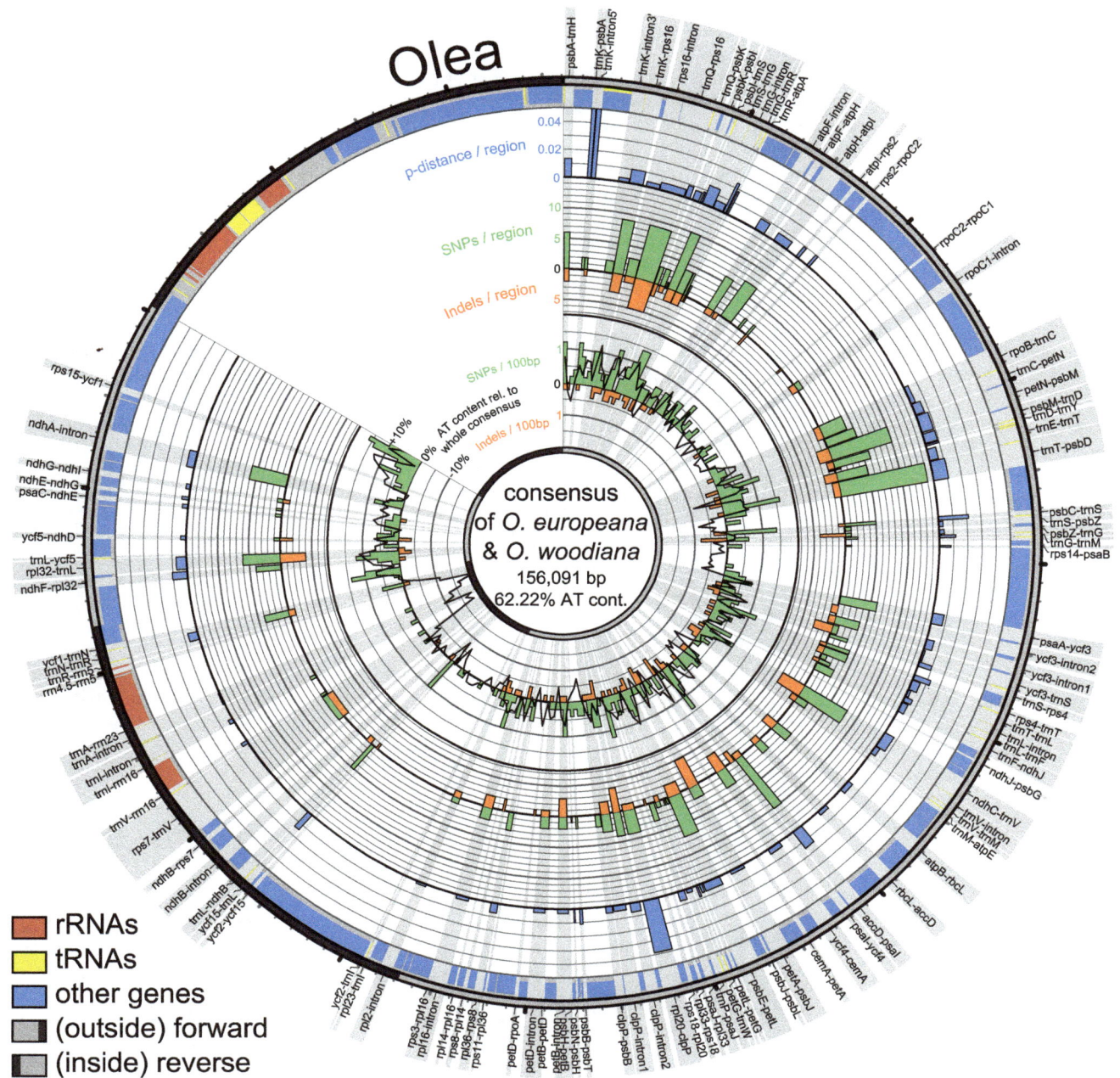

**Figure 4. Circular representation of plastid genome pairs in *Olea*.** Shown are consensus sequences of compared species pairs of *Olea europaea* and *O. woodiana* with their differing *p*-distances, numbers of SNPs and indels across the consensus. Radial grey highlights show the regions in focus of study with their names. Circular graphs from outside to inside: outermost circle with ticks for every 1,000 bp (small) and 10,000 bp (big) indicates part of genome, single copy regions in light grey and inverted repeats in dark grey; bands show locations of genes (blue), tRNAs (yellow) and rRNAs (red); the three outermost histograms display *p*-distances (blue), number of SNPs (green) and indels (orange) per spacer region; innermost graph shows number of SNPs (green histogram), indels (orange histogram), and AT content relative to the whole consensus (black line graph) of 500 bp long parts of the whole consensus.

has the highest number of PICs whereas it is only at rank 8 in *Pyrus*. The *trnR–atpA* spacer, which has the second-highest number of PICs in *Pyrus*, was not at all reported. However, the ranking of Shaw et al. may not be that comparable because the authors "normalized" their PICs with the aim to reduce the influence of different evolutionary rates or genetic distances. They divided the number of PICs within a region from a certain

taxonomic lineage by the total sum of PICs within the same lineage. Therefore, their results do not directly show lineage-specific differences in marker variability, although the absolute variability of a given genomic region is the only relevant fact in any analysis.

Low genetic distances in *Pyrus* have been pointed out in two earlier studies of *Pyrus* plastid genomes [27,39]. These studies were motivated by the horticultural importance of *Pyrus*, and focused on Asian species and cultivars. Katayama and Uematsu [39] provided a physical map of the plastid genome of *Pyrus ussuriensis* var. *hondoensis* and ran an RFLP analysis on cpDNAs from 11 accessions of five *Pyrus* and two *Prunus* species. However, there were no sequence data to support their conclusions. Terakami et al. [27] aligned the three plastid genomes of *Pyrus pyrifolia*, *Malus* × *domestica*, and *Prunus persica*. The authors calculated the proportion of mutational events using the same formula as Shaw et al. [8] for 89 noncoding regions, and ranked the compared regions according to their variability comparing *Pyrus* with *Malus* and *Prunus* (ingroup and outgroup were not specifically defined). While the *ndhC–trnV* and *trnR–atpA* spacers depict the highest sequence divergence in both, Terakami et al. and our work presented here, the overall rankings are strongly different. Terakami et al. found the spacers *rpl33–rps18*, *psbI–trnS*, and *rpl14–rpl16* from the third to fifth rank. In our *Pyrus* ranking, these spacers are at positions 22, 2, and 12 (based on *p*-distances) and 43, 22, and 41 (based on PICs), respectively. These differences may be explained by the much greater distance between the *Pyrus* and *Malus* plastid genomes than our two *Pyrus* genomes. The crown group of *Pyrus* diversified 27–33 mya while the crown group of *Malus* was inferred to have diversified 34–46 mya [40].

Various plastid regions have also been sequenced for a large number of samples in *Pyrus*. Katayama et al. [41] sequenced the *rps16–trnQ* and *accD–psaI* spacers and reconstructed a network based on 25 different haplotypes including 21 species of *Pyrus* and multiple individuals of *P. pyrifolia* and *P. ussurienis*, respectively. The authors found both spacers to contain highly variable AT-rich mutational hotspots and concluded that these regions are "hypervariable", while their remaining *Pyrus* sequences showed hardly any variation. The authors argued that their results confirmed their earlier hypothesis of strong sequence conservation in the plastid genomes of *Pyrus* [39]. No explanation, however, was given why particularly the *rps16–trnQ* and *accD–psaI* spacers had been chosen and not one of the highest ranked ones in terms of variability. The authors noted that the frequency of micro-structural mutations in both spacers studied was markedly higher than of substitutions and that haplotypes were mostly defined by indels. Such a dominance of microstructural mutations over substitutions is typical of AT-rich sequence elements that constitute terminal stem-loops of introns and transcribed spacers which are often unique to small lineages of plants [11]. At the same time such sequence elements often exhibit high levels of homoplasy. Thus, the exclusive application of these elements to calculate networks or trees may potentially lead to wrong conclusions. Wuyun et al. [42] sequenced the *rps16–trnQ* and *accD–psaI* spacers to reconstruct a phylogenetic network of *Pyrus ussuriensis* in China, which was largely based on the presence or absence of indels in the two spacers. Compared with our results, the two regions used by Katayama et al. [47] and Wuyun et al. [48] are also not the most variable plastid regions in *Pyrus*: the *trnQ–rps16* spacer ranks at place 24 for *p*-distances and at place 5 for PICs. The *accD–psaI* spacer ranks at place 18 for *p*-distances and at place 20 for PICs.

## Plastid markers proposed for *Pyrus*

Four intergenic spacers of 900 to 1000 bp and the *rpl16* group II intron (ca. 1000 bp) are proposed here to be sequenced for evolutionary studies in *Pyrus* (Table 4). They were selected from the most variable genomic regions (Table 3) considering an efficient sequencing strategy (see methods section).

Among the regions with a minimum size of 500 bp, the *ndhC–trnV* and *trnR–atpA* spacers rank 3rd and 4th according to *p*-distances, and *ndhC–trnV* has the highest number of PICs. Both can be sequenced with just one primer (either forward or reverse). Thus, these spacers are especially useful if large sample numbers need to be analysed. The *ndhF–rpl32* spacer (ranked 3rd of the regions >500 bp in Table 4) was not considered further because there are two large microsatellites. This fragment can therefore not be sequenced with two primers. The same problem occurs in the *rps16–trnK* spacer (ranked 4th of the regions >500 bp in Table 4) where two poly G and one poly T are likely to cause sequencing problems with pherograms unreadable after the homonulceotide stretches. The *trnQ–rps16* and *psbM–trnD* spacers follow in the ranking. Both also have polyA/T microsatellites. While they can be covered with two primer reads that overlap at the microsatellite, they may not be as efficiently sequenced than the *ndhC–trnV* and *trnR–atpA* spacers for large sample numbers. The *rpl16* intron (ranked at 7th position of the regions >500 bp in Table 4), is particularly recommended because it was shown to also possess a high phylogenetic structure *R* in different angiosperm sequence data sets [43–45]. Multiple *rpl16* sequence alignments can therefore be expected to yield well-resolved and well-supported trees also in *Pyrus*. The intron can be co-amplified with the *rpl14–rpl16* spacer. The use of the reverse primer PYR-rpl16R (Table 4) will allow to sequence the whole intron with one read. The *rpl16* intron contains a polyA/T stretch of variable length in different species of *Pyrus* (see also Fig. 5c), what implies that an additional forward primer read may be necessary to cover the whole intron in some samples.

Primers were newly designed for *trnR–atpA* as this region to our knowledge has never been used in any evolutionary study so far. For *ndhC–trnV*, primers were available [46] but we designed a new *Pyrus*-specific reverse primer in order to completely cover the spacer-exon boundary. For *trnQ-rps16*, the universal primers designed by Shaw & al. [9] work for *Pyrus* as well. Available primers for *psbM–trnD* [47] were re-designed for *Pyrus* to avoid mismatches in the forward and then to obtain a similar melting temperature in the reverse primer. For the *rpl16* intron, primers were also adapted to *Pyrus* following the general amplification strategy of [43] and [44] with a forward primer that anneals to the *rps3* exon. This ensures that the *rpl16* intron can be amplified and sequenced completely. The universal reverse primer rpl16R [48] was replaced by a *Pyrus*-specific primer that anneals further downstream to cover the intron-exon boundary.

**Comparison of plastid genomes with low *p*-distances in angiosperms.** In addition to *Pyrus*, we explored variability patterns in plastid genome pairs of *Oenothera argillicola* and *O. parviflora* (Onagraceae), *Olea europaea* and *O. woodiana* (Oleaceae), and *Cymbidium sinense* and *C. tortisepalum* (Orchidaceae) which have comparable low *p*-distances (Table 2). The variability patterns of all four genome pairs are illustrated using a Circos-plot (Figs. 2–4). Each genome pair has different regions with highest *p*-distances and highest numbers of PICs, resulting in a genome pair-specific ranking (Table 3). The results of the pairwise comparisons of individual introns and spacers for each genome pair are provided in Table S2.

The SNPs and indels are almost evenly spread across the LSC and the SSCs in *Olea*. In *Cymbidium*, SNPs and indels are more clustered. The plastid genomes of *Pyrus* and *Oenothera* exhibit strong variation in certain areas, e.g. between *trnT* and *rpoB* (Figs. 1, 3) but alsoalso homogeneously distributed mutations across their genomes. The *Olea* genome stands out by many more SNPs than indels, while the other genomes have almost as many indels as SNPs.

**Table 4.** Genomic regions proposed for evolutionary analyses in *Pyrus* and primers for their amplification.

| Region | Amplified fragment | Primer name | Primer sequence | Reference |
|--------|--------------------|-------------|-----------------|-----------|
| *ndhC–trnV* | 900 bp | ndhC–F | TGCCAAAATAGGAATAACAC | Goodson et al. [46] |
| | | PYRtrnV–150R | CCACATAATGAATCAGAGCAC | this study |
| *trnR–atpA* | 1000 bp | trnR–F | GTCTAATGGATAGGACAGAGG | this study |
| | | atpA–180R | GGAACRAACGGYTATCTTGATTC | this study |
| *psbM–trnD* | 1350 bp | PYRpsbM–F | CCTTGGCTGACTGTTTTTACG | this study |
| | | PYRtrnD–R | GAGCACCGCCCTGTCAAGG | this study |
| *trnQ–rps16* | 900 bp | trnQ (UUG) | GCGTGGCCAAGTGGTAAGGC | Shaw et al. [9] |
| | | rps16x1 | GTTGCTTTCTACCACATCGTTT | Shaw et al. [9] |
| *rpl16* intron | 1300 bp | PYR–rps3F | GATTATTGTTCCTATGCAG | this study |
| | | PYR–rpl16R | GCTTGAAGAGCATATCTAC | this study |

In our summary of the 30 most variable genomic regions including all four genome pairs, 77 different regions appear in total (Table 3). It is noteworthy that only two spacers, *ndhF–rpl32* and *trnK–rps16*, are consistently placed among the 30 most variable regions. Eight spacers appear three times: *atpI–rps2*, *psaA–ycf3*, *psbB–psbT*, *rps4–trnT*, *trnQ–psbK*, *trnS–trnG*, *trnT–psbD*, and *trnT–trnL*.

**Earlier comparisons of plastid genomes in angiosperms for marker selection.** In an approach to explore hitherto unused plastid regions as phylogenetic markers, Shaw et al. [9] in 2007 compared whole plastid genomes in a comprehensive way. They analysed genome pairs from three different lineages of angiosperms [*Atropa* and *Nicotiana* (Solanaceae) for the asterids, *Lotus* and *Medicago* (Fabaceae) for the rosids, and *Oryza* and *Saccharum* (Poaceae) for the monocots]. They found nine previously unexplored plastid regions with high levels of variation based on the numbers of PICs: *rpl32–trnL*, *trnQ–rps16*, *ndhC–trnV*, *ndhF–rpl32*, *psbD–trnT*, *psbJ–petA*, *rps16–trnK*, *atpI–atpH*, and *petL–psbE*. As noted before, we were interested to compare the distance levels of these genomes to the genome pairs examined here, as we expected considerable differences. The *p*-distances were indeed much higher and are here calculated as follows: *Lotus japonicus*/*Medicago truncatula* $p = 0.17603$, *Nicotiana tabacum*/*Atropa belladonna* $p = 0.01363$, *Saccharum* hybrid/*Oryza sativa* $p = 0.04879$.

Another comparative study of plastid genomes was carried out by Dong et al. [13] five years later. They looked at 14 angiosperm genera for which more than one plastid genome was available, again with the goal of finding markers for phylogeny reconstruction and DNA barcoding. They concluded that *ycf1*, *psbA–trnH*, *rpl32–trnL*, *trnQ–rps16*, *ndhC–trnV*, *trnK*/*matK*, and *trnS–trnG* are best-suited.

Next generation sequencing has resulted in an increased availability of plastid genome data in recent years (Table 5) that were used to find markers for various phylogenetic analyses in certain angiosperm lineages, to recover promising regions for haplotype studies or to differentiate closely related species and cultivars [14,21,22,24–27,49–52]. None of the authors addressed more general patterns of plastid genome mutational dynamics and molecular evolution. As noted before, the studies span an enormous range of different genetic distances in the genomes compared. The compared economically important asterids (e.g., *Solanum*, *Nicotiana*, *Lactuca*) are well represented while studies on other taxa are still scarce. Moreover, the approaches and methods applied in these studies differ. Most of them calculated

some kind of sequence variability, while others additionally or solely reconstructed phylogenetic trees based on small taxon sets to assess the phylogenetic utility of these regions. A spectrum of 37 plastid loci was reported as "highly variable" in the studies cited above. Most commonly mentioned were *rpl32–trnL* (7x), *trnQ–rps16* (5x) *trnK–rps16* (4x), and *ndhC–trnV* (4x). Nevertheless, the question remains how representative the earlier pairwise genome comparisons are, and to what extent their conclusions are also valid for other families and genera of flowering plants.

Shaw et al. [8] assumed a high universality of their results. But Daniell et al. [52], who compared plastid genomes of Solanaceae, found spacers with higher sequence divergence not mentioned in [8]. Timme et al. [49] analysed Asteraceae and indicated that their ranking of most variable regions barely overlapped with the ranking of Shaw et al., and suspected that "each family or major lineage will most likely have a unique set of variable regions" [43]. Shaw et al. [9] in 2007 found no less than 11 new highly variable markers not considered in their 2005 study therefore pointed to the need of a test-wise screening of the "universal" regions to find the most suitable one for a given lineage. Likewise, Dong et al. [13] stated that markers useful for one group may not be useful for another and recommended evaluating markers in detail before selecting them for further use. With the aim of resolving the species tree in the huge genus *Solanum*, Särkinen and George [14] found that the average amount of variable characters differs within subclades of the genus. In their view, the degree to which the utility of a marker can be extended to more inclusive clades would then also be clade-specific.

In summary, lineage specific differences in variability and phylogenetic utility of plastid genomic regions were reported in various cases in flowering plants although there was never any standardized comparative approach to better understand this issue. Moreover, none of the previous studies explicitly addressed phylogenetic signal as being different from similarity-based variability, or looked at any molecular evolutionary characteristics.

**Molecular evolution and lineage specific variability of genomic regions.** Lineage-specific differences in variability are often explained by patterns of molecular evolution. It has been exemplarily demonstrated for regions such as *psbA–trnH* [53] or *trnL–trnF* [54] that variability is strongly influenced by structural constraints. Empirical analysis of *petD* group II intron sequences has further shown that increased length correlates with increased AT strongly influenced byal constraints Empirical analysis of *petD* group II intron sequences has further shown that increased length correlates with increased AT content [12]. Figure 5 shows the AT

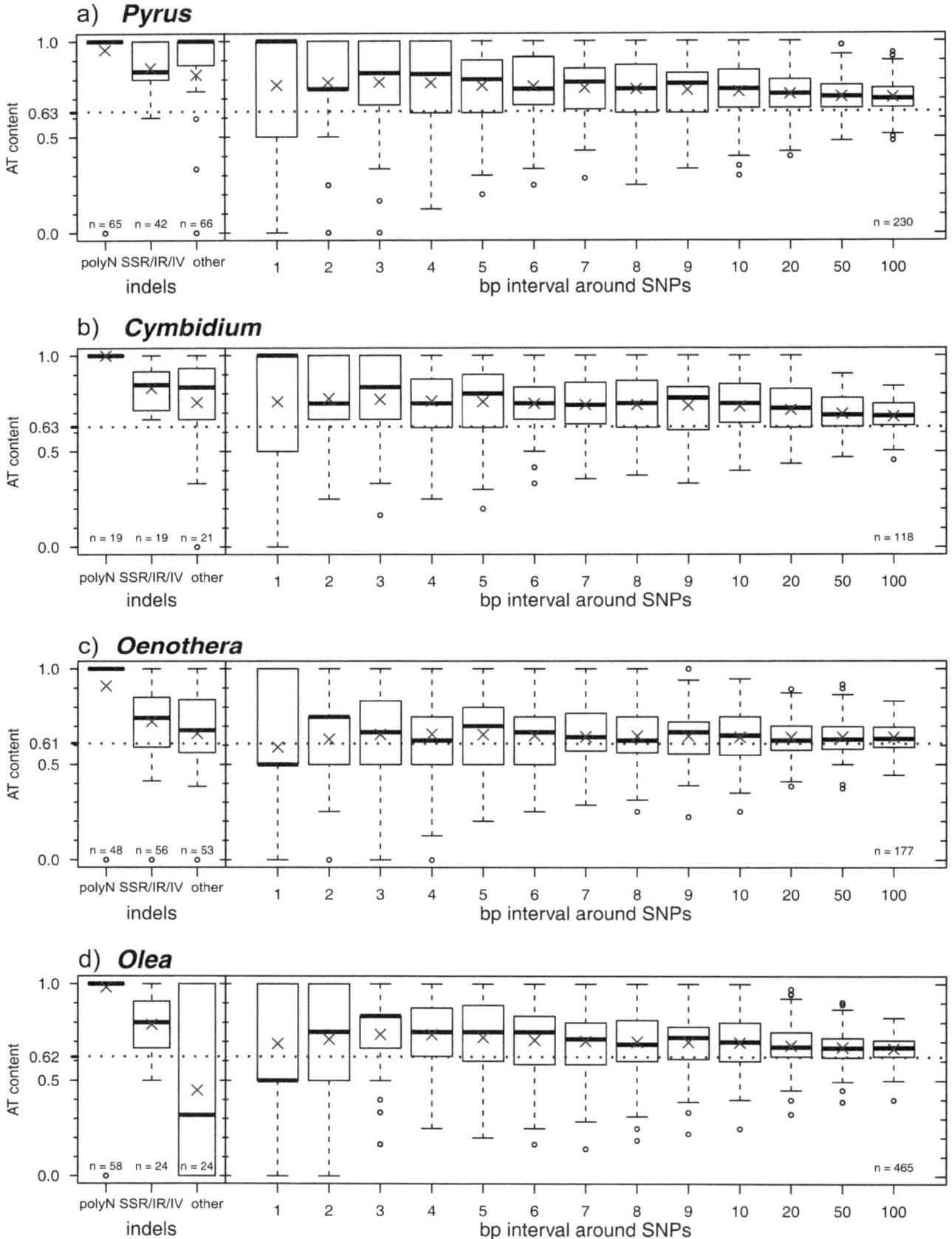

a) *Pyrus*

b) *Cymbidium*

c) *Oenothera*

d) *Olea*

**Figure 5. Mutational dynamics in group II introns.** a) Schematic consensus structure of plastid group II introns based on Michel et al. (1989). Roman numbers indicate the six domains. B) Alignment and predicted RNA secondary structure for domain IV of the *atpF* intron in *Cymbidium*, *Pyrus*, *Oenothera* and *Olea*. The apparently non-homologous sequence blocks are placed separately in the alignment. There are no substitutions or length mutations in *Pyrus* and *Cymbidium*, the structures shown are therefore identical in the two species compared. The shown secondary structures of *Oenothera* and *Olea* are consensus structures. Two conserved nucleotide blocks at the 3' and 5' ends, indicated by thick blue bars, are conserved across all taxa and homologous in primary sequence and secondary structure. These conserved sequence blocks form the stem of the domain while variation occurs in the terminal stem-loops part of the domain. c) Alignment and predicted secondary RNA structures of domain IV of the *rpl16* intron. For clarity, only the part of the domain with positions variable within genera are shown; "[-]" mark the omitted stem-loop elements. The apparently non-homologous sequence blocks are placed separately in the alignment. Those positions where variation occurs within a genus are marked with arrows. See text for more explanation.

contents of three types of indels (left side) and around SNPs (right side) in intervals of increasing size of each of our genome pairs. AT content distributions are displayed in boxplots with the cross showing the mean and the thick line referring to the median. Respective boxplots arranged along the *x*-axis then depict maximum distances of the intervals in each direction of the SNP. Apart from rare exceptions the surroundings of SNPs are distinctly more AT-rich than the whole genome (Fig. 6), indicating that substitutions occur predominantly in AT-rich stretches. The AT contents of the consensus sequences are displayed as dotted lines. Looking at indels, considerable differences are apparent in the frequency of different kinds among the four plant lineages. In *Olea*, length-variable polyA/T stretches are most common. In *Oenothera*, all three kinds of indels occur with almost equal frequency, while in *Cymbidium* and *Pyrus* indels without a clear motif predominate.

The AT content is significantly increased in sequence elements affected by microstructural changes (Fig. 6), both in SSRs and in the non-SSR indels. The SSRs are generally AT-rich, so the templates for these SSRs must be AT-rich as well. And therefore, their frequency is also significantly higher in AT-rich sequence elements. It can thus be suggested that mutational dynamics is

increased in AT-rich sequence. A strong correlation between high AT content and high substitutional rates was also recently demonstrated in plastid genomes of Lentibulariaceae [55].

Comparative studies of the molecular evolution of group II introns showed substitutions, length-variable homonucleotide stretches and indels to predominantly occur in domains I, III and IV. These domains are also the most variable with respect to size and experience less strong functional constraints compared to the other domains [12,56,57]. Furthermore, considerable variation occurs in sequence elements that are unique to certain lineages, where they have evolved through stepwise insertion processes connected to the formation of stable helical elements [11]. In our data set, this is for example evident in the *petD* and *rpl16* introns. They appear at strikingly different positions in the rankings of the respective genome pairs (Table 3 and S2). In both introns the variation between the sequences of a genome pair is mostly caused by length variable polyA/T stretches or AT-rich indels.

Domain IV of the *atpF* intron belongs to a conserved group II intron (Fig. 5a) with no variation between the *Cymbidium* and *Pyrus* sequences, two substitutions in *Olea* and a length-variable polyA-stretch in *Oenothera* (Fig. 5b). The alignment (Fig. 5b) illustrates two conserved sequence blocks that are homologous and

**Table 5.** Identification of most variable plastid regions based on pairwise genome comparisons across angiosperms.

| Reference | Taxa studied | Markers found as most variable |
|---|---|---|
| Daniell et al. [52] | Asterids: *Atropa belladonna*, *Nicotiana tabacum*, *Solanum bulbocastanum*, *S. lycopersicum* (Solanaceae) | *psbK–psbI*, *rps12–clpP*, *trnG–trnfM*, *trnK–rps16*, *trnQ–rps16* |
| Timme et al. [49] | Asterids: *Helianthus annuus*, *Lactuca sativa* (Asteraceae) | *ndhC–trnV*, *rpl32–trnL*, *rps12–clpP*, *trnE–rpoB*, *trnY–trnE* |
| Shaw et al. [9] | Angiosperms: Asterids: *Atropa belladonna*, *Nicotiana tabacum* (Solanaceae), Rosids: *Lotus*, *Medicago* (Fabaceae), Monocots: *Oryza*, *Saccharum* (Poaceae) | *rpl32–trnL* *trnQ–rps16* *ndhC–trnV*, *ndhF–rpl32*, *psbD–trnT*, *psbJ–petA*, *rps16–trnK*, *atpI–atpH*, *petL–psbE* |
| Doorduin et al. [50] | Asterids: *Jacobaea vulgaris*, *Helianthus anuus*, *Lactuca sativa*, *Parthenium argentatum*, *Guizotia abyssinica* (Asteraceae) | *ndhC–trnV*, *ndhC–atpE*, *rps18–rpl20*, *clpP*, *psbM–trnD* |
| Gargano et al. [51] | Asterids: *Solanum tuberosum* subsp. *tuberosum*, *S. bulbocastanum* (Solanaceae) | *ndhA intron*, *petN–psbM*, *rpl32–trnL*, *rps2–rpoC2*, *trnQ–rps16* |
| Yang et al. [24] | Monocots: *Cymbidium* (Orchidaceae) | *cemA–petA*, *clpP–psbB*, *ndhF–rpl32*, *petA–psbJ*, *psbA–trnK*, *rpl32–trnL*, *trnE–trnT*, *trnK–rps16*, *trnL–ccsA*, *trnP–psaJ*, *trnT–trnL* |
| Dong et al. [13] | Angiosperms: *Acorus* (Acoraceae), *Aethionema* (Brassicaceae), *Calycanthus* (Calycanthaceae), *Chimonanthus* (Calycanthaceae), *Eucalyptus* (Myrtaceae), *Gossypium* (Malvaceae), *Nicotiana* (Solanaceae), *Oenothera* (Onagraceae), *Oryza* (Poaceae), *Paeonia* (Paeoniaceae), *Populus* (Salicaceae), *Solanum* (Solanaceae) | *ycf1*, *trnH–psbA*, *rpl32–trnL*, *trnQ–rps16*, *ndhC–trnV*, *trnK/matK*, *trnS–trnG* |
| Ku et al. [26] | Asterids: *Catharanthus roseus* (Apocynaceae), *Asclepias syriaca* (Apocynaceae), *Coffea arabica* (Rubiaceae), *Solanum lycopersicon* (Solanaceae) | *ndhF–rpl32*, *rpl32–trnL*, *rps16–trnQ*, *trnE–trnT*, *trnK–rps16* |
| Ku et al. [25] | Asterids: *Ardisia polysticta* (Primulaceae – Myrsinioideae) *Panax ginseng* (Araliaceae) *Sesamum indicum* (Pedaliaceae) | *ccsA–ndhD*, *ndhG–ndhI*, *rpl14–rpl16*, *rpl32–trnL*, *trnK–rps16* |
| Särkinen & George [14] | Asterids: *Solanum tuberosum*, *S. bulbocastanum*, *S. lycopersicum* (Solanaceae) | *atpB–rbcL*, *clpP–psbB*, *ndhF*, *ndhF–rpl32*, *petL–psaJ*, *petN–psbM*, *rpl32–trnL*, *rpoC1–rpoB*, *trnA–trnI*, *trnK–rps16*, *ycf1* |

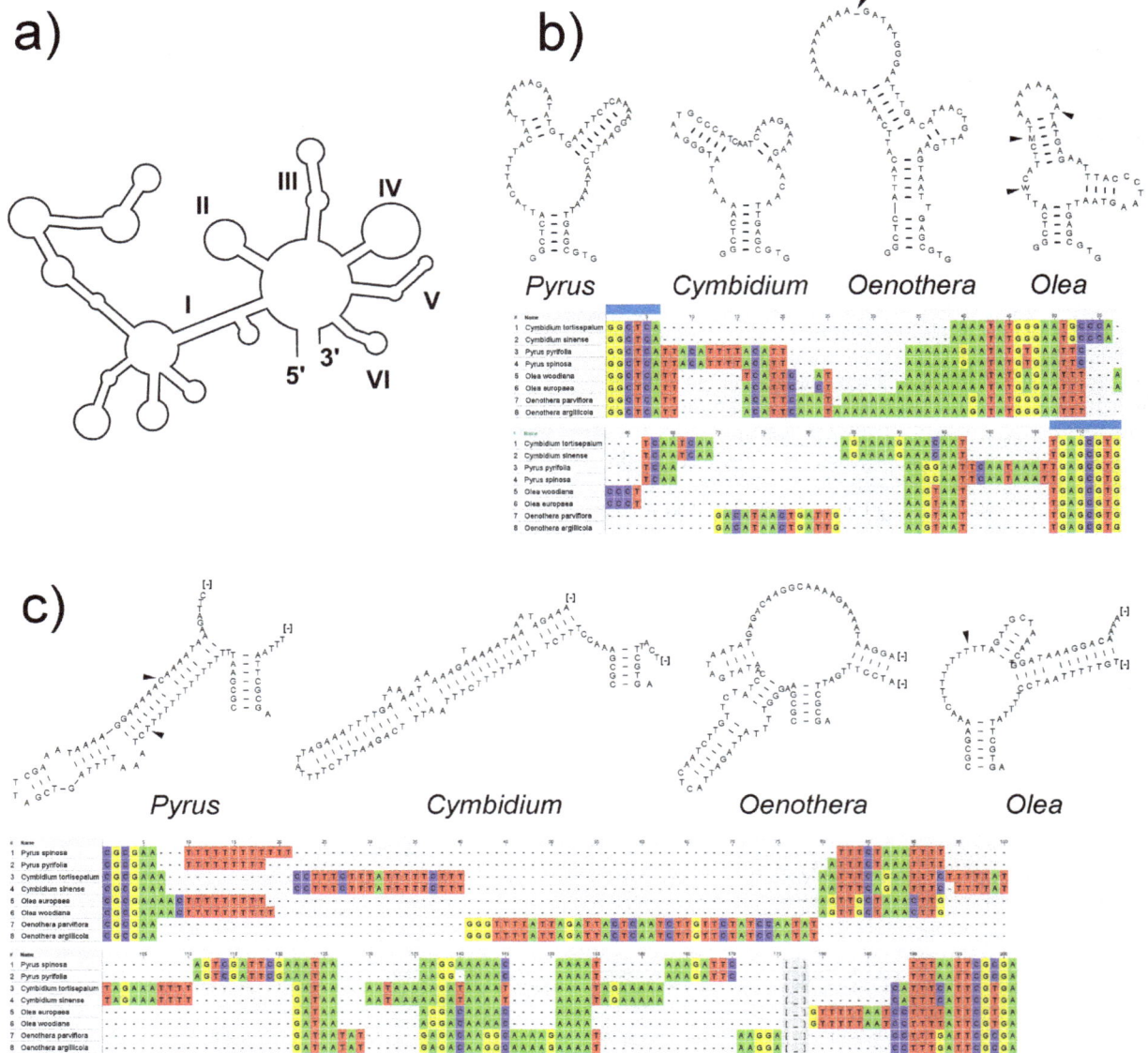

**Figure 6. AT content of indels and areas around substitutions.** Boxplot representation of the AT content in different types of indels (polyN, short sequence repeats (SSR) and other indels) on the left side and in areas with different sizes around all substitutions (SNPs) in the genome on the right side for a) *Pyrus spinosa* and *P. pyrifolia*), b) *Cymbidium tortisepalum* and *C. sinense*, c) *Oenonthera parviflora* and *O. argillicola* and d) *Olea europaea* and *O. woodiana*. The cross in each boxplot indicates the mean of the distribution, the thick line refers to the median. The dotted line shows the AT content of the whole consensus sequence.

conserved across all genera. They form the stem of the domain. Terminal parts of the domain such as the length-variable polyA-stretch in *Oenothera* have no structural constraints and therefore evolve rather freely. In *Olea*, there are two substitutions (indicated with ambiguity codes in the secondary structure) and one length variable polyA stretch. Again they occur in the terminal stem-loop and have no influence on the structure. The *rpl16* intron is more variable in *Pyrus* than in the other genome pairs. The polyT-stretch of *Olea* and *Pyrus* (beginning at position 10) is hypothe-sized as homologous in the alignment. But the predicted secondary structures (Fig. 5c) show that this polyT stretch forms different secondary structures caused by the different adjacent sequence elements. In *Olea*, it forms a bulge but in *Pyrus* it forms a stem-

element together with a complementary 'AAAACACAAAAAA' motif [12,54].

**Sequence variability versus phylogenetic signal.** It is important to note that sequence variability as such does not necessarily correlate with the amount of hierarchical phylogenetic signal in a multiple sequence matrix. Thus, *p*-distances and PICs_which are both measures of sequence variability and describe the similarity of sequences_will not necessarily indicate the phylogenetically most informative regions. The phylogenetic utility of genomic regions depends on the distribution and kind of character state transformations throughout the evolutionary history of the sequences. Several statistics have been proposed to measure the hierarchical phylogenetic signal (referring to the phylogenetic structure in a data set) that take into account the number of resolved nodes and the statistical support for these nodes [58,59]. Specifically, the statistics $R$, $B$, and $C$, have been defined by Müller et al. [59]. The most important one, $R$, measures the proportion of resolved clades and their support in a tree inferred from a given data set relative to the maximum possible resolution and support. If all nodes have maximum support, $R$ will get the value 1; if the phylogeny is completely unresolved (consists only of polytomies), $R$ will have the value 0.

The empirical evaluation of phylogenetic structure in a genomic region generally requires a multiple sequence alignment of a representatively sampled clade. From the datasets that have been evaluated in detail using the $R$ statistic [44,45,59], it is evident that at one hand higher variability often leads to more phylogenetic information (simply because there are more potentially informative characters). On the other hand, there are marked differences in the quality of hierarchical phylogenetic signal coming from the same number of variable positions in different kinds of genomic regions [45]. These can be explained by different molecular evolutionary patterns. The general trend across angiosperms is that high phylogenetic structure is found in intergenic spacers and group I and II introns, but not in protein-coding genes except *matK*. In our case of very closely related plastid genomes, the effects of multiple changes of the same site, eventually leading to saturation, or reversals, will probably not be very significant because these sequences are just starting to diverge. Nevertheless, it will be interesting to determine the phylogenetic structure in the top-ranked genomic regions in terms of variability once more extensive taxon sets will be available.

Moreover, highly variable regions will be needed to distinguish haplotypes (or species), even if they do not provide sufficient information about their phylogeny [44]. If haplotypes are used in the sense of individual alleles, the pure variability is most important. However, AT-rich sequence elements (often in stem-loops) can be highly homoplastic with respect to the evolution of microstructural mutations [60,61]. The most extreme causes of homoplasy are inversions [62,63]. Therefore, especially those markers that contain a single AT-rich mutational hotspot should be tested for congruence in signal with other plastid markers. Haplotype analyses often only use one or two markers, but experiences from other studies that have successfully reconstructed evolutionary relationships among closely related species indicate that the combination of four or five regions will be needed. An increased number of characters increases resolution and support also in network analyses [64,65].

**Implications for plastid marker development in angiosperms.** About 20–30 plastid spacers and introns are regularly sequenced for phylogenetic and haplotype analyses, for which universal amplification primers exist. Also, considerable progress has been made during recent years in predicting phylogenetic utility from molecular evolutionary patterns, revealing differences in phylogenetic structure of genes, group I and group II introns, and intergenic spacers [10–12,45,59]. In this way, markers with high versus low phylogenetic signal can be distinguished. For higher levels of genetic distance levels (e.g. distantly related species, genera, and families of flowering plants), a detailed evaluation of markers is therefore hardly necessary because sound predictions can be made. But is it worth to sequence whole plastid genomes when very closely related groups of species are to be studied?

Our comparison of genome pairs at comparable low distances shows that the mutational dynamics of plastid genomic regions may follow its own path in different lineages. While the variability in the respective unique sequence elements contributes the major proportion of the overall variability of a genomic region at that level, this contribution will be increasingly negligible at higher distance levels. The exploration of the plastid genome for the most variable and most suitable regions will therefore be a worthwhile investment when genetic distances are low.

It is of course possible to sequence all or at least most of the 30 promising plastid regions individually for a small taxon set in a given group. However, the effort needed is quite high. At least 60 individual fragments would need to be PCR-amplified and sequenced using many individual primers. Since only three to five loci are usually sequenced in evolutionary studies, a large part of these data would be wasted or deposited in GenBank as "unpublished". The sequencing and assembly of whole plastid genomes is still laborious, especially if critical areas of low coverage or homonucleotide stretches are verified by Sanger sequencing. Often overlooked costs have to be considered as well: this includes higher requirements for IT hardware and much increased time for sequence assembly and data management compared to traditional sequencing. Still, sequencing a complete plastid genome has many benefits over many single-marker PCRs. First, the complete genome sequence ensures that all genomic regions can be considered for marker development. And second, generating complete genomes allows for using the genome sequence for other studies, so that data are added in a complementary way to build proper information sources for the respective lineages (e.g., for comparative genomics, primer design, detection of plastid microsatellites, or extraction of regions for phylogenetic studies). We therefore conclude that whole plastid genome sequencing will remain a worthwhile approach for marker development in evolutionary studies of plants.

## Supporting Information

**Table S1  Primers for verification of sequence parts ambiguously read by the 454 sequencing.**

**Table S2  Ranking of all regions for the four genome pairs.**

**File S1  Pairwise alignment of the plastid genomes of *Pyrus spinosa* and *P. pyrifolia*.**

**File S2  Pairwise alignment of the plastid genomes of *Cymbidium tortisepalum* and *C. sinense*.**

**File S3  Pairwise alignment of the plastid genomes of *Oenothera parviflora* and *O. argillicola*.**

**File S4  Pairwise alignment of the plastid genomes of *Olea woodiana* and *O. europaea*.**

## Acknowledgments

The genomics work was done at the Berlin Center for Genomics in Biodiversity Research (BeGenDiv). We greatly acknowledge the lab assistance of Susan Mbedi and Virginia Duwe. We are also grateful to Felix Heeger for assistance with the genome assembly and Susan Wicke (Münster) for helpful advice on genome annotation.

This is publication number 11 of the Berlin Center for Genomics in Biodiversity Research (BeGenDiv). The study was carried out as part of the project "Developing tools for conserving the plant diversity of the Transcaucasus" funded by VolkswagenStiftung.

## Author Contributions

Conceived and designed the experiments: NK TB. Performed the experiments: NK HTV LN. Analyzed the data: NK LN HTV MA TB. Contributed reagents/materials/analysis tools: LN MA. Wrote the paper: NK LN TB.

## References

1. McCauley DE (1995) The use of chloroplast DNA polymorphism in studies of gene flow in plants. Trends Ecol Evol 10: 198–202.

2. Sang T, Crawford DJ, Stuessy TF (1997) Chloroplast DNA phylogeny, reticulate evolution, and biogeography of *Paeonia* (Paeoniaceae). Am J Bot 84: 1120–1136.

3. Sang T, Donoghue MJ, Zhang DM (1997) Evolution of alcohol dehydrogenase genes in peonies (*Paeonia*): Phylogenetic relationships of putative nonhybrid species. Mol Biol Evol 14: 994–1007.

4. Kurtto A (2009) Rosaceae (pro parte majore). Euro+Med Plantbase - the information resource for Euro-Mediterranean plant diversity. Available: http://www.emplantbase.org/home.html. Accessed February 2014 February 19.

5. Cuizhi G, Spongberg SA (2003) *Pyrus*. Flora of China: eFloras (2008). Published on the Internet http://www.efloras.org, Missouri Botanical Garden, St. Louis, MO & Harvard University Herbaria, Cambridge, MA. pp. 173–179.

6. Akopian JA (2007) O vidakh roda *Pyrus* L. (Rosaceae) v Armenii [On the *Pyrus* L. (Rosaceae) species in Armenia]. Fl Rast Rastitel Resurs Armenii 16: 15–26.

7. Fedorov AA (1954) Rod grusha *Pyrus* L. [The genus pear *Pyrus* L.]. In: Sokolov SJ, editor. Derev'a i kustarniki SSSR [The trees and shrubs of the USSR]. Moskva, Leningrad: Izdatel'stvo Akademii Nauk SSSR [National Science Academy of the U.S.S.R. publishing]. pp. 378–414.

8. Shaw J, Lickey EB, Beck JT, Farmer SB, Liu W, et al. (2005) The tortoise and the hare II: relative utility of 21 noncoding chloroplast DNA sequences for phylogenetic analysis. Am J Bot 92: 142–166.

9. Shaw J, Lickey EB, Schilling EE, Small RL (2007) Comparison of whole chloroplast genome sequences to choose noncoding regions for phylogenetic studies in angiosperms: the tortoise and the hare III. Am J Bot 94: 275–288.

10. Kelchner SA (2002) Group II introns as phylogenetic tools: Structure, function, and evolutionary constraints. Am J Bot 89: 1651–1669.

11. Borsch T, Quandt D (2009) Mutational dynamics and phylogenetic utility of noncoding chloroplast DNA. Plant Syst Evol 282: 169–199.

12. Korotkova N, Schneider J, Quandt D, Worberg A, Zizka G, et al. (2009) Phylogeny of the eudicot order Malpighiales: analysis of a recalcitrant clade with sequences of the *petD* group II intron. Plant Syst Evol 282: 201–228.

13. Dong W, Liu J, Yu J, Wang L, Zhou S (2012) Highly variable chloroplast markers for evaluating plant phylogeny at low taxonomic levels and for DNA barcoding. PLoS ONE 7: e35071.

14. Särkinen T, George M (2013) Predicting plastid marker variation: Can complete plastid genomes from closely related species help? PLoS ONE 8: e82266.

15. Jansen RK, Cai Z, Raubeson LA, Daniell H, dePamphilis CW, et al. (2007) Analysis of 81 genes from 64 plastid genomes resolves relationships in angiosperms and identifies genome-scale evolutionary patterns. Proc Natl Acad Sci USA 104: 19369–19374.

16. Moore MJ, Dhingra A, Soltis PS, Shaw R, Farmerie WG, et al. (2006) Rapid and accurate pyrosequencing of angiosperm plastid genomes. BMC Plant Biol 6.

17. Cronn R, Liston A, Parks M, Gernandt DS, Shen R, et al. (2008) Multiplex sequencing of plant chloroplast genomes using Solexa sequencing-by-synthesis technology. Nucleic Acids Res 36.

18. Parks M, Cronn R, Liston A (2009) Increasing phylogenetic resolution at low taxonomic levels using massively parallel sequencing of chloroplast genomes. BMC Biol 7.

19. Xu Q, Xiong G, Li P, He F, Huang Y, et al. (2012) Analysis of complete nucleotide sequences of 12 *Gossypium* chloroplast genomes: Origin and evolution of allotetraploids. PLoS ONE 7.

20. Njuguna W, Liston A, Cronn R, Ashman TL, Bassil N (2013) Insights into phylogeny, sex function and age of *Fragaria* based on whole chloroplast genome sequencing. Mol Phylogenet Evol 66: 17–29.

21. Mariotti R, Cultrera NG, Diez CM, Baldoni L, Rubini A (2010) Identification of new polymorphic regions and differentiation of cultivated olives (*Olea europaea* L.) through plastome sequence comparison. BMC Plant Biol 10: 211.

22. Ahmed I, Matthews PJ, Biggs PJ, Naeem M, McLenachan PA, et al. (2013) Identification of chloroplast genome loci suitable for high-resolution phylogeographic studies of *Colocasia esculenta* (L.) Schott (Araceae) and closely related taxa. Mol Ecol Resour: 929–937.

23. Jheng CF, Chen TC, Lin JY, Wu WL, Chang CC (2012) The comparative chloroplast genomic analysis of photosynthetic orchids and developing DNA markers to distinguish *Phalaenopsis* orchids. Plant Sci 190: 62–73.

24. Yang J-B, Tang M, Li H-T, Zhang Z-R, Li D-Z (2013) Complete chloroplast genome of the genus *Cymbidium*: lights into the species identification, phylogenetic implications and population genetic analyses. BMC Evol Biol 13: 84.

25. Ku C, Hu J-M, Kuo C-H (2013) Complete plastid genome sequence of the basal asterid *Ardisia polysticta* Miq. and comparative analyses of asterid plastid genomes. PLoS ONE 8: e62548.

26. Ku C, Chung W-C, Chen L-L, Kuo C-H (2013) The complete plastid genome sequence of Madagascar Periwinkle *Catharanthus roseus* (L.) G. Don: Plastid genome evolution, molecular marker identification, and phylogenetic implications in Asterids. PLoS ONE 8: e68518.

27. Terakami S, Matsumura Y, Kurita K, Kanamori H, Katayose Y, et al. (2012) Complete sequence of the chloroplast genome from pear (*Pyrus pyrifolia*): genome structure and comparative analysis. Tree Genet Genomes 8: 841–854.

28. Chevreux B, Wetter T, Suhai S (1999) Genome sequence assembly using trace signals and additional sequence information. Computer Science and Biology: Proceedings of the German Conference on Bioinformatics (GCB) 99: 45–56.

29. Wyman SK, Jansen RK, Boore JL (2004) Automatic annotation of organellar genomes with DOGMA. Bioinformatics 20: 3252–3255.

30. Müller J, Müller K, Neinhuis C, Quandt D (2005+) PhyDE: Phylogenetic Data Editor. – Available: www.phyde.de. Accessed 2014 March 3.

31. Huse SM, Huber JA, Morrison HG, Sogin ML, Welch DM (2007) Accuracy and quality of massively parallel DNA pyrosequencing. Genome Biol 8: R143.

32. Kelchner SA (2000) The evolution of non-coding chloroplast DNA and its application in plant systematics. Ann Missouri Bot Gard 87: 482–498.

33. Löhne C, Borsch T (2005) Molecular evolution and phylogenetic utility of the *petD* group II intron: A case study in basal angiosperms. Mol Biol Evol 22: 317–332.

34. Katoh K, Standley DM (2013) MAFFT multiple sequence alignment software version 7: improvements in performance and usability. Mol Biol Evol 30: 772–780.

35. Swofford DL (1998) PAUP*. Phylogenetic Analysis Using Parsimony (*and other Methods). Sunderland, Massachussets: Sinauer Associates.

36. Krzywinski MI, Schein JE, Birol I, Connors J, Gascoyne R, et al. (2009) Circos: An information aesthetic for comparative genomics. Genome Res.

37. Michel F, Umesono K, Ozeki H (1989) Comparative and functional anatomy of group II catalytic introns - a review. Gene 82: 5–30.

38. Mathews DH, Disney MD, Childs JL, Schroeder SJ, Zuker M, et al. (2004) Incorporating chemical modification constraints into a dynamic programming algorithm for prediction of RNA secondary structure. Proc Natl Acad Sci USA 101: 7287–7292.

39. Katayama H, Uematsu C (2003) Comparative analysis of chloroplast DNA in *Pyrus* species: physical map and gene localization. Theor Appl Genet 106: 303–310.

40. Lo EYY, Donoghue MJ (2012) Expanded phylogenetic and dating analyses of the apples and their relatives (Pyreae, Rosaceae). Mol Phylogenet Evol 63: 230–243.

41. Katayama H, Tachibana M, Iketani H, Zhang S-L, Uematsu C (2012) Phylogenetic utility of structural alterations found in the chloroplast genome of pear: hypervariable regions in a highly conserved genome. Tree Genet Genomes 8: 313–326.

42. Wuyun T, Ma T, Uematsu C, Katayama H (2013) A phylogenetic network of wild Ussurian pears (*Pyrus ussuriensis* Maxim.) in China revealed by hypervariable regions of chloroplast DNA. Tree Genet Genomes 9: 167–177.

43. Löhne C, Borsch T, Wiersema JH (2007) Phylogenetic analysis of Nymphaeales using fast-evolving and noncoding chloroplast markers. Bot J Linn Soc 154: 141–163.

44. Korotkova N, Borsch T, Quandt D, Taylor NP, Müller K, et al. (2011) What does it take to resolve relationships and to identify species with molecular markers? An example from the epiphytic Rhipsalideae (Cactaceae). Am J Bot 98: 1549–1572.

45. Barniske A-M, Borsch T, Müller K, Krug M, Worberg A, et al. (2012) Phylogenetics of early branching eudicots: Comparing phylogenetic signal across plastid introns, spacers, and genes. J Syst Evol 50: 85–108.

46. Goodson BE, Santos-Guerra A, Jansen RK (2006) Molecular systematics of *Descurainia* (Brassicaceae) in the Canary Islands: biogeographic and taxonomic implications. Taxon 55: 671–682.

47. Lee C, Wen J (2004) Phylogeny of *Panax* using chloroplast *trnC*–*trnD* intergenic region and the utility of *trnC*–*trnD* in interspecific studies of plants. Mol Phylogenet Evol 31: 894–903.

48. Campagna ML, Downie SR (1998) The intron in chloroplast gene *rpl16* is missing from the flowering plant families Geraniaceae, Goodeniaceae, and Plumbaginaceae. Trans Illinois State Acad Sci 91: 1–11.

49. Timme RE, Kuehl JV, Boore JL, Jansen RK (2007) A comparative analysis of the *Lactuca* and *Helianthus* (Asteraceae) plastid genomes: Identification of divergent regions and categorization of shared repeats. Am J Bot 94: 302–312.

50. Doorduin L, Gravendeel B, Lammers Y, Ariyurek Y, Chin-A-Woeng T, et al. (2011) The complete chloroplast genome of 17 individuals of pest species *Jacobaea vulgaris*: SNPs, microsatellites and barcoding markers for population and phylogenetic studies. DNA Res 18: 93–105.

51. Gargano D, Scotti N, Vezzi A, Bilardi A, Valle G, et al. (2012) Genome-wide analysis of plastome sequence variation and development of plastidial CAPS markers in common potato and related *Solanum* species. Genet Resour Crop Evol 59: 419–430.

52. Daniell H, Lee SB, Grevich J, Saski C, Quesada-Vargas T, et al. (2006) Complete chloroplast genome sequences of *Solanum bulbocastanum*, *Solanum lycopersicum* and comparative analyses with other Solanaceae genomes. Theor Appl Genet 112: 1503–1518.

53. Štorchová H, Olson MS (2007) The architecture of the chloroplast *psbA-trnH* non-coding region in angiosperms. Plant Syst Evol 268: 235–256.

54. Quandt D, Müller K, Stech M, Frahm J-P, Frey W, et al. (2004) Molecular evolution of the chloroplast *trnL-F* region in land plants. In: Goffinet B, Hollowell V, Magill R, editors. Molecular Systematics of Bryophytes. St Louis: Missouri Botanical Garden Press. pp. 13–37.

55. Wicke S, Schäferhoff B, dePamphilis CW, Müller KF (2014) Disproportional plastome-wide increase of substitution rates and relaxed purifying selection in genes of carnivorous Lentibulariaceae. Mol Biol Evol 31: 529–545.

56. Lehmann K, Schmidt U (2003) Group II introns: Structure and catalytic versatility of large natural ribozymes. Crit Rev Biochem Mol Biol 38: 249–303.

57. Pyle AM, Lambowitz AM (2006) Group II Introns: Ribozymes that splice RNA and invade DNA. In: Gesteland RF, Cech TR, Atkins JF, editors. The RNA World. 3rd ed. Cold Spring Harbor: Cold Spring Harbor Laboratory Press. pp. 449–505.

58. Källersjö M, Farris JS, Chase MW, Bremer B, Fay MF, et al. (1998) Simultaneous parsimony jackknife analysis of 2538 *rbcL* DNA sequences reveals support for major clades of green plants, land plants, seed plants and flowering plants. Plant Syst Evol 213: 259–287.

59. Müller K, Borsch T, Hilu KW (2006) Phylogenetic utility of rapidly evolving DNA at high taxonomical levels: Contrasting *matK*, *trnT-F*, and *rbcL* in basal angiosperms. Mol Phylogenet Evol 41: 99–117.

60. Tesfaye K, Borsch T, Govers K, Bekele E (2007) Characterization of *Coffea* chloroplast microsatellites and evidence for the recent divergence of *C. arabica* and *C. eugenioides* chloroplast genomes. Genome 50: 1112–1129.

61. Borsch T, Hilu KW, Wiersema JH, Lohne C, Barthlott W, et al. (2007) Phylogeny of *Nymphaea* (Nymphaeaceae): Evidence from substitutions and microstructural changes in the chloroplast *trnT-trnF* region. Int J Plant Sci 168: 639–671.

62. Quandt D, Müller K, Huttunen S (2003) Characterisation of the chloroplast DNA *psbT-H* region and the influence of dyad symmetrical elements on phylogenetic reconstructions. Plant Biol 5: 400–410.

63. Whitlock BA, Hale AM, Groff PA (2010) Intraspecific inversions pose a challenge for the *trnH-psbA* plant DNA barcode. PLoS ONE 5: e11533.

64. Fior S, Li M, Oxelman B, Viola R, Hodges SA, et al. (2013) Spatiotemporal reconstruction of the *Aquilegia* rapid radiation through next-generation sequencing of rapidly evolving cpDNA regions. New Phytol 198: 579–592.

65. Erixon P, Oxelman B (2008) Reticulate or tree-like chloroplast DNA evolution in Sileneae (Caryophyllaceae)? Mol Phylogenet Evol 48: 313–325.

66. Greiner S, Wang X, Herrmann RG, Rauwolf U, Mayer K, et al. (2008) The complete nucleotide sequences of the 5 genetically distinct plastid genomes of *Oenothera*, subsection *Oenothera*: II. A microevolutionary view using bioinformatics and formal genetic data. Mol Biol Evol 25: 2019–2030.

67. Greiner S, Wang X, Rauwolf U, Silber MV, Mayer K, et al. (2008) The complete nucleotide sequences of the five genetically distinct plastid genomes of *Oenothera*, subsection *Oenothera*: I. Sequence evaluation and plastome evolution. Nucleic Acids Res 36: 2366–2378.

68. Besnard G, Hernandez P, Khadari B, Dorado G, Savolainen V (2011) Genomic profiling of plastid DNA variation in the Mediterranean olive tree. BMC Plant Biol 11.

# Unusual Ratio between Free Thyroxine and Free Triiodothyronine in a Long-Lived Mole-Rat Species with Bimodal Ageing

**Yoshiyuki Henning[1]\*, Christiane Vole[1], Sabine Begall[1], Martin Bens[2], Martina Broecker-Preuss[3], Arne Sahm[2], Karol Szafranski[2], Hynek Burda[1], Philip Dammann[1,4]**

**1** Department of General Zoology, Faculty of Biology, University of Duisburg-Essen, Essen, Germany, **2** Genome Analysis, Leibniz Institute for Age Research - Fritz Lipmann Institute, Jena, Germany, **3** Department of Endocrinology and Metabolism and Division of Laboratory Research, University Hospital, University of Duisburg-Essen, Essen, Germany, **4** Central Animal Laboratory, University Hospital, University of Duisburg-Essen, Essen, Germany

## Abstract

Ansell's mole-rats (*Fukomys anselli*) are subterranean, long-lived rodents, which live in eusocial families, where the maximum lifespan of breeders is twice as long as that of non-breeders. Their metabolic rate is significantly lower than expected based on allometry, and their retinae show a high density of S-cone opsins. Both features may indicate naturally low thyroid hormone levels. In the present study, we sequenced several major components of the thyroid hormone pathways and analyzed free and total thyroxine and triiodothyronine in serum samples of breeding and non-breeding *F. anselli* to examine whether *a*) their thyroid hormone system shows any peculiarities on the genetic level, *b*) these animals have lower hormone levels compared to euthyroid rodents (rats and guinea pigs), and *c*) reproductive status, lifespan and free hormone levels are correlated. Genetic analyses confirmed that Ansell's mole-rats have a conserved thyroid hormone system as known from other mammalian species. Interspecific comparisons revealed that free thyroxine levels of *F. anselli* were about ten times lower than of guinea pigs and rats, whereas the free triiodothyronine levels, the main biologically active form, did not differ significantly amongst species. The resulting fT4:fT3 ratio is unusual for a mammal and potentially represents a case of natural hypothyroxinemia. Comparisons with total thyroxine levels suggest that mole-rats seem to possess two distinct mechanisms that work hand in hand to downregulate fT4 levels reliably. We could not find any correlation between free hormone levels and reproductive status, gender or weight. Free thyroxine may slightly increase with age, based on sub-significant evidence. Hence, thyroid hormones do not seem to explain the different ageing rates of breeders and non-breeders. Further research is required to investigate the regulatory mechanisms responsible for the unusual proportion of free thyroxine and free triiodothyronine.

**Editor:** Eliseo A. Eugenin, Rutgers University, United States of America

**Funding:** This work was supported by a grant of the German Research Foundation to PD (DFG-grant DA 992/3-1) (http://www.dfg.de/). The publication fee was funded by the "Publish Open Access"-Program of the University of Duisburg-Essen and the German Research Foundation. The funders had no role in study design, data collection and analysis, decision to publish, or preparation of the manuscript.

**Competing Interests:** The authors have declared that no competing interests exist.

\* Email: yoshiyuki.henning@uni-due.de

## Introduction

Most ageing theories assume a link between metabolism and ageing because of several inevitable side effects of metabolic processes that potentially impair somatic integrity in the long term. Examples of such side effects are the production of reactive oxygen species [1,2], formation of advanced glycation end products [3,4,5], and telomere shortening with every cell proliferation cycle [6].

Thyroid hormones (THs) play a major role in development, differentiation and metabolism in vertebrates and are therefore assumed to affect ageing, too [7,8]. Experimental as well as comparative studies on various mammal models support this assumption. For example, experimentally induced hypothyroidism increases lifespan in rats [9], whereas experimentally induced hyperthyroidism decreased lifespan in young and middle-aged rats

[10]. Additionally, *Ames dwarf mice* and *Snell dwarf mice*, which have extraordinary low levels of THs, and other hormones related to growth and development (e.g., somatotropin, insulin-like growth factor 1), live significantly longer than wild type mice [11]. Whereas the treatment with somatotropin does not have any effect on the lifespan of *Snell dwarf mice*, the administration of THs via food throughout adult life diminishes their lifespan, although it is still longer than in non-treated wild type mice [7,12]. Furthermore, longevity in vertebrate species is often associated with low metabolic rates, low TH levels, or both. For example, naked mole-rats (*Heterocephalus glaber*) are the longest-living rodent species (lifespan of >30 years) [13], showing only 79% of the allometrically expected resting metabolic rate of non-subterranean rodents [14] and very low levels of certain THs have been reported [15]. Also some long-lived bat species feature low

metabolic rates [16,17]. In humans, there is a significant correlation between low TH metabolism and longevity [7,18].

In all vertebrates, the main THs are thyroxine (T4) and triiodothyronine (T3). Both THs are derivatives of the amino acid tyrosine and are synthesised in the thyroid gland. Synthesis of T4 and T3 is stimulated by the thyroid-stimulating hormone (TSH), which is released from the pituitary gland. The structures of T4 and T3 are strongly conserved in all mammalian species studied thus far, whereas TSH is species-specific. TSH consists of an unspecific alpha-subunit (TSHA), and a beta-subunit (TSHB), which is responsible for biological specificity [19]. TSH is stimulated by the thyrotropin-releasing hormone (TRH), which is secreted by the hypothalamus [20]. This hypothalamic-pituitary-thyroid (HPT) axis is regulated by THs exerting a negative feedback control over the secretion of TRH and TSH [21]. In peripheral tissues, THs are actively transported through the plasma membrane mainly by the monocarboxylate transporters 8 and 10 (MCT8 and MCT10) and, at least in mice and rats, the organic anion-transporting polypeptide 1C1 [22]. In the cytoplasm, specific deiodinases type 1 and 2 (D1, D2) convert T4 into T3 by deiodination of the outer ring of the T4 molecule [23,24].

The main biologically active TH, namely T3, regulates gene expression in the nucleus at various loci by binding to two types of thyroid hormone receptors (THRA and THRB) [25]. In addition to these classical TH functions, some non-nuclear TH actions have been described recently [26]. T4 and T3 are typically secreted into the blood stream in a ratio of about 6:1 (T4:T3) in rats and 14:1 in humans [27,28]. Thus, circulating T4 levels are manifold higher than T3 levels in healthy organisms. After secretion, less than 1% of the THs are circulating as biologically active free hormones (fT4 and fT3), while the major amount of T4 and T3 is bound to transport proteins [29,30].

Ansell's mole-rats (*Fukomys anselli*) are subterranean rodents endemic to Zambia. They show some promising features for ageing and TH studies. Similar to naked mole-rats, which belong to the same family of African mole-rats (Bathyergidae), *F. anselli* live in eusocial families, in which reproduction is usually monopolized by a single breeding pair [31]. The species has an extraordinary maximum lifespan of more than 20 years, which is far more than expected based on their body weight (~60–150 g). Remarkably, reproductive individuals live about twice as long as non-reproductive animals, regarding both their average and maximum lifespan (breeders: mean ca. 10 years, max. >20 years; non-breeders: mean lifespan ca. 4.8 years, max. 11.1 years ([32], own unpublished data). This bimodal ageing pattern of Ansell's mole-rats and other species of the same genus [33,34] contradicts the classic model that assumes a trade-off between reproduction and somatic maintenance [35,36]. Until now, conflicts to this trade-off model have only been reported about eusocial insect species like ants or termites [37,38]. The mechanisms underlying the unusual ageing pattern of *Fukomys* mole-rats are largely unknown.

Several indications suggest that Ansell's mole-rats may be naturally hypothyroid. First, oxygen consumption of *F. anselli* is significantly lower than expected from allometric equations, suggesting a low metabolic rate [39]. Low resting metabolic rates are typical for bathyergid rodents and are probably a physiological adaptation to their subterranean, low-oxygen environment [40]. Second, the retinae of Ansell's mole-rats show a high density of short-wave sensitive S-cone opsin. This is untypical for rodents, as they usually show S-cone opsins, as well as middle-to-long-wave sensitive L-cone opsins in diverse arrangements [41]. Although the adaptive function of colour perception in mole-rats is not yet understood, the high S-cone opsin density in Ansell's mole-rats

would be in line with the expected low TH levels, as fT3 is essential for the expression of L-cone opsins during the prenatal development [42]. Moreover, in athyroid mice, opsin expression can be restored with postnatal T4 treatment [43].

In order to characterize the TH system of Ansell's mole-rats qualitatively and quantitatively, we first sequenced mRNA of several major components of TH related pathways in order to find out whether the TH system of Ansell's mole-rats is evolutionary conserved, or whether it contains qualitative peculiarities compared to other mammalian species. Then, we determined serum TH levels in individuals of different age, sex, and breeding status. Here, we focused mainly on the following two questions: i) Do Ansell's mole-rats have lower circulating TH levels than unrelated, euthyroid rodent species (rats, guinea pigs)? ii) Are there differences in free TH levels between the slowly ageing reproductive and the faster ageing non-reproductive individuals? We hypothesized that i) Ansell's mole-rats have naturally low TH levels in comparison to euthyroid rodents, and ii) lower free TH levels in breeding animals (slow ageing) compared to non-breeders (faster ageing) as a possible molecular trigger for the lifespan differences between these two cohorts.

## Materials and Methods

### Animals

All Ansell's mole-rats used in this study were born, raised and maintained at the animal facilities of the Department of General Zoology, University of Duisburg-Essen, Germany. The age of the animals ranged from 1.2–10.2 years in non-breeders and 5.4–13.5 years in breeders at the day of serum sampling. They were housed as family groups in glass terraria on horticultural peat and fed *ad libitum* with carrots and potatoes every day, apples every second day, and grain and lettuce once a week. Room temperature and humidity were kept constant at $24\pm1°C$ and $40\pm3\%$, respectively.

*Wistar-Unilever* rats aged 6–9 months (Central Animal Laboratory of the University Hospital Essen, Germany) and *Dunkin Hartley* guinea pigs aged 12–24 months (Charles River, Wilmington, MA, USA) served as healthy, fully grown euthyroid controls. We did not introduce major age variation within these groups because the intraspecific variation of THs in these species has been summarized and studied elsewhere [44,45] and was not the focus of our study. Both species were housed at $21\pm1°C$ and $55\pm5\%$ humidity in standard macrolon cages and were fed commercial, species-specific food pellets (ssniff).

### Ethics Statement

Maintenance and all treatments of the animals were approved by the North Rhine-Westphalia State Environment Agency (Permit number: 87–51.04.2010.A359). Blood sampling was the only invasive treatment and was performed under deep anesthesia (ketamine and xylazine or isoflurane), except for guinea pigs, since anaesthesia is not necessary for blood sampling via the *vena saphena* if the procedure is sufficiently quick and appropriate restraining is possible [46]. All efforts were made to minimize suffering.

### Sequencing and sequence analysis

The *F. anselli* transcriptome was characterized by high-throughput sequencing, as will be reported elsewhere. From the resulting data, we extracted transcripts of twelve thyroid-relevant genes. In detail, one male *F. anselli* was deeply anesthetized with isoflurane and killed by cervical translocation. Tissues from thyroid gland, ventral skin, adrenal gland, pancreas, testis, and brain stem were homogenized in a Tissue Lyser (Qiagen), and total mRNA was isolated using RNeasy (Qiagen, Valencia, CA, USA). Thereof, mRNA-Seq libraries were generated using platform-specific chemistry, according to the supplier's instructions (Illumina). Sequencing was performed using an Illumina Genome Analyzer IIx, resulting in a total of 88.7 million (9492 Mbp) single-end reads. Adapter clipping and trimming of low-quality 3′ ends (error probability of 0.5%) was performed with the programs cutadapt [47] and sickle [48], respectively. Reads shorter than 35 nt were removed, resulting in a total of 82.6 million (8,202 Mbp) reads. The transcriptome reads were pooled for a *de-novo* assembly with the Trinity software [49]. Resulting transcript contigs were labelled with gene symbols and were annotated for coding sequences (CDS) based on best bidirectional BLAST mapping against human protein coding genes (National Center for Biotechnology Information [NCBI], *H. sapiens* Annotation Release 104) using in-house scripts. The mRNA sequences were deposited in NCBI GenBank under accession numbers KJ958510-KJ958520 and KM676335. For species comparisons, *F. anselli* mRNA sequences were translated into proteins. These were aligned with orthologous protein sequences from 11 to 17 other mammalian species (RefSeq-database, NCBI) using CLUSTAL W. For genome-wide analysis of evolutionary selection trends we created five-species alignments of the CDS (human, dog, rat, mouse and naked mole-rat) using CLUSTAL W. For the thyroid target genes we additionally created the five-species multiple alignment with the Ansell's mole-rat sequence as the bathyergid representative. We used a parametric model of evolution implemented in the CodeML program of the PAML package [50,51] in performing the branch test on the mole-rat branch against all other branches as background (options CodonFreq = 2 and Kappa = 2). CodeML estimates the ratio of non-synonymous to synonymous mutations (Ka/Ks ratio), among several other parameters, separately for the mole-rat and all other branches. Additionally we used the "M0" model of CodeML to calculate the average Ka/Ks across the whole tree. We estimated the Ka/Ks ratios for the genome-wide orthologous gene set and used these to determine empirical probabilities ("percentiles") for particular Ka/Ks values, as well as Ka/Ks differences between the mole-rat branch and outer branches. This allowed to relate the Ansell's mole-rat Ka/Ks values of specific genes to the genome-wide Ka/Ks spectrum.

## Sample sizes and sampling protocols

We sampled a total of 32 Ansell's mole-rats (12 breeders and 20 non-breeders, sex-balanced), 4 male rats and 4 male guinea pigs for the free TH measurements.

Additionally, we sampled a subgroup of 12 mole-rats (sex-balanced; both reproductive groups) plus further 7 rats (4 males, 3 females) and 4 guinea pigs (sex-balanced) to determine total TH levels (free and protein-bound fractions in total; tT4 and tT3). Again, free T4 and free T3 were determined in these serum samples in order to calculate the ratios of the free: total fractions for each hormone.

For blood sampling, mole-rats were anesthetized according to a standard protocol for mole-rats with an intramuscular injection of a 6 mg/kg dose of ketamine (10%, Ceva GmbH) and 2.5 mg/kg xylazine (2%, Ceva GmbH) [52]. Rats were isoflurane anesthetized, while guinea pigs did not receive any anaesthesia. To avoid hypothermia, animals were kept under a heat lamp before and after the treatment.

The mole-rat and guinea pig blood samples were taken from the *vena saphen*a of the hind paw with a capillary (Servoprax, 100 μl) and transferred into a serum test tube (Multivette 600, Sarstedt). The rat blood samples were taken by orbita puncture. All samples

were taken at approximately the same daytime, in order to avoid any bias due to the circadian variation in TRH secretion [21,53]. After approximately 20 minutes, the blood samples were centrifuged (Biofuge Pico, Heraeus Instruments) at 900×g for 5 minutes [15] and the serum (the clear top layer) was stored at −80°C until use.

## Quantification of thyroid hormones

Free and total thyroxine and triiodothyronine levels were quantified by means of a solid phase competitive enzyme immunoassay (EIA) for human serum (DRG Instruments GmbH). The use of a human serum EIA is justifiable since the molecular structures of T3 and T4 are not species specific ([29], results of the present study]).

The accuracy of the fT3 and fT4 microplate EIA test system was confirmed by analyzing known hormone values, and by comparing the results with those of a reference method (radioimmunoassay). The correlation coefficient between the concentrations measured by the two methods was 0.95 (fT3) and 0.96 (fT4), which indicates a high accuracy of the test systems. The intra- and inter-assay variances are shown in Table S1. According to the manufacturer the cross-reactivities of the antibodies were as follows: *Triiodothyronine* − triiodothyronine: 100%; thyroxine: 0.02–0.37%; iodothyrosine, diiodothyrosine, phenylbutzone, sodium salicylate: 0.01–0.2%. *Thyroxine* − thyroxine 100%, triiodothyronine 3%, diiodothyronine, diiodotyrosine and iodotyrosine 0.01%. The assay sensitivities (i.e., detection limits) were 0.05 ng/dl (fT4), 8 nmol/l (tT4), 0.05 pg/ml (fT3), and 0.1 ng/ml (tT3).

## Statistical analyses

The statistical analyses of the thyroid hormone levels were conducted with the software SPSS Statistics v.20.0.0 (IBM Corp.). Normal distribution was tested with the Kolmogorov-Smirnov-test with Lilliefors correction. For interspecies comparisons, one-way ANOVA with Bonferroni post hoc tests were applied. For intraspecific comparisons, a generalized linear model (GLM) was run with sex, reproductive status, age, and weight as independent factors, and fT4 and fT3 values as dependent variables. We calculated *i*) the main effects of the independent factors alone and *ii*) a two factor model using the interaction of status × age as explaining variable.

In the present study, mole-rats proved to have very low fT4 levels in general: 8 out of 32 mole-rats showed fT4 levels below the detection limit (<0.05 ng/dl), and the majority of fT4 values fell relatively close to the detection limit of the assay. It was impossible to decide whether the 8 missing values were failures or represented fT4 levels lower than 0.05 ng/dl. Therefore, analyses of fT4 were run under two different scenarios: 1) treating missing values as failures (effective n = 24; "first scenario" henceforth), and 2) replacing the missing values with the value of the detection limit (0.05 ng/dl) (effective n = 32; "second scenario" henceforth).

## Results

### Molecular constituents of TH system are conserved in *F. anselli*

We first characterized the TH system of Ansell's mole-rats on the genetic level and compared protein sequences with those of several other species representing different mammalian subgroups. Starting with RNA-seq of five different *F. anselli* tissues, we obtained full protein sequences for the two TH receptors (THR alpha [THRA, Figure S1]; THR beta [THRB, Figure S2]), one member of the regulation cascade (TSH beta subunit [TSHB, Figure S3]), two metabolizing deiodinases (D1, Figure S4; D2,

Figure S5), two members of the synthesis pathway (thyroglobulin [TG, Figure S6]; thyroperoxidase [TPO, Figure S7]), two transporter proteins (transthyretin [TTR, Figure S8]; thyroxine-binding globulin [TBG, coded by *Serpina7*, Figure S9]), one TH-regulated protein (hypoxia-inducible factor 1 alpha [HIF1A, Figure S10]), one TH transporter (monocarboxylate transporter 8 [MCT8, coded by *Slc16a2*, Figure S11]) and the sodium/iodid symporter (NIS, coded by *SLC5a5*, Figure S12). In order to analyze the conservation status of these molecular markers for the *F. anselli* TH system in comparison to other mammals, we analyzed the ratio of non-synonymous versus synonymous nucleotide changes in the CDS (Ka/Ks; Table 1). In all genes, the Ka/Ks ratio across the selected species was within the 95% percentile of genome-wide Ka/Ks values ($\leq 0.443$), indicating average levels of purifying selection. For only one gene (*Serpina7*), the Ka/Ks ratio was outside the 90% percentile ($\leq 0.340$), which is well within bounds accounting for multiple testing on eleven genes. Furthermore, in nearly all cases, the branch-specific Ka/Ks ratio of *F. anselli* was close to the ratio of the background branches as well as the average across the tree (Table 1). Exceptions were *Slc16a2*, which showed a slightly higher purifying selection pressure, and *Ttr*, which showed a much weaker purifying selection in the *F. anselli* branch. Nevertheless, for *Ttr*, the absolute Ka/Ks difference between foreground and background branch is within the 80% percentile ($p = 0.201$, one-sided) that is seen genome-wide.

TSHB, an important upstream molecular target for the study of TH dynamics, shows three species-specific changes in *F. anselli* at positions that are conserved in all 16 other mammalian species (E46D, V112I and F157I; Figure S3). This is comparable to the rate of species-specific changes found in TSHB of other mammals, ranging from 0 to 9. In pairwise comparisons to the sequence of *F. anselli*, the number of amino acid differences varies from 10 (to thirteen-lined ground squirrel, *Ictidomys tridecemlineatus*) to 24 (guinea pig), which underlines that even closely related species (Ansell's mole-rats and guinea pigs) show high variations in the TSHB sequence. Finally, we analyzed the *F. anselli* TH receptors, particularly their ligand-binding domains, as specific markers for conservation of the TH molecules. The ligand-binding domain of *F. anselli* THRA (protein position 190–370) shows only one conservative change from glutamic acid to aspartic acid (E270D) when compared to the consensus of 14 other mammalian species (Figure 1a, Figure S1). The ligand-binding domain of THRB (position 222–464) does not show a single amino acid difference

between *F. anselli* and the sequence consensus of 17 other mammalian species (Figure 1b, Figure S2).

## Free TH levels: Ansell's mole-rats have low fT4, but normal fT3

Under both scenarios, Ansell's mole-rat fT4 levels ($0.18 \pm 0.08$ ng/dl [n = 24] and $0.15 \pm 0.09$ ng/dl [n = 32]; Table 2) were about 10 times lower than fT4 measured in rats ($2.11 \pm 0.67$ ng/dl) and guinea pigs ($2.25 \pm 0.25$ ng/dl; one-way ANOVA: $F$ [first/second scenario] $= 206.38/273.96$, $p < 0.0001$; Bonferroni post hoc comparisons in both scenarios: mole-rat vs. rat and mole-rat vs. guinea-pig: $p < 0.0001$; rat vs. guinea-pig: $p > 0.99$; Figure 2). By contrast, fT3 levels did not differ significantly among the three species (mole-rat: $2.24 \pm 0.96$ pg/ml; rat: $2.85 \pm 0.34$ pg/ml; guinea pig: $2.36 \pm 0.35$ pg/ml; one-way ANOVA: $F = 0.85$, $p = 0.44$; Figure 2).

## Total TH levels: Ansell's mole-rats have low tT4 and low tT3

Ansell's mole-rat tT4 levels ($20.07 \pm 6.32$ ng/ml) were significantly lower than those of rats ($47.05 \pm 10.39$ ng/ml) and guinea pigs ($42.01 \pm 14.77$ ng/ml; one-way ANOVA, $F = 19.51$, $p < 0.0001$; Bonferroni post hoc comparison: mole-rat vs. rat: $p < 0.0001$; mole-rat vs. guinea pig: $p = 0.003$; rat vs. guinea pig: $p > 0.99$; Figure 3). Levels of tT3 were significantly lower between mole-rats ($1.13 \pm 0.25$ ng/ml) and rats ($3.14 \pm 1.34$ ng/ml), but not guinea pigs ($1.80 \pm 1.03$ ng/ml; one-way ANOVA: $F = 11.13$, $p = 0.001$; Bonferroni post hoc comparison: mole-rat vs. rat: $p < 0.0001$; mole-rat vs. guinea pig: $p = 0.629$; rat vs. guinea pig: $p = 0.076$; Figure 3).

## Hormone ratios

We estimated the percentage of unbound hormone fractions by determining the proportion of fTH in relation to tTH per serum sample. The proportion of fT4 to tT4 was significantly lower in mole-rats ($0.02 \pm 0.01\%$) compared to rats ($0.04 \pm 0.01\%$) and guinea pigs ($0.05 \pm 0.02\%$; one-way ANOVA: $F = 9.60$, $p = 0.001$; Bonferroni post hoc comparison: mole-rats vs. rat: $p = 0.02$; mole-rats vs. guinea pig: $p = 0.002$; rat vs. guinea pig: $p = 0.496$; Figure 4). In contrast, the proportion of fT3 to tT3 was highest in mole-rats ($0.38 \pm 0.14\%$; rats: $0.17 \pm 0.09\%$; guinea pigs $0.23 \pm 0.08\%$), with differences being statistically significant in the comparison with rats and close to significance threshold in the

**Table 1.** K$_a$/K$_s$ ratios indicate persisting purifying selection on proteins of the TH system in *F. anselli*.

| Protein (Gene) | K$_a$/K$_s$ ratio across tree | K$_a$/K$_s$ ratio *F. anselli* branch | K$_a$/K$_s$ ratio background branches |
|---|---|---|---|
| THRA | 0.011 | 0.017 | 0.008 |
| THRB | 0.036 | 0.054 | 0.033 |
| TG | 0.312 | 0.336 | 0.305 |
| TPO | 0.127 | 0.088 | 0.108 |
| D1 (Dio1) | 0.229 | 0.227 | 0.198 |
| D2 (Dio2) | 0.144 | 0.158 | 0.149 |
| TSHB | 0.209 | 0.176 | 0.189 |
| TBG (Serpina7) | 0.342 | 0.358 | 0.333 |
| TTR | 0.226 | 0.531 | 0.202 |
| MCT8 (Slc16a2) | 0.076 | 0.007 | 0.098 |
| NIS (*Slc5a5*) | 0.074 | 0.051 | 0.068 |

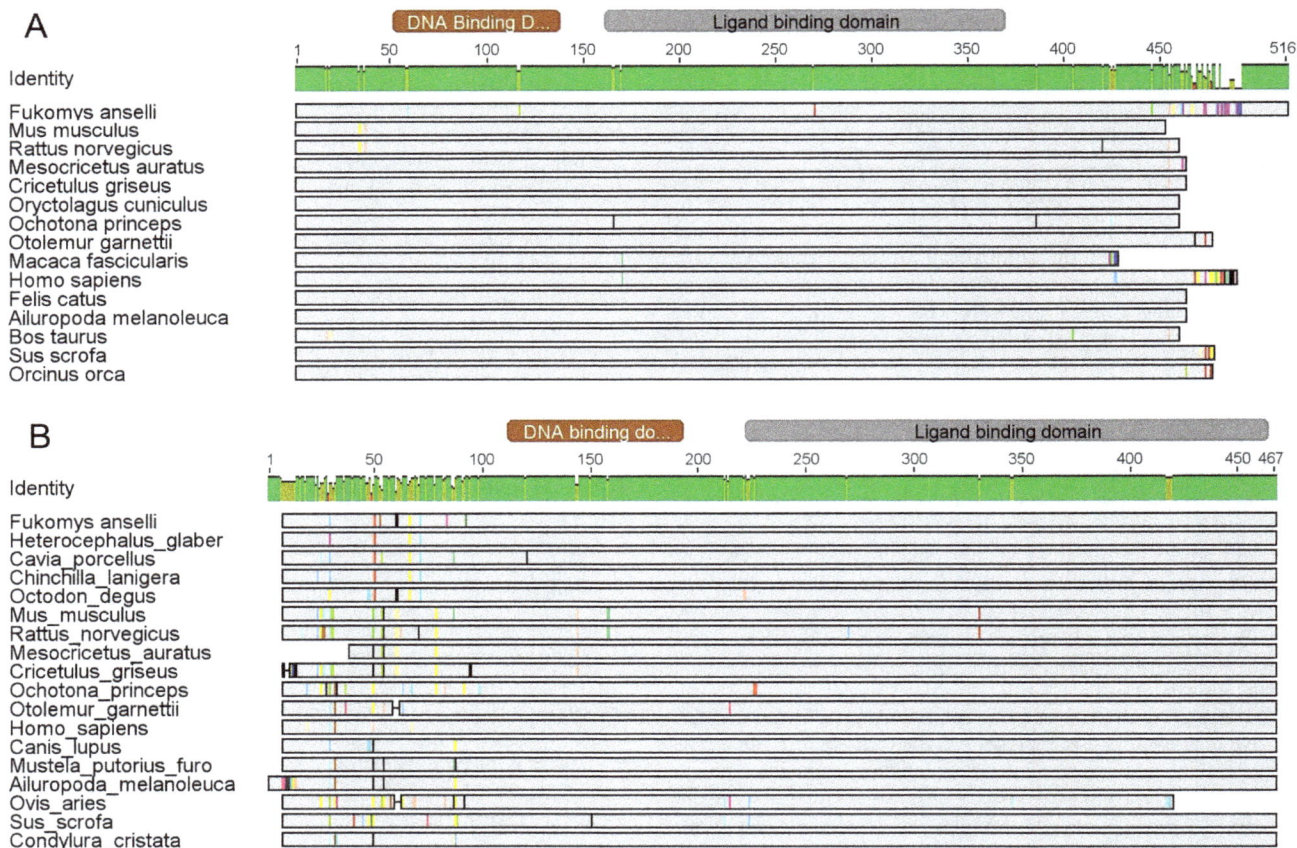

**Figure 1. Protein sequence alignment for (A) TH receptor α (THRA) and (B) TH receptor β (THRB) of different mammalian species.** *F. anselli* mRNA sequences were derived from RNA-seq and subsequently translated, the other sequences were retrieved from NCBI databases (accession numbers are given in Figures S1 and S2). Sequence differences are marked in gray. "Identity" shows the percentage of amino acid conformity for each position; the protein domain regions correspond to the human sequence entry. A fully resolved representation of the alignment is given as a supplement (Figures S1 and S2).

comparison with guinea pigs (one-way ANOVA: $F = 7.724$, $p = 0.004$; Bonferroni post hoc comparison: mole-rats vs. rats: $p = 0.004$; mole-rats vs. guinea pig: $p = 0.075$; rat vs. guinea pig: $p > 0.99$; Figure 4).

We also calculated the ratios of free T4:T3, and total T4:T3 in all three species (Table 3). Free T4:T3 ratios in mole-rats ($0.70 \pm 0.45$–$0.85 \pm 0.43$, depending on the scenario) were significantly lower than in both other species (rats: $7.28 \pm 1.87$, guinea pigs: $9.63 \pm 1.50$; one-way ANOVA [first/second scenario]: $F = 240.53/309.71$, $p < 0.0001$ each; Bonferroni post hoc comparison: mole-rat vs. rat: $p < 0.0001$ [both scenarios]; mole-rat vs. guinea pig: $p < 0.0001$; rat vs. guinea pig: $p = 0.002$). In contrast, tT4:T3 ratios did not differ between mole-rats and the other species (mole-rats: $18.60 \pm 6.98$, rats: $16.24 \pm 4.28$, guinea pigs: $26.43 \pm 9.51$; one-way ANOVA: $F = 3.00$, $p = 0.07$; Bonferroni post hoc comparison: mole-rat vs. rat: $p > 0.99$; mole-rat vs. guinea pig: $p = 0.17$; rat vs. guinea-pig: $p = 0.08$).

### Free thyroid hormone levels do not explain intraspecific ageing differences

Intraspecific comparisons revealed, that the fT4 levels of Ansell's mole-rats were not affected by any of the tested factors sex, reproductive status, age and weight alone in the first scenario ($n = 24$), and neither so by the interaction of reproductive status ×

age (Table 4). In the second scenario ($n = 32$), the fT4 levels were again not affected by sex, reproductive status and weight alone, nor by the interaction of status × age (Table 4). Levels fT4 increased significantly with age under this scenario ($p = 0.041$; Table 4).

Free T3 levels were not influenced by any of the tested factors or interactions in the GLM (Table 4).

### Discussion

The aim of our study was to characterize the TH system of Ansell's mole-rats and to determine T4 and T3 levels in this species to investigate if these animals are hypothyroid, and if their hormone concentrations correlate with their extraordinary lifespan and the bimodal ageing pattern of reproductive and non-reproductive animals.

### Molecular constituents of TH system are conserved in *F. anselli*

In *F. anselli*, the major molecular constituents of the mammalian TH system could be identified via their mRNAs, namely TG, TPO, TSHB, D1, D2, TTR, MCT8, TBG and NIS. The mRNAs show full protein-coding capacity, and sequence substitution patterns suggest that purifying selection acts on these molecules to a similar extent as found in other mammal species

**Table 2.** Mean (±SD) fT4 and fT3 values in Ansell's mole-rats.

| | fT4 (ng/dl; n=24) | | | fT4 (ng/dl; n=32) | | | fT3 (pg/ml; n=32) | | |
|---|---|---|---|---|---|---|---|---|---|
| | female | male | all | female | male | all | female | male | all |
| R[1] | 0.20±0.11 (n=5) | 0.17±0.08 (n=6) | 0.18±0.09 (n=11) | 0.18±0.12 (n=6) | 0.17±0.08 (n=6) | 0.17±0.09 (n=12) | 3.01±2.02 (n=6) | 2.24±0.49 (n=6) | 2.63±1.45 (n=12) |
| NR[2] | 0.18±0.07 (n=9) | 0.16±0.06 (n=4) | 0.17±0.06 (n=13) | 0.17±0.08 (n=10) | 0.09±0.07 (n=10) | 0.13±0.08 (n=20) | 1.97±0.23 (n=10) | 2.04±0.43 (n=10) | 2.01±0.34 (n=20) |

[1]NR = non-reproductive; R = reproductive.

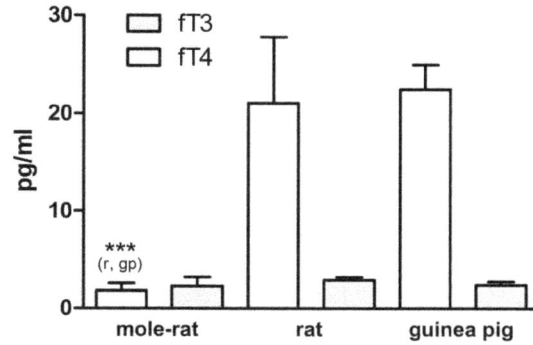

**Figure 2. Free T4 and free T3 levels of Ansell's mole-rats (n=24), rats (n=4) and guinea pigs (n=4).** Mean ± SD; all data expressed in pg/ml. One-way ANOVA, fT4: $F = 206.38$, $p < 0.0001$; fT3: $F = 0.85$, $p = 0.44$. Significant differences in the Bonferroni post hoc comparisons are indicated by asterisks coupled with the comparison species referred to (mr = mole-rat, r = rat, gp = guinea pig) in parenthesis. See "Results" section for statistical details and Table 3 for TH ratios obtained from these data. Mole-rat fT4 data refer to scenario 1; applying the second scenario (not depicted here) created essentially the same result because mean fT4 values of mole-rats were slightly lower.

(Table 1). However, the mutational drift for some proteins is such high that immunochemical detection methods, e.g. for TSHB, will require development of species-specific antibodies. In addition, preliminary work showed no results for *F. anselli* samples with commercially available TSH assays, both for human and guinea pig, which confirms the species-specificity of TSH as well.

Most notably, the two TH receptor isoforms, THRA and THRB, show a high level of sequence conservation in *F. anselli* compared to 15 mammalian species, especially within their hormone-binding domains (Figure 1, Figures S1 and S2). This finding supports the expectation that just as in other mammals, the structures of T4 and T3 are conserved in *F. anselli*, although these were not directly determined.

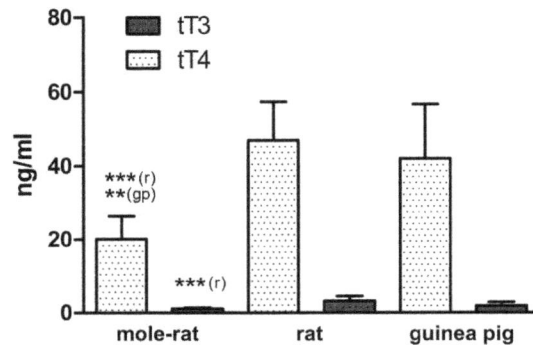

**Figure 3. Total T4 and total T3 levels of Ansell's mole-rats (n=12), rats (n=7) and guinea pigs (n=4).** Mean ± SD; all data expressed in ng/ml. One-way ANOVA, tT4: $F = 19.51$, $p < 0.0001$; fT3: $F = 11.13$, $p = 0.001$. Significant differences in the Bonferroni post hoc comparisons are indicated by asterisks coupled with the comparison species referred to (mr = mole-rat, r = rat, gp = guinea pig) in parenthesis. See "Results" section for statistical details and Table 3 for TH ratios obtained from these data.

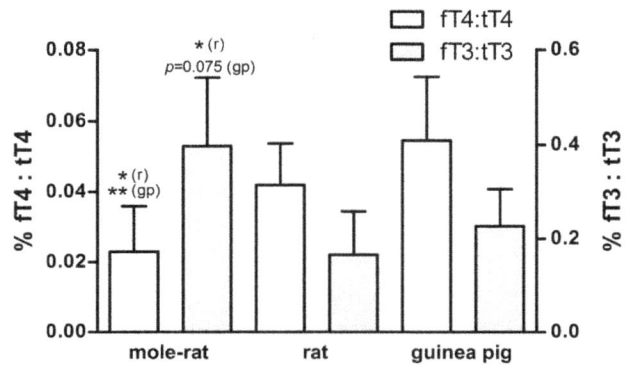

**Figure 4. fT4:tT4 and fT3:tT3 ratios in Ansell's mole-rats (n = 12), rats (n = 7) and guinea pigs (n = 4).** One-way ANOVA, fT4:tT4: $F = 9.60$, $p = 0.001$; fT3:tT3: $F = 7.724$, $p = 0.004$. Significant differences in the Bonferroni post hoc comparisons are indicated by asterisks coupled with the comparison species referred to (mr = mole-rat, r = rat, gp = guinea pig) in parenthesis. See "Results" section for statistical details.

## Interspecies comparison of TH hormone levels

The first striking result of our study is that serum fT4 levels were about 10 times lower in mole-rats than in guinea pigs and rats, regardless of the scenario applied. Low circulating T4 levels are often caused by iodine deficiency in the diet [54], but this explanations appears unlikely in Ansell's mole-rats, because carrots, which they receive *ad libitum*, contain more iodine than required [54]. Moreover, the sequence analysis of the natrium/iodide symporter (NIS; Figure S12) appears to be under strong purifying selection in Ansell's mole-rats as in other mammals (Table 1), thus reducing the possibility of an iodide deficiency in the thyroid gland. Note that typical symptoms of a lifelong iodide deficiency like e.g. goiters [55] have not been observed in *Fukomys* (own unpublished data) or *Heterocephalus* mole-rats [15] so far. Thus it is plausible to assume that the low fT4 and tT4 levels reported here reflect the natural status of these animals, which provides principal support for our first hypothesis that Ansell's mole-rats have naturally low TH levels in comparison to euthyroid rodents. As shown in various vertebrate species, long lifespan is often correlated with low T4 levels, low metabolic rates, or both [7,16–18]. Many long-lived bathyergid species, including *F. anselli*, have very low metabolic rates [39,40,56], and the only member of the family *Bathyergidae* in which TH levels have been quantified so far (the naked mole-rat) has also shown remarkably low fT4 [15]. The T4 levels observed in the present study are in good agreement with these findings.

Considering that the fT4 levels in mole-rats are by an order of magnitude lower than in rats and guinea pigs, it is interesting to

note that after 30 years of maintaining and breeding mole-rats, we have no indications for developmental or cognitive impairments of the progeny. This is noteworthy, because in other animal models (chicken, rats and mice), induction of even mild fT4 deficits in the mother during prenatal development affects brain development, potentially leading to significant cognitive and/or motoric impairments in the progeny [57–59]. In humans, maternal fT4 deficits during pregnancy are associated with an elevated risk of cognitive impairments in the child, including severe disorders like e.g. autism [60,61]. Preliminary own data suggests that female mole-rats do not elevate fT4 levels during pregnancy. Should this assumption be verified, it will be worthwhile to investigate the mechanisms that enable Ansell's mole-rats to deal with such low maternal fT4 levels during prenatal development without ontogenetic impairments.

The low fT4 values may, however, help to explain the unexpectedly high S-cone opsin concentration in the retina of Ansell's mole-rats [41], because THs are essentially involved in the expression of L-opsins in the mammalian retina by binding to a THRB isoform in the cones [42,43]. The adaptive value of colour perception for a strictly subterranean rodent is probably residual; studies by Kott et al. [62] suggest that while rods play an important role in the subterranean habitat, cones, especially S-cones, have no specific adaptive function. Our results provide the alternative explanation that the expression of these S-opsins could be a side effect of a natural state of low T4, which has evolved for other reasons (in this case potentially metabolism).

Total thyroxine levels (tT4) were also significantly lower in mole-rats than in the controls, but the differences were less pronounced than in the free hormone fractions; on average, mole-rat tT4 levels reached about 50% of those measured in rats and guinea pigs. Of note, we found that also the fT4/tT4 ratio is significantly lower in mole-rats than in the two control species (Figure 4). It hence appears that mole-rats do not only produce less T4 in their thyroid glands, but also recruit lesser proportions of their total T4 resources into the active form. Taken together, mole-rats seem to possess two distinct mechanisms that work hand in hand to downregulate fT4 levels reliably, which indicates an adaptive function of low T4 levels in these animals. We will discuss potential proximate mechanisms for their maintenance later in this manuscript.

Interestingly, and in sharp contrast to the low T4, fT3 levels were undistinguishable between *F. anselli* and the euthyroid controls (Figure 2). Total T3 levels of mole-rats were also not statistically different from those measured in guinea pigs, but significantly lower than in rats (Figure 3). The ratios between free and total T3 (Figure 4) suggest that mole-rats recruit significantly higher portions of the available T3 into the active unbound form than the other two species, counteracting the much lower T4 levels. Although these results should be treated with some caution because our rat tT3 levels appear atypically high (see e.g.

**Table 3.** Ratios ($\pm$ SD) of free and total T4:T3 in interspecies comparison.

| Species | Ratio fT4:fT3 | Ratio tT4:tT3 |
|---|---|---|
| Mole-rat | 0.70±0.45* 0.85±0.43[†] | 18.60±6.98 |
| Rat | 7.28±1.87 | 16.24±4.28 |
| Guinea pig | 9.63±1.50 | 26.43±9.51 |

fT4:fT3 ratios were obtained from the data shown in Figure 2 and tT4:tT3 ratios were obtained from data shown in Figure 3. See "Results" section for statistical details.
*: scenario 2 (n = 32).
[†]: scenario 1 (n = 24).

**Table 4.** Intraspecific fT4 (both scenarios) and fT3 differences in Ansell's mole-rats.

| Factor | fT4 first scenario (n = 24) | | fT4 second scenario (n = 32) | | fT3 (n = 32) | |
|---|---|---|---|---|---|---|
| | F | p | F | p | F | p |
| Sex | 2.47 | 0.13 | 4.00 | 0.056 | 0.90 | 0.35 |
| Reproductive status | 0.54 | 0.47 | 0.14 | 0.71 | 0.62 | 0.44 |
| Age | 2.48 | 0.13 | 4.58 | 0.041* | 0.63 | 0.44 |
| Weight | 2.04 | 0.17 | 0.59 | 0.45 | 0.32 | 0.58 |
| Reproductive status × age | 0.71 | 0.41 | 0.04 | 0.84 | 1.29 | 0.28 |

GLM main effects for all four factors alone and a GLM two factor model with reproductive status × age as independent variable. The p-values and the correlation coefficients F are shown.
* = significant (p<0.05).

[29,63,64] where rat tT3 levels between 0.8 ng/ml–1.62 ng/ml have been reported), there is little doubt that the overall T3 pattern differs quite clearly compared to T4.

The combination of low fT4 and "normal" fT3 resulted in a very uncommon fT4:fT3 ratio of only 0.70–0.85 (depending on the scenario applied) in mole-rats, compared to 7.28 and 9.63 in rats and guinea pigs, respectively (Table 3), the latter being in good agreement with published data [28,29]. This phenotype resembles that of hypothyroxinemia, a condition characterized by low levels of fT4 while TSH and often also fT3 are in a normal range or slightly elevated [57]. However, whether mole-rats are naturally hypothyroxinemic cannot be answered until TSH can be quantified reliably also in mole-rats.

Regardless of the terminology, our findings raise interesting questions about the proximate and ultimate mechanisms being responsible for this unusual and hitherto unreported hormone distribution. We have already discussed that mole-rats seem to recruit less T4 and more T3 from their respective resources than other rodents. Both may be linked to higher expression rates and/or binding affinities of the mole-rat TH binding proteins. In rodents, the main known binding proteins are albumin, TTR and TBG [65], and the combination of their expression rates and binding affinities have major influence on the half-life of circulating THs. TBG, for instance, has a high T4 binding affinity, but is expressed at very different levels across the lifetime of rats [66]. Specific expression rates and/or functional mutations affecting binding affinities of TBG and other involved proteins could potentially provide an explanation for the altered ratios between free and total TH fractions observed in Ansell's mole-rats and should therefore be focussed in future investigations. For instance, amino acid changes in TTR, at position 109 or 119, were shown to increase thyroxine affinity and decrease fT4/fT3 ratio in humans [67]. However, in the present study no such changes were observed in Ansell's mole-rats (Figure S8).

The observed TH pattern could be linked to alterations in deiodination rates in and/or efflux rates out of target cells. Deiodination of T4 to T3 takes place in the cytoplasm of target cells [24,68]. A higher D1 and D2 activity, both responsible for converting T4 to T3 [24], and/or a high efflux of T3 out of the cells could lead to a relatively high T3 concentration in the blood stream [69] and help compensate for low levels of T4. Therefore, expression rates of the regulatory components of the TH system as well as D1, D2 and D3 activities in the brain, the thyroid and peripheral organs should be determined in further studies.

The "normal" fT3 concentration is rather unexpected on the basis of the low metabolic rate of Ansell's mole-rats [39,40] and

the low L-opsin density in the retina [41]. Thus, alternative functions of T3 could help explain these contradictions: For instance, novel signalling pathways of T3, which imply indirect activation of transcription as a non-nuclear activity, are discussed. One such pathway initiated by THs is the activation of the transcription of the alpha subunit of hypoxia-inducible factor 1 (HIF1A). It is a transcription factor found in all mammalian cells and responsible for a wide range of cellular responses to hypoxia [70,71]. In human fibroblasts, HIF1A mRNA and protein concentration are upregulated by a pathway which is activated by T3 binding to THRB in the cytoplasm [72,73] without being transported into the nucleus.

Therefore, the maintenance of normal T3 levels despite low T4 levels may be an adaptive cellular mechanism of animals living in hypoxic environments to assure a more specific and continuous availability of HIF1. Of course, this is speculative at the moment. However, the importance of HIF1 in subterranean environments is supported by findings from another strictly subterranean mammal, *Spalax ehrenbergi*. In the skeletal muscles of these animals, the concentration of HIF1A mRNA is significantly higher than in rats [74].

Not surprisingly, a remarkably high concentration of HIF1A was also detected in the brain of old naked mole-rats [75,76]. This suggests that this kind of adaptation to a hypoxic environment is not restricted to *S. ehrenbergi* and may also be found in bathyergid species. Further research has to confirm whether these adaptations also occur in *F. anselli*.

## Intraspecies comparison of TH hormone levels

Intraspecific fT4 and fT3 comparisons suggest that THs are not the major determinants of the caste-specific ageing rates found in Ansell's mole-rats. In neither scenario was there a significant difference in hormone levels (fT4 or fT3) between non-breeders and breeders. Likewise, sex and weight of the animals did not have an influence on hormone levels. On the other hand fT4 levels did seem to increase with age, when applying the second scenario (Table 4). Age effects on TH levels are well-known, which is not surprising, because THs play a major role in development and metabolism. However, in other mammalian species, THs usually decline with age. In human for instance, fT3 levels usually decline with age, while fT4 levels remain more or less unchanged [77,78]. Guinea pigs do not show an alteration in serum fT4 levels as well [45].

In summary, our results indicate that in *F. anselli*, euthyroid fT3 levels are coupled with lower circulating levels of T4, which, in combination with their low metabolic rate, may represent a novel mechanism to cope with the hypoxic subterranean environment these animals have adapted to. However, THs do not seem to have a major influence on the intraspecific ageing rates in these mole-rats.

## Supporting Information

**Figure S1  Protein alignment of thyroid hormone receptor α (THRA) from different mammal species.** The mRNA sequence of *F. anselli* was obtained from RNA-seq and subsequently translated, other sequences were retrieved from NCBI databases with the following accession numbers: Mus musculus (CAA30576), Rattus norvergicus (NP_112396), Mesocricetus auratus (XP_005076008), Cricetulus_griseus (XP_003510526), Oryctolagus cuniculus (XP_002719397), Ochotona princeps(XP_004591220), Otolemur garnettii (XP_003786450), Macaca fascicularis (NP_001270601),Homo sapiens (NP_003241),Felis catus (XP_003996845), Ailuropoda melanoleuca (XP_002924975), Bos Taurus (NP_001039794), Sus scrofa (O97716), Orcinus orca (XP_004282800).

**Figure S2  Protein alignment of thyroid hormone receptor β (THRB) from different mammal species.** The mRNA sequence of *F. anselli* was obtained from RNA-seq and subsequently translated, other sequences were retrieved from NCBI databases with the following accession numbers: Heterocephalus glaber (XP_004892721), Cavia porcellus (XP_005008354), Chinchilla lanigera (XP_005387539), Octodon degus (XP_004634245), Mus musculus (P37242), Rattus norvegicus (P18113), Mesocricetus auratus (XP_005081037), Cricetulus griseus (ERE87100), Ochotona princeps (XP_004588346), Otolemur garnettii (XP_003781795), Homo sapiens (P10828), Canis lupus (XP_862690), Mustela putorius furo (XP_004786805), Ailuropoda melanoleuca (XP_002928077), Ovis aries (Q28571), Sus scrofa (XP_001928500), Condylura cristata (XP_004692336).

**Figure S3  Protein alignment of thyreotropin β subunit (TSHB) from different mammal species.** The mRNA sequence of *F. anselli* was obtained from RNA-seq and subsequently translated, other sequences were retrieved from NCBI databases with the following accession numbers: Cavia porcellus (XP_003479306), Octodon degus (XP_004641725), Mus musculus (NP_001159412), Rattus norvegicus (NP_037248), Ictidomys tridecemlineatus (XP_005334966), Otolemur garnettii (XP_003793878), Macaca fascicularis (XP_001111873), Nomascus leucogenys (XP_003268073), Gorilla gorilla (XP_004026450), Pan paniscus (XP_003805682), Pan troglodytes (XP_001160337), Homo sapiens (AAB30828), Bos taurus (XP_005204060), Orcinus orca (XP_004263279), Echinops telfairi (XP_004714911).

**Figure S4  Protein alignment of Type I iodothyronine deiodinase (D1) from different mammal species.** The mRNA sequence of *F. anselli* was obtained from RNA-seq and subsequently translated, other sequences were retrieved from NCBI databases with the following accession numbers: Heterocephalus glaber (XP_004908861), Cavia porcellus (NP_001244903), Octodon degus (XP_004642620), Mus musculus (Q61153), Rattus norvegicus (CAA41063), Cricetulus griseus (NP_001243688), Ochotona princeps (XP_004588749), Otolemur garnettii (XP_003793192), Macaca mulatta (NP_001116124), Pan troglodytes (NP_001116123), Homo sapiens (NP_000783), Canis lupus (NP_001007127), Felis catus (NP_001009267), Bos taurus

(NP_001116065), Sus scrofa (NP_001001627), Equus caballus (NP_001159924), Orcinus orca (XP_004273874).

**Figure S5  Protein alignment of Type II iodothyronine deiodinase (D2) from different mammal species.** The mRNA sequence of *F. anselli* was obtained from RNA-seq and subsequently translated, other sequences were retrieved from NCBI databases with the following accession numbers: Heterocephalus glaber (XP_004900438), Chinchilla lanigera (XP_005390287), Octodon degus (XP_004624767), Mus musculus (NP_034180), Rattus norvegicus (NP_113908), Ochotona princeps (XP_004584413), Homo sapiens (AAC95470), Canis lupus (NP_001116117), Ovis aries (XP_004011138), Sus scrofa (NP_001001626), Equus caballus (NP_001159927), Orcinus orca (XP_004262346), Condylura cristata (XP_004681708), Echinops telfairi (XP_004698804).

**Figure S6  Protein alignment of thyroglobulin (TG) from different mammal species.** The mRNA sequence of *F. anselli* was obtained from RNA-seq and subsequently translated, other sequences were retrieved from NCBI databases with the following accession numbers: Cavia porcellus (XP_003467392), Chinchilla lanigera (XP_005398080), Octodon degus (XP_004642544), Mus musculus (AAB53204), Rattus norvegicus (BAL14775), Ochotona princeps (XP_004580794), Otolemur garnettii (XP_003792914), Macaca mulatta (EHH28780), Pan troglodytes (XP_003311969), Homo sapiens (AAC51924), Canis lupus (XP_005627864), Felis catus (XP_004000173), Sus scrofa (NP_001161890), Equus caballus (XP_001916622), Orcinus orca (XP_004265356), Echinops telfairi (XP_004697442).

**Figure S7  Protein alignment of thyroperoxidase (TPO) from different mammal species.** The mRNA sequence of *F. anselli* was obtained from RNA-seq and subsequently translated, other sequences were retrieved from NCBI databases with the following accession numbers: Cavia porcellus (XP_003464975; patched), Octodon degus (XP_004644658), Mus musculus (EDL36934), Rattus norvegicus (EDM03234), Cricetulus griseus (XP_003501455), Ochotona princeps (XP_004582879), Otolemur garnettii (XP_003798602), Macaca mulatta (XP_001117795), Homo sapiens (XP_005264756), Canis lupus (Q8HYB7), Felis catus (XP_003984594), Bos taurus (XP_603356), Sus scrofa (P09933), Equus caballus (XP_001918216), Orcinus orca (XP_004274968), Echinops telfairi (XP_004709888).

**Figure S8  Protein alignment of transthyretin (TTR) from different mammal species.** The mRNA sequence of *F. anselli* was obtained from RNA-seq and subsequently translated, other sequences were retrieved from NCBI databases with the following accession numbers: Heterocephalus glaber (XP_004905241), Chinchilla lanigera (XP_005372800), Octodon degus (XP_004623610), Rattus norvegicus (AAA41801), Mesocricetus auratus (XP_005065406), Cricetulus griseus (XP_003510202), Ictidomys tridecemlineatus (XP_005337518), Oryctolagus cuniculus (XP_002713532), Chlorocebus aethiops (BAL44398), Homo sapiens (CAG33189), Equus caballus (XP_001495232), Echinops telfairi (XP_004702987).

**Figure S9  Protein alignment of thyroxine-binding globin (TBG) from different mammal species.** The mRNA sequence of *F. anselli* was obtained from RNA-seq and subsequently translated, other sequences were retrieved from

NCBI databases with the following accession numbers: Heterocephalus glaber (EHB09876), Octodon degus (XP_004646260), Mus musculus (P61939), Rattus norvegicus (AAA42205), Cricetulus griseus (ERE65740), Otolemur garnettii (XP_003801681), Gorilla gorilla (XP_004064693), Pan troglodytes (NP_001009109), Homo sapiens (NP_783866), Canis lupus (XP_538128), Bos taurus (AAI03464), Ovis aries (NP_001094390), Sus scrofa (Q9TT35), Equus caballus (XP_001493492), Orcinus orca (XP_004285286), Echinops telfairi (XP_004710081).

## Figure S10 Protein alignment of hypoxia-induced factor (HIF1A) from different mammal species.

The mRNA sequence of *F. anselli* was obtained from RNA-seq and subsequently translated, other sequences were retrieved from NCBI databases with the following accession numbers: Heterocephalus glaber (XP_004837489), Octodon degus (XP_004624861), Mus musculus (CAA70305), Rattus norvegicus (O35800), Ochotona princeps (XP_004597684), Otolemur garnettii (XP_003794480), Pan troglodytes (XP_001168972), Homo sapiens (NP_001521), Canis lupus (XP_003639249), Felis catus (XP_003987765), Bos taurus (NP_776764), Sus scrofa (NP_001116596), Orcinus orca (XP_004262152).

## Figure S11 Protein alignment of Monocarboxylate transporter 8 (MCT8) from different mammal species.

The mRNA sequence of *F. anselli* was obtained from RNA-seq and subsequently translated, other sequences were retrieved from NCBI databases with the following accession numbers: Heterocephalus glaber (XP_004905154), Cavia porcellus (XP_005004229), Octodon degus (XP_004648070), Mus musculus (AAC40078), Rattus norvegicus (EDM07172), Ochotona princeps (XP_004592817), Otolemur garnettii (XP_003802275), Macaca mulatta (XP_001096017), Homo sapiens (NP_006508), Bos taurus (NP_001193868), Orcinus orca (XP_004283857). Suggestions of NCBI for translation starts of Octodon degus (XP_004648070), Ocho-

tona princeps (XP_004592817), Macaca mulatta (XP_001096017) and Bos taurus (NP_001193868) were changed to the position of the orthologous sequences.

## Figure S12 Protein alignment of Natrium-Iodid-Symporter (NIS) from different mammal species.

The mRNA sequence of *F. anselli* was obtained from RNA-seq and subsequently translated, other sequences were retrieved from NCBI databases with the following accession numbers: Heterocephalus glaber (XP_004873534), Cavia porcellus (XP_003465226), Octodon degus (XP_004646953), Mus musculus (NP_444478), Rattus norvegicus (Q63008), Otolemur garnettii (XP_003796602), Macaca mulatta (EHH29802), Pan troglodytes (XP_524154), Homo sapiens (NP_000444), Canis lupus (XP_541946), Sus scrofa (NP_999575), Orcinus orca (XP_004277608).

## Table S1 Intra- and inter-assay variances of the Enzyme Immunoassays for fT3, fT4, tT3 and tT4.

Shown are the coefficients of variances (%CV) of the assays used in the present study, according to the manufacturer (DRG Instruments GmbH).

## Acknowledgments

The authors thank Christine Krueger, Gero Hilken and Andreas Wissmann for their help at blood sampling, and Lilia Kufeld and Andrea Jaeger for conducting the hormone measurements.

## Author Contributions

Conceived and designed the experiments: YH CV SB MB MBP AS KS HB PD. Performed the experiments: YH CV AS KS PD. Analyzed the data: YH CV SB MB AS KS PS. Contributed reagents/materials/analysis tools: SB HB MBP KS PD. Wrote the paper: YH CV SB AS KS HB PS.

## References

1. Harman D (1956) Aging - A theory based on free-radical and radiation-chemistry. Journals of Gerontology 11: 298–300.
2. Balaban RS, Nemoto S, Finkel T (2005) Mitochondria, oxidants, and aging. Cell 120: 483–495.
3. Monnier VM (1989) Toward a Maillard reaction theory of aging. Progress in clinical and biological research 304: 1–22.
4. Baynes JW (2001) The role of AGEs in aging: causation or correlation. Experimental Gerontology 36: 1527–1537.
5. Dammann P, Sell DR, Begall S, Strauch C, Monnier VM (2011) Advanced Glycation End-Products as Markers of Aging and Longevity in the Long-Lived Ansell's Mole-Rat (Fukomys anselli). Journals of Gerontology Series a-Biological Sciences and Medical Sciences 67: 573–583.
6. Harley CB, Vaziri H, Counter CM, Allsopp RC (1992) The telomere hypothesis of cellular aging. Experimental Gerontology 27: 375–382.
7. Bowers J, Terrien J, Clerget-Froidevaux MS, Gothié JD, Rozing MP, et al. (2013) Thyroid Hormone Signaling and Homeostasis During Aging. Endocrine Reviews 34: 556–589.
8. Zhang J, Lazar MA (2000) The mechanism of action of thyroid hormones. Annu Rev Physiol 62: 439–466.
9. Ooka H, Fujita S, Yoshimoto E (1983) Pituitary-thyroid activity and longevity in neonatally thyroxine-treated rats. Mechanisms of Ageing and Development 22: 113–120.
10. Ooka H, Shinkai T (1986) Effects of chronic hyperthyroidism on the lifespan of the rat. Mechanisms of Ageing and Development 33: 275–282.
11. Brown-Borg HM, Borg KE, Meliska CJ, Bartke A (1996) Dwarf mice and the ageing process. Nature 384: 33–33.
12. Vergara M, Smith-Wheelock M, Harper JM, Sigler R, Miller RA (2004) Hormone-treated snell dwarf mice regain fertility but remain long lived and disease resistant. Journals of Gerontology Series a-Biological Sciences and Medical Sciences 59: 1244–1250.
13. Liang S, Mele J, Wu Y, Buffenstein R, Hornsby PJ (2010) Resistance to experimental tumorigenesis in cells of a long-lived mammal, the naked mole-rat (Heterocephalus glaber). Aging Cell 9: 626–635.
14. Lovegrove BG (1986) The metabolism of social subterranean rodents: adaptation to aridity. Oecologia 69: 551–555.
15. Buffenstein R, Woodley R, Thomadakis C, Daly TJM, Gray DA (2001) Cold-induced changes in thyroid function in a poikilothermic mammal, the naked mole-rat. American Journal of Physiology-Regulatory Integrative and Comparative Physiology 280: R149–R155.
16. Willis CKR, Brigham RM, Geiser F (2006) Deep, prolonged torpor by pregnant, free-ranging bats. Naturwissenschaften 93: 80–83.
17. Becker NI, Encarnacao JA, Tschapka M, Kalko EKV (2013) Energetics and life-history of bats in comparison to small mammals. Ecological Research 28: 249–258.
18. Rozing MP, Houwing-Duistermaat JJ, Slagboom PE, Beekman M, Frolich M, et al. (2010) Familial Longevity Is Associated with Decreased Thyroid Function. Journal of Clinical Endocrinology & Metabolism 95: 4979–4984.
19. Shupnik MA, Chin WW, Habener JF, Ridgway EC (1985) Transcriptional regulation of the thyrotropin subunit genes by thyroid hormone. J Biol Chem 260: 2900–2903.
20. Kelly GS (2000) Peripheral metabolism of thyroid hormones: a review. Alternative medicine review: a journal of clinical therapeutic 5: 306–333.
21. Costa-e-Sousa RH, Hollenberg AN (2012) Minireview: The Neural Regulation of the Hypothalamic-Pituitary-Thyroid Axis. Endocrinology 153: 4128–4135.
22. Visser WE, Friesema EC, Visser TJ (2011) Minireview: thyroid hormone transporters: the knowns and the unknowns. Mol Endocrinol 25: 1–14.
23. Crantz FR, Larsen PR (1980) Rapid thyroxine to 3,5,3'-triiodothyronine conversion and nuclear 3,5,3'-triiodothyronine binding in rat cerebral-cortex and cerebellum. Journal of Clinical Investigation 65: 935–938.
24. Bianco AC, Larsen PR (2005) Cellular and structural biology of the deiodinases. Thyroid 15: 777–786.
25. Harvey CB, Williams GR (2002) Mechanism of thyroid hormone action. Thyroid 12: 441–446.
26. Moeller LC, Broecker-Preuss M (2011) Transcriptional regulation by nonclassical action of thyroid hormone. Thyroid research 4 Suppl 1: S6.

27. Gereben B, Zavacki AM, Ribich S, Kim BW, Huang SA, et al. (2008) Cellular and Molecular Basis of Deiodinase-Regulated Thyroid Hormone Signaling. Endocrine Reviews 29: 898–938.

28. Ma C, Xie J, Huang X, Wang G, Wang Y, et al. (2009) Thyroxine alone or thyroxine plus triiodothyronine replacement therapy for hypothyroidism. Nucl Med Commun 30: 586–593.

29. Davies DT (1993) Assessment of rodent thyroid endocrinology - Advantages and pit-falls. Comparative Haematology International 3: 142–152.

30. Schussler GC (2000) The thyroxine-binding proteins. Thyroid 10: 141–149.

31. Burda H, Honeycutt RL, Begall S, Locker-Grutjen O, Scharff A (2000) Are naked and common mole-rats eusocial and if so, why? Behavioral Ecology and Sociobiology 47: 293–303.

32. Dammann P, Burda H (2006) Sexual activity and reproduction delay ageing in a mammal. Current Biology 16: R117–R118.

33. Dammann P, Sumbera R, Massmann C, Scherag A, Burda H (2011) Extended Longevity of Reproductives Appears to be Common in Fukomys Mole-Rats (Rodentia, Bathyergidae). Plos One 6.

34. Schmidt CM, Jarvis J. U. M.; Bennett NC (2013) The Long-Lived Queen: Reproduction and Longevity in Female Eusocial Damaraland Mole-Rats (Fukomys damarensis). African Zoology 48: 193–196.

35. Kirkwood TB (1977) Evolution of ageing. Nature 270: 301–304.

36. Edward DAaC, T. (2011) Mechanisms underlying reproductive trade-offs: Costs of reproduction. In: Flatt THA, editor. Mechanisms of Life History Evolution - The Genetics and Physiology of Life History Traits and Trade-Offs New York Oxford University Press Inc. pp. 137–152.

37. Keller L, Genoud M (1997) Extraordinary lifespans in ants: a test of evolutionary theories of ageing. Nature 389: 958–960.

38. Corona M, Velarde RA, Remolina S, Moran-Lauter A, Wang Y, et al. (2007) Vitellogenin, juvenile hormone, insulin signaling, and queen honey bee longevity. Proceedings of the National Academy of Sciences of the United States of America 104: 7128–7133.

39. Marhold S, Nagel A (1995) The energetics of the common mole-rat Cryptomys, a subterranean eusocial rodent from Zambia. Journal of Comparative Physiology B-Biochemical Systemic and Environmental Physiology 164: 636–645.

40. Zelová J, Sumbera R, Sedlácek F, Burda H (2007) Energetics in a solitary subterranean rodent, the silvery mole-rat, Heliophobius argenteocinereus, and allometry of RMR in African mole-rats (Bathyergidae). Comparative biochemistry and physiology Part A, Molecular & integrative physiology 147: 412–419.

41. Peichl L, Němec P, Burda H (2004) Unusual cone and rod properties in subterranean African mole-rats (Rodentia, Bathyergidae). European Journal of Neuroscience 19: 1545–1558.

42. Glaschke A, Gloesmann M, Peichl L (2010) Developmental Changes of Cone Opsin Expression but Not Retinal Morphology in the Hypothyroid Pax8 Knockout Mouse. Investigative Ophthalmology & Visual Science 51: 1719–1727.

43. Glaschke A, Weiland J, Del Turco D, Steiner M, Peichl L, et al. (2011) Thyroid Hormone Controls Cone Opsin Expression in the Retina of Adult Rodents. The Journal of Neuroscience 31: 4844–4851.

44. Choksi NY, Jahnke GD, St Hilaire C, Shelby M (2003) Role of thyroid hormones in human and laboratory animal reproductive health. Birth Defects Research Part B-Developmental and Reproductive Toxicology 68: 479–491.

45. Mueller K, Mueller E, Klein R, Brunnberg L (2009) Serum thyroxine concentrations in clinically healthy pet guinea pigs (Cavia porcellus). Veterinary Clinical Pathology 38: 507–510.

46. Hoff J RL (2000) Methods of blood collection in the mouse. Lab animals 29: 47–53.

47. Martin M (2011) Cutadapt removes adapter sequences from high-throughput sequencing reads. 2011 17.

48. Joshi NA FJ (2011) Sickle: A sliding-window, adaptive, quality-based trimming tool for FastQ files (Version 1.21).

49. Grabherr MG, Haas BJ, Yassour M, Levin JZ, Thompson DA, et al. (2011) Full-length transcriptome assembly from RNA-Seq data without a reference genome. Nat Biotechnol 29: 644–652.

50. Yang Z (1997) PAML: A program package for phylogenetic analysis by maximum likelihood. Comput Appl Biosci 13: 555–556.

51. Yang Z (2007) PAML 4: Phylogenetic Analysis by Maximum Likelihood. Molecular Biology and Evolution 24: 1586–1591.

52. Garcia Montero A, Burda H, Begall S (2014) Chemical restraint of African mole-rats (Fukomys sp.) with a combination of ketamine and xylazine. Veterinary Anaesthesia and Analgesia: doi:10.1111/vaa.12180.

53. Campos-Barros A, Musa A, Flechner A, Hessenius C, Gaio U, et al. (1997) Evidence for circadian variations of thyroid hormone concentrations and type II 5′-iodothyronine deiodinase activity in the rat central nervous system. Journal of Neurochemistry 68: 795–803.

54. Negro R, Soldin O, Obregon M-J, Stagnaro-Green A (2011) Hypothyroxinemia and Pregnancy. Endocrine Practice 17: 422–429.

55. Bauch K, Meng W, Ulrich FE, Grosse E, Kempe R, et al. (1986) Thyroid status during pregnancy and post partum in regions of iodine deficiency and endemic goiter. Endocrinol Exp 20: 67–77.

56. Buffenstein R (2005) The naked mole-rat: a new long-living model for human aging research. J Gerontol A Biol Sci Med Sci 60: 1369–1377.

57. Opazo MC, Gianini A, Pancetti F, Azkcona G, Alarcón L, et al. (2008) Maternal Hypothyroxinemia Impairs Spatial Learning and Synaptic Nature and Function in the Offspring. Endocrinology 149: 5097–5106.

58. Dong H, You S-H, Williams A, Wade MG, Yauk CL, et al. (2014) Transient Maternal Hypothyroxinemia Potentiates the Transcriptional Response to Exogenous Thyroid Hormone in the Fetal Cerebral Cortex Before the Onset of Fetal Thyroid Function: A Messenger and MicroRNA Profiling Study. Cerebral Cortex.

59. Darras VM, Van Herck SLJ, Geysens S, Reyns GE (2009) Involvement of thyroid hormones in chicken embryonic brain development. General and Comparative Endocrinology 163: 58–62.

60. Haddow JE, Palomaki GE, Allan WC, Williams JR, Knight GJ, et al. (1999) Maternal Thyroid Deficiency during Pregnancy and Subsequent Neuropsychological Development of the Child. New England Journal of Medicine 341: 549–555.

61. Román GC, Ghassabian A, Bongers-Schokking JJ, Jaddoe VWV, Hofman A, et al. (2013) Association of gestational maternal hypothyroxinemia and increased autism risk. Annals of Neurology 74: 733–742.

62. Kott O, Moritz RE, Šumbera R, Burda H, Němec P (2014) Light propagation in burrows of subterranean rodents: tunnel system architecture but not photoreceptor sensitivity limits light sensation range. Journal of Zoology 294: 67–75.

63. Moreno M LA, Lombardi A GF (1997) How the thyroid controls metabolism in the rat: different roles for triiodothyronine and diiodothyronines. J Physiol 505(Pt 2): 529–538.

64. Hood A, Liu YP, Gattone VH 2nd, Klaassen CD (1999) Sensitivity of thyroid gland growth to thyroid stimulating hormone (TSH) in rats treated with antithyroid drugs. Toxicol Sci 49: 263–271.

65. Kaneko JJ (2008) Thyroid Function. In: Kaneko JJH, John W.; Bruss, Michael L., editor. Clinical Biochemistry of Domestic Animals. 6 ed. Amsterdam; Boston: Elsevier Academic Press. pp. 623–634.

66. Savu L, Vranckx R, Rouaze-Romet M, Nunez EA (1992) The pituitary control of rat thyroxine binding globulin. Acta Med Austriaca 19 Suppl 1: 88–90.

67. Refetoff S, Marinov VS, Tunca H, Byrne MM, Sunthornthepvarakul T, et al. (1996) A new family with hyperthyroxinemia caused by transthyretin Val109 misdiagnosed as thyrotoxicosis and resistance to thyroid hormone–a clinical research center study. J Clin Endocrinol Metab 81: 3335–3340.

68. Pohlenz J, Maqueem A, Cua K, Weiss RE, Van Sande J, et al. (1999) Improved radioimmunoassay for measurement of mouse thyrotropin in serum: Strain differences in thyrotropin concentration and thyrotroph sensitivity to thyroid hormone. Thyroid 9: 1265–1271.

69. Visser WE, Friesema EC, Jansen J, Visser TJ (2008) Thyroid hormone transport in and out of cells. Trends Endocrinol Metab 19: 50–56.

70. Weidemann A, Johnson RS (2008) Biology of HIF-1 alpha. Cell Death and Differentiation 15: 621–627.

71. Otto T, Fandrey J (2008) Thyroid hormone induces hypoxia-inducible factor 1 alpha gene expression through thyroid hormone receptor beta/retinoid X receptor alpha-dependent activation of hepatic leukemia factor. Endocrinology 149: 2241–2250.

72. Moeller LC, Dumitrescu AM, Refetoff S (2005) Cytosolic action of thyroid hormone leads to induction of hypoxia-inducible factor-1 alpha and glycolytic genes. Molecular Endocrinology 19: 2955–2963.

73. Storey NM, Gentile S, Ullah H, Russo A, Muessel M, et al. (2006) Rapid signaling at the plasma membrane by a nuclear receptor for thyroid hormone. Proceedings of the National Academy of Sciences of the United States of America 103: 5197–5201.

74. Avivi A, Shams I, Joel A, Lache O, Levy AP, et al. (2005) Increased blood vessel density provides the mole rat physiological tolerance to its hypoxic subterranean habitat. The FASEB Journal 19: 1314–1316.

75. Edrey YH, Park TJ, Kang H, Biney A, Buffenstein R (2011) Endocrine function and neurobiology of the longest-living rodent, the naked mole-rat. Experimental Gerontology 46: 116–123.

76. Kim EB, Fang X, Fushan AA, Huang Z, Lobanov AV, et al. (2011) Genome sequencing reveals insights into physiology and longevity of the naked mole rat. Nature 479: 223–227.

77. Peeters RP (2008) Thyroid hormones and aging. Hormones-International Journal of Endocrinology and Metabolism 7: 28–35.

78. Gesing A, Lewinski A, Karbownik-Lewinska M (2012) The thyroid gland and the process of aging; what is new? Thyroid Research 5: 1–5.

# The GCKIII Kinase Sps1 and the 14-3-3 Isoforms, Bmh1 and Bmh2, Cooperate to Ensure Proper Sporulation in *Saccharomyces cerevisiae*

**Christian J. Slubowski, Scott M. Paulissen, Linda S. Huang***

Department of Biology, University of Massachusetts Boston, Boston, Massachusetts, United States of America

## Abstract

Sporulation in the budding yeast *Saccharomyces cerevisiae* is a developmental program initiated in response to nutritional deprivation. Sps1, a serine/threonine kinase, is required for sporulation, but relatively little is known about the molecular mechanisms through which it regulates this process. Here we show that *SPS1* encodes a bona-fide member of the GCKIII subfamily of STE20 kinases, both through phylogenetic analysis of the kinase domain and examination of its C-terminal regulatory domain. Within the regulatory domain, we find Sps1 contains an invariant ExxxPG region conserved from plant to human GCKIIIs that we call the EPG motif; we show this EPG motif is important for *SPS1* function. We also find that Sps1 is phosphorylated near its N-terminus on Threonine 12, and that this phosphorylation is required for the efficient production of spores. In Sps1, Threonine 12 lies within a 14-3-3 consensus binding sequence, and we show that the *S. cerevisiae* 14-3-3 proteins Bmh1 and Bmh2 bind Sps1 in a Threonine 12-dependent fashion. This interaction is significant, as *BMH1* and *BMH2* are required during sporulation and genetically interact with *SPS1* in sporulating cells. Finally, we observe that Sps1, Bmh1 and Bmh2 are present in both the nucleus and cytoplasm during sporulation. We identify a nuclear localization sequence in Sps1 at amino acids 411–415, and show that this sequence is necessary and sufficient for nuclear localization. Taken together, these data identify regions within Sps1 critical for its function and indicate that *SPS1* and 14-3-3s act together to promote proper sporulation in *S. cerevisiae*.

**Editor:** Stefanie Pöggeler, Georg-August-University of Göttingen Institute of Microbiology & Genetics, Germany

**Funding:** This work was supported by a Goranson Research Award to CJS, an University of Massachusetts Boston Dissertation Support Grant to CJS and by the National Science Foundation (MCB-0544160) to LSH. The funders had no role in study design, data collection and analysis, decision to publish, or preparation of the manuscript.

**Competing Interests:** The authors have declared that no competing interests exist.

* Email: Linda.Huang@umb.edu

## Introduction

Yeast deprived of a fermentable carbon source and nitrogen undergo sporulation [1]. Sporulation begins with meiosis, which results in the production of four haploid nuclei from a single diploid cell. These four nuclei are encapsulated by the prospore membrane, which acts as the template for spore wall deposition. The spore wall differs from the vegetative cell wall, and contains the spore-specific chitosan and dityrosine layers that protect the spores during times of harsh environmental stress. Sporulation is a highly regulated process, and *SPS1*, which encodes a STE20 family serine/threonine kinase, is essential for sporulation [2].

STE20 family kinases are highly conserved from yeast to mammals and are divided into two subgroups, the p21-activated kinases (PAKs) and the germinal center kinases (GCKs) [3,4]. These two subgroups are distinguished both by the phylogenetic relationships among their kinase domains and by their domain architectures: In PAKs, the kinase domain is C-terminal to the regulatory domain, and this is reversed in GCKs [5]. Within the GCKs, the GCKIII subfamily of kinases includes the mammalian kinases MST3, MST4, and YSK1/SOK1/STK25 [3], which have been implicated in processes such as apoptosis [6] and axon

outgrowth [7], and may be involved in diseases such as Alzheimer's [8], type 2 diabetes [9], Parkinson's disease [10], and cerebral cavernous malformations [4].

In *S. cerevisiae*, *SPS1* is required for proper sporulation. In particular, previous work has shown that *SPS1* is required for the proper localization of the Gsc2, Chs3, and Gas1 enzymes involved in the construction of the spore wall [2,11,12]. In addition, Sps1 may play a role in histone modification [13], although whether this role is direct is currently unclear. *SPS1* has also been shown to regulate yeast replicative lifespan [14].

14-3-3 proteins are phosphopeptide binding proteins found in all eukaryotes [15]. There are seven 14-3-3 isoforms in mammals, at least thirteen in plants, and two in yeasts [16]. 14-3-3 family proteins function in a diverse range of biological processes and are implicated in human diseases [17–27].

At the molecular level, 14-3-3 proteins are acidic, readily form dimers and bind other proteins using a conserved binding groove [28]. Binding by 14-3-3 proteins has been shown to affect protein function through multiple mechanisms which include acting as a scaffold to facilitate interaction between proteins, modulating protein degradation rate, and altering protein subcellular localization [29]. 14-3-3 binding to substrates in a phosphorylation dependent manner was first shown between 14-3-3ζ and a serine-phosphorylated Raf-1 peptide [30]. Subsequently three different consensus sequences for 14-3-3 binding have been identified: RSX(pS/pT)XP, RXXX(pS/pT)XP [31] and (pS/pTX)(1–2)-COOH [32] (where pS/pT indicates a phosphoserine or phosphothreonine respectively and X represents any amino acid).

The *Saccharomyces cerevisiae* 14-3-3 homologs are encoded by *BMH1* and *BMH2*. Both Bmh1 and Bmh2 are expressed in vegetatively growing cells, although Bmh1 is the major isoform [33–35]. In most strain backgrounds, 14-3-3s are essential [36]. However, both *BMH1* and *BMH2* can be removed in the Σ1278b background, a strain in which they have been shown to bind to the kinase, Ste20, and regulate MAPK signaling during pseudohyphal growth [37]. Other 14-3-3 functions in *S. cerevisiae* include: cell cycle regulation [38], DNA replication [39], TOR-signaling [40], PKA signaling [41], transcription [42], cation homeostasis [43], Golgi function [44], lifespan regulation [45], rapamycin-mediated transcription [46], and the spindle position checkpoint [47].

In this study, we use phylogenetic analysis to determine the relationship of Sps1 to other Ste20 kinases, and demonstrate that Sps1 is a bona-fide member of the GCKIII family of STE20 kinases. Our comparative analyses also identify a C-terminal region in GCKIII kinases that is conserved from yeast to mammal to plant, and we show that this region is important for Sps1 function. To obtain insight into the regulatory interactions of Sps1, we map phosphorylation sites on Sps1 and identify threonine 12 (T12) as a residue important for Sps1 function and efficient sporulation. We show that Sps1-T12 is required for the physical interaction between Sps1 and the 14-3-3 proteins Bmh1 and Bmh2. We describe a role for 14-3-3 proteins in sporulation, and demonstrate that the relative levels of Bmh1 and Bmh2 change during sporulation. We show that Sps1 and 14-3-3 proteins are present in both the nucleus and cytoplasm during sporulation, and we identify a nuclear localization signal for Sps1. Because we see both a physical and genetic interaction between 14-3-3 proteins and Sps1, we propose that Bmh1, Bmh2, and Sps1 act together during sporulation to regulate spore formation.

## Materials and Methods

### Plasmids used in this study

All plasmids used in this study can be found in Table S1 and all primers in Table S2. Construction details are described below. All plasmid inserts amplified using PCR were verified by sequencing.

pCS22 (pRS426-$P_{TEF2}$-*GFP-SPS1*) was constructed by amplifying the *SPS1* coding sequence from genomic SK1 DNA using primers OLH1128 and OLH1129 and then cutting both the amplified DNA and pRS426-$P_{TEF2}$-*GFP-SPO71(1–1245)* [48] with HindIII and XhoI restriction enzymes. The *SPS1* ORF was then ligated into the GFP containing plasmid so that GFP was N-terminally fused to *SPS1*.

pCS20 (pRS426-$P_{TEF2}$-*SBP-SPS1*) was created by amplifying SBP (Streptavidin Binding Peptide) from plasmid pMK33-CTAP(SG) [49] using primers OLH1132 and OLH1133. The PCR product as well as plasmid pCS22 (pRS426-$P_{TEF2}$-*GFP-SPS1*) were cut with restriction enzymes EcoRI and HindIII. This allowed ligation of SBP in place of GFP.

pCS65 (pRS426-$P_{TEF2}$*GFP-sps1-ggaga*) and pCS130 (pRS426-$P_{TEF2}$*GFP-sps1-arappa*) were constructed by site-directed mutagenesis of pCS20 (pRS426-$P_{TEF2}$-*SBP-SPS1*) using primers OLH1182/OLH1183 and OLH1362/OLH1363 respectively. The mutagenized ORFs were then cloned into pCS22 (pRS426-$P_{TEF2}$-*GFP-SPS1*) in place of *SPS1* using the HindIII and XhoI restriction sites.

pCS28 (pRS426-$P_{TEF2}$-*SBP-sps1-K47R*) was constructed by site-directed mutagenesis of pCS22 (pRS426-$P_{TEF2}$-*GFP-SPS1*) and subsequent cloning into pCS20 (pRS426-$P_{TEF2}$-*SBP-SPS1*) using restriction sites HindIII and XhoI.

pCS75 (pRS426-$P_{TEF2}$-*GFP-GST-URA3*) was constructed by first amplifying GST out of plasmid pGEX-4T-3 using the primer combination OLH1258/OLH1259, cutting both the PCR product (which lacked 9 amino acids from the C-terminal due to an endogenous XhoI site) and pCS22 (pRS426-$P_{TEF2}$-*GFP-SPS1*) with HindIII and XhoI.

pCS60 (pRS426-$P_{TEF2}$-*GFP-GST-SPS1(387–438)*) and pCS78 (pRS426-$P_{TEF2}$-*GFP-GST-sps1-ggaga-(387–438)*) were constructed in the same manner as pCS75 (pRS426-$P_{TEF2}$-*GFP-GST-URA3*) except the primer combination OLH1258/OLH1260 was used to generate a GST product without a stop codon. OLH1261 and OLH1262 were then used to amplify the *SPS1* NLS region from pCS22 (pRS426-$P_{TEF2}$-*GFP-SPS1*) and pCS65 (pRS426-$P_{TEF2}$-*GFP-sps1-ggaga*) respectively. Overlap PCR was performed using the PCR products generated in the above reactions with primers OLH1258 and OLH1262. Overlap PCR resulted in products: *GST-SPS1(387–438)* and *GST-sps1-ggaga-(387–438)* respectively. These products, as well as pCS22 (pRS426-$P_{TEF2}$-*GFP-SPS1*), were then cut with the restriction enzymes HindIII and XhoI and ligated into pCS22 (pRS426-$P_{TEF2}$-*GFP-SPS1*).

pCS96 (pRS316-$P_{TEF2}$-*SBP-SPS1*) was created by cutting pCS20 (pRS426-$P_{TEF2}$-*SBP-SPS1*) with SacI and KpnI and ligating the resulting product into the identically cut pRS316.

pCS107 (pRS316-$P_{TEF2}$-*SBP*) was constructed by first amplifying SBP from pMK33-CTAP(SG) using the primer combination OLH1132/OLH1226 and then cutting the resulting product, as well as pCS96 (pRS316-$P_{TEF2}$-*SBP-SPS1*), with EcoRI and XhoI followed by ligation.

pCS98 (pRS316-$P_{TEF2}$-*SBP-sps1-T12A*) was constructed by mutagenic PCR of pCS20 (pRS426-$P_{TEF2}$-*SBP-SPS1*) that changed the codon for T12 using primers OLH1281 and OLH1282. The mutagenized *sps1* ORF was then excised using HindIII and XhoI restriction sites and ligated into pCS96 (pRS316-$P_{TEF2}$-*SBP-SPS1*).

pCS99 (pRS316-$P_{SPS1}$-*SBP-SPS1*) was created by first amplifying the promoter region of *SPS1* using the primer combination OLH1230/OLH1257. The PCR product and pCS96 (pRS316-$P_{TEF2}$-*SBP-SPS1*) were cut with SacI and EcoRI and the *SPS1* promoter was ligated in place of the *TEF2* promoter.

pCS100 (pRS316-$P_{SPS1}$-*SBP-sps1-T12A*) was created by cutting *sps1-T12A* out of pCS98 (pRS316-$P_{TEF2}$-*SBP-sps1-T12A*) using HindIII and XhoI and ligating it in place of *SPS1* in pCS99 (pRS316-$P_{SPS1}$-*SBP-SPS1*), which was also cut with the same enzymes.

pCS145 (pRS316-$P_{SPS1}$-*sfGFP-sps1-T12A*) was constructed by first PCR amplifying sfGFP from pDHL1029 using primers OLH1416 and OLH1417 and then cutting the resulting PCR product as well as pCS100 (pRS316-$P_{SPS1}$-*SBP-sps1-T12A*) with EcoRI and HindIII. sfGFP was then ligated in place of SBP.

pCS146 (pRS316-$P_{SPS1}$-*sfGFP-SPS1*) was constructed by cutting pCS99 (pRS316-$P_{SPS1}$-*SBP-SPS1*) and pCS145 (pRS316-$P_{SPS1}$-*sfGFP-sps1-T12A*) with the restriction enzymes HindIII and XhoI and ligating *SPS1* in place of *sps1-T12A*.

pCS47 (pBSIIKS+: $_{-137}$ *SPS1* $_{+297}$) was constructed by amplifying genomic DNA using primers OLH778 and OLH1195. The resulting product, as well as pBSIIKS+, were cut with the restriction enzymes ClaI and SpeI followed by ligation.

pCS159 was constructed by PCR mutagenesis of pCS47 (pBSIIKS+; $_{-137}$ $SPS1_{+297}$) using primers OLH1466 and OLH1467.

## Strains used in this study

All strains in this study were derived from the SK1 background [50] (list of strains, Table S3). C-terminal tagging, gene disruptions, and gene deletions were accomplished by PCR mediated recombination [51,52]. All of the above genome changes were confirmed by diagnostic PCR.

The *dit1::TRP1* and *bmh1::TRP1* alleles were created using primers OLH608/OLH609 and OLH1309/OLH1328, respectively, in conjunction with pCgW.

The *sps1::LEU2* and *sps1::HIS3* alleles were created using primers OLH131/OLH132 in conjunction with pLEU2 and pHIS3, respectively. The *SPS1-zz-URA3* and *sps1ΔEPG-zz-URA3* alleles were created using primers OLH391/OLH392 and OLH487/OLH488, respectively, with pBS1365. The *SPS1-13xmyc-TRP1* allele was created using primers OLH389/OLH390 in conjunction with pFA6a-13Myc-TRP1. The *BMH1-GFP-TRP1* and *BMH2-GFP-TRP1* alleles were created using primers OLH1305/OLH1327 and OLH1276/OLH1277, respectively, along with pFA6a-GFP(S65T)-TRP1. The *bmh2::URA3*, *URA3-SPS1* and *SBP-sps1(S345::URA3)* alleles were created using primers OLH1311/OLH1312, OLH826/OLH827 and OLH1459/OLH1460, respectively, in conjunction with pURA3. The DNA product from OLH1459/OLH1460 and pURA3 was then transformed into a strain carrying the *SBP-SPS1* allele (described below).

The alleles: *SBP-SPS1*, *SBP-sps1-T12A*, *sfGFP-SPS1*, and *sfGFP-sps1-T12A* were created by replacement of the *URA3* in the *URA3-SPS1* allele (described above) by PCR amplified DNA specific to each allele (see below). Counter-selection on plates containing 5-fluoroorotic acid (5FOA; Zymo Research) was used to screen for successful replacement of *URA3* as previously described [50]. The primers OLH780/OLH513 were used with pCS99, pCS100, pCS146 and pCS145 to amplify DNA for the creation of the alleles: *SBP-SPS1*, *SBP-sps1-T12A*, *sfGFP-SPS1*, and *sfGFP-sps1-T12A*, respectively. The allele *SBP-sps1-S345A* was created by 5FOA counter-selection as described above using the *SBP-sps1(S345::URA3)* allele and DNA generated using primers OLH507/OLH796 and pCS159. All PCR mediated alleles were verified afterward by DNA sequencing.

## Fluorescence microscopy

Microscopy was done using a $100\times$ (NA 1.45) objective on a Zeiss Axioskop Mot2. Images were taken using an Orca-ER cooled charge-coupled device camera (Hamamatsu) using Openlab 4.04 (Perkin Elmer) software. All imaging was done using live cells.

## Yeast growth conditions

Yeast cells were induced to sporulate as described previously [50]. In brief, yeast cells were grown to saturation in YPD (2% Peptone, 1% Yeast Extract, 2% Dextrose), and transferred to pre-sporulation media YPA (2% Peptone, 1% Yeast Extract, 1% Potassium Acetate). Cells were grown in pre-sporulation media overnight, and then shifted to sporulation media (2% Potassium Acetate). Cells grown in log phase were grown in either YPD or the appropriate selective medium to approximately $OD_{600}$ 0.8.

## Scoring of sporulation phenotypes

Spore efficiency was measured using liquid sporulation cultures from a minimum of three biological replicates. Meiotic progression was determined by examining Htb2-mCherry [48]. Cultures were allowed to sporulate in liquid media for 24 hours. Yeast cells were scored for sporulation if at least one refractile spore was formed; a minimum of 200 cells was counted for each culture.

To determine the number of spores formed in each ascus, three biological replicates were tested for each strain. Cultures were sporulated in liquid sporulation media for 24 hours. Yeast cells were scored for number of refractile spores formed per ascus. Cells that did not form refractile spores were not counted.

The dityrosine fluorescence assay was performed as described [53]. In short, cells were grown on YPD plates for 24 hours then transferred to SPO plates with a nitrocellulose filter, incubated for another 24 hours and then exposed to UV light. The nitrocellulose membrane was then imaged using a digital camera.

Spore viability was assayed by the dissection of 25 tetrads per strain [54]. Dissected spores were allowed to grow on YPD plates for 48 hours.

Spore wall permeability assays were carried out as previously described [55]. In brief, yeast strains were transformed with pRS424-ssGFP, induced to sporulate, and visualized for GFP in the ascal cytoplasm or trapped within the extracellular space between the plasma membrane and the spore wall. For each strain 100 cells were counted from three biological replicates for a total of 300 total cells counted per strain. Only cells with refractile spores were counted.

## Protein immunoblotting

Protein lysates for immunoblotting were prepared using trichloroacetic acid (TCA) denaturation, as described [31]. After TCA precipitation, proteins were re-suspended in SDS-PAGE sample buffer [50] boiled for 5 minutes and separated by SDS-PAGE. Proteins were transferred onto polyvinylidene fluoride (GE Healthcare) and probed with the following antibodies: rabbit pre-immune anti-sera (gift from K. Benjamin) at 1:1000 to detect the zz epitope (which contains two copies of the z-domain of Protein A [52]); mouse monoclonal antibody 9E10 (Covance) at 1:1000 to detect the Myc epitope; mouse monoclonal JL-8 (BD Living Colors) at 1:1000 to detect GFP, mouse monoclonal SB19-C4 (Santa Cruz Biotechnology) at 1:1000 to detect SBP, mouse monoclonal 22C5D8 (abcam) at 1:1000 to detect Pgk1, rabbit polyclonal anti-Bmh [35] at 1:10000 to detect Bmh1 and Bmh2 and rabbit anti-Ndt80 [56] at 1:1000 to detect Ndt80.

Blots visualized on the Kodak Image Station 4000R with Kodak Molecular Imaging Software v4.0.4 were stained with the appropriate anti-rabbit or anti-mouse HRP conjugated secondary antibodies at 1:10000 (Jackson ImmunoResearch). Secondary antibodies detected using Supersignal West Dura Extended Duration Substrate (Pierce).

Blots visualized on the Odyssey CLx Infrared Imaging System (LI-COR Biosciences) were blocked using Odyssey blocking buffer (LI-COR) and Goat Anti-Mouse IR Dye 800 CW (LI-COR) at 1:10000 as a secondary antibody. Protein bands were quantified using Image Studio v3.1 (LI-COR).

## Immunoprecipitation, phosphatase assay and mass spectrometry

Lysates for immunoprecipitation were prepared from 60 $OD_{600}$ of cells, lysed in a Mini Bead Beater8 (Biospec) at 4°C with glass beads in IP buffer (329.2 mM NaCl, 0.0823 mM EDTA, 16.46 mM HEPES, 1.2345 mM $MgCl_2$, 0.823% Nonidet P-40)

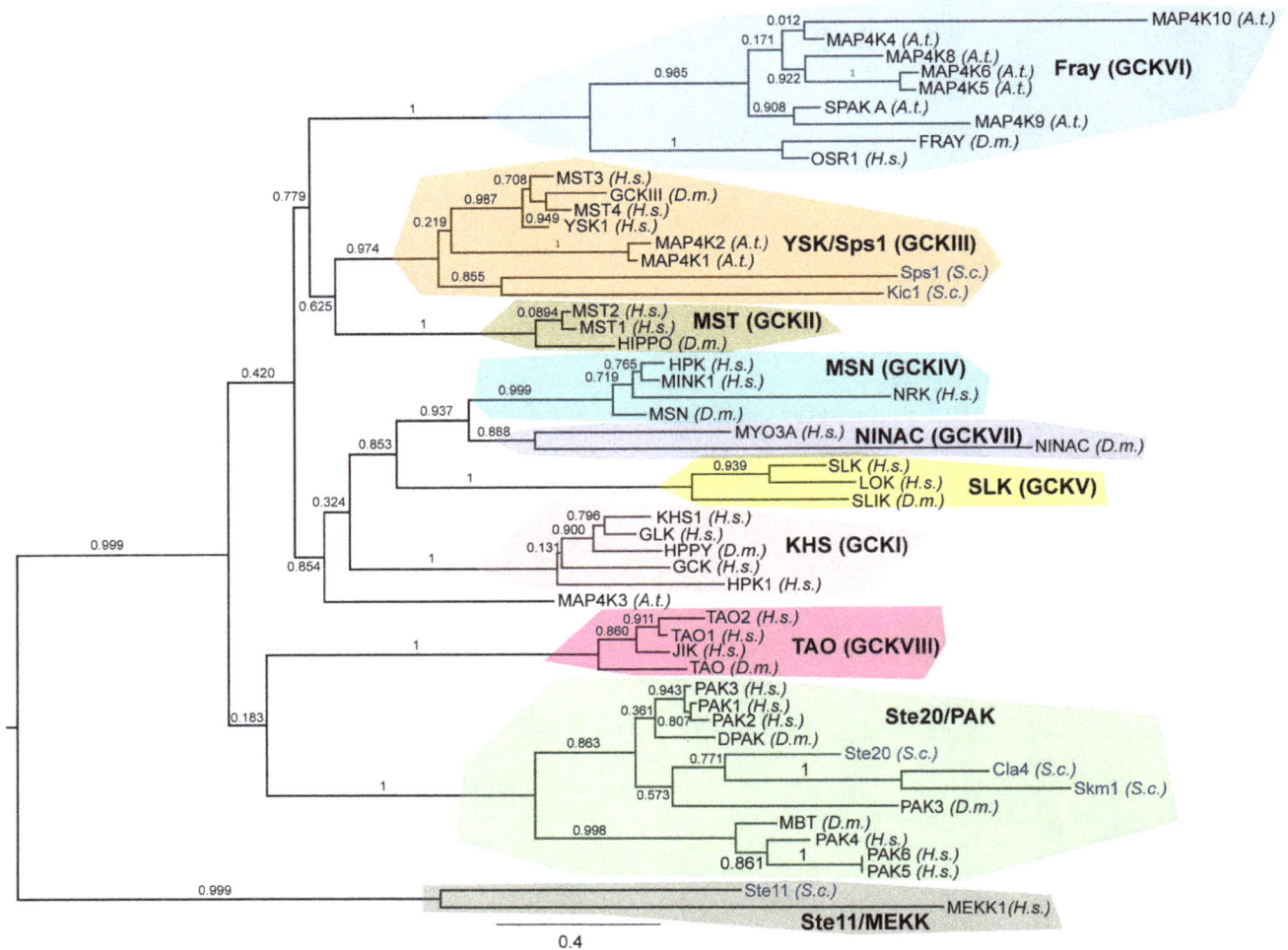

**Figure 1. Sps1 is a GCKIII family member.** Phylogenetic analysis using maximum likelihood-based methods. Branch support Ps are noted where relevant. (*A.t.*: *Arabidopsis thaliana*, *D.m.*: *Drosophila melanogaster*, *H.s.*: *Homo sapiens*, *S.c.*: *Saccharomyces cerevisiae*.) The GCK subfamilies, the PAK subfamily, and the MEKK outgroup are highlighted in different colors. The *S. cerevisiae* protein names are in blue; all other kinase names are in black.

and a protease and phosphatase inhibitor cocktail as described [56]. Sps1-zz was immunoprecipitated using IgG Sepharose (GE Healthcare) and SBP-Sps1 was immunoprecipitated using Streptavidin Plus Ultralink Resin (Pierce).

Sps1-zz immune complexes were washed 4 times in IP buffer and then re-suspended in SDS-PAGE sample buffer, boiled for 5 minutes and separated by SDS- PAGE. SBP-Sps1 immune complexes were washed 4 times in IP buffer and then SBP-Sps1 was eluted using biotin solution (2 mM biotin (Sigma), 0.28% ammonium hydroxide (Sigma)). Eluate was then TCA prepped as described above. The TCA precipitated protein was then resuspended in SDS-PAGE sample buffer, boiled for 5 minutes, and separated by SDS-PAGE.

Phosphatase assays were performed as described [56]. Briefly, Sps1-zz was immunoprecipitated, washed 4 times in IP buffer and then once in pre-phosphatase buffer (50 mM Tris-HCl at pH 7.5, 0.1 mM EDTA, 2 mM MnCl2, 0.1 mg/mL BSA) re-suspended in 50 μL of reaction buffer (50 mM Tris-HCl at pH 7.5, 0.1 mM EDTA, 5 mM DTT, 2 mM MnCl2, 0.01% Brij35, 50 U of λ protein phosphatase (New England Biolabs)). Control reactions omitted phosphatase or added phosphatase inhibitors (50 mM NaF, 50 mM β-glycerophosphate, 2 mM sodium orthovanadate).

Reactions were incubated for 30 minutes at 30°C. Beads were then resuspended in SDS-PAGE sample buffer, boiled for 5 minutes and separated by SDS-PAGE.

Samples for mass spectrometry phosphorylation analysis were immunoprecipitated and then resuspended in SDS-PAGE sample buffer, boiled for 5 minutes and separated by SDS-PAGE. Gel bands were excised and sent to the Taplin Mass Spectrometry Facility (Harvard Medical School). In brief, samples underwent a modified in-gel trypsin digestion [57], followed by peptide extraction procedures, separation on a reverse-phase HPLC capillary column [58], electrospray ionization and peptides then entered into an LTQ-Orbitrap mass spectrometer (Thermo Fisher, San Jose, CA). Peptide sequences were determined by matching protein or translated nucleotide databases with the acquired fragmentation pattern by the software program, Sequest [59]. To identify phosphorylated residues, the modification of 79.9663 mass units to serine, threonine, and tyrosine was included in the database searches to determine phosphopeptides and each phosphopeptide that was identified by the Sequest program was also inspected manually.

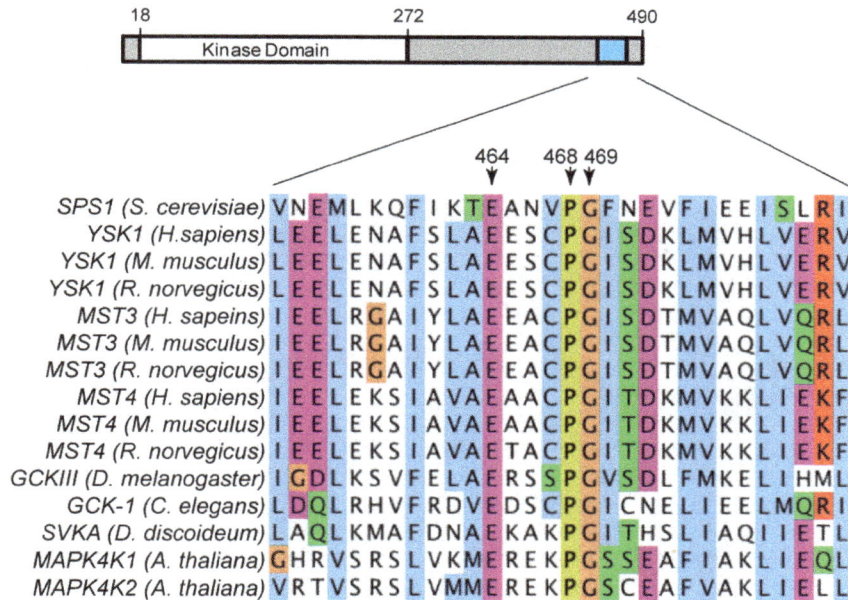

**Figure 2. The C-terminal region of Sps1 contains a conserved EPG motif.** Alignment of Sps1 residues 453–482 with the relevant regions from the human, mouse, rat, fly, nematode, slime mold, and mustard weed homologs. Amino acids with similar properties are colored using the ClustalX default color scheme [86]: magenta = acidic, red = basic, blue = hydrophobic, green = hydrophilic, orange = glycine, yellow = proline.

## Bioinformatics

ScanSite, http://scansite.mit.edu/ [60], was used to predict the 14-3-3 binding site in Sps1. ScanSite was run using the high stringency setting.

For phylogenetic analysis, kinase domains were identified based on PFAM kinase domain (PF00069) and aligned using MUSCLE 3.7 [61]. Multiple sequence alignments were then used for maximum likelihood phylogeny, performed using PhyML v3.0, [62] with LG substitution matrix [63] and Shimodaira-Hasegawa-like (SH-like) approximate likelihood ratio test (aLRT) branch supports [64]. Phylogenetic trees were displayed using FigTree (http://tree.bio.ed.ac.uk/software/).

## Results

### Sps1 is a GCKIII family member

Sps1 has been considered an outlier among Ste20 kinases, as initial analysis of the evolutionary relationships among Ste20 family kinases suggested it belonged to neither GCK nor PAK subfamilies [5]. However, we noticed that when compared to human Ste20 kinases, the kinase domain of Sps1 shares ~50% amino acid identity with the GCKs (e.g., 52% identity with the human GCKIII Ysk1), but only ~40% with the PAKs (e.g., 40% with the human Pak1). This raised the possibility that rather than being an outlier among Ste20 kinases, Sps1 might belong to the GCK subfamily, and prompted us to revisit the evolutionary relationship of the Sps1 kinase domain to other Ste20 family members.

Phylogenetic analysis of Sps1 was performed to examine relationships among the kinase domain sequences of all five *S. cerevisiae* Ste20 family members (Cla4, Kic1, Skm1, Sps1, and Ste20), as well as all human, *Drosophila melanogaster* and *Arabidopsis thaliana* Ste20 family members. Two MAP Kinase Kinases (MEKKs), *S. cerevisiae* Ste11 and human MEKK1, were used to root the tree, as MEKK and Ste20 kinases form related but distinct gene families (Figure 1). Maximum likelihood-based

phylogenetic analysis found, with strong branch support, that Sps1 resides in a monophyletic clade with human Mst3, Mst4, and Ysk1, as well as *Drosophila* GCK-III. As Mst3, Mst4, Ysk1, and GCK-III are canonical GCKIII family members [3–5], these data indicate that Sps1 is not an outlier among Ste20 kinases. Our analysis shows that Sps1 is a GCK, and specifically, a member of the GCKIII subfamily. Phylogenetic analysis performed using a different reconstruction approach (the less computationally intense Neighbor Joining method) also indicated Sps1 belongs to the GCKIII family. Consistent with this view, the Sps1 domain architecture also resembles GCK architecture, with the kinase domain located at the amino-terminus. Interestingly, two *Arabidopsis* MAP4Ks also fall within the GCKIII clade, suggesting that the GCKIII kinase lineage diverged from other Ste20 kinases prior to the separation of the plant and opisthokont (yeast/animal) lineages ~1 billion years ago. The other yeast Ste20 family member with a GCK-like domain architecture, Kic1, is also a member of the GCKIII clade.

As a member of the GCKIII family, Sps1 could share amino acid similarities outside the kinase domain. We performed sequence alignments with Sps1 and other GCKIIIs and found Sps1 contains a conserved region at its C-terminus (Figure 2). This region extends from amino acid 453 to 482 in Sps1, and it includes three amino acids conserved between Sps1 and the animal and plant GCKIIIs (Figure 2). In Sps1, these residues are glutamic acid (E) 464, proline (P) 468, and glycine (G) 469. We call this invariant ExxxPG region the EPG motif. Taken together, the phylogenetic evidence, domain architecture, and C-terminal sequence similarity all support the identification of Sps1 as a member of the GCKIII subfamily of Ste20 kinases.

### The C-terminal EPG motif containing region is required for Sps1 function

To test whether the C-terminal EPG motif containing region is required for Sps1 function, we created the *sps1ΔEPG-zz* allele by

**Table 1.** The EPG motif containing region is required for *SPS1* function.

| strain name | relevant genotype | % spores (mean ± S.D.) |
|---|---|---|
| LH902 | wild type | 80.7±2.1 |
| LH960 | SPS1-zz | 81.7±2.8 |
| LH966 | sps1::LEU2 | 4.0±1.8 |
| LH961 | sps1ΔEPG-zz | 6.8±2.9 |

200 cells/culture and a minimum of 3 cultures were tested for each strain.

removing the last 38 amino acids (and thus deleting the ExxxPG region). The deletion starts from the conserved valine (V) 453 to the C-terminus; this sequence is replaced with the zz (two tandem z domains from Protein A) epitope [52]. We compared a strain carrying the *sps1ΔEPG-zz* allele to strains carrying the alleles: *SPS1-zz*, *sps1::LEU2* as well as wild type for their ability to form refractile spores (summarized in Table 1). We found that wild type and *SPS1-zz* cells sporulate at 80.7% and 81.7% respectively, while *sps1ΔEPG-zz* reduces sporulation similar to the *sps1* null

allele (6.8% and 4.0%, respectively) (Student's *t* test comparison of LH960 and LH961 gives a *P*<0.01). This reduction in sporulation is not due to a reduction in protein level, as the *sps1ΔEPG-zz* mutation does not grossly disrupt steady state levels of Sps1 protein (Figure S1A).

## Sps1 is a phosphoprotein

We examined Sps1 during sporulation and saw that the level of Sps1 is induced during sporulation and peaks around 6 to 8 hours (Figure 3A), when cells are completing meiosis and starting spore morphogenesis. We see that Sps1 runs as a doublet, suggesting that Sps1 is post-translationally modified.

To test whether this post-translational modification was due to phosphorylation, we carried out a phosphatase assay by examining Sps1 purified from cells 8 hours into sporulation (Figure 3B). When λ phosphatase was added to the immunoprecipitated Sps1 we see that the more slowly migrating band is no longer detectable. When λ phosphatase and phosphatase inhibitors were added together, the higher mobility band can be seen more easily than in the phosphatase only treated sample (Figure 3B). These results indicate that Sps1 is phosphorylated (Figures 3A and 3B). Differences in the intensity of the more slowly migrating phosphorylated band in the phosphatase treatment experiment (Figure 3B) compared to the examination of cell lysates (Figure 3A) may be due to the use of different sample preparations (more native conditions for the phosphatase assay in Figure 3B compared to denaturing conditions for the cell lysates in Figure 3A) or the different epitope tags used in the two different experiments.

To identify potential phosphorylation sites on Sps1, we used mass spectrometric analysis on Sps1 purified from vegetatively growing log-phase cells that expressed Sps1 from a multi-copy plasmid under the *TEF2* promoter and from sporulating cells. Kinase dead *sps1* was also purified from log-phase cells and subjected to mass spectrometric analysis. We found that Sps1 was phosphorylated on threonine (T) 12 and serine (S) 345 in all three experiments. In addition to T12 and S345, four novel phosphorylation sites were found only when the kinase dead version of Sps1 was purified: S19, S22, S312, and S329 (Figure 3C). Thus, Sps1 is phosphorylated on multiple serine and threonine residues, and the phosphorylation of T12 and S345 are not simply due to autophosphorylation by Sps1. Whether S19, S22, S312, or S329 are biologically relevant is unclear, given that we only see these phosphorylation sites on the kinase dead version of Sps1 that is ectopically expressed in log phase cells.

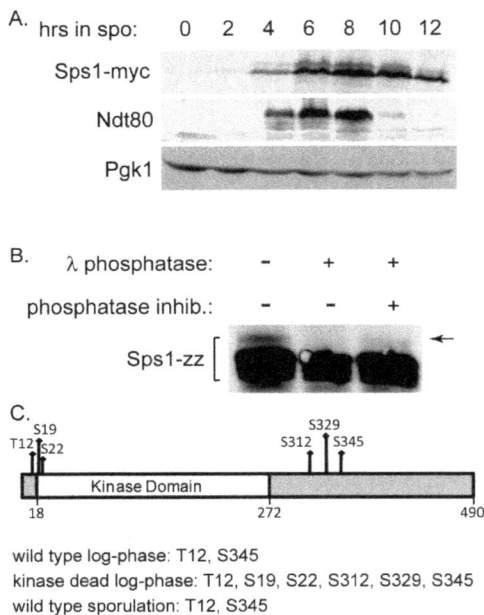

**Figure 3. Sps1 is a phosphoprotein.** (A) Lysates from LH875 (*SPS1-myc*) were collected throughout sporulation, run on an immunoblot and probed with anti-Myc antibody. Ndt80 was detected using an anti-Ndt80 antibody and used to monitor progression through sporulation [56]. Pgk1 was used as a loading control. (B) Sps1 was immunoprecipitated from lysates collected from LH791 (*SPS1-zz*) harvested 8 hrs into sporulation. The immunoprecipitate was divided and treated with (+) or without (−) λ phosphatase and with (+) or without (−) phosphatase inhibitors. Reactions were run on an immunoblot and then probed with rabbit pre-immune anti-sera, to detect the zz epitope. (C) Mass spectrometry phosphorylation analysis of Sps1 identified several phosphorylation sites. For log-phase cultures, LH988 (containing *SPS1* [pRS426-SBP-Sps1]) and LH989 (containing kinase dead *sps1* [pRS426-SBP-sps1-K47R]) were grown in selective media and collected during logarithmic phase growth (0.5–0.8 OD$_{600}$). For sporulating cultures, LH954 (*SBP-SPS1*) cells were collected 8 hours into sporulation. Sps1 was immunoprecipitated using the SBP epitope and analyzed by mass spectrometry to identify phosphorylation sites.

## sps1-T12A displays reduced sporulation efficiency

To test the importance of Sps1 T12 or S345 phosphorylation during sporulation, we replaced the *SPS1* locus with either *SBP-sps1-T12A* or *SBP-sps1-S345A*. We did not see any sporulation defect with the *SBP-sps1-S345A* mutant and did not study the

**Table 2.** *sps1-T12A* has a sporulation defect.

| strain name | relevant genotype | % spores (mean ± S.D.) |
|---|---|---|
| LH902 | wild type | 80.7±2.1 |
| LH962 | SBP-SPS1 | 85.8±5.0 |
| LH967 | SBP-sps1-S345A | 83.7±2.7 |
| LH968 | SBP-sps1-T12A | 65.7±3.3 |
| LH969 | SBP-sps1-T12A/+ | 82.8±3.8 |

200 cells/culture and a minimum of 3 cultures were tested for each strain.

effects of this site further. In contrast, the *SBP-sps1-T12A* did affect spore formation. We found that the *sps1* null forms spores 4.0% of the time, while wild type and the *SBP-SPS1* strain sporulate at 80.7% and 85.8% of the time respectively. *SBP-sps1-T12A* forms spores 65.7% of the time, a statistically significant decrease in sporulation efficiency compared to wild type ($P<0.01$; $t$ test), although not as pronounced as in an *sps1* null (Table 2).

To examine whether the T12A mutation disrupted Sps1 protein levels or altered its temporal expression profile, we compared wild type SBP-Sps1 and SBP-Sps1-T12A expression during sporulation and found comparable protein levels and similar temporal expression patterns throughout sporulation (Figure S1B).

### *sps1-T12A* is a reduction-of-function mutation defective in spore packaging

Because the *SBP-sps1-T12A* allele did not have a phenotype as severe as the null allele, we wanted to determine the manner in which is disrupts *SPS1* function. First, we tested whether *SPB-sps1-T12A* acts as a dominant negative mutant by creating a heterozygous strain. If *sps1-T12A* were acting as a dominant negative mutation, we would expect a reduction in spore efficiency

in heterozygous cells compared to wild type cells. However, we found that there was no reduction in spore formation between wild type (80.7%) and *SBP-sps1-T12A/+* (82.8%; $P=0.4$, Student's $t$ test).

To test whether *SBP-sps1-T12A* acts as a reduction-of-function allele, we varied the dosage of *sps1*. Consistent with *sps1-T12A* acting as a partial loss-of-function allele, the spore efficiency of *SBP-sps1-T12A/sps1Δ* was reduced compared to *SBP-sps1-T12A/SBP-sps1-T12A* (53.0% compared to 65.67%; $P<0.05$; $t$ test).

To determine whether the spores that are produced in *sps1-T12A* allele show defects, we examined the number of spores formed per ascus when at least one refractile spore was formed in an ascus (Figure S2A), the ability of the outer spore wall layer (the dityrosine layer) to fluoresce [53] (Figure S2B), as well as spore wall permeability using a signal sequence fused to GFP (pRS424-ssGFP; [55]) (Figure S2C). We saw no differences in the spores formed by cells homozygous for the *sps1-T12A* allele compared to wild type cells. We also examined the germination efficiency in *SBP-sps1-T12A* cells by dissecting 25 tetrads each of *SBP-sps1-T12A* and *SBP-SPS1* (only cells that made four refractile tetrads in an ascus were dissected) and did not see any differences in the

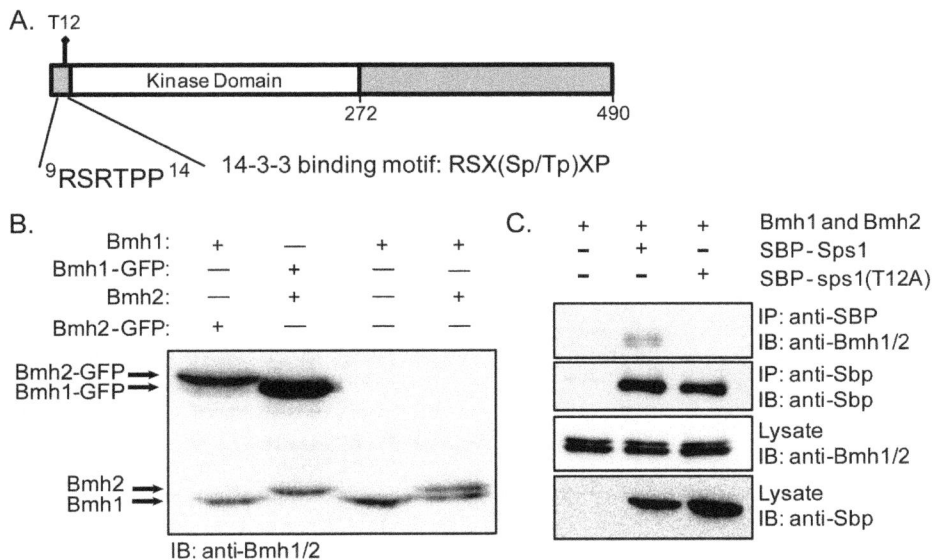

**Figure 4. Sps1 can physically interact with Bmh1 and Bmh2.** (A) The 14-3-3 consensus sequence within Sps1. (B) Immunoblot probed with anti-Bmh antibody; strains from left to right: LH958 (*BMH2-GFP*), LH957 (*BMH1-GFP*), LH959 (*bmh2Δ*) and LH177 (*WT*) (C) Co-immunoprecipitation between Sps1 and the 14-3-3 homologs, Bmh1 and Bmh2. LH994 (pRS426-*SBP*), LH993 (pRS426-*SBP-SPS1*) and LH992 (pRS426-*SBP-sps1-T12A*), were grown in selective media and harvested during log-phase growth. SBP immunoprecipitation (IP) was performed, and samples were analyzed by probing immunoblots (IB) with antibodies as indicated.

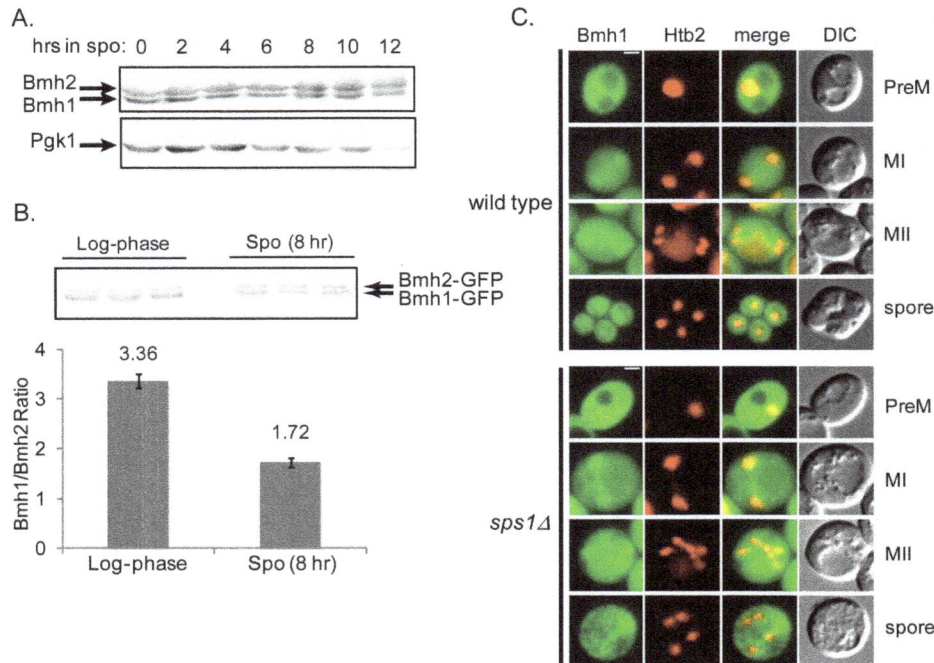

**Figure 5. Bmh1 and Bmh2 protein expression and localization during sporulation.** (A) Immunoblot probed with anti-Bmh antibody, showing Bmh1 and Bmh2 expression during sporulation using LH177 (*wild type*). Pgk1 was used as a loading control. (B) Quantification of Bmh isoform ratio during log-phase growth and sporulation. LH971 (*BMH1-GFP BMH2-GFP*) was sampled during log-phase growth and at 8 hours into sporulation. Immunoblot was probed with anti-GFP antibody and band intensities were measured. (C) Localization of Bmh1-GFP in a wild type background (LH972; top) and in an *sps1Δ* background (LH974; bottom) during sporulation. Htb2-mCherry is used as a nuclear marker. PreM: Pre-meiosis, MI: Meiosis I, MII: Meiosis II, Spore: mature spore. Scale bar = 2 μ.

ability of haploid spores to germinate. Of 100 spores dissected, 98 germinated in the *SPB-SPS1* containing strain, while 96 germinated in the *SBP-sps1-T12A* containing strain. Taken together, these results suggest that while *sps1-T12A* reduces the efficiency of spore packaging, the spores that are packaged and form refractile structures show no obvious defects.

## Sps1 has a consensus 14-3-3 binding site and can physically interact with Bmh1 and Bmh2

When we examined the region surrounding T12 within the Sps1 protein, we found that this threonine resides within a potential 14-3-3 phosphopeptide binding consensus sequence. The sequence of Sps1 starting from arginine (R) 9 to proline (P) 14 closely matches the 14-3-3 consensus sequence: R-S-X-(pS/pT)-X-P [31] (Figure 4A).

To examine Bmh1 expression in the SK1 strain, we used an anti-Bmh antibody [35]. To test the specificity of the anti-Bmh antibody in the SK1 strain, we examined its ability to recognize Bmh1 and Bmh2 using wild type, *bmh2Δ*, *BMH1-GFP*, and *BMH2-GFP* containing strains, which allowed us to distinguish between the two 14-3-3 isoforms, as appending GFP at the C-terminus of the protein shifts the migration of an isoform (Figure 4B). We found that this antibody can detect both isoforms of 14-3-3 proteins in our strain background, with the more slowly migrating isoform corresponding to Bmh2 while the faster migrating isoform is Bmh1, as expected by their predicted protein sizes (31 kDa for Bmh2 and 30 kDa for Bmh1).

We next asked if Sps1 could bind to Bmh1 and Bmh2, and found that we can co-immunoprecipitate Bmh1 and Bmh2 with SBP-Sps1 (Figure 4C). However, we do not detect an interaction

of Bmh1 and Bmh2 with SBP-Sps1-T12A (Figure 4C). These results indicate that Sps1 interacts with Bmh1 and Bmh2 and that T12 within Sps1 is required for the interaction.

## Bmh1 and Bmh2 are expressed during sporulation and are present in the nucleus and the cytoplasm

We examined Bmh1 and Bmh2 expression during sporulation in wild type cells, and found that both are expressed throughout sporulation (Figure 5A). As Bmh1 was previously reported to be the major isoform in vegetative growth [35,65], we compared the levels of expression of Bmh1 and Bmh2 during log-phase growth and sporulation. We used the strain *BMH1-GFP BMH2-GFP* and examined the protein levels using an anti-GFP antibody to avoid any bias in isoform detection by the anti-Bmh antibody. We found that the ratio of Bmh1 to Bmh2 in log-phase growth was 3.4±0.2 (mean ± S.D.), decreasing to 1.7±0.2 (mean ± S.D.) in sporulating cells. This suggests that Bmh1 is more abundant than Bmh2 during log-phase growth, but that Bmh2 levels increase relative to Bmh1 in sporulating cells (Figure 5B).

We also examined Bmh1 and Bmh2 localization during sporulation. High-throughput studies have examined Bmh1 and Bmh2 in haploid log-phase cells and found both cytoplasmic and nuclear localization [66]. We examined Bmh1-GFP localization in diploids during both log-phase growth and sporulating cells and saw both nuclear and cytoplasmic localization (Figure 5C, *top*). We see similar results for Bmh2-GFP (Figure S3, *left*). We next asked if 14-3-3 localization is affected in an *sps1Δ* background. We see both nuclear and cytoplasmic localization in the absence of *SPS1* (Figure 5C, *bottom* and Figure S3, *middle*), suggesting that *SPS1* is not required for proper 14-3-3 localization. Finally, we

**Table 3.** *BMH1* and *BMH2* are required for sporulation.

| strain name | relevant genotype | % spores (mean ± S.D.) |
|---|---|---|
| LH902 | wild type | 80.7±2.1 |
| LH980 | *bmh1Δ/bmh1Δ* | 71.0±3.8 |
| LH981 | *bmh2Δ/bmh2Δ* | 70.8±4.5 |
| LH979 | *bmh1Δ/+ bmh2Δ/+* | 79.7±4.8 |
| LH982 | *bmh1Δ/bmh1Δ bmh2Δ/+* | 55.8±8.3 |
| LH983 | *bmh1Δ/+ bmh2Δ/bmh2Δ* | 50.0±2.5 |

200 cells/culture and a minimum of 3 cultures were tested for each strain.

asked if 14-3-3 localization was dependent on the presence of both isoforms. To answer this question, we examined *BMH1-GFP* in a *bmh2Δ* background and we saw no obvious difference in localization in sporulating cells (Figures S3 and S4).

## Yeast deficient in 14-3-3 isoforms exhibit reduced sporulation efficiency

Because we see a defect in sporulation efficiency in cells carrying the *sps1-T12A* allele, we asked if 14-3-3s play a role in sporulating cells. We examined strains that lack either *BMH1* or *BMH2* (LH980: *bmh1Δ/bmh1Δ* and LH981: *bmh2Δ/bmh2Δ*) and saw that both had slight reductions in sporulation efficiency (Table 3). Cells lacking *BMH1* form spores at 71.0%, while cells lacking *BMH2* form spores 70.8% of the time, compared to a wild type value of 80.6% (P<0.02 and P<0.03 respectively compared to wild type, Student's *t* test). To examine whether the slight sporulation efficiency defect in each *bmh* mutant was due to the missing 14-3-3 isoform, or the reduction in the total number of 14-3-3 genes, we examined sporulation in the doubly heterozygous LH979 strain (*bmh1Δ/+ bmh2Δ/+*), which sporulated at 79.6%, similar to wild-type (Table 3).

We also examined the effects of further reductions in 14-3-3 gene dosage. Although a strain lacking both *bmh1Δ* and *bmh2Δ* was viable in the SK1 strain background, this strain displayed severe growth defects and was unable to enter meiosis (likely because of an inability to successfully grow in YPA pre-sporulation medium, as 14-3-3s have been shown to be required for growth in media containing acetate [33]), precluding our ability to assay the sporulation efficiency of the *bmh1Δ bmh2Δ* double mutant strain. However, we were able to examine sporulation in LH982 (*bmh1Δ/bmh1Δ bmh2Δ/+*) and LH983 (*bmh1Δ/+ bmh2Δ/bmh2Δ*), and found that the spore efficiency defect became more severe as the gene dosage of 14-3-3 was further reduced. Specifically, we see that *bmh1Δ/bmh1Δ bmh2Δ/+* and *bmh1Δ/+ bmh2Δ/bmh2Δ*

produced 55.8% and 50.0% spores respectively (both P<0.01 compared to wild type, Student's *t* test). Taken together, these results suggest a role for 14-3-3 proteins in sporulation efficiency (Table 3).

## SPS1 genetically interacts with the 14-3-3 isoforms, *BMH1* and *BMH2*

Given the physical interaction between Sps1 and the 14-3-3 proteins, and as both display sporulation efficiency defects, we tested whether *SPS1* genetically interacts with *BMH1* and *BMH2*. Consistent with a genetic interaction between the *14-3-3*'s and *SPS1*, LH902 wild type (80.6%), LH979 *bmh1Δ/+ bmh2Δ/+* (79.6%), and LH965 *sps1Δ/+* (81.0%) all show similar levels of sporulation, while LH984 (*bmh1Δ/+ bmh2Δ/+ sps1Δ/+*) showed a reduction in spore formation, forming spores 67.3% of the time (P<0.03 compared to wild type, Student's *t* test) (Tables 3 and 4).

If mutation of threonine 12 to alanine was sufficient to completely abrogate the interaction of the 14-3-3 proteins and Sps1, then LH985 (*bmh1Δ/+ bmh2Δ/+ SBP-sps1-T12A/sps1Δ*) should have a spore efficiency defect equivalent to LH970 (*SBP-sps1-T12A/sps1Δ*), since the lack of the T12 should eliminate the Sps1 14-3-3 interaction. Instead, we found that LH985 (*bmh1Δ/+ bmh2Δ/+ SBP-sps1-T12A/sps1Δ*) sporulates at an efficiency of 22.5%, which is less efficient than LH970 (*SBP-sps1-T12A/sps1Δ*), which forms spores at an efficiency of 53.0% (P<0.01 compared to *SBP-sps1-T12A/sps1Δ*, Student's *t* test). This result suggests that T12 of Sps1 is not the sole mediator of its interaction with Bmh1 and Bmh2 (Table 4).

## Localization of Sps1 and Sps1-T12A during sporulation

Having shown that Sps1, Bmh1 and Bmh2 are all expressed at the same time during sporulation and that they are capable of physically interacting, we investigated if Sps1 localization resembled that of the 14-3-3s. While a previous report suggested an N-

**Table 4.** *SPS1* genetically interacts with *BMH1* and *BMH2*.

| strain name | relevant genotype | % spores (mean ± S.D.) |
|---|---|---|
| LH902 | wild type | 80.7±2.1 |
| LH984 | *bmh1Δ/+ bmh2Δ/+ sps1Δ/+* | 67.3±7.8 |
| LH965 | *sps1Δ/+* | 81.0±5.4 |
| LH985 | *bmh1Δ/+ bmh2Δ/+ SBP-sps1-T12A/sps1Δ* | 22.5±3.9 |
| LH970 | *SBP-sps1-T12A/sps1Δ* | 53.0±6.5 |

200 cells/culture and a minimum of 3 cultures were tested for each strain.

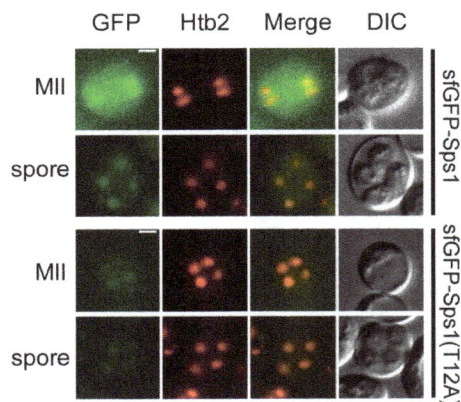

**Figure 6. Localization of Sps1 and Sps1-T12A during sporulation.** Localization of sfGFP-Sps1 (LH986; top) and sfGFP-Sps1-T12A (LH987; bottom) during sporulation. Htb2-mCherry is used as a nuclear marker. MII: Meiosis II, Spore: mature spore. Scale bar = 2 μ.

terminal GFP-Sps1 fusion contained within strain Y5050 [11] localizes to the prospore membrane, we were unable to detect such localization. Amplification and sequencing of the *GFP-SPS1* locus from Y5050 revealed that the start codon (ATG) was absent from the construct, indicating that it does not produce the intended protein product (Figure S5).

To determine Sps1 localization, we constructed an N-terminal *sfGFP-SPS1* fusion by integrating a monomeric variant of the fast-folding *superfolderGFP* (*sfGFP*) [67–69] at the *SPS1* locus. We see sfGFP-Sps1 in both the cytoplasm and the nucleus (Fig. 6 *top*). As spores matured and became refractile, nuclear localization became more distinct and cytoplasmic localization was reduced, compared to cells earlier in the sporulation process (Fig. 6 *top*). As 14-3-3 proteins are known to affect the localization of their target proteins, we wanted to determine if mutation of threonine 12 to alanine affected Sps1 localization. Using *sfGFP-sps1-T12A*, we saw a similar pattern of localization as we did in *sfGFP-SPS1* (Figure 6 *bottom*), suggesting that 14-3-3 proteins do not affect Sps1 localization.

## Sps1 has a nuclear localization sequence

As we observed sfGFP-Sps1in the nucleus, we examined the amino acid sequence of Sps1 and found two putative nuclear localization sequences (NLS) based on their similarity to the classical SV40 monopartite NLS [70] (Figure 7A). We compared the localization of GFP-Sps1 with GFP-tagged alleles in which each putative NLS was mutated (pRS426-*GFP-SPS1*, pRS426-*GFP-sps1-arappa* and pRS426-*GFP-sps1-ggaga*). We visualized these different fusion proteins in log-phase cells expressing Sps1 and mutant variants ectopically, using the *TEF2* promoter on a multicopy vector (Fig. 7B). We found that changing residues 231–236 KRKPPK to ARAPPA had no effect on nuclear localization (Fig. 7B, *middle row*). However, mutation of residues 411–415 KKHKK to GGAGA prevented distinct nuclear localization (Fig. 7B, *bottom row*). This experiment demonstrated the necessity of residues 411–415 (KKHKK) for proper nuclear localization.

We then asked whether this sequence was sufficient to drive the nuclear localization of a protein that is normally not present in the nucleus. A GFP-GST fusion was created, in which the addition of GST prevents the passive diffusion into the nucleus of GFP [71]. We observed an increase in nuclear localization of the fusion protein containing an intact NLS in comparison to those with just

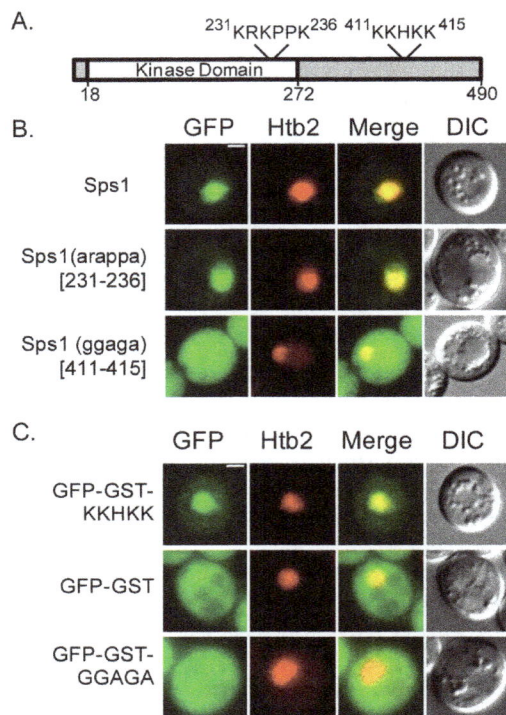

**Figure 7. Sps1 has a nuclear localization sequence.** (A) Diagram of Sps1 indicating potential nuclear localization signals. The first KRKPPK (231–236) lies within the kinase domain and the second, KKHKK (411–415) is located toward the C-terminus. (B) The amino acid sequence, KKHKK (411–415) is necessary for proper nuclear localization of Sps1. LH995 (*pGFP-SPS1*), top, LH997 (*pGFP-sps1-arappa (213–236)*), middle, and LH996 (*pGFP-sps1-ggaga (411–415)*), bottom, were visualized during log-phase growth; Htb2-mCherry is used as a nuclear marker. (C) The NLS of Sps1 (KKHKK (411–415)) is sufficient to localize a GFP-GST construct to the nucleus. LH999 (*pGFP-GST-SPS1(387–438)*), top, LH998 (*pGFP-GST*), middle, and LH1000 (*pGFP-GST-sps1-ggaga-(387–438)*), bottom, were visualized during log-phase growth; Htb2-mCherry is used as a nuclear marker. Scale bar = 2 μ.

the GFP-GST fusion or a GFP-GST fusion with mutated NLS (Fig. 7C). Therefore, Sps1 has a functioning NLS that is both necessary and sufficient for nuclear localization.

## Discussion

Sps1 is a serine/threonine kinase that is expressed during sporulation and functions in the process of spore formation. Interestingly, genome-wide studies have identified targets of yeast 14-3-3 proteins [16,72], although none of these studies have identified Sps1 as a target for either Bmh1 or Bmh2.

In this study, we use phylogeny to show that Sps1 is a bona-fide GCKIII kinase, and identify three regions important for its function: a conserved C-terminal motif containing the invariant ExxxPG at residues 464–469, a phosphorylation site T12, and a nuclear localization signal from 411 to 415 (KKHKK). We also show that T12 is part of a 14-3-3 binding site, and is required for the physical interaction between Sps1 and the 14-3-3 isoforms Bmh1 and Bmh2. We demonstrate that 14-3-3s are important for sporulation, as cells lacking in *BMH1* and *BMH2* have defects in sporulation efficiency and show a genetic interaction with *SPS1*. Because we see both a genetic and physical interaction, we propose that *SPS1* and the 14-3-3s act together to regulate sporulation.

## Sps1 is a GCKIII family member

While an initial analysis suggested Sps1 was an outlier with respect to the Ste20 family [5], our bioinformatic analyses indicate that Sps1 belongs to the GCKIII family of Ste20 kinases. The identification of Sps1 as a GCKIII emerged using two different phylogenetic reconstruction strategies, supporting the robustness of the determination. The phylogenetic results also indicate that Sps1 is present in a distinct clade of animal and plant GCKIII kinases, indicating that the GCKIII kinases are an ancient and distinct subfamily of Ste20 kinases, likely present in the common eukaryotic ancestor ~1 billion years ago.

Our comparison of Sps1 with other GCKIII kinases led to the identification of a previously unappreciated conserved region found in all GCKIII kinases. This region is found at the C-terminus of the regulatory domain, and we call this region the EPG motif. Our experimental results demonstrate the importance of this region and suggest that the EPG motif may play an important role in other GCKIII kinases.

We also identify a nuclear localization sequence on Sps1, which is both necessary and sufficient for directing nuclear localization (residues 411–415). Other GCKIII proteins have also been reported to localize to both the nucleus and cytoplasm, and the nuclear localization domain of Mst3 has been mapped to residues 278–294 [73]; the location of this nuclear localization signal is conserved in the mammalian GCKIIIs and in *Drosophila*, but has changed in Sps1.

## Role of *BMH1* and *BMH2* in sporulation

Here we show that *BMH1* and *BMH2* are important for the efficient formation of spores in the SK1 background. Previous studies have not identified a role in sporulation for *BMH1*, including a genome-wide study using the yeast deletion collection in the S288c background [74]. It is possible that loss of *BMH1* and *BMH2* does not have a sporulation defect in the S288c background, or it could be that the less efficiently sporulating S288c strain background made it more difficult to detect the mild sporulation defect we see in *bmh1* and *bmh2* mutants.

In most yeast strain backgrounds examined, the *bmh1 bmh2* double mutant is inviable. In the SK1 strain, the *bmh1 bmh2* mutant is viable, but grows very slowly and produces cells of abnormal morphology, precluding the ability to examine the double mutant during sporulation. Because of this, we examined sporulation in a *bmh1/+ bmh2/bmh2* strain and see that this strain sporulates at 50% (compared to wild type levels of about 81%), a more severe phenotype than that seen with either single mutant. Our results suggest that if we were able to examine the sporulation defect in a *bmh1 bmh2* double mutant, we may see a more severe sporulation efficiency of less than 50%.

Why are there two 14-3-3 isoforms in yeast? In higher eukaryotes, there are several isoforms of 14-3-3 proteins, and these different isoforms are hypothesized to play different roles [28,75]. The yeast Bmh1 and Bmh2 are considered paralogs, and likely arose during a whole genome duplication event [76]. The yeast 14-3-3 isoforms appear to be largely redundant in terms of function, although specific phenotypes have been associated with the loss of only a single isoform. For instance, the loss of *BMH1* causes an increase in glycogen accumulation [77] whereas loss of *BMH2* results in abnormal accumulation of polyphosphate [78].

We see that 14-3-3 isoform levels shift during sporulation. In vegetatively growing cells, we see that Bmh1 is more prevalent compared to Bmh2, as previously described [65]. However, during sporulation, Bmh2 levels rise in comparison to Bmh1 levels. We do not know whether this relative increase in Bmh2 level is important for 14-3-3 function or is merely a consequence of transcriptional regulation changes that occur during sporulation.

## 14-3-3 regulation of Sps1 during sporulation

Our data suggests that Bmh1 and Bmh2 are important for the positive regulation of Sps1 function during sporulation, since the phenotypes of the strains containing *sps1-T12A* are similar to those lacking *bmh1*, and *bmh2*. As Bmh1 and Bmh2 physically interact with Sps1 in a manner that depends on the T12 within Sps1, these data raise the possibility that this interaction is functionally significant.

We propose that the interaction with 14-3-3s modulates Sps1 function but is not absolutely required for Sps1 activity, because an *sps1* null allele has a much more severe phenotype than either the *sps1-T12A* or the *bmh1* or *bmh2* mutants. In comparison to cells carrying *sps1Δ* allele, which rarely produce refractile spores, cells with the *sps1-T12A* allele display a less severe phenotype, sometimes producing spores that appear normal with respect to their ability to form the outer layers of the spore wall, to package the appropriate number of spores within an ascus, and to germinate.

Although we cannot directly assay the phenotype of the *bmh1 bmh2* double mutant, we anticipate that it would be more severe than the *sps1-T12A* phenotype because cells carrying *sps1-T12A* allele sporulate more efficiently than the *bmh1/+ bmh2/bmh2* mutant cells (Tables 2 and 3). The more severe defect seen in the *bmh1/+ bmh2/bmh2* mutant along with more severe defect seen in *bmh1Δ/+ bmh2Δ/+ SBP-sps1-T12A/sps1Δ* compared to *SBP-sps1-T12A/sps1Δ* (Tables 3 and 4) suggests that Sps1 may not be the only relevant binding partner of 14-3-3 proteins during sporulation. Alternatively, it is possible that T12 is not the only residue on Sps1 that can mediate the Sps1-14-3-3 interaction. It is possible that the weak phenotype seen with *sps1-T12A* compared to the *bmh1/+ bmh2/bmh2* mutant is due to the presence on Sps1 of other 14-3-3 interaction sites.

## 14-3-3 regulation of other GCKIII family kinases

The modulation of GCKIII kinases by 14-3-3s may be evolutionarily conserved. First, a high-throughput screen identified a physical interaction between the mammalian GCKIII Mst4 and 14-3-3ε [79]. Second, both MST4 and YSK1/SOK1 have predicted 14-3-3 binding motifs C-terminal to their kinase regions (Scansite). Third, the brain specific isoform of MST3 has been shown to be phosphorylated on threonine 18 [80], which is within a 14-3-3 consensus binding site, though binding with 14-3-3s has yet to be demonstrated.

Other interactions with 14-3-3 proteins and GCKIII kinases have been identified. 14-3-3ζ is a substrate of YSK1/SOK1, and its phosphorylation has been shown to be important for Golgi positioning [81] and to disrupt the binding of the proapoptotic factor ASK to 14-3-3ζ during apoptosis [82].

Interestingly, there may be a role for 14-3-3s in modulating Ste20 family kinases in general. In the Σ1278 background, 14-3-3s have been found to bind directly to the C-terminal kinase-containing portion (from amino acid 494 to 939) of the Ste20 kinase, a PAK family member [37]. Since the domain architecture of PAKs and GCKs are reversed, and since Bmh1 and Bmh2 interacts with the C-terminal half of Ste20 and the N-terminal half of Sps1 (which are both STE20 family kinases), it is tempting to speculate that Bmh1 and Bmh2 may modulate pseudohyphal-development directly through the modulation of Ste20 activity, especially because our analysis of amino acids 494–939 of Ste20 using Scansite predicts a 14-3-3 binding site of medium stringency surrounding threonine 546, which has been found to be phosphorylated by a number of different studies [83–85].

## Supporting Information

**Figure S1   Analysis of Sps1 protein.** (A) Lysates from LH791 (*SPS1*-zz) and LH953 (*sps1ΔEPG*-zz) were collected throughout sporulation. Immunoblots were probed with rabbit antisera. (B) Lysates from LH954 (*SBP-SPS1*) and LH955 (*SBP-sps1-T12A*) were collected throughout sporulation. Immunoblots were probed with anti-SBP antibody. Pgk1 was used as a loading control in both (A) and (B).

**Figure S2   Analysis of *sps1-T12A* spores.** (A) Quantification of the number of spores formed per ascus for (left to right): LH902 (*WT*), LH962 (*SBP-SPS1*) and LH968 (*SBP-sps1-T12A*) and LH970 (*SBP-sps1-T12A/sps1Δ*). (B) Dityrosine assay examining the outer spore wall layer. Visible light image of nitrocellulose membrane with yeast cell patches, *left*; UV light image of the same membrane, *right*. Strains shown are LH956 (*dit1Δ*), LH177 (*WT*), LH872 (*sps1Δ*), LH955 (*SBP-sps1-T12A*) and LH954 (*SBP-SPS1*). (C) Spore wall permeability assay. Impermeable spore with ssGFP correctly localized to the spore wall, *left*, permeable spore that has incorrectly allowed ssGFP to disperse between the spore wall and ascal membrane, *right*. LH954 (*SBP-SPS1*) and LH955 (*SBP-sps1-T12A*) were transformed with pRS424-ssGFP. The permeability of the spore was scored for each strain.

**Figure S3   Bmh2 localization during sporulation does not depend on *SPS1* or *BMH1*.** Localization during sporulation of Bmh2, as seen in LH973 (*BMH2-GFP*), *left*, LH975 (*BMH2-GFP sps1Δ*), *middle*, and LH978 (*BMH2-GFP bmh1Δ*), *right*. Htb2-mCherry is used as a nuclear marker. PreM: Premeiosis, MI: Meiosis I, MII: Meiosis II, Spore: mature spore. Scale bar = 2 μ.

**Figure S4   Bmh1 localization during sporulation does not depend on *BMH2*.** Localization during sporulation of Bmh1, as seen in LH977 (*BMH1-GFP bmh2Δ*). Htb2-mCherry is used as a nuclear marker. PreM: Pre-meiosis, MI: Meiosis I, MII: Meiosis II, Spore: mature spore. Scale bar = 2 μ.

**Figure S5   Sequencing of Y5050 reveals no start codon.** Sequencing results of the Sps1 locus from Y5050 [11] reveals the lack of a start codon before GFP.

**Table S1   Plasmids used in this study.**

**Table S2   Primers used in this study.**

**Table S3   *S. cerevisiae* strains used in this study.**

## Acknowledgments

We would like to thank Sandra Lemmon for the Bmh antibody, Aaron Neiman for pRS424-ssGFP, Kirsten Benjamin for the rabbit antisera and the Ndt80 antibody, Dirk Landgraf for mSuperfolderGFP, Alexey Veraksa for the SBP construct and for technical assistance, Paul Garrity for his assistance with the phylogenetic analysis, and Paul Garrity, Katherine Gibson, Mitch McVey, Joe Roesner, Ross Tomaino, and Alexey Veraksa for helpful comments on the manuscript.

## Author Contributions

Conceived and designed the experiments: CJS SMP LSH. Performed the experiments: CJS SMP LSH. Analyzed the data: CJS SMP LSH. Contributed reagents/materials/analysis tools: CJS SMP LSH. Wrote the paper: CJS LSH.

## References

1.  Neiman AM (2011) Sporulation in the budding yeast *Saccharomyces cerevisiae*. Genetics 189: 737–765.
2.  Friesen H, Lunz R, Doyle S, Segall J (1994) Mutation of the *SPS1*-encoded protein kinase of *Saccharomyces cerevisiae* leads to defects in transcription and morphology during spore formation. Genes Dev 8: 2162–2175.
3.  Delpire E (2009) The mammalian family of sterile 20p-like protein kinases. Pflugers Arch 458: 953–967.
4.  Sugden PH, McGuffin LJ, Clerk A (2013) SOcK, MiSTs, MASK and STicKs: the GCKIII (germinal centre kinase III) kinases and their heterologous protein-protein interactions. Biochem J 454: 13–30.
5.  Dan I, Watanabe NM, Kusumi A (2001) The Ste20 group kinases as regulators of MAP kinase cascades. Trends Cell Biol 11: 220–230.
6.  Wu H-Y, Lin C-Y, Chen T-C, Pan S-T, Yuan C-J (2011) Mammalian Ste20-like protein kinase 3 plays a role in hypoxia-induced apoptosis of trophoblast cell line 3A-sub-E. Int J Biochem Cell Biol 43: 742–750.
7.  Irwin N, Li Y-M, O'Toole JE, Benowitz LI (2006) Mst3b, a purine-sensitive Ste20-like protein kinase, regulates axon outgrowth. Proc Natl Acad Sci U S A 103: 18320–18325.
8.  Matsuki T, Zaka M, Guerreiro R, van der Brug MP, Cooper JA, et al. (2012) Identification of Stk25 as a genetic modifier of Tau phosphorylation in Dab1-mutant mice. PLoS One 7: e31152.
9.  Nerstedt A, Cansby E, Andersson CX, Laakso M, Stančáková A, et al. (2012) Serine/threonine protein kinase 25 (STK25): a novel negative regulator of lipid and glucose metabolism in rodent and human skeletal muscle. Diabetologia 55: 1797–1807.
10. Zach S, Felk S, Gillardon F (2010) Signal transduction protein array analysis links LRRK2 to Ste20 kinases and PKC zeta that modulate neuronal plasticity. PLoS One 5: e13191.
11. Iwamoto MA, Fairclough SR, Rudge SA, Engebrecht J (2005) *Saccharomyces cerevisiae* Sps1p regulates trafficking of enzymes required for spore wall synthesis. Eukaryot Cell 4: 536–544.

12. Rolli E, Ragni E, de Medina-Redondo M, Arroyo J, de Aldana CRV, et al. (2011) Expression, stability, and replacement of glucan-remodeling enzymes during developmental transitions in *Saccharomyces cerevisiae*. Mol Biol Cell 22: 1585–1598.
13. Krishnamoorthy T, Chen X, Govin J, Cheung WL, Dorsey J, et al. (2006) Phosphorylation of histone H4 Ser1 regulates sporulation in yeast and is conserved in fly and mouse spermatogenesis. Genes Dev 20: 2580–2592.
14. Managbanag JR, Witten TM, Bonchev D, Fox LA, Tsuchiya M, et al. (2008) Shortest-path network analysis is a useful approach toward identifying genetic determinants of longevity. PLoS One 3: e3802.
15. Aitken A, Collinge DB, van Heusden BP, Isobe T, Roseboom PH, et al. (1992) 14-3-3 proteins: a highly conserved, widespread family of eukaryotic proteins. Trends Biochem Sci 17: 498–501.
16. Van Heusden GPH (2009) 14-3-3 Proteins: insights from genome-wide studies in yeast. Genomics 94: 287–293.
17. Luo J, Feng J, Lu J, Wang Y, Tang X, et al. (2010) Aberrant methylation profile of 14-3-3 sigma and its reduced transcription/expression levels in Chinese sporadic female breast carcinogenesis. Med Oncol 27: 791–797.
18. Zurita M, Lara PC, del Moral R, Torres B, Linares-Fernández JL, et al. (2010) Hypermethylated 14-3-3-sigma and ESR1 gene promoters in serum as candidate biomarkers for the diagnosis and treatment efficacy of breast cancer metastasis. BMC Cancer 10: 217.
19. Li D-J, Deng G, Xiao Z-Q, Yao H-X, Li C, et al. (2009) Identifying 14-3-3 sigma as a lymph node metastasis-related protein in human lung squamous carcinoma. Cancer Lett 279: 65–73.
20. Chohan G, Pennington C, Mackenzie JM, Andrews M, Everington D, et al. (2010) The role of cerebrospinal fluid 14-3-3 and other proteins in the diagnosis of sporadic Creutzfeldt-Jakob disease in the UK: a 10-year review. J Neurol Neurosurg Psychiatry 81: 1243–1248.
21. Zhao J, Meyerkord CL, Du Y, Khuri FR, Fu H (2011) 14-3-3 proteins as potential therapeutic targets. Semin Cell Dev Biol 22: 705–712.

22. Watanabe K, Thandavarayan RA, Gurusamy N, Zhang S, Muslin AJ, et al. (2009) Role of 14-3-3 protein and oxidative stress in diabetic cardiomyopathy. Acta Physiol Hung 96: 277–287.

23. Kilani RT, Maksymowych WP, Aitken A, Boire G, St-Pierre Y, et al. (2007) Detection of high levels of 2 specific isoforms of 14-3-3 proteins in synovial fluid from patients with joint inflammation. J Rheumatol 34: 1650–1657.

24. Alexander JS, Minagar A, Harper M, Robinson-Jackson S, Jennings M, et al. (2007) Proteomic analysis of human cerebral endothelial cells activated by multiple sclerosis serum and IFNbeta-1b. J Mol Neurosci 32: 169–178.

25. Ametzazurra A, Matorras R, García-Velasco JA, Prieto B, Simón L, et al. (2009) Endometrial fluid is a specific and non-invasive biological sample for protein biomarker identification in endometriosis. Hum Reprod 24: 954–965.

26. Wang DY, Ray A, Rodgers K, Ergorul C, Hyman BT, et al. (2010) Global gene expression changes in rat retinal ganglion cells in experimental glaucoma. Invest Ophthalmol Vis Sci 51: 4084–4095.

27. Favier FB, Costes F, Defour A, Bonnefoy R, Lefai E, et al. (2010) Downregulation of Akt/mammalian target of rapamycin pathway in skeletal muscle is associated with increased REDD1 expression in response to chronic hypoxia. Am J Physiol Regul Integr Comp Physiol 298: R1659–66.

28. Aitken A (2011) Post-translational modification of 14-3-3 isoforms and regulation of cellular function. Semin Cell Dev Biol 22: 673–680.

29. Obsil T, Obsilova V (2011) Structural basis of 14-3-3 protein functions. Semin Cell Dev Biol 22: 663–672.

30. Muslin AJ, Tanner JW, Allen PM, Shaw AS (1996) Interaction of 14-3-3 with signaling proteins is mediated by the recognition of phosphoserine. Cell 84: 889–897.

31. Yaffe MB, Rittinger K, Volinia S, Caron PR, Aitken A, et al. (1997) The structural basis for 14-3-3:phosphopeptide binding specificity. Cell 91: 961–971.

32. Ganguly S, Weller JL, Ho A, Chemineau P, Malpaux B, et al. (2005) Melatonin synthesis: 14-3-3-dependent activation and inhibition of arylalkylamine N-acetyltransferase mediated by phosphoserine-205. Proc Natl Acad Sci U S A 102: 1222–1227.

33. Van Heusden GP, Wenzel TJ, Lagendijk EL, de Steensma HY, van den Berg JA (1992) Characterization of the yeast BMH1 gene encoding a putative protein homologous to mammalian protein kinase II activators and protein kinase C inhibitors. FEBS Lett 302: 145–150.

34. Van Heusden GP, Griffiths DJ, Ford JC, Chin-A-Woeng TF, Schrader PA, et al. (1995) The 14-3-3 proteins encoded by the BMH1 and BMH2 genes are essential in the yeast Saccharomyces cerevisiae and can be replaced by a plant homologue. Eur J Biochem 229: 45–53.

35. Gelperin D, Weigle J, Nelson K, Roseboom P, Irie K, et al. (1995) 14-3-3 proteins: potential roles in vesicular transport and Ras signaling in Saccharomyces cerevisiae. Proc Natl Acad Sci U S A 92: 11539–11543.

36. Van Heusden GPH, Steensma HY (2006) Yeast 14-3-3 proteins. Yeast 23: 159–171.

37. Roberts RL, Mösch HU, Fink GR (1997) 14-3-3 proteins are essential for RAS/MAPK cascade signaling during pseudohyphal development in S. cerevisiae. Cell 89: 1055–1065.

38. Grandin N, Charbonneau M (2008) Budding yeast 14-3-3 proteins contribute to the robustness of the DNA damage and spindle checkpoints. Cell Cycle 7: 2749–2761.

39. Engels K, Giannattasio M, Muzi-Falconi M, Lopes M, Ferrari S (2011) 14-3-3 Proteins regulate exonuclease 1-dependent processing of stalled replication forks. PLoS Genet 7: e1001367.

40. Bertram PG, Zeng C, Thorson J, Shaw AS, Zheng XF (1998) The 14-3-3 proteins positively regulate rapamycin-sensitive signaling. Curr Biol 8: 1259–1267.

41. Lee P, Paik S-M, Shin C-S, Huh W-K, Hahn J-S (2011) Regulation of yeast Yak1 kinase by PKA and autophosphorylation-dependent 14-3-3 binding. Mol Microbiol 79: 633–646.

42. Walter W, Clynes D, Tang Y, Marmorstein R, Mellor J, et al. (2008) 14-3-3 interaction with histone H3 involves a dual modification pattern of phosphoacetylation. Mol Cell Biol 28: 2840–2849.

43. Zahrádka J, van Heusden GPH, Sychrová H (2012) Yeast 14-3-3 proteins participate in the regulation of cell cation homeostasis via interaction with Nha1 alkali-metal-cation/proton antiporter. Biochim Biophys Acta 1820: 849–858.

44. Demmel L, Beck M, Klose C, Schlaitz A-L, Gloor Y, et al. (2008) Nucleocytoplasmic shuttling of the Golgi phosphatidylinositol 4-kinase Pik1 is regulated by 14-3-3 proteins and coordinates Golgi function with cell growth. Mol Biol Cell 19: 1046–1061.

45. Wang C, Skinner C, Easlon E, Lin S-J (2009) Deleting the 14-3-3 protein Bmh1 extends life span in Saccharomyces cerevisiae by increasing stress response. Genetics 183: 1373–1384.

46. Trembley MA, Berrus HL, Whicher JR, Humphrey-Dixon EL (2014) The yeast 14-3-3 proteins Bmh1 and Bmh2 differentially regulate rapamycin-mediated transcription. Biosci Rep 34(2): e00099.

47. Caydasi AK, Micoogullari Y, Kurtulmus B, Palani S, Pereira G (2014) The 14-3-3 protein Bmh1 functions in the spindle position checkpoint by breaking Bfa1 asymmetry at yeast centrosomes. Mol Biol Cell 25: 2143–2151.

48. Parodi EM, Baker CS, Tetzlaff C, Villahermosa S, Huang LS (2012) SPO71 mediates prospore membrane size and maturation in Saccharomyces cerevisiae. Eukaryot Cell 11: 1191–1200.

49. Kyriakakis P, Tipping M, Abed L, Veraksa A (2008) Tandem affinity purification in Drosophila: the advantages of the GS-TAP system. Fly (Austin) 2: 229–235.

50. Huang LS, Doherty HK, Herskowitz I (2005) The Smk1p MAP kinase negatively regulates Gsc2p, a 1,3-beta-glucan synthase, during spore wall morphogenesis in Saccharomyces cerevisiae. Proc Natl Acad Sci U S A 102: 12431–12436.

51. Longtine MS, McKenzie A, Demarini DJ, Shah NG, Wach A, et al. (1998) Additional modules for versatile and economical PCR-based gene deletion and modification in Saccharomyces cerevisiae. Yeast 14: 953–961.

52. Puig O, Rutz B, Luukkonen BG, Kandels-Lewis S, Bragado-Nilsson E, et al. (1998) New constructs and strategies for efficient PCR-based gene manipulations in yeast. Yeast 14: 1139–1146.

53. Briza P, Breitenbach M, Ellinger A, Segall J (1990) Isolation of two developmentally regulated genes involved in spore wall maturation in Saccharomyces cerevisiae. Genes Dev 4: 1775–1789.

54. Guide to yeast genetics and molecular biology. (1991). Methods Enzymol 194: 1–863.

55. Suda Y, Rodriguez RK, Coluccio AE, Neiman AM (2009) A screen for spore wall permeability mutants identifies a secreted protease required for proper spore wall assembly. PLoS One 4: e7184.

56. Benjamin KR, Zhang C, Shokat KM, Herskowitz I (2003) Control of landmark events in meiosis by the CDK Cdc28 and the meiosis-specific kinase Ime2. Genes Dev 17: 1524–1539.

57. Shevchenko A, Wilm M, Vorm O, Mann M (1996) Mass spectrometric sequencing of proteins silver-stained polyacrylamide gels. Anal Chem 68: 850–858.

58. Peng J, Gygi SP (2001) Proteomics: the move to mixtures. J Mass Spectrom 36: 1083–1091.

59. Eng JK, McCormack AL, Yates JR (1994) An approach to correlate tandem mass spectral data of peptides with amino acid sequences in a protein database. J Am Soc Mass Spectrom 5: 976–989.

60. Obenauer JC, Cantley LC, Yaffe MB (2003) Scansite 2.0: Proteome-wide prediction of cell signaling interactions using short sequence motifs. Nucleic Acids Res 31: 3635–3641.

61. Edgar RC (2004) MUSCLE: multiple sequence alignment with high accuracy and high throughput. Nucleic Acids Res 32: 1792–1797.

62. Guindon S, Dufayard J-F, Lefort V, Anisimova M, Hordijk W, et al. (2010) New algorithms and methods to estimate maximum-likelihood phylogenies: assessing the performance of PhyML 3.0. Syst Biol 59: 307–321.

63. Le SQ, Gascuel O (2008) An improved general amino acid replacement matrix. Mol Biol Evol 25: 1307–1320.

64. Hall BG, Salipante SJ (2007) Measures of clade confidence do not correlate with accuracy of phylogenetic trees. PLoS Comput Biol 3: e51.

65. Garrels JI, Futcher B, Kobayashi R, Latter GI, Schwender B, et al. (1994) Protein identifications for a Saccharomyces cerevisiae protein database. Electrophoresis 15: 1466–1486.

66. Tkach JM, Yimit A, Lee AY, Riffle M, Costanzo M, et al. (2012) Dissecting DNA damage response pathways by analysing protein localization and abundance changes during DNA replication stress. Nat Cell Biol 14: 966–976.

67. Pédelacq J-D, Cabantous S, Tran T, Terwilliger TC, Waldo GS (2006) Engineering and characterization of a superfolder green fluorescent protein. Nat Biotechnol 24: 79–88.

68. Zacharias DA, Violin JD, Newton AC, Tsien RY (2002) Partitioning of lipid-modified monomeric GFPs into membrane microdomains of live cells. Science 296: 913–916.

69. Shaner NC, Steinbach PA, Tsien RY (2005) A guide to choosing fluorescent proteins. Nat Methods 2: 905–909.

70. Kalderon D, Roberts BL, Richardson WD, Smith AE (1984) A short amino acid sequence able to specify nuclear location. Cell 39: 499–509.

71. Süel KE, Chook YM (2009) Kap104p imports the PY-NLS-containing transcription factor Tfg2p into the nucleus. J Biol Chem 284: 15416–15424.

72. Kakiuchi K, Yamauchi Y, Taoka M, Iwago M, Fujita T, et al. (2007) Proteomic analysis of in vivo 14-3-3 interactions in the yeast Saccharomyces cerevisiae. Biochemistry 46: 7781–7792.

73. Lee W-S, Hsu C-Y, Wang P-L, Huang C-YF, Chang C-H, et al. (2004) Identification and characterization of the nuclear import and export signals of the mammalian Ste20-like protein kinase 3. FEBS Lett 572: 41–45.

74. Enyenihi AH, Saunders WS (2003) Large-scale functional genomic analysis of sporulation and meiosis in Saccharomyces cerevisiae. Genetics 163: 47–54.

75. Uhart M, Bustos DM (2014) Protein intrinsic disorder and network connectivity. The case of 14-3-3 proteins. Front Genet 5: 10.

76. Byrne KP, Wolfe KH (2005) The Yeast Gene Order Browser: combining curated homology and syntenic context reveals gene fate in polyploid species. Genome Res 15: 1456–1461.

77. Wilson WA, Wang Z, Roach PJ (2002) Systematic identification of the genes affecting glycogen storage in the yeast Saccharomyces cerevisiae: implication of the vacuole as a determinant of glycogen level. Mol Cell Proteomics 1: 232–242.

78. Freimoser FM, Hürlimann HC, Jakob CA, Werner TP, Amrhein N (2006) Systematic screening of polyphosphate (poly P) levels in yeast mutant cells reveals strong interdependence with primary metabolism. Genome Biol 7: R109.

79. Kristensen AR, Gsponer J, Foster LJ (2012) A high-throughput approach for measuring temporal changes in the interactome. Nat Methods 9: 907–909.

80. Zhou TH, Ling K, Guo J, Zhou H, Wu YL, et al. (2000) Identification of a human brain-specific isoform of mammalian STE20-like kinase 3 that is regulated by cAMP-dependent protein kinase. J Biol Chem 275: 2513–2519.

81. Preisinger C, Short B, De Corte V, Bruyneel E, Haas A, et al. (2004) YSK1 is activated by the Golgi matrix protein GM130 and plays a role in cell migration through its substrate 14-3-3zeta. J Cell Biol 164: 1009–1020.

82. Zhou J, Shao Z, Kerkela R, Ichijo H, Muslin AJ, et al. (2009) Serine 58 of 14-3-3zeta is a molecular switch regulating ASK1 and oxidant stress-induced cell death. Mol Cell Biol 29: 4167–4176.

83. Chi A, Huttenhower C, Geer LY, Coon JJ, Syka JEP, et al. (2007) Analysis of phosphorylation sites on proteins from *Saccharomyces cerevisiae* by electron transfer dissociation (ETD) mass spectrometry. Proc Natl Acad Sci U S A 104: 2193–2198.

84. Breitkreutz A, Choi H, Sharom JR, Boucher L, Neduva V, et al. (2010) A global protein kinase and phosphatase interaction network in yeast. Science 328: 1043–1046.

85. Soulard A, Cremonesi A, Moes S, Schütz F, Jenö P, et al. (2010) The rapamycin-sensitive phosphoproteome reveals that TOR controls protein kinase A toward some but not all substrates. Mol Biol Cell 21: 3475–3486.

86. Thompson JD, Gibson TJ, Plewniak F, Meanmougin F, Higgins DG (1997) The CLUSTAL_X windows interface: flexible strategies for multiple sequence alignment aided by quality analysis tools. Nucleic Acids Res 25: 4876–4882.

# Discrimination of *Escherichia coli* O157, O26 and O111 from Other Serovars by MALDI-TOF MS Based on the *S10*-GERMS Method

Teruyo Ojima-Kato[1]*, Naomi Yamamoto[2], Mayumi Suzuki[2], Tomohiro Fukunaga[3], Hiroto Tamura[2]*

1 Hub of Knowledge Aichi, Aichi Science and Technology Foundation, Yakusa, Toyota, Aichi, Japan, 2 School of Agriculture, Meijo University, Shiogamaguchi, Tenpaku-ku, Nagoya, Aichi, Japan, 3 Japan Food Research Laboratories, Osu, Naka-ku, Nagoya, Aichi, Japan

## Abstract

Enterohemorrhagic *Escherichia coli* (EHEC), causes a potentially life-threatening infection in humans worldwide. Serovar O157:H7, and to a lesser extent serovars O26 and O111, are the most commonly reported EHEC serovars responsible for a large number of outbreaks. We have established a rapid discrimination method for *E. coli* serovars O157, O26 and O111 from other *E. coli* serovars, based on the pattern matching of mass spectrometry (MS) differences and the presence/absence of biomarker proteins detected in matrix-assisted laser desorption/ionization time-of-flight MS (MALDI-TOF MS). Three biomarkers, ribosomal proteins S15 and L25, and acid stress chaperone HdeB, with MS *m/z* peaks at 10138.6/10166.6, 10676.4/10694.4 and 9066.2, respectively, were identified as effective biomarkers for O157 discrimination. To distinguish serovars O26 and O111 from the others, DNA-binding protein H-NS, with an MS peak at *m/z* 15409.4/15425.4 was identified. Sequence analysis of the O157 biomarkers revealed that amino acid changes: Q80R in S15, M50I in L25 and one mutation within the start codon ATG to ATA in the encoded HdeB protein, contributed to the specific peak pattern in O157. We demonstrated semi-automated pattern matching using these biomarkers and successfully discriminated total 57 O157 strains, 20 O26 strains and 6 O111 strains with 100% reliability by conventional MALDI-TOF MS analysis, regardless of the sample conditions. Our simple strategy, based on the *S10-spc-alpha* operon gene-encoded ribosomal protein mass spectrum (*S10*-GERMS) method, therefore allows for the rapid and reliable detection of this pathogen and may prove to be an invaluable tool both clinically and in the food industry.

**Editor:** Muna Anjum, Animal Health and Veterinary Laboratories Agency, United Kingdom

**Funding:** This work was financially supported by Aichi Science and Technology Foundation (Japan, http://www.astf.or.jp/). The funder had no role in study design, data collection and analysis, decision to publish, or preparation of the manuscript.

**Competing Interests:** The authors have declared that no competing interests exist.

* Email: teruyo.ojima@gmail.com (TO); hiroto@meijo-u.ac.jp (HT)

## Introduction

Matrix-assisted laser desorption/ionization time-of-flight mass spectrometry (MALDI-TOF MS) is a robust approach for the rapid identification of microorganisms. The identification mechanism is based on the protein MS pattern obtained by MALDI-TOF MS matching microbial sequence data in available databases, the so-called fingerprinting method. This method has been rapidly developed and expanded, and has been successfully applied to the clinical field because it offers a stable, rapid and cost-effective system for microbial identification.

In using MALDI-TOF MS for the identification of microorganisms, the majority of the high-intensity MS peaks detected is derived from ribosomal proteins encoded in the *S10-spc-alpha* operon, where at least half of the ribosomal subunit proteins are encoded. This operon is highly conserved among eubacterial genomes [1–5]. These peaks can be reliable biomarkers with which to discriminate bacteria at a strain or pathovar level because strain-specific peaks can be predicted and verified from the DNA sequence information before measurement [6,7]. This methodology, known as the '*S10*-GERMS (*S10-spc-alpha* operon gene-encoded ribosomal protein mass spectrum) method', offers

theoretically calculated *m/z* ion peaks of ribosomal proteins that are species- or strain-specific. An accurate database can then be constructed by comparing the experimentally observed *m/z* values with the theoretical values. The *S10*-GERMS method has been effectively employed in the identification of serovars of *Pseudomonas syringae* [7] and strains of *Lactobacillus casei* [8]. Strain typing by direct bacterial profiling has increasingly been studied as a method for bacterial species identification in recent years [9–11].

Shiga toxin-producing *Escherichia coli*, known as enterohemorrhagic *E. coli* (EHEC), causes bloody diarrhea, hemorrhagic colitis and life-threatening hemolytic-uremic syndrome. Serovar O157:H7 is the most commonly reported EHEC serovar causing many outbreaks and significantly threatening human life worldwide. Serovars O26 and O111 are also responsible for a large number of EHEC outbreaks.

Attempts to classify EHEC serovars by MALDI-TOF MS have been reported [12]; however, the results are dependent on sample preparation conditions and the biomarker proteins are not assigned. To allow this method to be practically applied in the field, it needs to be versatile and reliable. In another study, *E. coli* O157:H7-specific biomarkers HdeA, HdeB, CspC, YbgS, YjbJ

**Table 1.** *E. coli* strains used in this study.

| Strain | Characteristics | Source |
|---|---|---|
| GTC 03904 | O157:H7, VT–Shiga toxin negative | NBRP |
| GTC 14513 | O157:H7:VT2 | NBRP |
| GTC 14535 | O157:H7:VT1&2 | NBRP |
| GTC 14536 | O157:H7:VT1&2 | NBRP |
| GTC 14537 | O157:H7:VT2 | NBRP |
| GTC 14544 | O157:H7:VT1&2 | NBRP |
| GTC 14545 | O157:H7:VT1&2 | NBRP |
| GTC 14546 | O157:H7:VT2 | NBRP |
| GTC 14547 | O157:H7:VT2 | NBRP |
| GTC 14550 | O157:H7:VT2 | NBRP |
| GTC 14551 | O157:H7:VT1&2 | NBRP |
| GTC 14552 | O157:H7:VT1&2 | NBRP |
| GTC 14553 | O157:H7:VT2 | NBRP |
| GTC 14507 | O111:H-:VT1&2 | NBRP |
| GTC 14517 | O111:H-:VT1 | NBRP |
| GTC 14515 | O26:H11:VT1&2 | NBRP |
| GTC 14516 | O26:H11:VT1 | NBRP |
| GTC 14538 | O26:H-:VT1 | NBRP |
| GTC 14539 | O26:H11:VT1&2 | NBRP |
| GTC 14540 | O26:H11:VT1 | NBRP |
| GTC 14548 | O26:H-:VT1 | NBRP |
| GTC 14549 | O26:H11:VT1 | NBRP |
| GTC 14557 | O26:H11:VT1 | NBRP |
| GTC 14558 | O26:H11:VT1 | NBRP |
| GTC 14530 | O121:H19:VT2 | NBRP |
| GTC 14601 | O121:H19:VT2 | NBRP |
| GTC 14602 | O121:H19:VT2 | NBRP |
| GTC 14518 | O115:H10:VT1 | NBRP |
| GTC 14529 | O119:H2:VT1 | NBRP |
| GTC 14559 | O63:H6:VT2 | NBRP |
| GTC 14603 | O128:H-:VT1&2 | NBRP |
| NBRC 12713 | Genome sequenced K-12 strain. The alias of W3110. | NITE |
| ATCC 47076 | Genome sequenced K-12 strain. The alias of MG1655. | ATCC |
| NBRC 13893 | | NITE |
| NBRC 15034 | | NITE |
| NBRC 14237 | | NITE |
| NBRC 13891 | | NITE |
| NBRC 3301 | K-12 strain. | NITE |
| NBRC 3972 | | NITE |
| NBRC 12062 | | NITE |
| NBRC 13168 | | NITE |
| NBRC 3548 | | NITE |
| NBRC 12734 | | NITE |
| JCM16574 | Genome sequenced strain, O152:H28 | JCM |
| ATCC BAA-1743 | Genome sequenced strain | ATCC |
| JCM16575 | Genome sequenced strain, O150:H5 | JCM |
| NBRC 3991 | | NITE |
| WT-141 | O157: VT- Shiga toxin Negative, isolated from human | |
| WT-351 | O157: VT- Shiga toxin Negative, isolated from cattle | |
| WT-352 | O157: VT- Shiga toxin Negative, isolated from cattle | |

**Table 1.** Cont.

| Strain | Characteristics | Source |
|--------|----------------|--------|
| jfrl 01 | O157:H7:VT2, isolated from pork in 1998 | |
| jfrl 02 | O157:H7:VT2, isolated from beef in 1996 | |
| jfrl 03 | O157:H7:VT1&2, isolated from beef in 1998 | |
| jfrl 04 | O157:H7:VT1&2, isolated from beef in 1996 | |
| jfrl 05 | O157:H7:VT2, isolated from welsh onion in 1996 | |
| jfrl 06 | O157:VT1&2, isolated from beef in 2003 | |
| jfrl 07 | O157:VT 2, isolated from beef in 1999 | |
| jfrl 08 | O157: VT1&2, isolated from beef in 1999 | |
| jfrl 09 | O157:VT2, isolated from beef in 2010 | |
| jfrl 10 | O157:VT2, isolated from beef in 2010 | |
| jfrl 11 | O157:VT2, isolated from beef in 2010 | |
| jfrl 12 | O26, VT1, isolated from beef in 2010 | |
| A11-1 | O157:H7, VT1&2 | APIPH |
| A11-85 | O157:HUT, VT1&2 | APIPH |
| A11-87 | O157:H7, VT1&2 | APIPH |
| A11-88 | O157:H7, VT1&2 | APIPH |
| A11-89 | O157:H7, VT1&2 | APIPH |
| A11-90 | O157:H7, VT1&2 | APIPH |
| A11-161 | O157:H7, VT2 | APIPH |
| A11-163 | O157:H7, VT2 | APIPH |
| A11-168 | O157:H7, VT1 | APIPH |
| A11-169 | O157:H7, VT1&2 | APIPH |
| A11-175 | O157:H7, VT1 | APIPH |
| A11-176 | O157:H7, VT1 | APIPH |
| A11-177 | O157:H7, VT1 | APIPH |
| A11-225 | O157:H7, VT2 | APIPH |
| A11-234 | O157:H7, VT2 | APIPH |
| A12-154 | O157:H7, VT2 | APIPH |
| A12-163 | O157:H7, VT1&2 | APIPH |
| A12-164 | O157:H7, VT2 | APIPH |
| A12-166 | O157:H7, VT1&2 | APIPH |
| A12-167 | O157:H7, VT1&2 | APIPH |
| A12-183 | O157:H7, VT2 | APIPH |
| A12-185 | O157:HUT, VT1&2 | APIPH |
| A12-190 | O157:H7, VT1&2 | APIPH |
| A12-191 | O157:HUT, VT1&2 | APIPH |
| A12-193 | O157:HUT, VT1&2 | APIPH |
| A12-201 | O157:H7, VT2 | APIPH |
| A12-209 | O157:H7, VT2 | APIPH |
| A12-212 | O157:H7, VT1&2 | APIPH |
| A12-222 | O157:H7, VT2 | APIPH |
| A12-223 | O157:H7, VT2 | APIPH |
| A12-97 | O26:H11, VT1 | APIPH |
| A12-98 | O26:H11, VT1 | APIPH |
| A12-99 | O26:H11, VT1 | APIPH |
| A12-100 | O26:H11, VT1 | APIPH |
| A12-147 | O26:H11, VT1 | APIPH |
| A13-137 | O26:H11, VT1 | APIPH |
| A13-138 | O26:H11, VT1 | APIPH |

**Table 1.** Cont.

| Strain | Characteristics | Source |
|--------|-----------------|--------|
| A13-154 | O26:H11, VT2 | APIPH |
| A13-155 | O26:H11, VT2 | APIPH |
| A13-165 | O26:H11, VT1 | APIPH |
| A12-152 | O111:HUT, VT1 | APIPH |
| A12-161 | O111:H21, VT1 | APIPH |
| A12-162 | O111:H21, VT1&2 | APIPH |
| A12-200 | O111:HUT, VT1 | APIPH |
| A12-216 | O121:H19, VT2 | APIPH |

and YbgO were identified using MALDI-TOF/TOF-MS/MS [13], in which only 1 Da difference was sufficient to distinguish *E. coli* O157 from other serovars.

Here, we report the discrimination of *E. coli* O157, O26 and O111 serovars with four specific biomarker proteins based on the *S10*-GERMS method by MALDI-TOF MS. These biomarker peaks that are assigned and validated by DNA sequence analysis are detected under any of the sample conditions tested, with high reproducibility, using conventional MALDI-TOF MS analysis.

## Materials and Methods

### Bacterial strains and growth conditions

Thirty EHEC strains, 4 shiga toxin non-producing O157 strains and 16 non-EHEC strains were used for the construction of a theoretical mass database (Table 1). They were purchased from the National BioResource Project (NBRP; a division of pathogenic microbe, Gifu University, Gifu, Japan), the American Type Culture Collection (ATCC; Rockville, MD, USA), the Japan Collection of Microorganisms, RIKEN BRC (JCM, Tsukuba, Japan), which is participating in the National BioResource Project of the Ministry of Education, Culture, Sports, Science and Technology, Japan, and the Biological Resource Center at the National Institute of Technology and Evaluation (NITE, Kisarazu, Japan). Three shiga toxin non-producing O157 (strains WT-141, WT-351 and WT-352) were kindly provided by Dr. Hiroshi Asakura (National Institute of Health Sciences, Japan). Nutrient broth (Becton Dickinson, Franklin Lakes, NJ, USA), tryptone soya agar (Thermo Scientific, Waltham, MA), or Luria–Bertani broth (Nacalai, Kyoto, Japan) were used for cultivation. For the blind test, another 57 *E. coli* strains, namely 12 *E. coli* strains (strains jfrl 01–12), that were isolated from food samples through 1996 to 2010 and identified as O157 or O26 by the antisera coagglutination test (Denka Seiken, Tokyo, Japan), and 45 *E. coli* strains that were kindly provided from Aichi Prefectural Institute of Public Health (APIPH) were used (Table 1). The production of verotoxin in these strains was also checked by the coagglutination test (VTEC-RPLA, Denka Seiken).

### Construction of the protein mass database

The amino acid sequences of ribosomal subunit proteins and biomarker candidates of genome sequenced strains were obtained from the National Center for Biotechnology Information (NCBI) database. The theoretical ionized mass of each protein was calculated using a Compute pI/Mw tool on the ExPASy proteomics server (http://web.expasy.org/compute_pi/), considering the N-terminal rule. For the non-genome-sequenced strains,

the DNA sequence of the ribosomal proteins encoded in the *S10-spc-alpha* operon and biomarker candidates were analyzed as described previously [14]. In brief, respective regions of ribosomal protein-encoding genes ($\approx$5 kbp) or genes encoding biomarker proteins were amplified using high-fidelity DNA polymerase, KOD plus (Toyobo, Osaka, Japan), and primers designed against the consensus DNA sequences up- and down-stream of the target regions in the *E. coli* genome sequences in the NCBI database. Sequencing reactions were carried out using a BigDye ver. 3.1 Cycle Sequencing Kit (Applied Biosystems, Foster City, CA, USA). DNA primers used for PCR and sequence analysis are listed in Table 2.

### MALDI-TOF MS analysis for the evaluation of the mass database

Bacterial colonies grown on agar plate were picked and placed directly onto a measurement steel plate, while bacteria from liquid culture were harvested by centrifugation then washed with TMA-I buffer (10 mM Tris-HCl pH 7.8, 30 mM $NH_4Cl$, 10 mM $MgCl_2$ and 6 mM 2-mercaptoethnol). Approximately $10^7$ cfu were mixed well with 1 µL of matrix solution consisting of 20 mg/mL sinapic acid (Wako Pure Chemical) or saturated α-cyano-4-hydroxycinnamic acid (CHCA), and 1% (v/v) trifluoroacetic acid (Wako Pure Chemical) in 50% (v/v) acetonitrile. The mixture was spotted onto the MALDI sample plate and air dried. MALDI-TOF MS analysis was performed using an AXIMA micro-organism identification (Shimadzu/Kratos, Kyoto, Japan) as described previously, with minor modifications [7]. Briefly, the sample was measured in the positive linear mode in the spectrum range of $m/z$ 2000–20000. Data were obtained from the sum of 100 individual laser shots and calibrated with the *E. coli* strain DH5α using the peaks at $m/z$ 4365.4, 7274.5, 10300.1, 12770.6 and 14365.6, corresponding to ribosomal proteins L36, L29, S19, L18 and L17, respectively. After calibrating manually, each sample was automatically calibrated with the same internal peaks as DH5α. Theoretical and measured masses were matched with 500 ppm tolerance. The actual masses in the MALDI-TOF MS spectra were matched with the theoretical values and corrected appropriately.

### Automated MALDI-TOF MS analysis for validation

Samples prepared from colonies were automatically analyzed to verify the effectiveness and reproducibility of selected biomarkers. Four analytes per strain were prepared as described above. To evaluate the effects of culture medium on the masses of selected biomarkers, typical selective media for *Enterobacteriaceae* or O157, desoxycholate agar (Nissui Pharmaceutical, Tokyo, Japan),

**Table 2.** Primers used in this study.

| Name | Sequence (5′ – 3′) | Purpose |
| --- | --- | --- |
| EcW3110-S10-F | AAGAACGGTTACACTCTCCC | amplification of *S10* region |
| EcW3110-S10-R | ACACCGCTTCAAGGATATGG | amplification of *S10* region |
| EcW3110-S10-1 | AATCGTAATGGGTCTGAGGAG | sequencing |
| EcW3110-S10-2 | AAGCTGGCCACTTCGCTAAAG | sequencing |
| EcW3110-S10-3 | TGCTGAAGTAACTGGTTCCGG | sequencing |
| EcW3110-S10-4 | AAGCTGCTGTGCAGAAACTG | sequencing |
| EcW3110-S10-5 | CATAACGTAGAAATGAAACCAGG | sequencing |
| EcW3110-S10-6 | ACGTTCCGGTATTTGTAACCG | sequencing |
| EcW3110-S10-7 | TCAGTACCTGACTAAGGAAC | sequencing |
| EcW3110-S10-8 | AGCGTCGCTGATGTTACAAC | sequencing |
| EcW3110-S10-9 | AGCAAGTGCGTCGCGATGTCG | sequencing |
| EcW3110-S10-10 | GCTGGCATGATTCGTGAAGAACG | sequencing |
| EcW3110-spc-F | AACGGCTCAGAAATGAGCCG | amplification of *spc* region |
| EcW3110-spc-R | AGCAGTCTGCGTTTCAGCTC | amplification of *spc* region |
| EcW3110-spc-1 | TCTACCCATATCCTTGAAGC | sequencing |
| EcW3110-spc-2 | ATTGTTGAAGGTATCAACCTG | sequencing |
| EcW3110-spc-3 | TCGTGGTAACTACAGCATG | sequencing |
| EcW3110-spc-4 | ACCATGCCTTCCTCCAAGCT | sequencing |
| EcW3110-spc-5 | TTGGTGTAGGTTACCGTGCAG | sequencing |
| EcW3110-spc-6 | ATGCTGCCCGTGAAGCTGGC | sequencing |
| EcW3110-spc-7 | ATCGGTCGTCTGCCGAAACAC | sequencing |
| EcW3110-spc-9 | GTCACCATGCCTTCCTCCAAG | sequencing |
| EcW3110-spc-1r | GATGATGTCGCCTACGCCTGC | sequencing |
| EcW3110-spc-2r | TTACCGGTTAACACGATAAC | sequencing |
| EcW3110-alpha-F | AGTGCCAAAGGTGGCTTAGGC | amplification of *alpha* region |
| EcW3110-alpha-R | ACAGCTATTGTAGATAAGTGG | amplification of *alpha* region |
| EcW3110-alpha-1 | TGCCCATACTATCGAGCAAGC | sequencing |
| EcW3110-alpha-2 | TCACTGCTTATCGTTGTTGTC | sequencing |
| EcW3110-alpha-3 | TGTCGTTGAAGGTGATCTGCG | sequencing |
| EcW3110-alpha-4 | AATGGCAAGATATTTGGGTC | sequencing |
| EcW3110-alpha-5 | TGCGGACATTAACGAACACCTG | sequencing |
| EcW3110-alpha-6 | TGCCTACAATGTTGAAGCAGCG | sequencing |
| EcW3110-alpha-7 | AGCTGCGCCGCGTAGTTGAGC | sequencing |
| EcW3110-alpha-1r | AGCTGGATAATGATCGACGC | sequencing |
| EcW3110-L25-F | TTCGAGCAGCTTTTTATCCGCC | amplification of L25 |
| EcW3110-L25-R | AAGGCTACGAACTGGAAGAGAGC | amplification of L25 |
| EcW3110-L25-1 | ATACGCGCACACCGGGCATC | sequencing |
| EcW3110-L25-1r | AGACCGTAGCACACTGCGTCAG | sequencing |
| EcW3110-S15-F | TACGAACGATCGGATTAAGCAATG | amplification of S15 |
| EcW3110-S15-R | TTACTTGATCCATTACTGATGCC | amplification of S15 |
| EcW3110-S15-1 | GGATTAAGCAATGTAATATCC | sequencing |
| EcW3110-S15-1r | ATTACTGATGCCAATGGACAGTCC | sequencing |
| Ec_HdeB-F | GATATGTAATTCCGGGAATGC | amplification and sequencing of HdeB |
| Ec_HdeB-R | AAGGAGCAGCAAGATGGCTCAAC | amplification of HdeB |
| Ec_YdaQ-F | TCATAGCTGATTATTAATAATC | amplification of YdaQ |
| Ec_YdaQ-R | ATGAACCAGATGCGAATGTAT | amplification of YdaQ |
| Ec_HNS-F | TGAATTCCTTACATTCCTGGC | amplification and sequencing of H-NS |
| Ec_HNS-R | AGCTTATTCTTATTAAATTGTC | amplification of H-NS |

**Table 3.** Theoretical masses of selected biomarker proteins for *E. coli* discrimination.

| Protein | Coded operon | Group of mass pattern | | | | | | | | | | | | | | | |
|---|---|---|---|---|---|---|---|---|---|---|---|---|---|---|---|---|---|
| | | A | B | C | D | E | F | G | H | I | J | K | L | M | N | O | P |
| | | O157 | O157 | O157 | O26 O111 | O26 | O121, O128, O152, - | O115 | O119 | O63 | K12 | - | - | - | - | - | O150 |
| L23 | S10 | 11200.1 | 11200.1 | 11200.1 | 11200.1 | 11200.1 | 11200.1 | 11147.1 | 11200.1 | 11200.1 | 11200.1 | 11200.1 | 11200.1 | 11200.1 | 11200.1 | 11200.1 | 11200.1 |
| L24 | spc | 11186.0 | 11186.0 | 11186.0 | 11186.0 | 11186.0 | 11186.0 | 11186.0 | 11186.0 | 11216.0 | 11186.0 | 11186.0 | 11216.0 | 11186.0 | 11186.0 | 11216.0 | 11216.0 |
| S14 | spc | 11450.3 | 11450.3 | 11450.3 | 11450.3 | 11450.3 | 11450.3 | 11450.3 | 11450.3 | 11450.3 | 11450.3 | 11450.3 | 11450.3 | 11450.3 | 11464.3 | 11450.3 | 11450.3 |
| L15 | spc | 14967.4 | 14967.4 | 14967.4 | 14967.4 | 14967.4 | 14967.4 | 14981.4 | 14945.0 | 14967.4 | 14981.4 | 14981.4 | 14967.4 | 14981.4 | 14967.4 | 14967.4 | 14967.4 |
| S11+Me | alpha | 13728.8 | 13728.8 | 13728.8 | 13728.8 | 13728.8 | 13728.8 | 13728.8 | 13728.8 | 13728.8 | 13728.8 | 13728.8 | 13728.8 | 13728.8 | 13728.8 | 13728.8 | 13756.8 |
| YdaQ | | 8325.6 | - | 8325.6 | 8325.6 | - | 8325.6 | 8325.6 | 8325.6 | 8325.6 | 8325.6 | 8325.6 | - | - | 8325.6 | - | |
| S15 | | **10166.6** | **10166.6** | 10138.6 | 10138.6 | 10138.6 | 10138.6 | 10138.6 | 10138.6 | 10138.6 | 10138.6 | 10137.6 | 10138.6 | 10138.6 | 10138.6 | 10138.6 | 10138.6 |
| L25 | | **10676.4** | **10676.4** | 10694.4 | 10694.4 | 10694.4 | 10694.4 | 10694.4 | 10694.4 | 10693.5 | 10694.4 | 10694.4 | 10693.5 | 10694.4 | 10694.4 | 10693.5 | 10693.5 |
| HdeB | | - | - | - | 9066.2 | 9066.2 | 9066.2 | 9066.2 | 9066.2 | 9066.2 | 9066.2 | 9066.2 | 9066.2 | 9066.2 | 9066.2 | 9066.2 | 9066.2 |
| H-NS | | 15409.4 | 15409.4 | 15409.4 | **15425.4** | **15425.4** | 15409.4 | 15409.4 | 15409.4 | 15409.4 | 15409.4 | 15882.0 | 15409.4 | 15409.4 | 15409.4 | 15409.4 | 15409.4 |

Theoretical mass values (*m/z* [M+H]+) of possible biomarkers for discrimination of *E. coli* strains are shown. The database was constructed by validated *E. coli* strains available on public collections and three isolated strains. Groups A to P indicate the classification based on mass patterns. – in the group column indicates the O-antigen is not determined. – in MS column means the peaks are absent. *E. coli* strains belong to the groups A to P are as bellows; A: O157-351, GTC 14545, GTC 14546, GTC 14552, O157-141, GTC 14513, GTC 14535, GTC 14536, GTC 14537, GTC 14544, GTC 14547, GTC 14551 and GTC 03904; B: O157-352; C: GTC 14550 and GTC 14553; D: GTC 14517, GTC 14507, GTC 14516, GTC 14538, GTC 14540, GTC 14549, GTC 14557 and GTC 14558; E: GTC 14515, GTC 14539 and GTC14548; F: GTC 14530, GTC 14601, GTC 14602, GTC 14603, JCM16574, NBRC 12574, NBRC 13168, NBRC 12734 and NBRC 3991; G: GTC 14518; H: GTC 14518; I: GTC 14529; J: NBRC 12713, ATCC 47076, NBRC 3301, NBRC 3972; K: NBRC 13893; L: NBRC 15034 and NBRC 14237; M: NBRC 13891; N: NBRC 3548; O: ATCC BAA-1743; P: JCM 16575.

CT-SMAC (Kyokuto Pharmaceutical Industrial, Tokyo, Japan), Chromagar X-gal (Chromagar, Paris, France), and crystal violet neutral red bile lactose agar (VRBL, Thermo scientific) were tested in addition to the normal growth media such as nutrient broth, tryptone soya agar or Luria–Bertani broth.

## Cluster analysis

Fingerprints of protein mass patterns were analyzed with SARAMIS (Spectral Archive and Microbial Identification System, AnagnosTec, Postdam-Golm, Germany) to construct binary matrices of biomarkers. The data were imported into the PAST software (http://folk.uio.no/ohammer/past/, Natural History Museum, Oslo University, Norway) to calculate distance matrices using the neighbor-joining method with Kimura algorithm. A phylogenetic tree was constructed using the FigTree ver. 1.4.0 software (http://tree.bio.ed.ac.uk/software/figtree/) as described previously [16].

## Blind test using isolated wile-type E. coli strains

To evaluate the discrimination method using our selected biomarkers, 57 E. coli strains, individually isolated from food (such as beef, pork and Welsh onions) or humans and identified as serovars O157, O26, O111 or O121 by antisera testing, were analyzed by MALDI-TOF MS. Semi-automated classification was demonstrated according to the mass patterns of selected four biomarker proteins.

## Nucleotide sequence accession numbers

The nucleotide sequences of ribosomal proteins encoded in the S10-spc-alpha operon, biomarker proteins S15 and L25, acid stress chaperon HdeB and DNA-binding protein H-NS, of E. coli strains determined in this study were deposited in the DNA data bank of Japan (DDBJ, http://www.ddbj.nig.ac.jp) with accession numbers from AB903039 to AB903902 and AB915955 to AB916334.

## Results and Discussion

### Construction of the protein mass database

In this study we have attempted to employ the S10-GERMS method for the discrimination of major serovars of EHEC O157, O26 and O111 from the others. The theoretical masses of ribosomal proteins encoded by the S10-spc-alpha operon were calculated based on the sequence analysis and genome sequence information (Table 3). The mass values were compared with the actual analytical results of MALDI-TOF MS and manually validated. The masses of the S10-spc-alpha operon-encoded ribosomal proteins not shown in Table 3, namely S10, L3, L4, L23, L2, S19, L22, S3, L16, L29, S17, L14, L5, S14, S8, L6, L18, S5, L30, L36, S13, S11, S4 and L17, were all identical respectively in all of the E. coli strains used for database construction. Whereas, L24, S5 and S13, thought to be biomarker candidates from their calculated masses, gave unclear peaks because of small differences in masses or high molecular weights (Table 3). The S10-GERMS method has successfully been employed for Pseudomonas sp., Bacillus sp. and Lactobacillus sp. in previous studies [7,8,14,17]. However, in the case of E. coli, strain or serovar typing using ribosomal proteins encoded in the S10-spc-alpha operon appears to be more challenging due to a less diversity of the masses. Although the ribosomal proteins encoded in the S10-spc-alpha operon were not suitable as biomarkers for serovars O157, O26 and O111, the other strains which are classified into group G to P in Table 1 show unique mass patterns of ribosomal proteins in the operon. It helps the strain level discrimination of E. coli using these biomarkers.

Otherwise, unique and clear mass shifts of the ribosomal proteins S15 and L25 were observed specific in E. coli O157 compared with the other E. coli serovars (Fig. 1, Table 3). Sequence analysis revealed that a point mutation, A239G, on ribosomal protein S15 caused an amino acid residue change, Q80R, resulting in a MS shift of m/z 10138.6 to 10166.6. Similarly, the O157-specific mutation G150A in the gene encoding L25, resulting in an amino acid substitution, M50I, led to a mass shift of m/z 10694.4 to 10676.4. These two ribosomal proteins also showed mass shifts in the theoretical masses of E. coli strains GTC 14559, NBRC 15034, NBRC 14237, ATCC BAA-1743 and JCM16575 (group K, L, O and P in Table 3), although the differences were too small to distinguish in actual MALDI-TOF MS analysis. Exceptionally, two E. coli O157 strains, GTC 14550 and GTC 14553, showed the same mass patterns as most of the other strains except for the absence of m/z 9066.2.

To our knowledge, this is the first report that ribosomal proteins S15 and L25, H-NS would be important biomarkers for O157 in MALDI-TOF MS analysis, a finding overlooked by others [12]. The mass differences of ribosomal proteins greatly contribute to strain classification owing to their variability. The great abundance of these proteins in cells is also advantageous because their mass peaks are always detected as stable biomarkers under any analytical conditions (regardless of variables such as the method of sample preparation, the type of matrix or the MALDI system). In fact, the peak intensity and sharpness for proteins S15 and L25 in O157 serovars were sufficient to distinguish them from other E. coli serovars (Fig. 1). The same was possible using either sinapic acid or CHCA, whether the sample was a colony or a liquid extracted with formic acid (data not shown). Compared with the previously reported system that required time-consuming and complex sample preparation [12], our method is more applicable for routine MALDI-TOF MS analysis because it can be performed directly from a single colony.

The mass spectrum of the acid stress chaperone HdeB in non-EHEC strains was previously reported by Fagerquist et al [13]. Likewise we identified HdeB at m/z 9066.2 [M+H]$^+$ in non-EHEC strains, and a loss of this peak was observed in all O157 serovars used in this study with complete reproducibility, as reported by Carter et al [15] (Fig. 1, Table 3). Sequence analysis of the hdeB gene confirmed that the putative start codon, ATG, had a point mutation (ATA) in all O157 strains, while in all other E. coli strains of other serovars ATG was observed. This strongly supported the suggestion that this mutation correlates to the lack of the HdeB peak in O157 strains [15].

The peak at m/z 6040 has been reported as a biomarker specifically present in O157 strains [12]. However, in our study, the intensity of the peak at m/z 6040 was too low to be detectable and in more than half of the O157 strains used for the mass database (namely GTC 14513, GTC 14535, GTC 14536, GTC 14537, GTC 14544, GTC 14547, GTC 14551 and GTC 03904) the peak was absent (data not shown), suggesting that the presence/absence of suspicious biomarker proteins of low intensity is insufficient as a method for discrimination at the strain or serovar level.

In this study, the identification of other prevalent EHEC strains (O26 and O111) was considered. O26 and O111 strains could be distinguished from other E. coli strains by the peak at m/z 15409.4/15425.4 [M+H]$^+$ (Table 3, Fig. 1). From the sequence analysis, an amino acid change (A81S) in the DNA-binding protein H-NS in strains O26 and O111 was observed. A previous report had suggested that the protein corresponding to m/z

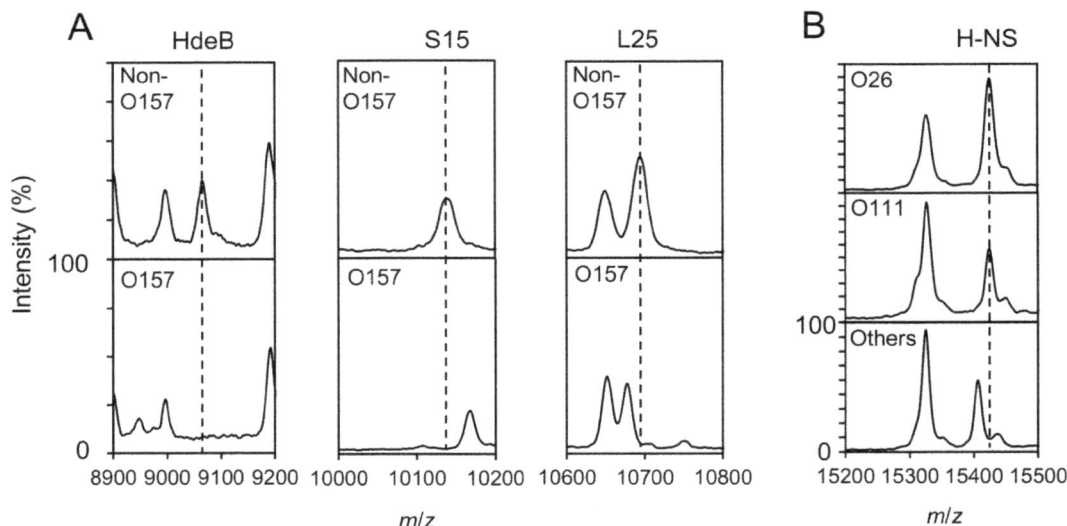

**Figure 1. Typical mass spectra of four biomarker proteins in _E. coli_.** A) MALDI mass spectra of non-EHEC _E. coli_ strain NBRC12713 and EHEC _E. coli_ strain O157 GTC 14513. Three biomarker peaks, HdeB (_m/z_ 9066.2 [M+H]$^+$), ribosomal protein S15 (_m/z_ 10138.6/10166.6 [M+H]$^+$) and L25 (_m/z_ 10676.4/10694.4 [M+H]$^+$), measured using sinapic acid as the matrix, are shown. B) Biomarker peak of DNA binding protein H-NS (_m/z_ 15409.4/ 15425.4 [M+H]$^+$) for strains O26, O111 and other _E. coli_.

15409.4 [M+H]$^+$ using K-12 strain (accession number P0ACF8) may be DNA-binding protein H-NS [18]; however, this required further confirmation. In our study, we first assigned the H-NS mass peaks and corresponding DNA sequences, and identified a specific mass shift in the H-NS protein in strains O26 and O111. Using the _S10_-GERMS method, in which the theoretical masses of biomarker proteins are confirmed, a valid and reliable mass database could be provided.

An additional peak at _m/z_ 8326 [M+H]$^+$ appeared to be another potential biomarker for _E. coli_ classification (Table 3). Its mass was identical to the theoretical mass of hypothetical protein YdaQ in a TagIdent tool search (http://web.expasy.org/tagident/). PCR analysis of this gene was performed and a target band was detected in three out of the eight strains tested that did not show the peak at _m/z_ 8326 (data not shown), suggesting that at least five strains may lack the _ydaQ_ gene in their genome causing a loss of this peak in MALDI-TOF MS analysis. The expression level of YdaQ in the three strains possessing the corresponding gene may be low. Investigations into the identification of these biomarker proteins are now in progress.

### Effects of culture media

In species level discrimination by MALDI-TOF MS, growth condition often affect the expression pattern of proteins thus causes less reproducibility of mass spectra [19]. Here, colonies grown on various selective media were analyzed by MALDI-TOF MS and evaluated whether the important biomarker peaks HdeB, S15 and L25 for O157, and H-NS for O26 and O111 work well for their discrimination. As a result, ribosomal protein S15 and L25 whose mass shifts are characteristic to O157 were not affected by any growth medium in accordance with the previous report that the impact of growth conditions on ribosomal proteins were minimum [20]. Similarly the masses of DNA-binding protein H-NS were not affected by culture medium in any _E. coli_ strain. On the other hand, mass intensity of HdeB in some strains of serovar O111 and O26 was decreased when grown on chromagar X-gal or VRBL, but nonetheless the peaks were enough to be detected in a default threshold. Therefore in the case of discrimination of O157, O26

and O111 from the others in colony directed MALDI-TOF MS analysis, the normal growth media and selective media such as desoxycholate agar, CT-SMAC, chromagar X-gal or VRBL will be available for the pre-selection of _E. coli_.

### Cluster analysis

All of the strains analyzed were correctly identified as _E. coli_ by SARAMIS. Cluster analysis based on the theoretical mass database of 10 biomarker proteins listed in Table 3, in which the mass patterns were classified into groups A to P, was performed using the actually detected peaks in semi-automated MALDI-TOF MS analysis. As mentioned above, small mass shifts of around 1 Da in the S15 and L25 proteins observed in the theoretical database in group K, I, L, O and P in Table 1 were difficult to detect in actual MALDI-TOF MS analysis, and therefore differences in these proteins were not reflected in the cluster profiling summarized in Table 4. In a phylogenetic tree illustrated based on this profiling, all the O157 strains were correctly classified into groups A, B, or C namely the 'O157 group' (Fig. 2). In addition, strains O26 and O111 belonged to the same cluster, groups D and E, owing to a mass difference in the peak at _m/z_ 15425.4, which was observed at _m/z_ 15409.4 in the other _E. coli_ strains tested (Fig. 1). Although high molecular weight proteins over 10000 Da are less detectable in many cases in MALDI-TOF MS [12,17], they could be powerful biomarkers, as reported for _Salmonella_ serovar identification [21]. To distinguish the small mass differences of S15 and L25 in group K, I, L, O and P from the others, MALDI-TOF MS analysis has mechanical limitation therefore MALDI-TOF/TOF-MS/MS analysis will be required to utilize such biomarkers in _E. coli_. Instead, the presence or absence of the _m/z_ 8326 peak made it possible to apply more detailed grouping.

### Discrimination of isolated wild-type _E. coli_ strains

The discrimination method proposed in this study was verified by performing blind tests using 12 _E. coli_ strains (O157 and O26) named as jfrl 01–12, and 45 wild-type strains (O157, O26, O111

**Table 4.** Peak pattern of *E. coli* strains in actual MALDI-TOF MS analysis.

| Group of mass pattern | | | | | | | | | | | | | | | |
| | A | B | C | D | E | F | G | H | I | J | K | L | M | N | O | P |
|---|---|---|---|---|---|---|---|---|---|---|---|---|---|---|---|---|
| L23 | 1 | 1 | 1 | 1 | 1 | 1 | 2 | 1 | 1 | 1 | 1 | 1 | 1 | 1 | 1 | 1 |
| L24 | 1 | 1 | 1 | 1 | 1 | 1 | 1 | 1 | 2 | 1 | 1 | 2 | 1 | 1 | 1 | 2 |
| S14 | 1 | 1 | 1 | 1 | 1 | 1 | 1 | 1 | 1 | 1 | 1 | 1 | 1 | 2 | 1 | 1 |
| L15 | 1 | 1 | 1 | 1 | 1 | 1 | 2 | 3 | 1 | 2 | 2 | 1 | 2 | 1 | 1 | 1 |
| S11+Me | 1 | 1 | 1 | 1 | | 1 | 1 | 1 | 1 | 1 | 1 | 1 | 1 | 1 | 1 | 1 |
| YdaQ | 1 | 0 | 1 | 1 | 0 | 1 | 1 | 1 | 1 | 1 | 1 | 0 | 0 | 1 | 0 | 0 |
| S15 | 2 | 2 | 1 | 1 | 1 | 1 | 1 | 1 | 1 | 1 | 1 | 1 | 1 | 1 | 1 | 1 |
| L25 | 2 | 2 | 1 | 1 | 1 | 1 | 1 | 1 | 1 | 1 | 1 | 1 | 1 | 1 | 1 | 1 |
| HdeB | 0 | 0 | 0 | 1 | 1 | 1 | 1 | 1 | 1 | 1 | 1 | 1 | 1 | 1 | 1 | 1 |
| H-NS | 1 | 1 | 2 | 2 | 2 | 1 | 1 | 1 | 1 | 1 | 3 | 1 | 1 | 1 | 1 | |

The actual mass of the peaks detected in MALDI-TOF MS has been replaced with number 1, 2 and 3 to represent the peak mass; 0 indicates no peak. Ribosomal protein S15 in group K and L25 in group I, L O and P were not distinguished from the others due to small mass differences in actual analysis with 500 ppm tolerance.

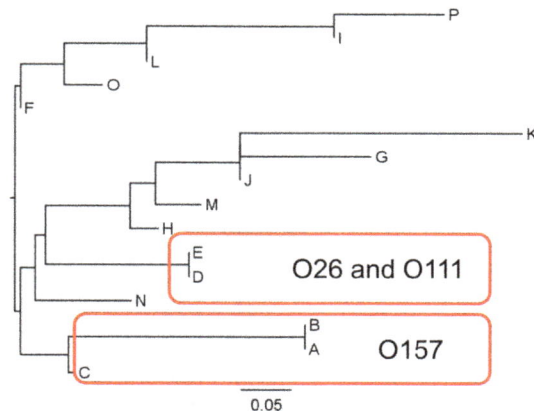

**Figure 2. Cluster analysis for *E. coli* strains with selected biomarkers.** Phylogenetic tree according to the binary graph made by actual analysis in Table 4 is shown. A to P indicate the *E. coli* groups classified by the theoretical mass patterns of biomarker protein peaks in Table 3.

and O121) kindly provided from provided from APIPH (Table 1). Among total 41 O157 strains, all of them showed typical mass shifts of the ribosomal proteins S15 and L25 and an absence of the peak at *m/z* 9066.2 with MALDI-TOF MS analysis. The ribosomal protein L23 and L24 were eliminated from the biomarkers because their mass peaks were not clear. Nevertheless they were correctly classified into the O157 group as categorized in Table 3 and Fig. 2. In detail, 39 strains were classified into group A and jfrl 01 and 07 were classified into group B due to a loss of the *m/z* 8326 peak. This result indicates that the variety of mass patterns in our database may be sufficient for serovar level discrimination of wild-type O157 strains regardless of the place or date of isolation. It should be noted that among the genome or partial sequence available strains of *E. coli* O157 (total 126 strains in the NCBI database), 119 (94.4%) strains including Sakai, FRIK2000, EC4206, EC4045, EC4196, EC4076, EC4113, EC4486, EC869, EC4501, EC508, EC4024, FRIK966, EC4115, EC4401, EC4486, EC4501, TW14588, TW14359, EDL933 and EC4042, have the same theoretical mass patterns for the biomarker proteins S15 and L25 as group A, that is typical 'O157 group' in our experiments. Only 7 strains in the database, namely G5101, 493–89, H 2687, LSU-61, 2010C-4979C1 and 98–3133, show the same masses of the other *E. coli* strains as classified in group C type O157 in Table 3. These findings suggest that a mass shift of ribosomal proteins S15 and L25 is common in most of O157 strains in the database, indicating that our discrimination approach that focuses on the mass shifts of S15 and L25 with the combination of a loss of HdeB peak, could be universally applied for O157 strain discrimination worldwide.

Similarly 11 O26 strains and 4 O111 strains for blind test were correctly classified into group E in Table 3 and Fig. 2 due to specific masses of H-NS. Although we could not isolate another O26 and O111 samples, the sequence available O26 strains in database (strain 11368, CVM9942, CVM10026, CVM10224, CVM10021, CVM9952, CVM10030, CFSAN001629, 2010C-4347, 2010C-4788, 05-3646, 06-3464, 03-3500, 2010C-4430, 2010C-4819, 2010C-4834, 2010C-5028, 2011C-3270, 2010EL-1699, 2011C-3387, 2011C-3282, 2011C-3506, 2011C-3655, 2009C-3612, 2009C-3689, 2009C-3996, 2009C-4760, 2009C-4826, 2010C-3051, 2010C-3871, 2010C-3472, 2010C-3902, 2010C-4244 and 2009C-4747) and O111 strains (strain 11128, CVM9534,

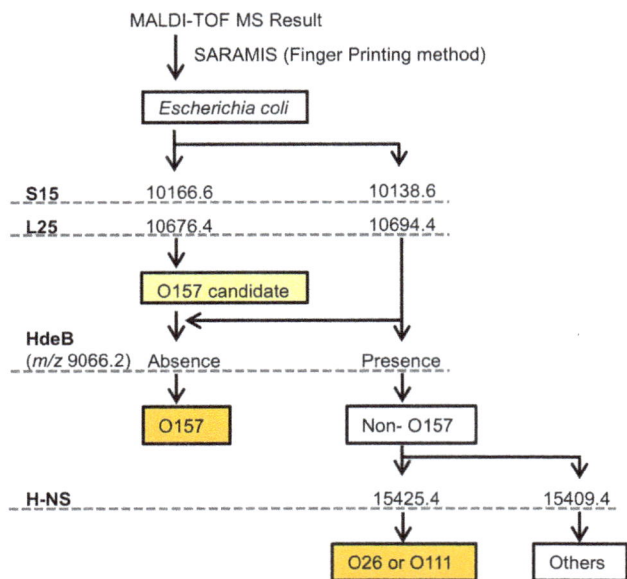

**Figure 3. Strategy for distinguishing between _E. coli_ strains O157, O26, O111 and the others using four biomarker peaks in MALDI-TOF MS.**

CVM9574, CVM9570, CVM9545, CVM9602, CVM9634, CVM9455, CVM9553, CFSAN001632, CFSAN001630, 2010C-4221, 2010C-3977, 2010C-4086, 08-4487, 2011C-3632, 2011C-3679, 2011C-3573, 2011C-3362, 2011C-3170, 2010C-4818, 2010C-4799, 2010C-4746, 2010C-4735, 2010C-4715, 2010C-4622, 2010C-4592, 04-3211, 03-3484, F6627, K6723, K6728, K6722, K6890, K6895, K6897, K6904, K6898, K6908, K6915, 2009C-4006, 2009C-4052, 2010C-3053, 2009EL-2169, 2009C-4126 and 2011C-3453) have the same theoretical mass values for DNA binding protein H-NS as group E, namely 'O26 and O111 group'. Therefore our findings for O26 and O111 will be also applicable for the isolates regardless of the date or place. A remaining wild-type strain, O121 was classified into group F.

To define O157, O26 and O111 from the others in a routine MALDI-TOF MS analysis in laboratory, masses of four selected biomarkers must be characteristic to these serovars. Although we have tested more than 11 different serovars in the present study for the construction of database and confirmed that at least no other serovars showed the same mass patterns as either O157, O26 or O111 in this experimental scale (Table 1, Table 3) and the biomarkers were effective in the blind test using 57 wild-type

strains including O121, the examination with wider variety of serovars is desired. Since it was difficult to obtain _E. coli_ strains with various serovars, the probability of our database was validated _in silico_ by checking the theoretical masses of biomarkers in various types of _E. coli_ serovars. Out of more than thousands of _E. coli_ strains available in the NCBI database, the theoretical masses of ribosomal protein S15 and L25 in all non-O157 _E. coli_ strains were calculated as _m/z_ 10138.6 and 10694.4, respectively. They are completely equal to that of group C to P shown in Table 3, namely 'non-O157 group', indicating the database constructed in this study will work well for screening of O157 from various types of serovars.

On the other hand, the specific biomarker, H-NS (_m/z_ 15425.4 [M+H]⁺) for O26 and O111 were observed in another few strains, O118 (strain 08-3651, 06-3612, 06-3256, 2009C-4446, 07-4255), O69 (strain 07-4281, 06-3325, 08-4661, 2009C-3601, 07-3763), O123 (strain 2009C-3307) and O103 (strain 2010C-3214). As they may be classified into the same group with O26 and O111 in our discrimination system using H-NS as biomarker, the extra biomarkers will be required for more detailed identification. Nevertheless it is promising that candidates for serovar O26 and O111 could be found in our system.

The major EHEC serovars O157, O26 and O111 present a great risk for human life, and therefore not only will the rapid discrimination of this strain from other _E. coli_ strains aid diagnostics, but it is also vital in ensuring clinical security and food safety. We propose a possible strategy for the effective discrimination of strains O157, O26 and O111 using specific four biomarkers by MALDI-TOF MS as shown in Fig. 3. Our _S10_-GERMS-based discrimination method uses the arbitrary selected masses of established biomarkers that are confirmed from the approaches of both genomics and proteomics. For automated processing and clustering of the data generated by MALDI-TOF MS, the analytical software 'Strain Solution' (Shimadzu, Kyoto, Japan) could be employed to realize the _S10_-GERMS approach. Our discrimination method will be an important screening tool clinically and in the dairy industry to ensure food safety.

## Acknowledgments

We thank MD., PhD. Hiroko Minagawa (Aichi Prefectural Institute of Public Health) for kindly supplying us wild-type _E. coli_ strains.

## Author Contributions

Conceived and designed the experiments: TO NY HT. Performed the experiments: TO NY MS. Analyzed the data: TO NY. Contributed reagents/materials/analysis tools: TF. Wrote the paper: TO NY HT.

## References

1. Watanabe H, Mori H, Itoh T, Gojobori T (1997) Genome plasticity as a paradigm of eubacteria evolution. J Mol Evol 44: S57–64.
2. Itoh T, Takemoto K, Mori H, Gojobori T (1999) Evolutionary instability of operon structures disclosed by sequence comparisons of complete microbial genomes. Mol Biol Evol 16: 332–346.
3. Coenye T, Vandamme P (2005) Organisation of the S10, spc and alpha ribosomal protein gene clusters in prokaryotic genomes. FEMS Microbiol Lett 242: 117–126.
4. Victoria B, Ahmed A, Zuerner RL, Ahmed N, Bulach DM, et al. (2008) Conservation of the _S10-spc-α_ locus within otherwise highly plastic genomes provides phylogenetic insight into the genus _Leptospira_. PLoS One 3: e2752.
5. Barloy-Hubler F, Lelaure V, Galibert F (2001) Ribosomal protein gene cluster analysis in eubacterium genomics: homology between _Sinorhizobium meliloti_ strain 1021 and _Bacillus subtilis_. Nucleic Acids Res 29: 2747–2756.
6. Teramoto K, Sato H, Sun L, Torimura M, Tao H, et al. (2007) Phylogenetic classification of _Pseudomonas putida_ strains by MALDI-MS using ribosomal subunit proteins as biomarkers. Anal Chem 79: 8712–8719.

7. Tamura H, Hotta Y, Sato H (2013) Novel accurate bacterial discrimination by MALDI-time-of-flight MS based on ribosomal proteins coding in _S10-spc-alpha_ operon at strain level _S10_-GERMS. J Am Soc Mass Spectrom 24: 1185–1193.
8. Sato H, Torimura M, Kitahara M, Ohkuma M, Hotta Y, et al. (2012) Characterization of the _Lactobacillus casei_ group based on the profiling of ribosomal proteins coded in _S10-spc-alpha_ operons as observed by MALDI-TOF MS. Syst Appl Microbiol 35: 447–454.
9. Josten M, Reif M, Szekat C, Al-Sabti N, Roemer T, et al. (2013) Analysis of the matrix-assisted laser desorption ionization-time of flight mass spectrum of _Staphylococcus aureus_ identifies mutations that allow differentiation of the main clonal lineages. J Clin Microbiol 51: 1809–1817.
10. Khot PD, Fisher MA (2013) Novel approach for differentiating _Shigella_ species and Escherichia coli by MALDI-TOF mass spectrometry. J Clin Microbiol 51: 3711–3716.
11. Benagli C, Demarta A, Caminada AP, Ziegler D, Petrini O, et al. (2012) A rapid MALDI-TOF MS identification database at genospecies level for clinical and environmental _Aeromonas_ strains. Plos One 7: e48441

12. Clark CG, Kruczkiewicz P, Guan C, McCorrister SJ, Chong P, et al. (2013) Evaluation of MALDI-TOF mass spectroscopy methods for determination of *Escherichia coli* pathotypes. J Microbiol Methods 94: 180–191.

13. Fagerquist CK, Garbus BR, Miller WG, Williams KE, Yee E, et al. (2010) Rapid identification of protein biomarkers of *Escherichia coli* O157:H7 by matrix-assisted laser desorption ionization-time-of-flight-time-of-flight mass spectrometry and top-down proteomics. Anal Chem 82: 2717–2725.

14. Hotta Y, Teramoto K, Sato H, Yoshikawa H, Hosoda A, et al. (2010) Classification of genus *Pseudomonas* by MALDI-TOF MS based on ribosomal protein coding in *S10-spc-alpha* operon at strain level. J Proteome Res 9: 6722–6728.

15. Carter MQ, Louie JW, Fagerquist CK, Sultan O, Miller WG, et al. (2012) Evolutionary silence of the acid chaperone protein HdeB in enterohemorrhagic *Escherichia coli* O157:H7. Appl Environ Microbiol 78: 1004–1014.

16. Ziegler D, Mariotti A, Pflüger V, Saad M, Vogel G, et al. (2012) In situ identification of plant-invasive bacteria with MALDI-TOF mass spectrometry. PloS One 7: e37189.

17. Hotta Y, Sato J, Sato H, Hosoda A, Tamura H (2011) Classification of the genus *Bacillus* based on MALDI-TOF MS analysis of ribosomal proteins coded in S10 and spc operons. J Agric Food Chem 59: 5222–5230.

18. Momo RA, Povey JF, Smales CM, O'Malley CJ, Montague GA, et al. (2013) MALDI-ToF mass spectrometry coupled with multivariate pattern recognition analysis for the rapid biomarker profiling of *Escherichia coli* in different growth phases. Anal Bioanal Chem 405: 8251–8265.

19. Wieme AD, Spitaels F, Aerts M, De Bruyne K, Van Landschoot A, et al. (2014) Effects of growth medium on matrix-assisted laser desorption-ionization time of flight mass spectra: a case study of acetic acid bacteria. Appl Environ Microbiol 80: 1528–1538.

20. Valentine N, Wunschel S, Wunschel D, Petersen C, Wahl K (2005) Effect of Culture Conditions on Microorganism Identification by Matrix-Assisted Laser Desorption Ionization Mass Spectrometry. Appl Environ Microbiol 71: 58–64

21. Dieckmann R, Malorny B (2011) Rapid screening of epidemiologically important *Salmonella enterica* subsp. *Enterica* serovars by whole-cell matrix-assisted laser desorption ionization-time of flight mass spectrometry. Appl Environ Microbiol 77: 4136–4146.

# Is There Still Room for Novel Viral Pathogens in Pediatric Respiratory Tract Infections?

Blanca Taboada[1], Marco A. Espinoza[1], Pavel Isa[1], Fernando E. Aponte[1], María A. Arias-Ortiz[2], Jesús Monge-Martínez[2], Rubén Rodríguez-Vázquez[2], Fidel Díaz-Hernández[2], Fernando Zárate-Vidal[2], Rosa María Wong-Chew[3], Verónica Firo-Reyes[4], Carlos N. del Río-Almendárez[5], Jesús Gaitán-Meza[6], Alberto Villaseñor-Sierra[7], Gerardo Martínez-Aguilar[8], Ma. del Carmen Salas-Mier[8], Daniel E. Noyola[9], Luis F. Pérez-Gónzalez[10], Susana López[1], José I. Santos-Preciado[3], Carlos F. Arias[1]*

1 Instituto de Biotecnología, Universidad Nacional Autónoma de México, Cuernavaca, Morelos, Mexico, 2 Colegio de Pediatría del Estado de Veracruz, Veracruz, Mexico, 3 Facultad de Medicina, Universidad Nacional Autónoma de México, México D.F., Mexico, 4 Hospital General de México, México D.F., Mexico, 5 Hospital Pediátirco de Coyoacán, México D.F., Mexico, 6 Nuevo Hospital Civil de Guadalajara "Dr. Juan I. Menchaca", Guadalajara, Jalisco, Mexico, 7 Centro de Investigación Biomédica de Occidente, IMSS, Guadalajara, Jalisco, Mexico, 8 Unidad de Investigación Biomédica IMSS, Durango, Durango, Mexico, 9 Universidad Autónoma de San Luis Potosí, San Luis Potosí, Mexico, 10 Hospital Central "Dr. Ignacio Morones Prieto", San Luis Potosí, Mexico

## Abstract

Viruses are the most frequent cause of respiratory disease in children. However, despite the advanced diagnostic methods currently in use, in 20 to 50% of respiratory samples a specific pathogen cannot be detected. In this work, we used a metagenomic approach and deep sequencing to examine respiratory samples from children with lower and upper respiratory tract infections that had been previously found negative for 6 bacteria and 15 respiratory viruses by PCR. Nasal washings from 25 children (out of 250) hospitalized with a diagnosis of pneumonia and nasopharyngeal swabs from 46 outpatient children (out of 526) were studied. DNA reads for at least one virus commonly associated to respiratory infections was found in 20 of 25 hospitalized patients, while reads for pathogenic respiratory bacteria were detected in the remaining 5 children. For outpatients, all the samples were pooled into 25 DNA libraries for sequencing. In this case, in 22 of the 25 sequenced libraries at least one respiratory virus was identified, while in all other, but one, pathogenic bacteria were detected. In both patient groups reads for respiratory syncytial virus, coronavirus-OC43, and rhinovirus were identified. In addition, viruses less frequently associated to respiratory infections were also found. Saffold virus was detected in outpatient but not in hospitalized children. Anellovirus, rotavirus, and astrovirus, as well as several animal and plant viruses were detected in both groups. No novel viruses were identified. Adding up the deep sequencing results to the PCR data, 79.2% of 250 hospitalized and 76.6% of 526 ambulatory patients were positive for viruses, and all other children, but one, had pathogenic respiratory bacteria identified. These results suggest that at least in the type of populations studied and with the sampling methods used the odds of finding novel, clinically relevant viruses, in pediatric respiratory infections are low.

**Editor:** Amit Kapoor, Columbia University, United States of America

**Funding:** This work was supported by grants 153639 (to JIS-P) and "Influenza 2009" (to CFA) from the National Council for Science and Technology—Mexico. www.conacyt.mx. FEA was recipient of a scholarship from the National Council for Science and Technology—Mexico. The funders had no role in study design, data collection and analysis, decision to publish, or preparation of the manuscript.

**Competing Interests:** The authors have declared that no competing interests exist.

* Email: arias@ibt.unam.mx

## Introduction

Acute respiratory infections (ARIs) are the most common illnesses in humans and are associated with significant morbidity and mortality in young children in developing countries and elderly people in developed countries. In children, 156 million episodes of pneumonia are recorded annually worldwide, of which more than 95% are reported in developing countries [1,2]. In 2008, 1.6 million children younger than 5 years died from pneumonia [3]. To try to reduce child mortality due to ARIs, is important to perform a more accurate diagnosis of the pathogens associated with those deaths in children younger than 5 years of age [1].

Introduction of PCR-based diagnostic methods has increased the ability to detect respiratory viruses, which are responsible for most ARIs in young children [4,5,6]. Several respiratory viruses, such as influenza, parainfluenza virus, adenovirus, respiratory syncytial virus (RSV) and coronavirus (HCoV) have been known for some time as etiological agents of lower tract respiratory infections (LRTI). More recently, with the improvement of diagnostic methods, rhinovirus (RV), which had been thought to be mostly associated with mild-to-moderate upper respiratory tract infections (URTI) was also found to be associated with severe respiratory infections [7,8] and, in the last decade, several new respiratory viruses have been identified, such as human metapneumovirus (hMPV), HCoV-NL63 and -HKU1, human bocavirus (HBoV), parechovirus (HPeV), polyomavirus KI and WU, and enterovirus 104 and 109 [9,10,11,12,13,14,15]. In this regard, the fact that even with state-of-the-art diagnostic tools in most studies a virus is detected in only 50% to 80% of upper and lower ARIs [4,5,6,16,17,18,19] a wonder is if there are more respiratory viruses associated to ARIs than those currently known [20].

In this work, we analyzed by next generation sequencing (NGS) nasopharyngeal samples from children with LRTI and URTI that had been found negative for a panel of 21 respiratory pathogens (15 viruses and 6 bacteria) using commercial multiplex PCR methods. This study contributes to the description of the viral and bacterial populations present in nasopharyngeal samples from children with lower and upper ARIs using a metagenomic approach, which so far has been employed in limited studies [21,22,23], and suggests that the current diagnostic methods likely miss known respiratory pathogens, which might explain the relatively high proportion of undiagnosed cases.

## Materials and Methods

### Study populations and clinical samples

Two pediatric populations with symptomatic respiratory tract infections were included in this study. The first consisted of children with LTRI that required hospital admission due to clinical or radiological signs or symptoms of pneumonia in four different states of Mexico. Nasal washings with 1.5 ml of saline solution were collected from 250 children (male:female ratio, 1.43; age range, 1–76 months) between March 2010 and April 2011. The second population was composed of patients with symptomatic URTI that attended the private consult in five different cities of the state of Veracruz, Mexico. Nasopharyngeal swabs (rayon-tipped, BD BBL) were collected from 526 children (male:female ratio, 1.27; age range, 0–191 months) from September 2011 to April 2012. All samples were placed in vials containing viral transport medium (1:1 in the case of nashal washings; Microtest M4-RT, Remel) and sent frozen in blue ice either to the Institute of Biotechnology in Cuernavaca (URTI samples) or to the School of Medicine in Mexico City (LRTI samples) and stored at −70°C until analyzed. All children were previously healthy, not diagnosed with tuberculosis or signs of malnutrition, and not immunocompromised. Administration of antibiotics before hospital admission was not registered; in outpatients no antibiotics were administered before sample collection. The children included in the study were those that arrived consecutively at the collection places during the study period, with no further selection. The study (project 186) was approved by the institutional review boards of the School of Medicine and the Institute of Biotechnology of the National University of Mexico and from the institutional review board and ethics committee of each participant hospital. Written informed consent was obtained from each parent or guardian prior to enrollment.

### Pathogen detection

The respiratory specimens from hospitalized and outpatient children were previously screened for viruses using the xTAG Bioplex respiratory Viral Panel (Abbott, Rungis, France) (JI Santos et al., in preparation) and the Seeplex RV15 ACE detection kit (Seegene, Seoul, Korea) (Wong-Chew et al., in preparation), respectively. The virus-negative samples from both groups of patients were screened in this work by a multiplex PCR (Seeplex

Pneumobacter ACE detection kit, Seegene, Seoul, Korea) for the presence of six bacteria commonly associated to respiratory infections: *Streptococcus pneumoniae*, *Haemophilus influenzae*, *Chlamydophila pneumoniae*, *Legionella pneumophila*, *Bordetella pertussis*, and *Mycoplasma pneumoniae*.

### Nucleic acid extraction, amplification and barcode labeling

Genetic material from clinical samples was extracted with the PureLink Viral RNA/DNA kit according to the manufacturer's instructions (Invitrogen, Waltham, MA). Before extraction, samples (200 μl) were treated with Turbo DNAse (Ambion, Waltham, MA) and RNAse (Sigma, St. Louis, MO) for 30 min at 37°C and immediately chilled on ice. Nucleic acids were eluted in nuclease-free water, aliquoted, quantified in NanoDrop ND-1000 (Nano-Drop Technologies, Waltham, MA), and stored at −70°C until further use. Sample random primer-amplification of nucleic acids was performed essentially as described previously [24]. Briefly, reverse transcription was done using SuperScript III Reverse Transcriptase (Invitrogen, Waltham, MA) and primer-A (5'-GTTTCCCAGTAGGTCTCN$_9$-3'). Complementary DNA (cDNA) strand was generated by two rounds of synthesis with Sequenase 2.0 (USB, USA). The cDNA obtained was then amplified with KlenTaq polymerase (Sigma, St. Louis, MO) using the primer-B (5'-GTTTCCCAGTAGGTCTC-3') and 20 cycles of the following program: 30 sec at 94°C, 1 min at 50°C, 1 min at 72°C. After cleaning the PCR products with the DNA Clean & Concentrator-5 kit (Zymo Research, Irvine, CA), DNA was digested with the GsuI restriction enzyme (Fermentas Waltham, MA) for 2 h at 30°C to remove sequences corresponding to PCR primers. After digestion, samples were purified again and used as starting material to prepare 300 bp-sized libraries using Illumina's Genomic DNA sample Prep Kit with multiplex primers as suggested by the manufacturer (Illumina, San Diego, CA). Libraries were loaded in a flow cell (4 or 5 libraries per lane) and sequencing was performed by 72 cycles of nucleotide extension followed by acquisition of multiplex code in a Genome Analyzer IIx. The datasets generated by the GAIIx were deposited in the European Nucleotide Archive, with study accession numbers PRJEB7390 and PRJEB7391 for URTI and LRTI samples respectively.

### Deep sequencing and sequence analysis

Image analysis and base calling were performed with the Illumina GAPipeline program (version 1.3.0) using standard parameters. To separate the samples, the pooled data from each lane were binned by barcode. In-house scripts were developed for the sequence analysis, including the following steps:

*i) Preprocessing.* For each read, the adapter and 5' and 3' bases with no-call sites (N residues) and low-quality (Phred-like scores < 20) were trimmed. Then, low complexity reads and less than 35 bases long were removed. Finally, identical reads were collapsed into a single representative sequence to optimize analysis time. Only reads passing the preprocessing step were considered valid.

*ii) Removal of host sequences.* The program SMALT (Wellcome Trust Sanger Institute, 2012) was used to align the reads against mitochondrial, human genome, and bacterial ribosomal RNA to remove them, using 90% coverage and 90% identity.

*iii) Taxonomic identification.* To minimize CPU time, valid reads were aligned to bacteria, fungi and viruses nt NCBI databases, using SMALT with 70% coverage and identity. Then, the reads that mapped were aligned with standalone BLASTn against the databases described above, using an E-value of 1e–03.

To avoid misclassification, the first 100 hits were obtained for each sequence. Reads that did not map were considered as unidentified.

*iv) Taxonomic classification.* To assign reads to the most appropriate taxonomic level the software MEGAN 4.70.4 was used, which assigns a read to the lowest common taxonomic ancestor of the organisms corresponding to the set of significant hits.

*v) Assembly.* Reads assigned to the same virus family level were subsequently used for *de novo* assembly with Velvet 1.1.04 to increase the accuracy of classification. Each assembly contig was aligned against BLASTn database.

*vi) Detection of novel viruses.* All unidentified sequences unaligned using SMALT nucleotide alignment were assembled *de novo* with Metavelvet modified by us to improve the assembly efficiency. First, we conducted exploratory assemblies of the reads using multiple hash lengths (k = 17–35). Then, additional assembly of all unused reads from the exploratory assemblies was done (k = 21). Finally, we assembled all contigs obtained from all exploratory assemblies and the unused reads assembly by using the program VelvetOptimiser. From this final assembly, contigs that were greater than 180 nt were directly compared with NCBI nr (non- redundant protein) database using BLASTx with an E-value of 100 in an attempt to identify novel viruses.

## Phylogenetic tree inference

Metagenomic contigs from specific viruses that were at least 150 nt-long, were phylogenetically characterized. The analysis required a different approach compared to full-length genomes due to the fact that metagenomics sequences are fragmentary and not completely overlapping. Therefore, for each virus, a database of complete genomes was first created using all sequences available in GenBank until January 2014. Then, a reference alignment was done with sequences of this database by using MUSCLE method. Next, we combined metagenomics contigs into a single large alignment by using the software MAFFT with the option align fragment sequences to reference alignment. Finally, maximum likelihood trees were generated with 1000 repetitions bootstrap using the MEGA program.

## Results

### Pathogen detection

In previous studies we screened by RT-PCR the presence of 15 respiratory viruses in nasal washings from 250 hospitalized children with clinical diagnosis suggestive of viral pneumonia and in 526 nasopharyngeal samples from pediatric children with URTI (see Materials and Methods). Table 1 shows the frequency of the different viruses found in both types of samples. Among the viruses detected, considering both single and multiple infections, RSV-A and rhinovirus showed the highest frequency in both LRTI and URTI. At least one virus was detected in 71.2% (178/250) of LRTI (Santos et al., manuscript in preparation) and 71.5% (376/526) of URTI (Wong-Chew et al., manuscript in preparation). In 40 of the 250 LRTI samples (16%) a viral coinfection was found. Thirty-four of these samples had a dual infection, with the combination of RSV-A/RV and RSV-A/AdV being the more frequent, while 6 children had triple virus infections. In the case of URTI, 73 of the 526 samples (13.9%) showed a viral coinfection. Sixty-three of these samples had a dual infection, with the combination of AdV/EV and RV/CoV 229/N63 being the more frequent. Eight children had triple virus infections, and two were infected simultaneously with four viruses.

The virus-negative samples were screened by a multiplex PCR for the presence of six bacteria commonly associated to respiratory infections. In 64.7% (46/71, LRTI) and 68.7% (103/150, URTI) of the virus-negative samples at least one bacterial pathogen was found. The most frequent bacteria detected in children in both types of populations were *S. pneumoniae* (36 LRTI, 88 URTI) and *H. influenzae* (24 LRTI, 47 URTI); in a few cases *C. pneumoniae* (9 URTI) and *M. pneumoniae* (2 LRTI, 2 URTI) were also detected. In 37 children with URTI two different bacteria were found, and in 3 children 3 bacteria were detected. In the case of LRTI, 8 children had a mixed infection. It is important to have in mind that bacterial colonization, frequently at lower bacterial colony counts, may be detected by very sensitive laboratory tests, and even more frequently than viruses, these bacteria may not be associated with acute disease.

After screening for common respiratory viruses and bacteria, 90% of children with LRTI and 91.3% with URTI had at least one pathogen identified. The remaining 25 (10%) hospitalized and 46 (8.7%) outpatient children remained negative for all the tested pathogens and were then characterized by next-generation sequencing (NGS).

## Next-generation sequencing of negative samples

To search for either known or novel respiratory pathogens in the double-negative (virus and bacteria) samples, the nucleic acids in these samples were isolated, amplified by PCR, and sequenced using the Illumina platform, as described in Materials and Methods. The 25 samples from children with LRTI were sequenced individually (listed in Table 2). In the case of the URTI samples, 9 were sequenced individually, while the amount of DNA isolated from the other 37 samples was too low to be analyzed independently, thus, they were used to prepare 16 pools for sequencing: 13 pools of two samples, 1 pool of three samples, and 2 pools of four samples (Table 3).

The total number of DNA reads and the valid unique reads obtained from each sample after passing the quality controls are shown in Tables 2 and 3. The valid reads were analyzed for the presence of sequences from human, bacterial, fungal, or viral origin. As expected, the most abundant reads were from human origin, representing 70% and 80% of LRTI and URTI patients, respectively (Fig. 1). Bacterial sequences made up the second largest data set, representing 15.2% of the sequence reads in LRTI and 8.5% in URTI. Viral sequences represented 0.56% and 0.57% of valid reads in LRTI and URTI, respectively, and only 0.05% of reads corresponded to fungi (Fig. 1). Finally, approximately 13% and 10% of the sequences in both LRTI and URTI could not be classified since no homolog was found (E-value 1e–03) or there were contradicting database hits. This category is referred to as 'undefined' in Figure 1. Of interest, despite the fact that the samples from LRTI and URTI were collected by different methods (nasal washings vs. swabs), and from children with different clinical syndromes and varying severity of respiratory disease, the proportion of sequences from different origins was very similar.

The undefined sequence reads from all samples were assembled, and contigs ≥180 nt were compared with non-redundant protein database of GenBank (E-value 100) to find sequences that could be distantly related to known viral sequences and could thus represent novel viruses. Indeed, short sequences are less likely than long sequences to retrieve statistically significant similarities in Blast searches, and sequence assembly into longer contigs is helpful to overcome this difficulty. As result of this, all filtered contigs aligned either to bacterial or human proteins during BLASTx runs. An analysis revealed that the contigs that map to bacteria showed only 60–80% nucleotide identity to their best-matching reference, indicating that they most likely represent novel species within their

**Table 1.** Frequency of viral pathogens in children with URTI and LRTI.

| Virus | URTI (%) [a] | LRTI (%) |
|---|---|---|
| Respiratory syncytial virus-A | 96 (18.3) | 77 (30.8) |
| Rhinovirus | 92 (17.5) | 62 (24.8) |
| Influenza virus A | 48 (9.1) | 4 (1.6) |
| Adenovirus | 38 (7.2) | 14 (5.6) |
| Enterovirus | 31 (5.9) | 2 (0.8) |
| Metapneumovirus | 28 (5.3) | 19 (7.6) |
| Coronavirus 229E/NL63 | 28 (5.3) | 2 (0.8) |
| Coronavirus OC43 | 18 (3.4) | 4 (1.6) |
| Parainfluenza virus 3 | 18 (3.4) | 27 (10.8) |
| Parainfluenza virus 1 | 15 (2.9) | 8 (3.2) |
| Bocavirus | 13 (2.5) | 8 (3.2) |
| Parainfluenza virus 4 | 13 (2.5) | 2 (0.8) |
| Parainfluenza virus 2 | 9 (1.7) | 7 (2.8) |
| Influenza virus B | 7 (1.3) | 4 (1.6) |
| Respiratory syncytial virus-B | 7 (1.3) | 2 (0.8) |

[a]The number of viruses include those present in single and mixed infections. The percentage refers to the total number of viruses detected.

**Table 2.** DNA reads obtained after NGS sequencing of LRTI samples.

| Sample | No. of reads | [a]No. of valid reads (%) |
|---|---|---|
| 11 | 7,336,101 | 6,243,183 (85.8) |
| 17 | 13,118,032 | 10,877,350 (83.5) |
| 24 | 8,522,571 | 580,599 (7.0) |
| 28 | 4,210,763 | 3,400,829 (81.3) |
| 47 | 9,051,977 | 1,474,628(16.3) |
| 64 | 10,626,262 | 1,125,214 (10.7) |
| 66 | 11,937,236 | 713,915 (6.1) |
| 67 | 10,779,916 | 696,529 (6.6) |
| 86 | 16,500,963 | 1,806,051 (11.1) |
| 111 | 13,312,546 | 941,951 (7.3) |
| 124 | 17,270,372 | 1,943,449 (11.4) |
| 125 | 10,053,798 | 627,927 (6.4) |
| 147 | 6,229,459 | 464,760 (7.5) |
| 151 | 14,208,710 | 1,248,513(8.9) |
| 206 | 2,881,815 | 137,274 (9.1) |
| 210 | 16,684,541 | 9,969,067 (60.2) |
| 211 | 10,784,070 | 1,099,819 (10.4) |
| 213 | 9,137,653 | 775,688 (8.6) |
| 214 | 11,503,832 | 4,321,038 (37.8) |
| 225 | 18,787,796 | 1,534,587 (8.4) |
| 227 | 19,294,597 | 1,483,230 (7.9) |
| 233 | 14,731,213 | 3,628,367 (24.9) |
| 236 | 13,712,286 | 974,744 (7.2) |
| 237 | 17,095,111 | 2,780,835 (16.8) |
| 238 | 12,373,402 | 3,277m334 (26.7) |

[a]Valid DNA reads after discarding those that did not pass the quality filter, and removing repeated reads (see Methods).

**Table 3.** DNA reads obtained after NGS sequencing of URTI samples.

| [a]Individual and pooled samples | No. of reads | [b]No. of valid reads (%) |
|---|---|---|
| C06, C55, T78, V24 | 4,323,309 | 1,707,160 (39.5) |
| C16, C61 | 2,113,065 | 1,134,387 (53.7) |
| C27, C01 | 2,951,784 | 1,474,233 (49.9) |
| C29, M40, T41, V39 | 3,325,315 | 1,235,867 (37.2) |
| C41, T50 | 3,765,099 | 1,708,531 (45.4) |
| C46, P54 | 4,289,979 | 2,278,068 (53.1) |
| M23, M44 | 3,607,011 | 1,982,487 (55.0) |
| M28 | 7,294,104 | 3,292,828(45.1) |
| P06, P150 | 5,010,364 | 2,387,550 (47.7) |
| P108 | 4,609,150 | 1,021,900 (22.2) |
| P147, P191 | 3,454,018 | 1,540,283 (44.6) |
| P149, P153 | 4,381,981 | 2,534,733 (57.8) |
| P151, P181 | 4,638,675 | 2,207,800 (47.6) |
| P173 | 2,974,252 | 613,487 (20.6) |
| P176, P186, P213 | 4,210,620 | 2,176,224 (51.7) |
| P183 | 1,654,081 | 272,720 (16.5) |
| P19, P88 | 8,727,411 | 5,067,242 (58.1) |
| P69 | 3,072,261 | 631,373 (20.6) |
| T33, T39 | 4,863,936 | 3,308,371 (68.0) |
| T36 | 1,228,687 | 446,932 (36.4) |
| T38 | 3,309,781 | 1,420,555 (42.9) |
| T43, T44 | 3,183,181 | 2,325,025 (73.0) |
| T65 | 78,467 | 30,443 (38.8) |
| V26 | 3,469,556 | 1,496,608 (43.1) |
| P131, V61 | 3,939,137 | 2,288,528 (58.1) |

[a]The samples that were pooled for sequencing are indicated.
[b]Valid DNA reads after discarding those that did not pass the quality filter, and removing repeated reads (see Methods).

corresponding genera and thus could not be classified during alignments with BLASTn. Nonetheless, the vast majority of reads (50% to 90%) were not assembled into contigs. The unassembled reads were low complexity sequences or library artifacts as adapter chimeras, suggesting that it is unlikely that they correspond to novel viruses. A remaining small amount of sequences could not be assembled due to non-uniform read depth because of a non-uniform species abundance distribution.

## Viruses detected by NGS in double-negative samples

DNA sequence reads from at least one virus commonly associated to respiratory infections was found in 20 out of the 25 double-negative samples of LRTI patients (Table 4): 5 samples were positive for RSV reads, 11 samples for HCoV-OC-43, and 9 for RV. In addition, 5 samples contained HBoV and in 12 samples anelloviruses (torque teno -TTV-, torque teno mini -TTMV-, or torque teno midi viruses -TTMDV) were also detected; rotavirus, papillomavirus, and herpesvirus sequences were identified once in the samples, and reads from several viruses from both animal (bat picornavirus, bovine viral diarrheal virus, bovine kobovirus) and plant origin (potato virus Y, pepper mild mottle virus), as well as various bacteriophages were also found (Table 4). Regarding bacteria, DNA sequence reads from *S. pneumoniae* were the most frequent, being present in all but one of the 25 samples sequenced, and *M. catarrhalis*, *L. pneumoniae*, and *H. influenzae* were less

frequently found. DNA reads from other bacteria less commonly associated with respiratory infections were also detected (Table 4). Some of the samples had sequence reads corresponding to up to 8 different viruses or 15 different bacteria. Of interest, including the NGS results, 79.2% (198/250) of the samples had a respiratory virus detected, and in the remaining 52 samples at least one bacteria was found, such that all 250 samples from children with LRTI had a respiratory pathogen identified.

DNA reads from one to five typical respiratory viruses were detected in 22 of the 25 sequenced double-negative individual and/or pooled samples from children with URTI (Table 5): The virus most frequently detected was RV, which was found in 19 of the pooled and/or individual samples; some of the samples had more than one type of virus, such that we found sequence reads from 4 RV subtype A, 4 subtype B, and 19 subtype C. One sample was positive for RSV, 3 for HCoV-OC43, 3 for human enterovirus A71, and 3 samples had HBoV. Of interest, 5 of the samples had DNA reads from Saffold virus, a virus recently described to be associated to respiratory infections. Also, among these samples we identified 3 containing herpesvirus, 5 papillomavirus, 2 human astrovirus, 4 rotavirus, and 10 anelloviruses (TTV, TTMV, TTMDV). Similar to what was found in LRTI, in children with URTI DNA reads of viruses from animal (white spot syndrome and bat picornavirus) and plant origin (Okra mosaic virus, capsicum chlorosis virus, cucumber mosaic virus, pepper

## Valid LRTI reads

Bacteria
15.25%

Undefined
13.70%

Fungi
0.05%

Viruses
0.56%

Human
70.44%

## Valid URTI reads

Bacteria
8.89%

Undefined
9.92%

Fungi
0.05%

Viruses
0.57%

Human
80.57%

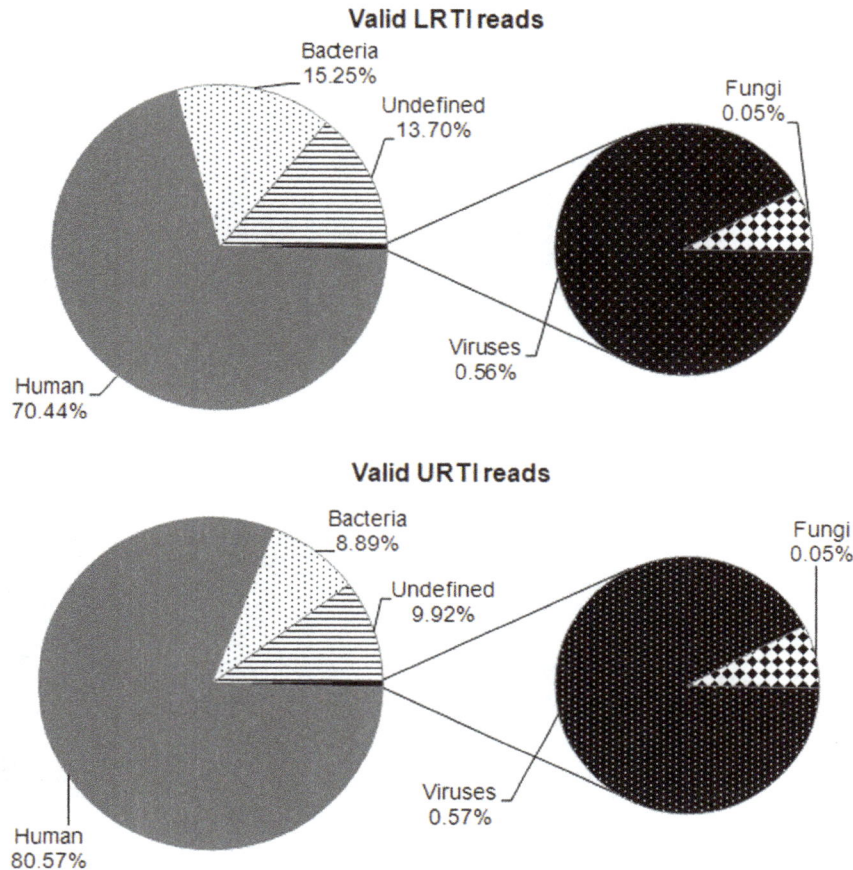

**Figure 1. Taxonomic classification of the generated DNA sequencing reads.** Valid DNA reads obtained by NGS of LRTI and URTI clinical samples were split into human, bacterial, fungal, and viral origin. Those reads not present in the four previous categories were classified as "undefined". Average values for all LRTI and URTI samples are shown.

mild mottle virus, and tomato mosaic virus) were also detected. In the majority of samples bacteriophages were found. After NGS, all pooled samples had DNA sequences from at least one respiratory virus, while three individual (not pooled) samples remained negative for common respiratory viruses (Table 5). In two of these samples (P183 and T38) *M. catarrhalis* was detected and *S. pneumoniae* was additionally present in one of them. The third sample (T65) remained negative for both viruses (including phages) and bacteria. Some of the samples from URTI had sequence reads corresponding to up to 11 different viruses or 18 different bacteria. Considering the NGS results, and assuming that in the pooled samples the identified viruses were each present in a different sample, 76.6% (403/526) of the children had a common respiratory virus detected, while in all but one of the other 129 children a respiratory bacteria was identified.

### Genome assembly and phylogenetic analyses

To estimate the sequence coverage of the NGS-identified viruses, the sequence reads were assembled *de novo*, and all contigs were used to estimate the extent of the virus genome coverage. A significant coverage was obtained for several of the detected viruses. In patients with LRTI, the genome of 14 viruses was assembled with coverage higher than 20% (Table 6). As indication of the sensitivity of NGS and of the relative abundance of some viruses not detected by the conventional PCR, we could assemble more than 95% of the genome of two RV strains, 98% of one

HBoV, and 99.8% of one HCoV-OC43 strain. In the case of children with URTI, at least 50% of the genome was covered for 15 viruses, including RV, HEV, Saffold virus, and TTV (Table 7), with 8 of them having a genome coverage of more than 90%. For the DNA reads of the animal viruses identified in both types of children populations the coverage ranged between 0.57 and 4.9%, and for plant viruses between 0.13 and 7.15% in the case of tomato mosaic virus (Tables 5A and B).

The assembled sequences from HCoV-OC43, RV, HBoV, Saffold, and anelloviruses were used to construct phylogenetic trees to determine the genetic similarity of the viruses characterized in this work with those in databases (see Materials and Methods). All viruses from LRTI and URTI grouped with clades formed by previously reported virus sequences (not shown). Anelloviruses could be readily classified as TTV, TTMV, or TTMDV, with some samples containing the three genera (Tables 4 and 5). All HCoV detected belong to species HCoV-OC43 and grouped with other known betacoronaviruses. HBoVs were all genotype 1, while the Saffold viruses detected in this work belonged to either genotype 2 or 3. Of interest, in the case of some RV, HBoV, Saffold, and anelloviruses, different contigs mapped to different clades, suggesting that recombination events are common in this type of viruses, as has been reported for RV [25].

**Table 4.** Pathogens identified by NGS in children with lower respiratory tract infections.

| Pathogen detected | 47 | 86 | 124 | 111 | 28 | 206 | 210 | 211 | 213 | 214 | 225 | 227 | 233 | 236 | 237 | 238 | 17 | 24 | 11 | 125 | 147 | 15 | 164 | 66 | 67 |
|---|---|---|---|---|---|---|---|---|---|---|---|---|---|---|---|---|---|---|---|---|---|---|---|---|---|
| **Samples** | | | | | | | | | | | | | | | | | | | | | | | | | |
| **VIRUSES** | | | | | | | | | | | | | | | | | | | | | | | | | |
| Respiratory syncytial virus | 3[a] | 372 | - | - | - | - | - | - | - | - | - | - | - | - | - | - | 12 | - | - | - | - | 7 | - | 318 | - |
| Coronavirus OC43 | - | - | - | - | 5 | - | 45 | 4 | 3 | 1222 | - | - | - | - | - | - | - | - | 392 | 600 | - | - | - | - | 3 |
| Rhinovirus A | 5 | - | 66 | - | 2 | - | - | 16009 | - | - | - | - | 206 | - | 239049 | 16 | 1672 | 2 | - | - | - | 3 | - | - | - |
| Rhinovirus C | - | - | 121 | - | - | - | - | 63 | - | 87209 | 3 | - | 38475 | - | - | 39 | - | - | - | - | - | - | - | - | - |
| Human bocavirus | - | - | - | - | - | 162 | - | - | - | 6 | - | - | 322 | - | - | - | 483 | - | 46 | - | - | - | - | - | - |
| Torque teno virus (TTV) | - | - | - | - | - | - | 5 | - | 616 | 7 | 1007 | - | 125 | 15 | - | - | 7 | 32 | - | - | 2 | 9 | - | 160 | - |
| Torque teno mini virus (TTMV) | - | - | - | - | - | - | 46 | - | 6 | - | - | - | 3 | - | - | - | - | 7 | - | - | - | 2 | - | 54 | - |
| Torque teno midi virus (TTMDV) | - | - | - | - | - | - | - | - | - | - | - | - | 9 | - | - | - | - | - | - | - | - | - | - | 6 | - |
| Human herpes virus | - | - | - | - | - | - | - | - | - | - | - | - | 8 | - | - | - | - | - | 2 | - | - | - | - | - | - |
| Human papilloma virus | - | - | - | - | - | - | - | - | - | - | - | - | - | - | - | - | - | - | - | - | - | - | - | - | - |
| Rotavirus | - | - | - | - | - | - | 3 | - | - | - | - | 31 | - | - | - | - | - | - | - | - | - | - | - | - | - |
| Bat picornavirus | - | - | - | - | - | - | 4 | - | - | - | - | - | - | - | - | - | - | - | - | - | - | - | - | - | - |
| Bovine kobuvirus | - | - | - | - | - | 6 | 130 | - | - | - | - | - | - | - | - | - | - | - | - | - | - | - | - | - | - |
| Bovine viral diarrhea virus | - | - | - | - | - | 5 | 20 | - | - | - | - | - | - | - | - | - | - | - | - | - | - | - | - | - | - |
| Pepper mottle virus | - | - | - | - | - | - | - | - | - | - | - | - | - | - | - | - | - | - | - | - | - | - | - | - | - |
| Potato virus Y | - | - | - | 7 | - | - | - | - | - | - | - | - | - | - | - | - | - | - | - | - | - | - | - | - | - |
| Phages | - | 236 | 294 | 955 | - | 33 | - | 109 | 83 | 9 | 82 | 394 | 32 | 117 | 25 | 17 | 335 | 2 | 2384 | 569 | - | 383 | 67 | 1201 | 1345 |
| **BACTERIA** | | | | | | | | | | | | | | | | | | | | | | | | | |
| Streptococcus pneumoniae | 3 | 10 | 14 | 7 | - | 57 | 4 | 34 | 25 | 37 | 3 | 22 | 25 | 6 | 94 | 418 | 4 | 25 | 23 | 27 | 420 | 48 | 186 | 104 | 3,964 |
| Haemophilus influenzae | 4 | 130 | 41 | 210 | - | - | - | 5 | 35 | 17 | - | 13 | - | - | - | - | - | 41 | - | - | - | - | 105 | 15 | - |
| Moraxella catarrhalis | 110 | 11,845 | 380 | - | - | - | - | 23 | 2,213 | 12 | 1,237 | 3 | 1,909 | 5 | 70 | - | 58 | 6 | - | 10 | - | 3 | - | 6 | 13 |
| Legionella pneumophila | - | 5 | 5 | - | - | 3 | - | 21 | - | - | 3 | 24 | - | - | - | - | - | - | 3 | - | - | 118 | - | - | - |
| Klebsiella pneumoniae | - | - | 37 | - | 5 | - | 5 | 5 | 6 | 2 | - | - | 4 | 4 | - | - | - | - | - | - | - | - | - | - | - |
| Staphylococcus aureus | - | - | - | - | - | 64 | - | 12 | 85 | 2 | 248 | - | 17 | 322 | 5 | - | 52 | - | - | 43 | 4 | 31 | - | - | 22 |
| Rahnella | - | 14 | - | 2,239 | - | - | - | - | - | - | 3 | - | - | - | - | - | 906 | - | - | 8 | - | - | - | 15 | 4 |
| Burkholderia cepacia | 3 | 228 | 242 | 1,753 | - | 133 | 29 | 475 | 236 | 22 | 420 | 1,135 | 98 | 148 | 6 | 16 | 325 | 25 | 163 | - | - | 15 | 255 | 1,168 | 106 |
| Acinetobacter baumannii | 67 | 1,419 | 1,092 | 4,622 | - | 685 | 152 | 1,491 | 1,660 | 57 | 1,252 | 1,793 | 579 | 46 | 37 | 54 | 2,248 | - | 2,210 | 180 | - | 350 | 347 | 783 | 2,425 |
| Pseudomonas aeruginosa | 14 | 398 | 222 | 1,044 | - | 531 | 61 | 282 | 430 | 9 | 719 | 1,525 | 123 | 914 | 26 | 10 | 1 | - | 965 | 63 | - | 199 | 307 | 1,175 | 369 |
| Mycobacterium tuberculosis | - | 7 | - | 11 | - | - | - | 3 | - | - | 7 | - | - | - | - | - | - | - | - | - | - | - | - | - | - |
| Actinomycetales | 15,563 | 25 | 52 | 316 | - | 38 | - | - | - | - | - | - | 17 | 19 | 32 | - | - | - | - | 33 | - | 10 | 15 | - | - |
| Burkholderia gladioli | 22 | 3,716 | 3,897 | 36,947 | - | 174 | 320 | 2,264 | 439 | 10,157 | 13,733 | 1,661 | 2,473 | 235 | 127 | 5 | 645 | 355 | - | - | - | 355 | 3,143 | 460 | 500 |
| Malassezia globosa | 35 | 154 | 270 | 447 | - | 1,554 | 48 | 785 | 208 | - | 146 | 848 | 214 | 620 | 1,915 | 1,062 | 21 | - | 392 | 141 | 102 | - | - | 142 | 87 |

**Table 4.** Cont.

| Pathogen detected | Samples | | | | | | | | | | | | | | | | | | | | | | | | |
|---|---|---|---|---|---|---|---|---|---|---|---|---|---|---|---|---|---|---|---|---|---|---|---|---|---|
| | 47 | 86 | 124 | 111 | 28 | 206 | 210 | 211 | 213 | 214 | 225 | 227 | 233 | 236 | 237 | 238 | 17 | 24 | 11 | 125 | 147 | 151 | 64 | 66 | 67 |
| Pseudomonas mendocina | - | 98 | 67 | 359 | - | 154 | 12 | 112 | 86 | - | 291 | 430 | 27 | 90 | 10 | - | - | 63 | - | 623 | 19 | 74 | 93 | 50 | 75 |
| Staphylococcus epidermidis | - | 88 | - | 2681 | - | 5,007 | - | - | 1,341 | - | - | - | - | - | 88 | - | - | - | - | 15 | - | 210 | - | 40 | 111 |
| Leifsonia xyli | 2,494 | - | - | - | - | - | - | - | - | - | - | - | - | - | - | - | - | - | - | - | - | - | - | - | - |
| Acidovorax | 69 | 1,280 | 885 | 6,704 | 808 | 222 | 1,325 | 2,042 | 106 | 3,814 | 5,079 | 543 | 551 | 59 | 40 | - | 17 | 7,074 | - | 7,704 | 2,057 | 995 | 771 | 5,345 | 5,061 |

[a]Number of valids DNA reads in the sample por the corresponding pathogen.

## Discussion

Improvements in diagnostic methods have increased the rate of identification of viral pathogens in different clinical conditions, such as gastrointestinal, respiratory, or neurologic infections. However, despite these advances there are still a significant number of cases (20–50%), in which the etiologic agents are believed to be viruses, but the agent is not identified [26]. Previously, we reported the presence of a respiratory virus in about 71% of nasal samples obtained from children with LRTI and URTI (Aponte et al., manuscript in preparation; see also the Pathogen Detection section above), using a PCR method able to detect 15 different respiratory viruses. After PCR screening the virus-negative samples for the presence of respiratory bacteria, 89.6% of children with LRTI and 91.1% with URTI had at least one potential pathogen identified. These percentages were raised to levels close to 80% for viruses in both patient populations after NGS analysis of the double-negative samples, and essentially 100% of the samples had either a common respiratory virus or bacteria identified (in only one of 526 URTI samples DNA reads from a potential pathogen was not identified). It is interesting that 6 of the 8 samples from both LRTI and URTI that were negative for viruses after NGS had less than one million valid reads (Tables 2 and 3). Since the number of sequence reads directly correlates with the amount of nucleic acids present in the original sample [21], the absence of virus detection in these samples could represent false-negative results; it is likely that with a larger amount of sample, or deeper sequencing, respiratory viruses could have also been detected.

The samples that resulted negative for viruses by PCR, and subsequently determined as PCR-positive for respiratory bacterial pathogens, were not characterized by NGS, but it is reasonable to assume that a high percentage of them could have also been positive for viruses by deep sequencing. It is difficult, however, to determine with confidence, which, if any, of the detected pathogens could be responsible for the clinical respiratory symptoms observed. The virus or bacteria detected by these methods could be present in the patient as an asymptomatic carrier state or as causal agents of asymptomatic infections. Studies comparing the presence of respiratory pathogens in nasal specimens from healthy children will help to resolve this issue. In addition, PCR and NGS are such sensitive techniques that the presence of small amounts of viral targets may not necessarily have clinical relevance. An additional limitation of this study is the limited number of samples analyzed. Exploring the possibility to define cutoff levels represents the next necessary step for diagnosing viral respiratory infections using molecular tests [27]. It is important to mention, however, that for several of the RV and HCoV detected in this study high genome sequence coverages were achieved. This observation indicates that a high number of DNA reads, and probably also of virus particles, were present in the samples. These viruses could have been undetected by PCR due to mismatches in the diagnostic primers used.

Of interest, the classes of viruses found by NGS in patients with LRTI and URTI were very similar, although their frequencies were different in the two study populations. RV was more frequently found in URTI (19 of 25 samples) vs. LRTI (9 of 25 samples), while coronavirus was more represented in LRTI (11/25) than in URTI (3/25). Only one of the 46 samples of children with URTI was positive for RSV, while in 5 of 25 samples from children with LRTI RSV was detected. Saffold viruses, members of the picornaviridae family and cardiovirus genus, were found only in children with URTI. Since their initial description in 2007, these viruses have been shown to circulate worldwide, occur early

**Table 5.** Pathogens identified by NGS in children with upper respiratory tract infections.

| Pathogen detected | C06, C55, T78, V24 | C16, C61 | C27, C01 | C29, M40, M41, V39 | C41, T50 | C46, P54 | M23, M44 | M28 | P06, P150 | P108 | P131, V61 | P147, P191 | P14,9 P153 | P151, P181 | P176, P186, P173 P213 | P19, P183 P88 | T33, T69 T39 | T36 T38 T44 | T43 | T65 V26 |
|---|---|---|---|---|---|---|---|---|---|---|---|---|---|---|---|---|---|---|---|---|
| **VIRUSES** | | | | | | | | | | | | | | | | | | | | |
| Respiratory syncytial virus | - | - | - | - | - | - | - | - | - | - | - | - | - | - | - | - | 3 | - | - | - |
| Coronavirus OC43 | - | - | - | - | - | - | - | - | - | - | - | 3 | 24 | - | 38 | - | - | - | - | - |
| Rhinovirus A | - | - | - | - | - | - | - | - | - | - | 49 | 59 | - | 1,052 | - | - | 2 | - | - | - |
| Rhinovirus B | - | - | - | - | - | - | - | - | 14 | - | 3,995 | 12 | - | 8 | - | - | - | - | - | - |
| Rhinovirus C | 8418[b] | 15 | 59 | 3,667 | 115,617 | 44 | 65,889 | 2 | 11,233 | - | 702 | 46 | 55 | 54 | 66 | 11,026 | 1,395 | 41 | 14,561 | 207 |
| Human enterovirus A | - | - | - | 15 | - | - | - | - | 15,236 | - | 78 | - | - | - | - | - | - | - | - | - |
| Human bocavirus | - | - | - | - | - | - | - | - | - | 12 | 8 | - | - | - | - | - | - | - | - | 98 |
| Saffold virus | 182 | - | - | 183 | - | - | - | - | - | - | 10 | - | 2,672 | - | 9,891 | - | - | - | - | - |
| Torque teno virus (TTV) | - | 9 | - | 7 | - | 187 | - | - | 12 | - | 212 | 14 | - | 7 | - | - | 3 | - | - | - |
| Torque teno mini virus (TTMV) | - | - | - | 3 | - | 100 | - | - | 22 | - | 3,408 | 9 | - | 17 | - | - | - | - | - | - |
| Torque teno midi virus (TTMDV) | 3 | 8 | - | 12 | - | 11 | - | - | - | - | 2,711 | 5 | - | 7 | 11 | - | - | - | - | - |
| Human herpes virus | - | - | - | - | 3 | - | - | - | - | 36 | - | - | - | - | - | 53 | - | - | - | - |
| Human papilloma virus | - | 3 | - | 3 | - | 6 | - | - | - | - | 3 | - | - | - | - | 5 | - | - | - | - |
| Rotavirus | - | - | - | 3 | - | 27 | - | - | - | - | - | - | 118 | - | 3 | - | 4 | - | - | - |
| Human astrovirus | 6 | - | - | 3 | - | - | - | - | - | - | - | - | - | - | 3 | - | - | - | - | - |
| Bat picornavirus | - | - | - | - | - | - | - | - | - | - | 4 | - | - | - | - | - | - | 7 | - | - |
| Capsicum chlorosis virus | - | - | - | - | - | - | - | - | - | - | - | - | - | - | 2 | - | - | - | - | - |
| Cucumber mosaic virus | 12 | - | - | - | - | 6 | - | - | - | - | - | - | - | - | - | - | - | - | - | - |
| Okra mosaic virus | - | 18 | - | - | - | - | - | - | - | - | - | - | - | - | - | 2 | - | - | - | - |
| Pepper mottle virus | - | - | 2 | - | - | - | - | - | - | - | - | - | - | - | - | - | - | - | - | - |
| Tomato mosaic virus | - | - | 12 | - | - | 4 | - | - | 3 | - | 4 | - | - | - | - | 8 | - | - | - | - |
| White spot syndrome virus | - | - | - | - | - | - | - | - | - | - | - | - | - | - | 31 | - | - | - | - | - |
| Phages | 10 | 28 | 31 | 54 | 46 | 102 | 60 | 6 | 12 | 11 | 16 | 327 | 36 | 139 | 10 | 568 | 6 | 21 | 8 | 43 |
| **BACTERIA** | | | | | | | | | | | | | | | | | | | | |
| Streptococcus pneumoniae | 13 | 80 | 103 | 14 | 3 | 180 | 12 | 51 | 42 | 3 | 15 | 21 | 51 | 83 | 20 | 274 | 9 | - | 4 | - |
| Haemophilus influenzae | 15 | - | 4 | 4 | - | 47 | 15 | - | 4 | - | 173 | - | 10 | 41 | 288 | 47 | 6 | - | 11 | 8 |
| Moraxella catarrhalis | 235 | 8 | 15 | 4 | 34 | 172 | 22 | 26 | 10 | - | 446 | 6 | 11 | 6 | 142 | - | 76 | - | 11 | 8 |
| Legionella pneumophila | 8 | 22 | 14 | - | - | 22 | 4 | 15 | - | - | - | 11 | - | 6 | - | 2 | - | - | - | - |
| Klebsiella pneumoniae | - | 9 | 3 | 3 | - | - | - | - | - | - | - | - | - | - | - | 3 | - | - | - | - |
| Staphylococcus aureus | 9 | 13 | 24 | 16 | - | 36 | - | 12 | - | - | - | 20 | - | 10 | 12 | 21 | 5 | - | 288 | - |

Samples[a]

**Table 5.** Cont.

| Pathogen detected | C06, C55, T78, V24 | C16, C61 | C27, C01 | C29, M40, M41, V39 | C41, T50 | C46, P54 | M23, M44 | M28 | P06, P150 | P108 | P131, V61 | P147, P191 | P14,9 P153 | P151, P181 | P173 | P176, P186, P213 | P183 P88 | P19, P88 | T69T39 | T33, T39 | T36T38T44 | T43, T44 | T65 V26 |
|---|---|---|---|---|---|---|---|---|---|---|---|---|---|---|---|---|---|---|---|---|---|---|---|
| | | | | | | | | | | | | | | | | | | | | | | | (Samples[a]) |
| Rahnella | 5 | - | - | - | - | - | - | - | - | - | - | - | 15 | - | - | - | - | - | - | - | - | - | - |
| Burkholderia cepacia | 65 | 46 | 16 | 9 | 7 | - | - | 14 | - | - | - | - | 4 | - | - | - | - | - | - | - | - | - | - |
| Acinetobacter baumannii | 1,783 | 1,754 | 825 | 1,301 | 778 | 2,390 | 5,204 | 402 | 202 | - | 654 | 2,107 | 208 | 1,544 | - | 1,347 | 556 | 179 | - | - | - | 179 | 50 |
| Pseudomonas aeruginosa | 81 | 37 | 32 | 134 | 16 | 115 | 97 | 29 | 131 | - | 61 | 167 | 163 | 211 | - | 196 | 274 | 17 | 16 | - | - | 7 | - |
| Mycobacterium tuberculosis | 6 | - | 4 | - | - | - | - | - | - | - | - | - | - | - | - | - | - | - | - | - | - | - | - |
| Actinomycetales | 1,017 | 410 | 2,283 | 1,141 | 7,369 | 575 | 692 | 357 | 1,600 | 64 | 482 | 1,049 | 1,106 | 779 | 13 | 753 | 122 | 13,679 | 201 | 2,159 | 13 | 341 | 117 |
| Burkholderia gladioli | - | 4 | - | 9 | - | - | - | - | - | - | - | - | - | - | - | - | - | - | - | - | - | - | - |
| Malassezia globosa | 21 | 10 | 87 | 17 | 145 | 103 | 34 | 267 | - | 43 | 578 | 12 | 111 | 48 | 123 | 102 | 136 | 31 | 77 | 11 | 75 | 57 | 42 |
| Pseudomonas mendocina | 124 | 106 | 53 | 242 | 56 | 109 | 160 | 46 | 146 | - | 73 | 201 | 137 | 205 | - | 214 | 370 | 50 | 17 | - | - | 3 | - |
| Rhodotorula glutinis | 265 | 396 | 74 | 202 | 134 | 69 | 510 | 11 | - | - | - | - | 10 | - | - | 8 | 19 | 7 | - | - | - | - | - |
| Staphylococcus epidermidis | 27 | 110 | 62 | 19 | 149 | 27 | 57 | 9 | 52 | - | 18 | - | 50 | 205 | - | 62 | - | 29 | 6 | - | - | 3 | - |
| Lysinibacillus sphaericus | 115 | 301 | 539 | 29 | 84 | 117 | 568 | 234 | - | - | - | - | - | - | - | - | 18 | - | - | - | - | - | - |
| Acidovorax | 508 | 190 | 117 | 356 | 76 | 80 | 315 | 124 | 30 | - | 163 | 136 | - | 189 | - | 207 | 58 | 31 | - | - | - | 9 | 12 |
| Fervidobacterium nodosum | 278 | 677 | 519 | 345 | 320 | 999 | 1,152 | 164 | 349 | - | 270 | 681 | 1,487 | 1,094 | - | 590 | - | - | 93 | 208 | - | 121 | - |

[a]When more than one sample were pooled for sequencing, the code for the various samples is mentioned.
[b]Number of valids DNA reads in the sample por the corresponding pathogen.

**Table 6.** Genome coverage for viruses present in LRTI samples.

| Sample | Virus[a] | No. of Contigs | No. of reads incorporated into contigs | Genome size | Genome Coverage (%) |
|--------|----------|----------------|----------------------------------------|-------------|---------------------|
| 86 | RSV | 22 | 213 | 15,191 | 6.75 |
| 124 | RV-A | 6 | 47 | 7,129 | 7.43 |
| | RV-C | 10 | 96 | 7,107 | 24.34 |
| 111 | PVY | 1 | 7 | 9,704 | 1.75 |
| 210 | HBoV | 16 | 74 | 5,299 | 33.97 |
| | HCoV | 3 | 10 | 30,578 | 5.89 |
| | TTMV | 2 | 14 | 2,912 | 8.24 |
| | BKV | 1 | 4 | 8,374 | 1.07 |
| | BVDV | 1 | 3 | 12,230 | 0.57 |
| 211 | RV-A | 1 | 13,418 | 7,129 | 94.60 |
| | RV-C | 11 | 47 | 7,107 | 17.03 |
| | BatPV | 1 | 3 | 7,753 | 0.97 |
| | BKV | 5 | 83 | 8,374 | 4.90 |
| | BVDV | 1 | 9 | 12,230 | 0.61 |
| | RV | 1 | 3 | 17,360 | 0.40 |
| 213 | TTV | 12 | 407 | 3,725 | 32.48 |
| 214 | HBoV | 4 | 53,548 | 5,299 | 97.62 |
| | HCoV | 80 | 848 | 30,578 | 39.96 |
| | TTMV | 1 | 3 | 2,912 | 2.40 |
| 225 | HBoV | 1 | 3 | 5,299 | 1.51 |
| | RV-C | 1 | 2 | 3,725 | 2.01 |
| 227 | TTV | 8 | 777 | 3,725 | 34.90 |
| | HPV | 1 | 12 | 7,466 | 0.94 |
| 233 | RV-A | 16 | 127 | 7,129 | 26.65 |
| | RV-C | 2 | 30712 | 7,107 | 97.88 |
| | TTV | 16 | 75 | 3,725 | 52.35 |
| 237 | RV-A | 1 | 3 | 7,129 | 0.98 |
| | RV-C | 6 | 35 | 7,107 | 7.46 |
| 238 | HBoV | 22 | 202 | 5,299 | 38.69 |
| | HCoV | 5 | 85766 | 30,578 | 99.81 |
| | TTV | 1 | 4 | 3,725 | 2.01 |
| 17 | HBoV | 32 | 279 | 5,299 | 61.80 |
| | HCoV | 112 | 1,021 | 30,578 | 20.52 |
| | RSV | 1 | 3 | 15,191 | 0.49 |
| 24 | TTV | 12 | 1 | 3,725 | 1.88 |
| 11 | HBoV | 6 | 45 | 5,299 | 12.08 |
| | HCoV | 15 | 291 | 30,578 | 5.72 |
| 125 | HCoV | 32 | 382 | 30,578 | 11.28 |
| 151 | RSV | 1 | 4 | 15,191 | 0.46 |
| 66 | RSV | 24 | 274 | 15,191 | 15.40 |
| | TTV | 3 | 107 | 3,725 | 9.93 |
| | TTMV | 2 | 23 | 2,912 | 4.46 |

[a]RSV, respiratory syncytial virus; RV-A, rhinovirus species A, RV-C; rhinovirus species C; HBoV, human bocavirus, HCoV, human coronavirus OC43; TTV, torque teno virus; TTMV, torque teno mini virus; BKV, bovine kobuvirus; BVDV, bovine viral diarrhea virus; BatPV, bat picornavirus; HPV, human papillomavirus; PVY, potato virus Y.

in life, and involve the respiratory and gastrointestinal tracts. The association of these viruses with clinical symptoms is under investigation and requires additional epidemiological studies to clarify their pathogenicity [28]. Anelloviruses (TTV, TTMV, TTMDV) were found in both LRTIs and URTIs. Members of this family ubiquitously infect humans and establish persistent infections, although causal disease associations are currently lacking [10]. It is interesting to note that common gastrointestinal viruses such as astrovirus and rotavirus were found in some of the samples. It is not surprising though, since rotaviruses have long being

**Table 7.** Genome coverage for viruses present in URTI samples.

| Sample | Virus[a] | No. of Contigs | No. of reads incorporated into contigs | Genome size | Genome Coverage (%) |
|---|---|---|---|---|---|
| C27, C01 | ToMV | 2 | 12 | 6,383 | 1.88 |
| | TTMDV | 2 | 5 | 3256 | 4.45 |
| | HPV | 1 | 2 | 7466 | 0.94 |
| | RV-C | 9 | 45 | 7107 | 11.3 |
| C41, T50 | TTMV | 2 | 5 | 2912 | 3.78 |
| | TTMDV | 2 | 10 | 3256 | 3.22 |
| | TTV | 1 | 3 | 3725 | 2.28 |
| | RV-C | 1 | 92130 | 7107 | 95.40 |
| C16, C61 | CMV | 4 | 10 | 3356 | 7.15 |
| | ToMV | 3 | 15 | 6386 | 2.98 |
| | RV-C | 2 | 10 | 7107 | 2.74 |
| M23, M44 | CMV | 1 | 3 | 3356 | 5.36 |
| | RV-C | 1 | 39840 | 7107 | 97.20 |
| P108 | HHV | 5 | 12 | 116114 | 0.28 |
| | HBoV | 2 | 5 | 5299 | 2.64 |
| P06, P150 | HEV-A | 2 | 12352 | 7,413 | 97.88 |
| | TTMV | 1 | 6 | 2912 | 4.12 |
| | RV-C | 4 | 8798 | 7107 | 92.74 |
| P149, P153 | HCoV | 3 | 11 | 30,578 | 0.78 |
| | SAFV | 5 | 1909 | 8,115 | 83.03 |
| | RV-C | 4 | 26 | 7107 | 4.50 |
| P173 | HCoV | 3 | 9 | 30,578 | 0.69 |
| P151, P181 | RV-A | 7 | 943 | 7,129 | 50.78 |
| | RV-B | 1 | 4 | 7,215 | 0.97 |
| | HPV | 9 | 45 | 7466 | 12.46 |
| | TTMV | 2 | 7 | 2912 | 4.98 |
| | TTV | 1 | 3 | 3725 | 0.87 |
| | RV-C | 5 | 17 | 7107 | 6.33 |
| P147, P191 | RV-A | 3 | 12 | 7,129 | 4.35 |
| | RV-B | 1 | 4 | 7,215 | 1.25 |
| | ToMV | 1 | 3 | 6,384 | 1.25 |
| | TTMV | 1 | 3 | 2912 | 3.43 |
| | TTMDV | 1 | 3 | 3256 | 2.15 |
| | TTV | 1 | 5 | 3725 | 2.01 |
| | RV-C | 5 | 15 | 7107 | 6.26 |
| P176, P186, P213 | WSSV | 6 | 31 | 292,967 | 0.13 |
| | SAFV | 5 | 8023 | 8,115 | 90.92 |
| | RV-C | 8 | 37 | 7107 | 9.01 |
| C46, P54 | RV | 3 | 12 | 17360 | 1.09 |
| | TTMV | 6 | 73 | 2912 | 15.45 |
| | TTMDV | 1 | 11 | 3256 | 2.15 |
| | TTV | 13 | 150 | 3725 | 50.74 |
| | RV-C | 3 | 9 | 7107 | 2.25 |
| P19, P88 | HHV | 3 | 12 | 116114 | 0.18 |
| | HPV | 0 | 0 | 7466 | 0.00 |
| | ToMV | 1 | 7 | 6,384 | 1.25 |
| | TTMDV | 1 | 6 | 3256 | 3.38 |
| | RV-C | 10 | 8726 | 7107 | 62.77 |
| T36 | RV-C | 5 | 21 | 7107 | 4.43 |
| T33, T39 | RV-C | 12 | 1319 | 7107 | 68.74 |

**Table 7.** Cont.

| Sample | Virus[a] | No. of Contigs | No. of reads incorporated into contigs | Genome size | Genome Coverage (%) |
|---|---|---|---|---|---|
| T43, T44 | RV-C | 3 | 14165 | 7107 | 92.92 |
| C06, C55, T78, V24 | HASTV | 2 | 5 | 6,759 | 2.52 |
| | CMV | 1 | 6 | 3356 | 2.09 |
| | SAFV | 2 | 126 | 8,115 | 2.42 |
| | RV-C | 3 | 7989 | 7107 | 92.66 |
| V26 | HBoV | 8 | 77 | 5299 | 15.30 |
| | RV-C | 7 | 143 | 7107 | 15.06 |
| C29, M40, M41, V39 | SAFV | 8 | 135 | 8,115 | 8.76 |
| | TTMV | 1 | 5 | 2912 | 2.75 |
| | RV-C | 11 | 3061 | 7107 | 63.89 |
| P131, V61 | BatPV | 1 | 2 | 7,753 | 0.90 |
| | HBoV | 1 | 4 | 5299 | 2.08 |
| | RV-A | 3 | 14 | 7,129 | 3.23 |
| | RV-B | 4 | 3145 | 7,215 | 90.33 |
| | HEV A | 3 | 13 | 7,413 | 3.10 |
| | TTMV | 9 | 1531 | 2912 | 47.12 |
| | TTMDV | 4 | 1881 | 3256 | 29.67 |
| | TTV | 13 | 126 | 3725 | 42.31 |
| | RV-C | 4 | 618 | 7107 | 19.77 |

[a]RSV, respiratory syncytial virus; RV-A, rhinovirus species A; RV-B; rhinovirus species B; RV-C; rhinovirus species C; HBoV, human bocavirus, HCoV, human coronavirus OC43; TTV, torque teno virus; TTMV, torque teno mini virus; TTMDV, torque teno midi virus; BKV, bovine kobuvirus; BVDV, bovine viral diarrhea virus; BatPV, bat picornavirus; HPV, human papillomavirus; PVY, potato virus Y, ToMV, tomato mosaic virus; CMV, cucumber mosaic virus; HHV, human herpes virus; HVE-A, human enterovirus A; SAFV, Saffold virus; WSSV, white spot syndrome virus; HASTV, human astrovirus.

suspected to reach the gastrointestinal tract via mouth and nose. In fact, some rotavirus infections have been associated with respiratory symptoms [29]. Finally, we found low amounts of DNA reads corresponding to animal and plant viruses. The number of these types of viruses is larger than previously reported in respiratory samples [21], although plant viruses have been found more abundantly in human feces [30]. Both, plant and animal viruses are thought to be derived either from consumed food or acquired from the environment.

The search for new viruses using NGS technologies in mammalian, avian, and in particular, human samples, has contributed to the identification of new viruses in animal reservoirs and in different conditions of disease [31]. However, the important effort invested in mammal and avian virus detection has only resulted in the discovery of variants of virus species, sister species to known viruses, and rarely genera. These observations contrast with the recent efforts to discover arthropod viruses, which have yielded widely divergent taxa that sometimes have even defined novel families [32]. Altogether, these observations and the presence of DNA sequence reads from common respiratory viruses or bacteria in essentially 100% of the samples collected

from children with LRTI and URTI, suggest there is limited potential for the discovery of so far undescribed, clinically relevant, viruses associated to pediatric respiratory disease at least in the type of populations studied and with the sampling and diagnostic methods employed.

## Acknowledgments

We thank Miguel L. García-León for his help in handling and organizing all pneumonia samples. We also thank the Instituto de Biotecnologia-UNAM for giving us access to its computer cluster and Jerome Verleyen for his computer support. This work was supported by grants 153639 (to JI Santos) and "Influenza 2009" (to CF Arias) from the National Council for Science and Technology-Mexico (CONACYT). F.E.A. is recipient of a scholarship from CONACYT.

## Author Contributions

Conceived and designed the experiments: BT PI CFA. Performed the experiments: BT MAE FEA PI. Analyzed the data: BT CFA. Contributed reagents/materials/analysis tools: MAA-O JM-M R-RV FD-H FZ-V RMW-C VF-R CNR-A JG-M AV-S GM-A MCS-M DEN LFP-G JIS-P. Wrote the paper: CFA BT SL JIS-P DEN GM-A RMW-C AV-S.

## References

1. Rudan I, O'Brien KL, Nair H, Liu L, Theodoratou E, et al. (2013) Epidemiology and etiology of childhood pneumonia in 2010: estimates of incidence, severe morbidity, mortality, underlying risk factors and causative pathogens for 192 countries. J Glob Health 3: 010401.

2. Ruuskanen O, Lahti E, Jennings LC, Murdoch DR (2011) Viral pneumonia. Lancet 377: 1264–1275.

3. Black RE, Cousens S, Johnson HL, Lawn JE, Rudan I, et al. (2010) Global, regional, and national causes of child mortality in 2008: a systematic analysis. Lancet 375: 1969–1987.

4. Chiu CY, Urisman A, Greenhow TL, Rouskin S, Yagi S, et al. (2008) Utility of DNA microarrays for detection of viruses in acute respiratory tract infections in children. J Pediatr 153: 76–83.

5. Ruohola A, Waris M, Allander T, Ziegler T, Heikkinen T, et al. (2009) Viral etiology of common cold in children, Finland. Emerg Infect Dis 15: 344–346.

6. van Gageldonk-Lafeber AB, Heijnen ML, Bartelds AI, Peters MF, van der Plas SM, et al. (2005) A case-control study of acute respiratory tract infection in general practice patients in The Netherlands. Clinical infectious diseases: an official publication of the Infectious Diseases Society of America 41: 490–497.

7. Gern JE (2010) The ABCs of rhinoviruses, wheezing, and asthma. J Virol 84: 7418–7426.
8. Heikkinen T, Jarvinen A (2003) The common cold. Lancet 361: 51–59.
9. Debiaggi M, Canducci F, Ceresola ER, Clementi M (2012) The role of infections and coinfections with newly identified and emerging respiratory viruses in children. Virol J 9: 247.
10. Jartti T, Jartti L, Ruuskanen O, Soderlund-Venermo M (2012) New respiratory viral infections. Curr Opin Pulm Med 18: 271–278.
11. Allander T, Tammi MT, Eriksson M, Bjerkner A, Tiveljung-Lindell A, et al. (2005) Cloning of a human parvovirus by molecular screening of respiratory tract samples. Proc Natl Acad Sci U S A 102: 12891–12896.
12. Harvala H, Simmonds P (2009) Human parechoviruses: biology, epidemiology and clinical significance. J Clin Virol 45: 1–9.
13. van den Hoogen BG, de Jong JC, Groen J, Kuiken T, de Groot R, et al. (2001) A newly discovered human pneumovirus isolated from young children with respiratory tract disease. Nat Med 7: 719–724.
14. van der Hoek L, Pyrc K, Jebbink MF, Vermeulen-Oost W, Berkhout RJ, et al. (2004) Identification of a new human coronavirus. Nat Med 10: 368–373.
15. Woo PC, Lau SK, Chu CM, Chan KH, Tsoi HW, et al. (2005) Characterization and complete genome sequence of a novel coronavirus, coronavirus HKU1, from patients with pneumonia. J Virol 79: 884–895.
16. Erdman DD, Weinberg GA, Edwards KM, Walker FJ, Anderson BC, et al. (2003) GeneScan reverse transcription-PCR assay for detection of six common respiratory viruses in young children hospitalized with acute respiratory illness. J Clin Microbiol 41: 4298–4303.
17. Gruteke P, Glas AS, Dierdorp M, Vreede WB, Pilon JW, et al. (2004) Practical implementation of a multiplex PCR for acute respiratory tract infections in children. J Clin Microbiol 42: 5596–5603.
18. Murdoch DR, Jennings LC, Bhat N, Anderson TP (2010) Emerging advances in rapid diagnostics of respiratory infections. Infect Dis Clin North Am 24: 791–807.
19. Syrmis MW, Whiley DM, Thomas M, Mackay IM, Williamson J, et al. (2004) A sensitive, specific, and cost-effective multiplex reverse transcriptase-PCR assay for the detection of seven common respiratory viruses in respiratory samples. J Mol Diagn 6: 125–131.
20. Hustedt JW, Vazquez M (2010) The changing face of pediatric respiratory tract infections: how human metapneumovirus and human bocavirus fit into the overall etiology of respiratory tract infections in young children. Yale J Biol Med 83: 193–200.
21. Lysholm F, Wetterbom A, Lindau C, Darban H, Bjerkner A, et al. (2012) Characterization of the viral microbiome in patients with severe lower respiratory tract infections, using metagenomic sequencing. PloS One 7: e30875.
22. Willner D, Furlan M, Haynes M, Schmieder R, Angly FE, et al. (2009) Metagenomic analysis of respiratory tract DNA viral communities in cystic fibrosis and non-cystic fibrosis individuals. PloS One 4: e7370.
23. Wylie KM, Mihindukulasuriya KA, Sodergren E, Weinstock GM, Storch GA (2012) Sequence analysis of the human virome in febrile and afebrile children. PloS One 7: e27735.
24. Sorber K, Chiu C, Webster D, Dimon M, Ruby JG, et al. (2008) The long march: a sample preparation technique that enhances contig length and coverage by high-throughput short-read sequencing. PloS One 3: e3495.
25. Waman VP, Kolekar PS, Kale MM, Kulkarni-Kale U (2014) Population structure and evolution of Rhinoviruses. PloS One 9: e88981.
26. Tang P, Chiu C (2010) Metagenomics for the discovery of novel human viruses. Future Microbiol 5: 177–189.
27. Jansen RR, Wieringa J, Koekkoek SM, Visser CE, Pajkrt D, et al. (2011) Frequent detection of respiratory viruses without symptoms: toward defining clinically relevant cutoff values. J Clin Microbiol 49: 2631–2636.
28. Himeda T, Ohara Y (2012) Saffold virus, a novel human Cardiovirus with unknown pathogenicity. J Virol 86: 1292–1296.
29. Estes MK, Greenberg HB (2013) Rotaviruses. In: Knipe DM, Howley, P.M., editor. Fields Virology. 6th ed. Philadelphia: Wolters Kluwer/Lippincott Williams & Wilkins. pp. 1347–1401.
30. Zhang T, Breitbart M, Lee WH, Run JQ, Wei CL, et al. (2006) RNA viral community in human feces: prevalence of plant pathogenic viruses. PLoS Biology 4: e3.
31. Chiu CY (2013) Viral pathogen discovery. Curr Opin Microbiol 16: 468–478.
32. Junglen S, Drosten C (2013) Virus discovery and recent insights into virus diversity in arthropods. Curr Opin Microbiol 16: 507–513.

# Pdsg1 and Pdsg2, Novel Proteins Involved in Developmental Genome Remodelling in *Paramecium*

**Miroslav Arambasic❾, Pamela Y. Sandoval❾, Cristina Hoehener, Aditi Singh, Estienne C. Swart, Mariusz Nowacki***

Institute of Cell Biology, University of Bern, Bern, Switzerland

## Abstract

The epigenetic influence of maternal cells on the development of their progeny has long been studied in various eukaryotes. Multicellular organisms usually provide their zygotes not only with nutrients but also with functional elements required for proper development, such as coding and non-coding RNAs. These maternally deposited RNAs exhibit a variety of functions, from regulating gene expression to assuring genome integrity. In ciliates, such as *Paramecium* these RNAs participate in the programming of large-scale genome reorganization during development, distinguishing germline-limited DNA, which is excised, from somatic-destined DNA. Only a handful of proteins playing roles in this process have been identified so far, including typical RNAi-derived factors such as Dicer-like and Piwi proteins. Here we report and characterize two novel proteins, Pdsg1 and Pdsg2 (*Paramecium* protein involved in <u>D</u>evelopment of the <u>S</u>omatic <u>G</u>enome 1 and 2), involved in *Paramecium* genome reorganization. We show that these proteins are necessary for the excision of germline-limited DNA during development and the survival of sexual progeny. Knockdown of *PDSG1* and *PDSG2* genes affects the populations of small RNAs known to be involved in the programming of DNA elimination (scanRNAs and iesRNAs) and chromatin modification patterns during development. Our results suggest an association between RNA-mediated trans-generational epigenetic signal and chromatin modifications in the process of *Paramecium* genome reorganization.

**Editor:** Giacomo Cavalli, Centre National de la Recherche Scientifique, France

**Funding:** This work was supported by Swiss National Science Foundation grant 31003A_129957 (http://www.snf.ch/); European Research Council grant "EPIGENOME" GA 260358 (http://erc.europa.eu/); European Cooperation in Science and Technology Action BM1102 (http://www.cost.eu/). The funders had no role in study design, data collection and analysis, decision to publish, or preparation of the manuscript.

**Competing Interests:** The authors have declared that no competing interests exist.

* Email: mariusz.nowacki@izb.unibe.ch

❾ These authors contributed equally to this work.

## Introduction

In ciliates, such as *Paramecium*, small RNAs participate in the programming of large-scale DNA deletion and genome organization during development. A characteristic feature of ciliates is the division of germline and somatic functions between two types of nuclei: the diploid micronucleus (MIC) and the polyploid macronucleus (MAC) respectively. New MIC and MAC develop from zygotic nuclei produced by fusion of haploid parental MIC during sexual cycle. In this process, a new MAC genome (72 Mb in *Paramecium tetraurelia*) matures from the MIC genome (97 Mb) by extensive editing [1] which is accompanied by polyploidization to an average copy number of ~800n. Genome correcting during development discards most non-genic DNA, transposable elements and other repeated sequences and numerous Internal Eliminated Sequences (IESs). While the new MAC is developing, the old MAC is fragmented and later eliminated from the cell.

*Paramecium tetraurelia* IESs are abundant (~45,000), short (frequently less than 30 bp long), single-copy, noncoding sequences that are precisely excised from MIC DNA to produce the mature MAC genome [1]. Each *Paramecium* IES is flanked by two 5′-TA-3′ dinucleotides (TA repeats). IES excision leads to the retention of a single TA dinucleotide. The ends of IESs have symmetrical inverted base frequencies, which can be crudely represented by the consensus sequence TAYAGYNR with no other known conserved motifs. IES ends are similar to the ends of Tc1/mariner transposons [2] but this consensus is recognized by the *Paramecium* IES excisase, a domesticated PiggyBac-related transposase (PiggyMac)[3]. The IES end sequence also appears to be important for the staggered double-strand cuts that initiate excision [4–6].

In *Paramecium*, both precise excision of IESs and imprecise elimination of genomic regions containing transposable elements can be controlled by maternal effects. This epigenetic dependency was demonstrated by microinjection of DNA, in the form of an IES sequence. The introduction of an IES into the old MAC can prevent the elimination of identical sequences from the progeny's somatic genome [7,8]. However; IESs are retained to different degrees depending on the quantity of injected sequence and the sensitivity of the IES. The retention is inherited in future generations and is true for a third of total *Paramecium* IESs assayed [8], known as maternally controlled IESs, or mcIESs.

These observations imply that trans-nuclear genome comparison occurs during development.

In ciliates, multiple RNA interference-related pathways exist. Post-transcriptional gene silencing can be induced by untranslatable transgenes [9,10] or by feeding cells with *E. coli* producing double-stranded RNAs [11]. In both cases, silencing of targeted genes generates complementary ~23 nt siRNAs produced by Dicer-related protein (Dcr1) [12–15]. In *Paramecium tetraurelia* and the related Oligohymenophorean ciliate *Tetrahymena thermophila*, a second RNAi-related pathway employs a distinctive class of sRNAs known as "scan RNAs" (scnRNAs) to perform the trans-nuclear comparison for the precise targeting of DNA elimination. In *Paramecium* and *Tetrahymena* Dicer-like proteins are responsible for producing scnRNAs in the meiotic MIC and Piwi proteins for binding and protecting scnRNAs [14,16–19]. High throughput sequencing of *Paramecium* sRNA has demonstrated that scnRNAs initially correspond to the entire germline genome, and become progressively enriched in IESs matching sequences [20,21]. This enrichment of IESs matching sequences is the result of the reduction of the total population of scnRNA and proposed to be due to a process known as "RNA scanning". scnRNAs produced from transcripts across the MIC genome are filtered by pairing to transcripts from the old MAC genome leaving only scnRNAs matching to the germline-limited sequences [14,22–24]. The remaining germline-specific scnRNAs are then transported to the developing new MAC where they target DNA elimination [25]. In *Paramecium* scnRNAs are produced by a pair of paralogous Dicer-like proteins (Dcl2 and Dcl3) [14] and become associated with a pair of Piwi-like protein paralogs (Ptiwi01p and Ptiwi09p) [16]. The mechanism by which scnRNAs trigger DNA elimination in the developing MAC is not completely understood. In *Tetrahymena* scnRNAs are responsible for trimethylation of lysine 27 and lysine 9 of histone H3 (H3K27me3 and H3K9me3 respectively) within chromatin destined to be eliminated [26–28]. It is therefore possible that scnRNA-mediated chromatin modification defines genomic regions to be targeted by DNA excision machinery.

In *Paramecium*, a third class of development-specific sRNAs is present, iesRNAs varies in length from ~21–31 nt (peaking at 27/28 nt) and are produced in the new MAC by a distinct Dicer-like protein, Dcl5 [21]. iesRNAs appear at late stages of development when IESs are being excised and match IESs exclusively, leading to the proposal that they may be derived from excised IESs [21]. Silencing of *DCL5* shows that iesRNAs are involved in targeting DNA excision in the developing MAC [21].

In *Paramecium* one of the key players in the RNA-mediated trans-nuclear crosstalk is the Nowa1 protein (Nowa1) [15]. Nowa1 is expressed specifically during sexual development and is required for the elimination of transposons and maternally controlled IESs [15]. It accumulates in the maternal MAC shortly before meiosis and later translocates to the developing MAC. The functions of Nowa1 in both maternal MAC and developing MAC are still under examination.

In this study we report the discovery of two proteins, Pdsg1 and Pdsg2, involved in *Paramecium* genome development. Both proteins are necessary for the excision of germline-limited DNA and for the survival of sexual progeny. Through high-throughput sequencing we show that *PDSG1* and *PDSG2* knockdowns affect *Paramecium*'s development-specific sRNAs. Together these results suggest that these proteins are involved in the epigenetic programming of DNA remodelling and in IES excision. In addition, the knockdowns affect chromatin modification patterns during development which suggests a link between DNA elimination and histone modifications in the process of *Paramecium* genome maturation.

## Materials and Methods

### *Paramecium* cultivation

*Paramecium* strain51, mating type 7 was used in all experiments. *Paramecium* cells were cultured in Wheat Grass Powder (WGP; Pines International, Lawrence, KS) medium bacterized with *Klebsiella pneumoniae*, and supplemented with 0.8 mg/l of β-sitosterol. Cultures used in all the experiments were grown at 27°C.

### Silencing experiments, survival test and IES retention PCR

For the silencing constructs different regions of the coding sequences of each candidate were selected and cloned into L4440 plasmid (list of specific primers can be found in Fig. S1). The plasmid was used for the transformation of HT1115 (DE3) *E.coli* strain. *Paramecium* cells were seeded into silencing medium at a density of 200 cells/ml and silencing was carried out as previously described [29]. Upon completion of development, single cells (n = 30) were isolated in fresh medium for the evaluation of survival of the progeny. Cells were monitored for 12 cell cycles after their isolation and categorized into three groups according to the observed phenotype (normal, sick or unviable). In parallel, 50 ml cultures were harvested and DNA extraction was performed by using GeneElute – Mammalian Genomic DNA MIniprep Kit (Sigma-Aldrich). IES PCR was done with GoTaq polymerase (Promega) standard protocols.

### Dot blot

Dot blot assays were conducted following standard protocols [30]. In detail, 3 µg of DNA of post-developmental cultures were fixed on a nylon membrane. Sardine and Thon transposons specific probes were labelled with α-32P dATP (3000 Ci/mmol) using RadPrime DNA Labeling System (Invitrogen). Probe against *ACTIN* gene targets the first 240 bp of the coding sequence. The signal was quantified with ImageJ 1.48e.

### Northern blot

10 µg of RNA were separated by electrophoresis in a denaturing gel and transferred to a nylon membrane. Full-length *PDSG1* gene probes were used for the specific detection. Probe against *PDSG2* targeted the 960 bp region, the same region that was used to clone the silencing fragment.

### GFP tagging, microinjection and GFP localization experiment

A set of specific primers (5′-TGATTTACAATTAAGGAT-TAGGAGTATTTTGA-3′ and (5′-CAGGCATTGATTGTAT-TTTAATTAATTTTAAATCT-3′) were used for the amplification of *PDSG1* gene including 175 bp upstream and 68 bp downstream of the coding region. Full length *PDSG2* gene along with 414 bp upstream and 343 bp downstream was amplified (5′-CGATAAAAGTTTGTTTTAATAAAATGATAATAAATCTC-ATAAAAGTG-3′ and 5′-GTATTTACTGCAGGTTTTTTTTT-GAATTGCATAAAC-3′). In case of *PDSG2*, GFP was inserted at the N-terminus. *PDSG1* was tagged at both ends but only the C-terminal tagged version was expressing a functional GFP. The constructs were linearized and microinjected into the MAC of the vegetative cells. Positively injected clones were selected by dot blot analysis. Cells were collected at different time points during sexual development and counterstained with DAPI (4,6-diamidino-2-

2phenylindole). Images for this experiment were acquired with a Leica microscope (Wetzlar, Germany).

## Immunocytochemistry and confocal microscopy

Cells were collected during sexual development and prepared for immunostaining according to standard protocol [31]. Anti-trimethyl-Histone H3 (Lys9) antibody (07-442, Millipore) and Anti-trimethyl-Histone H3 (Lys27) (07-449, Millipore) at 1:100 dilution were used. A FLUOVIEW FV1000 (Olympus) system with PLAPON 60× O SC NA 1.40 objective was used for imaging capture.

## Peptide competition assay

Total protein extract form an intermediate developmental stage was loaded in triplicate and proteins were separated by SDS-PAGE. Prior to immunoblotting, 0.66 ng of Anti-trimethyl-Histone H3 (Lys27) was diluted 1:3000 in 5% BSA supplemented with 0.1% Tween-20. Diluted antibody was incubated with dilution buffer only or with 0.02 ug of unmodified human histone H3 peptide (Ab2903, Abcam) or with 0.02 ug of human histone H3 (tri methyl K27) peptide (Ab1782, Abcam) for 3 h at the room temperature. Blot was incubated with secondary HRP conjugated antibody (Sc-2004, Santa Cruz Biotechnology).

## Small RNA analysis

*Paramecium* cells from 800 ml of culture were harvested and resuspended in 6 ml of TRI reagent BD (Sigma-Aldrich). Total RNA extraction was carried out following the TRI Reagent BD protocol. The enrichment of small RNA was performed using mirVana miRNA Isolation Kit (Ambion). Enriched samples were used for library preparation following the TruSeq Small RNA Sample Preparation protocol (Illumina). Reads were mapped with BWA [32] and uniquely mapping sRNAs selected by a custom Python script. To generate the sRNA size histograms, we normalized the number of sRNAs by the total number of mapped sRNAs (both MAC genome- and IES-matching).

## Accession Numbers

*PDSG1* and *PDSG2* sequences are available from the GenBank under the accessions: XM_001442856 and XM_001425883, respectively. *ACTIN* gene accession number: XM_001443584. Raw sRNA sequence data for the *PDSG1*-KD, *PDSG2*-KD, control early and control late time points can be obtained from the European Nucleotide Archive (ENA) under the accessions: PRJEB5853, PRJEB5867, SRR907874 and SRR907875, respectively.

## Protein domain prediction

Homology detection and structure prediction were estimated by open access Pfam (http://pfam.sanger.ac.uk/search) [33] and HHpred (toolkit.tuebingen.mpg.de/hhpred)[34] software.

# Results

## Selection of candidate genes involved in genome development in *Paramecium tetraurelia*

Only a handful of key factors playing role in *Paramecium* genome development have been described so far [3,15,16,21,35]. A common feature of all these genes is the transcriptional upregulation during sexual reproduction. We took this characteristic transcriptional upregulation as the main criterion for the selection of putative factors involved in MAC development and more specifically in the RNA-mediated genome reorganization.

A publicly accessible BioMart database [36] was used to query microarray data during the life cycle of *Paramecium* [37]. We chose twenty-eight genes that are highly upregulated during MAC development, all of which have no identifiable paralogs in the *Paramecium* genome (Fig. S2). The selected candidate genes were subdivided into two groups according to their expression profiles. Twelve candidates are early expressed genes and the peak of their expression is during meiosis of MIC or MAC fragmentation (Fig. S3A). The second group of 16 candidates included late expressed genes whose maximum expression occurs during the formation of the new MAC (Fig. S3B). Among the 28 selected candidates, the proteins of 10 contain predicted domains while the rest of the candidates have no homology to known proteins.

## Pdsg1 and Pdsg2 are essential for the generation of sexual progeny

To determine whether the developmentally upregulated gene candidates have an effect on the formation of fully functional progeny; we silenced each one of the genes independently during development and observed the effects on the offspring based on the ability of maintaining normal vegetative growth. The silencing was induced by feeding *Paramecium* cells with *E. coli* expressing dsRNA corresponding to the target genes (see Materials and Methods and supplementary information). *E. coli* producing dsRNA corresponding to the empty bacterial plasmid was used as negative control (empty vector control (EV)) and as a positive control we used a NOWA1 silencing construct which blocks the excision of maternally controlled IESs and is lethal to the sexual progeny [15].

The silencing of each of the 28 candidate genes was started 3 to 4 vegetative cell cycles before commitment to development. None of the silencing had a noticeable effect on cell growth during the vegetative cycle. To assess the survival of the sexual progeny, from each of the experiments 30 random post-developmental cells were monitored individually and scored for survival. Among the twenty-eight candidates only two showed impaired cell viability (Fig 1A). We cannot exclude the possibility that some of the remaining 26 candidates are also involved in developmentally specific processes, since the efficiency of each RNAi silencing construct has not been determined. The two genes essential for the generation of sexual progeny were named as *Paramecium protein involved in Development of the Somatic Genome 1 and 2 (PDSG1* and *PDSG2*). The dramatic effect on the survival of sexual progeny of *PDSG1*-KD and *PDSG2*-KD (90% and 93%) was comparable to Nowa1 depletion (used as control), suggesting that Pdsg1 and Pdsg2 are essential for adequate completion of developmental process in *Paramecium*. In addition to the survival test, silencing efficiency was assessed by Northern blot (Fig. 1B). Furthermore, no evident delay in the progression of developmental stages was noticed after cytological evaluation of major structures (MIC, old MAC and new MAC) in PDSG1, and PDSG2 silenced cultures. All the silencings were repeated 5 times and were highly reproducible (data not shown).

## Depletions of Pdsg1 and Pdsg2 impair elimination of germline-limited DNA in the developing macronucleus

In order to determine whether Pdsg1and Pdsg2 were involved in the *Paramecium* DNA elimination process we checked for retention of MIC-limited sequences in the new MAC genome upon the completion of development. As mentioned earlier, precise and imprecise mechanisms of DNA elimination determine the genome content in the new MAC. Both mechanisms may include common set of factors but some of them may be unique in

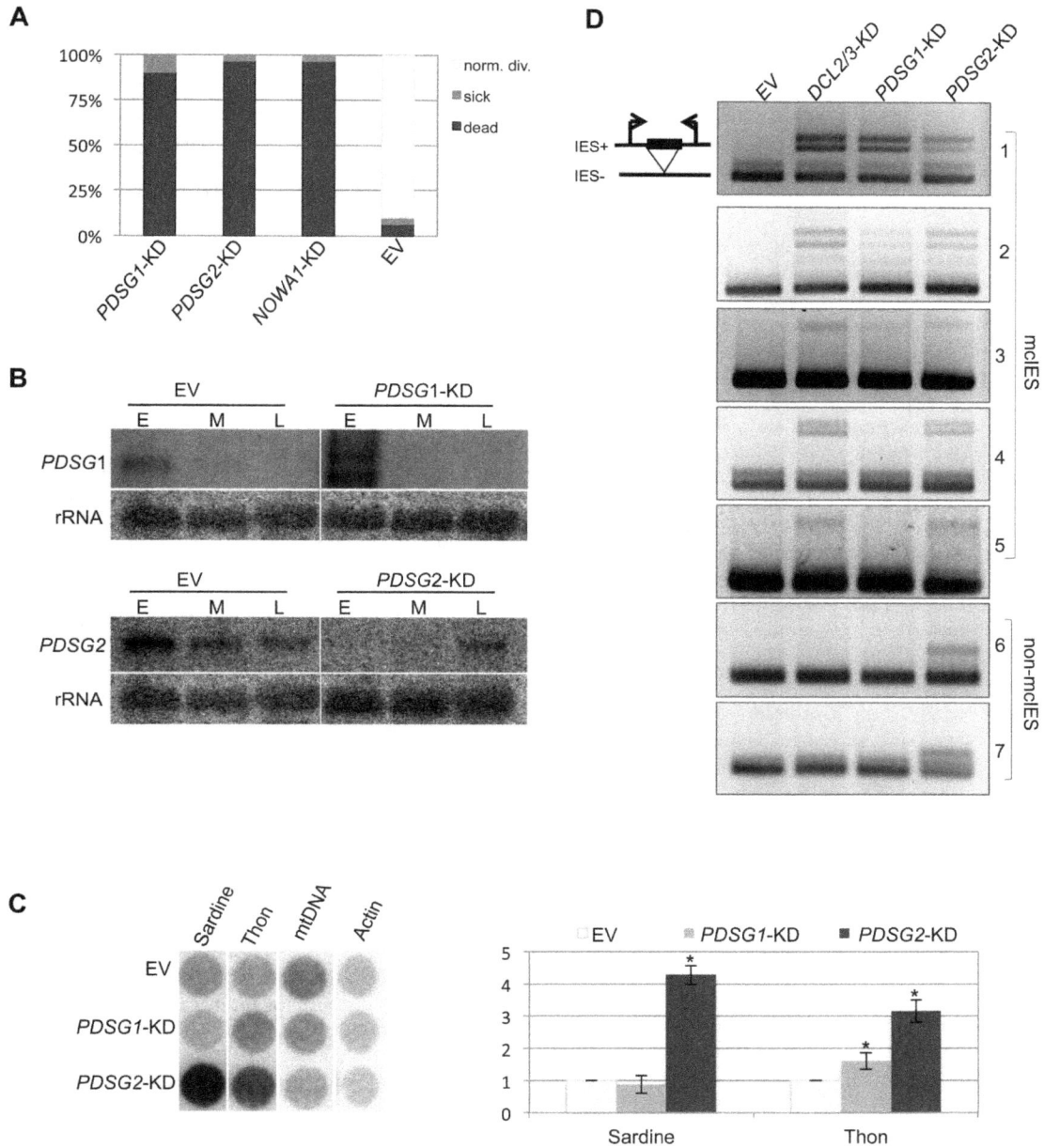

**Figure 1. Effects of PDSG1 and PDSG2 silencing on progeny survival and IES excision.** (A) Survival test. Graphic representation of percent of normally dividing (white), sick (grey) and dead (black) progeny cells. The silencing of PDSG1 and PDSG2 was lethal in 95% and 97% of cells, respectively. NOWA1-KD is a positive control. Empty vector (EV) is a negative control. (B) PDSG1 and PDSG2 silencing efficiency was assayed by Northern blot by comparing the control (EV, empty vector) and silenced cultures. Three developmental points were analysed during sexual cycles. Early developmental stage (E) includes cells undergoing meiosis and 50% of cells present fragmented old MAC. Middle developmental stage (M) presents 100% of the cells with fragmented old MAC. Late developmental stage (L) includes cells with fragmented old MAC and a substantial number of cells with evident new developing MAC. (C) Effect of PDSG1 and PDSG2 silencing on transposon elimination. Macronulcear DNA was extracted form PDSG1-KD and PDSG2-KD cultures and analyzed for the retention of Sardine and Thon transposons using specific probes. Quantification signal of two classes of transposons was normalized to the mitochondrial DNA probe (mtDNA). * A two-tailed Student's t-test was used to assess statistical significance of the differences in the mean (values of bars), and an asterisk is shown if the p-value from this test is <0.05. For Sardine elements p-values are: PDSG1-KD vs Ctrl: 0.054; PDSG2-KD vs Ctrl: 2.7e-3. For Thon elements p-values are: PDSG1-KD vs Ctrl: 8.8e-5; PDSG2-KD vs Ctrl: 2.3e-4. Probe against Actin gene was used as an additional loading control. (D) IES retention PCR. Excision of 5 mcIES (1–5) and 2 non-mcIESs (6–7) are shown. Upper band represents IES+, lower band represents DNA with excised IES (IES-). IES: 1 (mtA promoter IES); 2 (51G4404); 3 (51A6649); 4 (51A2591); 5 (51G2832); 6 (51G1413); 7 (51A1835).

order to determine the precision of the elimination that differ between both mechanisms.

Since repetitive sequences like minisatellites and transposable elements are imprecisely eliminated from the new MAC [38], we

checked for the correct removal of Sardine and Thon transposons. The level of retention of these transposons was measured in MAC genomic DNA samples collected from cells that have completed sexual development (progeny) from control (EV), PDSG1-KD, and *PDSG2*-KD cultures (Fig. 1C). Hybridization with specifics probes showed that the silencing of *PDSG1* has slight effect on the elimination of Thon of transposons but not on the elimination of Sardine since the levels of retention were similar to the levels observed in the control. However, depletion of Pdsg2 induces a strong retention of Sardine and Thon transposons. These results were corroborated by two independent experiments and suggest that only Pdsg2 is involved in the imprecise mechanism of DNA elimination.

Next, the precise elimination of IESs in the new MAC was assayed by PCR on total genomic DNA from cells that have completed development. We tested 7 different IESs; 5 maternally and 2 non-maternally controlled, with primers located in DNA regions flanking IESs (Fig. 1D and Fig. S4). Dcl2/3 silenced cells were used as a positive control for the retention of maternally controlled IESs. The depletion of both, *PDSG1*, and *PDSG2*, prevents the accurate excision of maternally controlled IESs (Fig. 1D, 1–5). *PDSG2* silencing seems to have a stronger effect in the retention of this group of IESs than *PDSG1*silencing since the later only affects three of the maternally controlled IESs. Furthermore, only the silencing of *PDSG2* affects the excision of non-maternally controlled IESs. This finding identifies *PDSG2* as one of the few known factors involved in a general mechanism of DNA elimination [39,40]. These results were reproduced in four independent experiments and are summarized in Fig. S5.

## Pdsg1 is present in both maternal and developing nuclei whereas Pdsg2 localizes exclusively to developing MAC

GFP fusion proteins were generated in order to monitor the subcellular localization of Pdsg1 and Pdsg2 during development (Fig. 2). The expression of Pdsg1-GFP and Pdsg2-GFP was undetectable during vegetative growth (Fig. 2A, F). Furthermore, both Pdsg1-GFP and Pdsg2-GFP signals were exclusively detected during development.

Pdsg1-GFP was detected in the parental MAC during early development (Fig. 2B) and the signal was persistent throughout development showing a dynamic localization from the parental MAC to the newly developing MAC (Fig. 2C, D, E). The Pdsg1-GFP signal gained intensity in the new MAC at late developmental stages. On the other hand, Pdsg2-GFP signal was not detected in the cells at early developmental stages and only appeared in the new MAC (Fig. 2G, H, I, J). Both transgenes do not alter either cell viability or the process of DNA elimination, which was confirmed by the survival test and IES retention PCRs, respectively (data not shown).

### *PDSG1* knockdown blocks RNA scanning process while *PDSG2* knockdown affects iesRNA population

To investigate possible effects of *PDSG1* and *PDSG2* knockdowns on developmental-specific small RNA, we examined sRNA-seq data from *PDSG1* and *PDSG2*-silenced cells. Early and late developmental stages were analyzed (correspond to the stages described in Fig. 1B). In cells fed with empty vector (EV, control), 25 nt sRNAs are mostly composed of scnRNAs (based on the high proportion of 5′UNG molecules [21] (Fig. 3D). As previously shown by absolute quantification of sRNAs from electrophoretic gels, there is a significant decrease in quantity of scnRNAs in the late developmental stage due to the elimination of MAC genome-matching scnRNAs during the scanning process (Fig. 3A). In *PDSG1*-KD cells the elimination of 25 nt scnRNAs was suppressed (Fig. 3B), indicating that the scanning process is altered. This suggests that *PDSG1* is involved, indirectly or directly, either in the transport of scnRNA into the parental MAC,

facilitating the interaction between scnRNAs and their targets in the parental MAC, or in the elimination of MAC genome-matching scnRNAs. In the case of *PDSG2* knockdown, no aberrant effect on scnRNA quantity was observed (Fig. 3C).

iesRNAs (peaking at 27–28 nt) appear in the late developmental stage in cells fed with empty vector (control) (Fig. 3A) as a consequence of Dcl5 cleavage of RNA which is likely transcribed from excised DNA [21]. This assumption is based on the fact that all iesRNAs map precisely to IES sequences with none of them overlapping MAC-IES junctions. In addition, the fact that iesRNAs originate only from excised IESs and not from non-excised IESs in the case of knockdown of proteins that are involved in IES excision but affect only a fraction of IESs. The inhibition of the scanning process in the parental MAC during the depletion of *PDSG1* prevents the normal excision of IESs during development. The low amount of iesRNAs seen in sRNA histograms of *PDSG1*-KD in late development (Fig. 3B) may thus be a direct consequence of a higher relative quantity of scnRNAs in these samples as well as massive IES retention that would prevent the production of iesRNAs. Depletion of *PDSG2* does not appear to inhibit RNA scanning (MAC genome-matching scnRNAs are reduced), but does prevent the production of iesRNAs at late developmental stage (Fig. 3C). The inhibition of iesRNAs can either be due to a complete retention of all IESs or due to a more direct involvement of Pdsg2 in iesRNA production.

## Pdsg1 is directly involved in the scanning process

In order to determine the role of Pdsg1 in the scanning process we monitored the retention of a specific IES (51A2591) during the depletion of Pdsg1. In wild type strain the IES-51A2591 is normally excised after development and its excision was not affected in PDSG1-KD strains (Fig. 1D, IES number 4). First, we challenged the system by injecting the IES-51A2591 into the maternal MAC, which will promote the retention of this specific sequence in the progeny MAC, as previously reported [8] (Fig. 4, lane 2). As a control, and to monitor the normal progression of development, both strains were subjected to feeding with *E. coli* producing RNA from empty vector construct (Fig. 4, lanes 3 and 4). Since empty vector control does not include target sequences for silencing, in the progeny of the wild type background IES-51A2591 was correctly excised (Fig. 4, lane 3). Conversely, under this condition in the injected strain the retention of IES-51A2591 was passed to the progeny as anticipated (Fig. 4, lane 4). Next, *PDSG1* was silenced in wild type strain and compared to the IES-51A2591 injected strain (Fig. 4, lanes 5 and 6 respectively). As expected, the depletion of Pdsg1 in wild type background did not promote the retention of IES-51A2591 (Fig. 4, lane 5). However, the silencing of *PDSG1* promoted complete excision of the IES-51A2591 in the injected strain (Fig. 4, lane 6). This suggests that the depletion of Pdsg1 has a negative effect on the scanning process, arguing that Pdsg1has a direct functional role in the scanning. Furthermore, elimination of the scnRNA production (Dcl2/3 silencing, Fig. 4, lanes 7 and 8) was used as a positive control of the disruption of the scanning process. In this case IES-51A2591 was retained in both strains due to the absence of scanning molecules (scnRNAs).

## Depletion of Pdsg2 affects the distribution of Dcl5 in the new MAC

Since both *PDSG1*- and *PDSG2*-KD affect the production of iesRNAs we decide to check the consequences of the depletion of Pdsg1 and Pdsg2 on Dcl5 localization. As previously described [21], Dcl5-GFP fusion protein in wild type cells forms distinct foci in the developing MACs (Fig. S6). The depletion of Pdsg1 had no

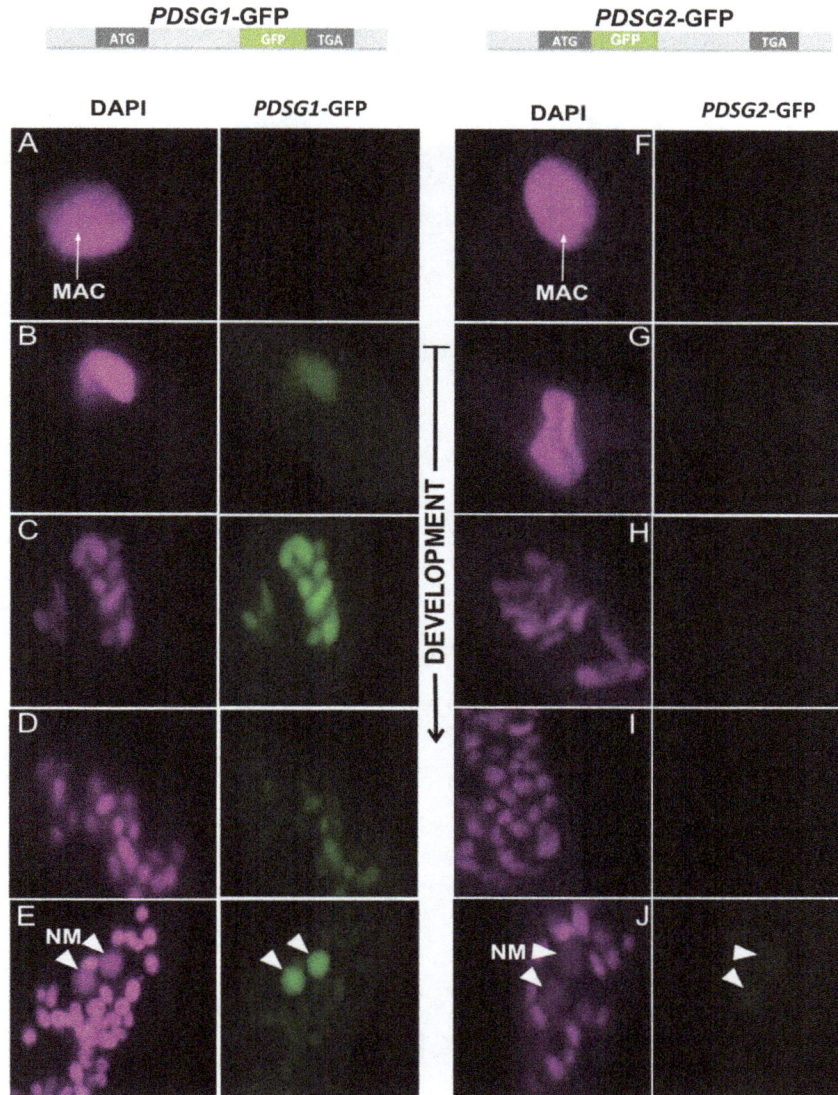

**Figure 2. Subcellular localization of Pdsg1-GFP and Pdsg2-GFP fusion proteins during development.** 1. Graphical representation of C-terminally tagged *PDSG1* with GFP. 2. Graphical representation of N-terminally tagged *PDSG2* with GFP. (A–E) Localization pattern of PDSG1-GFP. (F–J) Localization pattern of Pdsg2-GFP. (A, F) Vegetative cells with the intact MAC. (B, C, G, H) beginning of old MAC fragmentation that represents early development. (D, I) Middle stage of development with fragmented MAC. (E, J) Late development when new MAC is formed while the fragments of the old MAC are still present in the cytoplasm. Magenta: DAPI; green: GFP; white arrow: old macronucleus; arrowheads: new MAC (NM).

obvious effect on the normal expression or the foci formation of Dcl5. In contrast, the depletion of Pdsg2 disrupts Dcl5 foci organization. Although Dcl5 is still expressed only in the new MAC the number and size of Dcl5 foci is completely altered.

## Knockdowns of *PDSG1* and *PDSG2* affect histone methylation during development

Since H3K27me3 and H3K9me3 histone marks are associated with DNA elimination in *Tetrahymena* we decided to test if these decorations are also present in *Paramecium* and whether they are affected by the knockdowns of *PDSG1* and *PDSG2*.

Similar to *Tetrahymena*, both H3K27me3 and H3K9me3 are undetectable in the macronucleus of vegetative cells [26,28] (Fig. 5A, D, G and Fig. S7A, D and G). In cells fed with empty vector control, the staining for H3K27me3 became evident at

early stages of development (Fig. 5B) where it is present in the fragments of the old MAC. At later developmental stages, an H3K27me3 signal was also evident in the developing MAC (Fig. 5C) where the signal intensity was comparable to the one in the fragmented parental MAC. The H3K27me3 signal in the fragments of the parental MAC was slightly weaker in *PDSG1*-KD cells (Fig. 5E) when compared to cells fed with empty vector (control), but it was completely absent in the developing MACs (Fig. 5F). The lack of H3K27me3 in the developing MAC of *PDSG1*-KD cells suggests a link between RNA scanning in the parental MAC and histone modification in the developing MAC, both being required for epigenetic programming of the elimination of maternally controlled IESs.

The H3K27me3 signal in the fragments of the parental MAC was not removed by depletion of *PDSG2* (Fig. 5H). However, these cells had a considerably stronger H3K27me3 signal in the

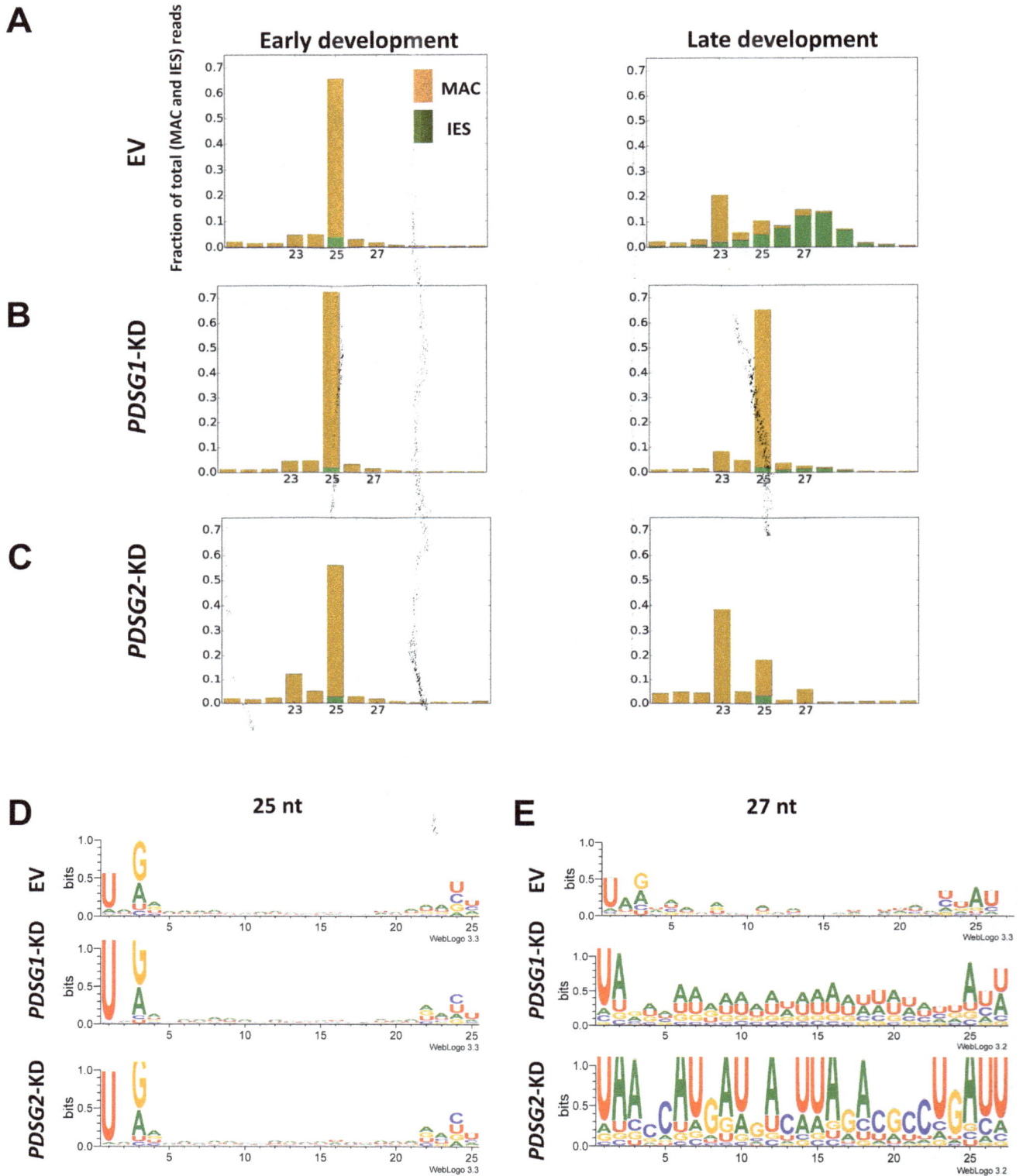

**Figure 3. Effects of *PDSG1* and *PDSG2* knockdowns on developmental-specific small RNA.** (A–C) Histograms of sRNAs size for MAC genome-matching (yellow) and IES- matching (green) reads of the early and late developmental stages for control and *PDSG1* and *PDSG2* knockdowns. (D, E) Representative sequence logos of 25 nt and 27 nt sRNAs for control and knockdown cells from the late developmental stage. Note that due to the low abundance of iesRNAs in the knockdowns in (E) a single sRNA is relatively abundant in the 27 nt sRNAs.

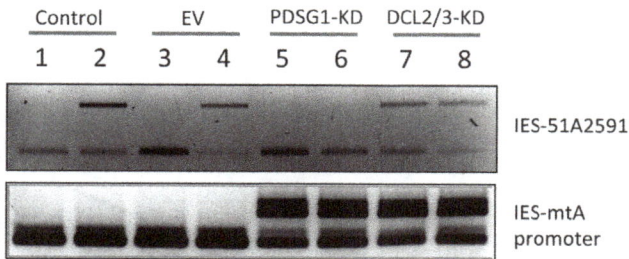

**Figure 4. Effect of PDSG1 depletion on the retention of injected IES (51A2591).** Lanes 1 and 2, control non-silenced cells. Lanes 3 and 4, control silencing with empty vector (EV). Lanes 5 and 6, PDSG1 depleted cells. Lanes 7 and 8, DCL2 and DCL3 depleted cells. Lanes 2, 4, 6 and 8, IES 51A2591 injected cells. IES-mtA promoter (control mcIES).

developing MACs (Fig. 5I). This may suggest an accumulation of H3K27me3-marked chromatin due to a complete block of DNA elimination caused by the silencing of *PDSG2*. Furthermore, similar result was obtained with H3K9me3-specific antibody (Fig. S7).

H3K27me3 antibody specificity was confirmed by a peptide competition assay (Fig. 5J). Only in case when antibody was pre-incubated with H3K27me3 peptide, specific band was not detectible due to a complete out-competition of the antibody.

## Discussion

In *Paramecium* sRNAs guide the elimination of germline-limited sequences during development. scnRNAs, that derive from the germline genome, target the elimination of DNA sequences subject to epigenetic control from developing MAC during its development[41]. A second class of sRNAs, iesRNAs, produced from excised germline-specific DNA in the new MAC complements scnRNAs and ensures complete DNA excision [21]. In this study we identify and characterize two novel proteins in *Paramecium* that affect both DNA excision and the development-specific sRNA populations during new MAC genome reorganization.

*PDSG1* reaches the highest level of expression at the beginning of sexual development and remains expressed at a high level until the new MAC is formed. The localization of *PDSG1* is very similar to that of Nowa1 protein, which was shown to be involved in the excision of maternally-controlled IESs [15]. Nowa1 is an Argonaute and RNA binding domain-containing protein [15], and its putative role is to facilitate the interactions between Piwi-bound small RNAs and their targets in the maternal MAC (RNA scanning process) and in the developing MAC (targeting DNA excision). We show that depletion of *PDSG1* also blocks the excision of some maternally-controlled IESs, which may suggest that both Pdsg1 and Nowa1 are components of the same molecular machinery (Fig. 6). Surprisingly, the silencing of *PDSG1* does not affect the elimination of Sardine and Thon transposons which is strongly affected by *NOWA1* silencing [15]. No other protein in *Paramecium* is known to affect only the

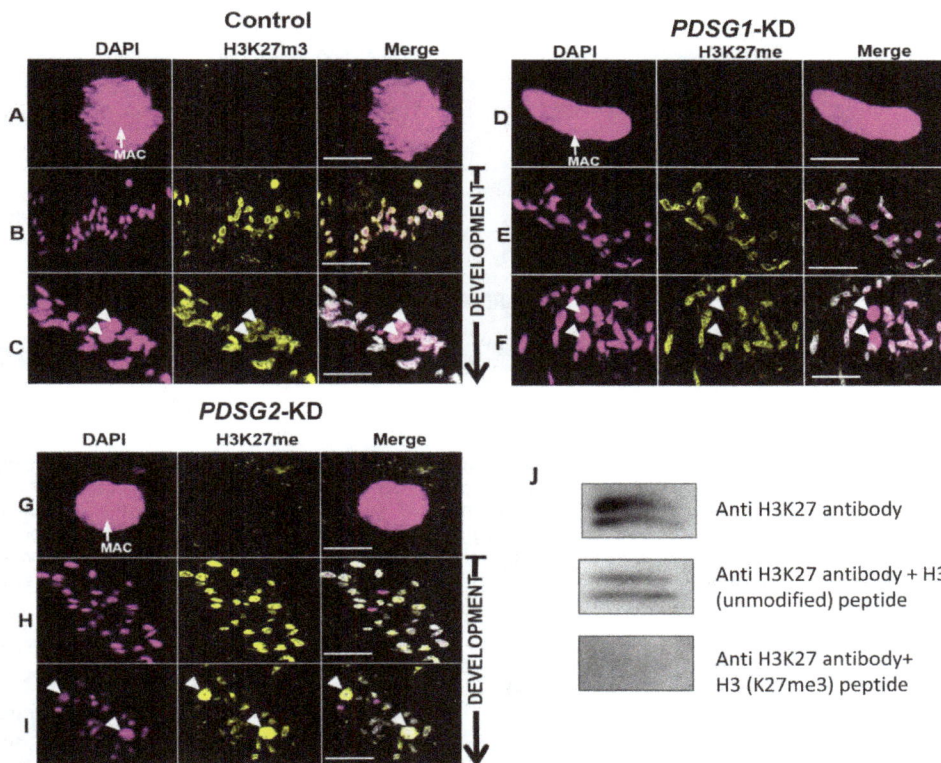

**Figure 5. Effects of Pdsg1 and Pdsg2 depletion on H3K27me3 histone modification.** (A, D, G) Maternal MAC in vegetative cells of control, *PDSG1*-KD and *PDSG2*-KD. (B, E, H) Middle stage of development with the fragments decorated with H3K27me3. (C, F, I) Late stage of development with histone modification present in the new MAC. J Peptide competition assay. Magenta: DAPI; yellow: H3K27me3; white arrow: MAC; arrowhead: new MAC. Scale bar 5 μm.

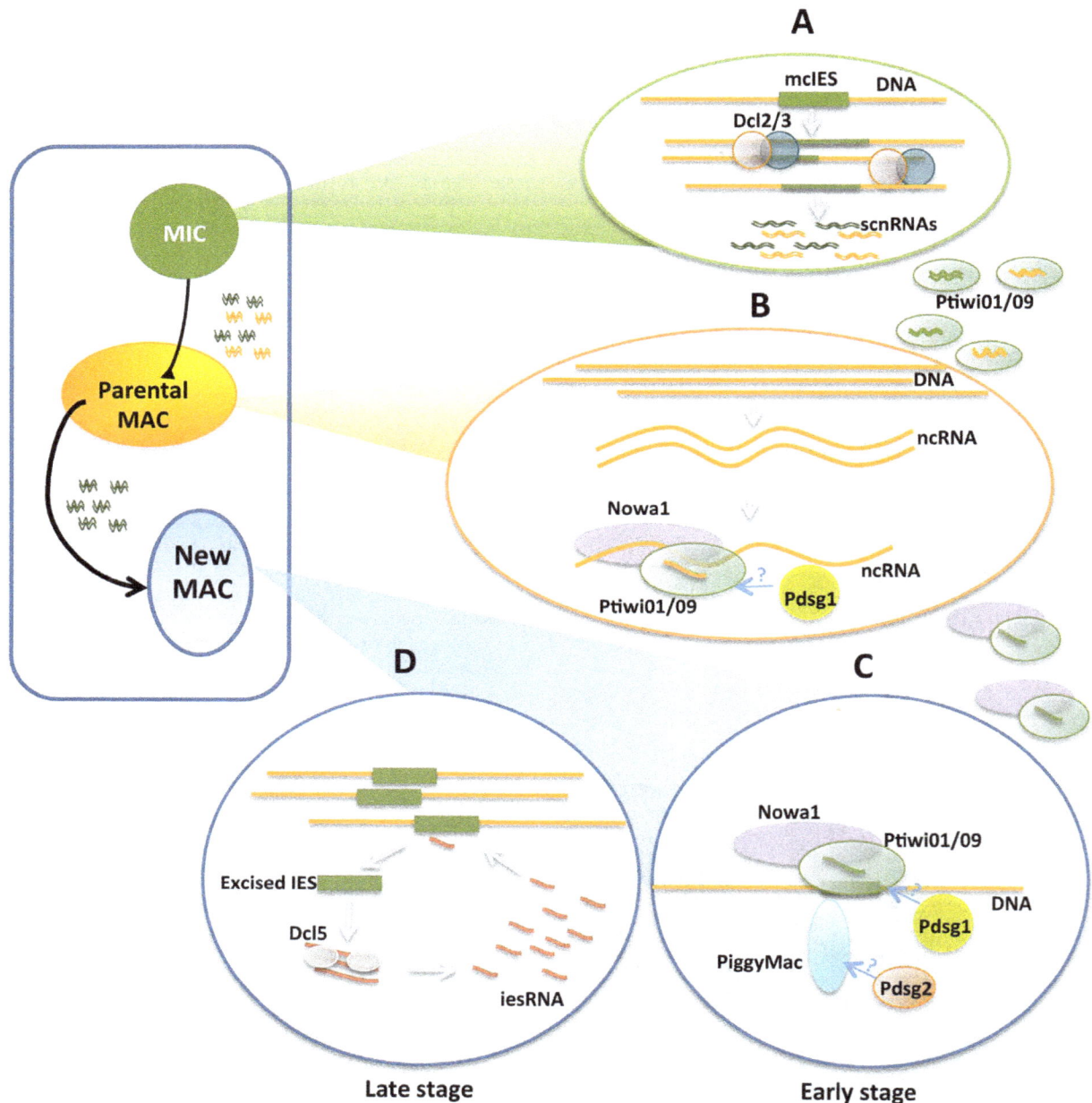

**Figure 6. Proposed model of Pdsg1 and Pdsg2 roles in elimination of mcIESs during MAC development.** (A) Early development, MIC germline genome is transcribed and transcripts are processed by Dcl2/3 into scnRNAs. (B) scnRNAs are transported from the MIC to parental MAC by Ptiwi01/09. Once in the parental MAC, scanning takes place filtering out the MAC genome-matching scnRNAs. We propose that the matching of scnRNAs to complementary sequences may be driven by a multiprotein complex that may include Pdsg1. (C) scnRNAs without matching sequences are transported to the new MAC where they target the excision of complementary sequences. This study suggests that Pdsg2 may be part of the DNA excision machinery. (D) After the excision, IESs are used as templates for iesRNA production to ensure reinforce the signal for an efficient targeting of IES excision.

excision of mcIESs but not the transposons. This difference may indicate that the machinery involved in the process of marking transposons for elimination differs from the one employed in the selection of mcIESs even though the two germline-specific DNA elements seem to require sRNA machinery for their recognition and/or excision.

Our results suggest that Pdsg1 may be involved in scnRNA transport between nuclei or in the interaction of scnRNAs with the complementary target sequences (Fig. 6). Pdsg1 depletion prevents

the degradation of MAC-matching scnRNAs and induces a massive accumulation of these sRNAs suggesting a direct role in the scanning process. In addition, we show that *PDSG1* knockdown prevents the retention of an IES that is normally retained in the developing MAC after microinjection of its sequence into the maternal MAC prior development (Fig. 4). Because retention of mcIESs is believed to be due to the sequestration of scnRNAs complementary to the injected IES in the parental MAC, our data suggests that Pdsg1 is involved in the scanning process. The reduced amount of iesRNAs in Pdsg1 knockdown at late developmental stage may be attributed to the reduced amount of excised IESs, which are prerequisites for transcription and iesRNA production.

Contrary to the effects of *PDSG1* knockdown, silencing of *PDSG2* does not perturb scnRNA quantities or the scanning. However, the production of iesRNAs is highly impaired by the depletion of Pdsg2 suggesting that iesRNA production is blocked because of the absence of excised IESs or that this protein might be involved in iesRNA production or stability (Fig. 6). The fact that Pdsg2 depletion affects the localization of Dcl5 may support the former idea. Furthermore, the substantial retention of mc- and non-mcIESs and the lethality of the progeny on PDSG2-KD indicate that Pdsg2 is essential for excision of IESs.

At the chromatin level, the methylation of histone H3 (H3K9me3 and H3K27me3), shown to be a necessary for scnRNA-dependent DNA elimination in *Tetrahymena* [26], is also present in *Paramecium*. In *Paramecium*, only the IESs that require sRNAs for excision may be dependent on histone modification since not all IESs require scnRNAs and iesRNAs for their excision [42]. This suggests that both ciliates present similar mechanism for targeting of DNA excision, however, as debated previously, in *Paramecium* this seems unlikely since some IESs are shorter than the length of DNA wrapped around a nucleosome [41]. The fact that *PDSG1* knockdown blocks the selection of scnRNAs, affects the methylation of histones and blocks the excision of maternally controlled IESs, may suggest that the these processes are associated with each other, but the precise mechanism still needs to be resolved. From the effects of the *PDSG2* knockdown it can be concluded that a block of IES excision induces an accumulation of H3K27me3 and H3K9me3. A future logical follow up to this study will be to determine potential binding partners of Pdsg1 and Pdsg2 to further elucidate their role in DNA elimination in *Paramecium*.

## Supporting Information

**Figure S1   Primers used for amplification of silencing constructs for all selected candidates.**

**Figure S2   Twenty-eight selected candidates with accession numbers and predicted protein domains.**

**Figure S3   Gene expression profiles of selected candidates.**

**Figure S4   Primers used for IES retention PCR.**

**Figure S5   Summarized results of IES retention PCR.** Seven tested IES are listed in the first column. IES retention is indicated with (+). Maternally controlled IESs are labelled as 'mc' and non-maternally controlled IESs as 'non-mc'.

**Figure S6   Depletion of Pdsg2 affect the normal distribution of Dcl5-GFP in the new MAC.** Localization of Dcl5-GFP to the new MAC in control, PDSG1-KD and PDSG2-KD strains. DAPI staining (blue, DNA), Dcl5-GFP (green) and merge images are shown. White arrows indicate developing MACs. (bottom panel) New MAC is shown in detail from the merge images.

**Figure S7   The effects of *PDSG1* and *PDSG2* knockdowns on H3K9me3 histone modification.** (A, D, G) Maternal MAC in vegetative cells of EV, *PDSG1*-KD and *PDSG2*-KD. (B, E, H) Middle stage of development with the fragments decorated with H3K27me3. (C, F, I) Late stage of development with histone modification present in new MAC. Magenta: DAPI; yellow: H3K27me3; white arrow: MAC; arrow head: new MAC. Scale bar 5 μm.

## Acknowledgments

We would like to thank Nasikhat Stahlberger and Adeel Manaf for their technical support.

## Author Contributions

Conceived and designed the experiments: MN. Performed the experiments: MA PS CH AS. Analyzed the data: MA ES MN. Contributed reagents/materials/analysis tools: MA PS CH AS ES. Wrote the paper: MA MN.

## References

1. Arnaiz O, Mathy N, Baudry C, Malinsky S, Aury JM, et al. (2012) The Paramecium germline genome provides a niche for intragenic parasitic DNA: evolutionary dynamics of internal eliminated sequences. PLoS Genet 8: e1002984.

2. Klobutcher LA, Herrick G (1995) Consensus inverted terminal repeat sequence of Paramecium IESs: resemblance to termini of Tc1-related and Euplotes Tec transposons. Nucleic Acids Res 23: 2006–2013.

3. Baudry C, Malinsky S, Restituito M, Kapusta A, Rosa S, et al. (2009) PiggyMac, a domesticated piggyBac transposase involved in programmed genome rearrangements in the ciliate Paramecium tetraurelia. Genes Dev 23: 2478–2483.

4. Betermier M (2004) Large-scale genome remodelling by the developmentally programmed elimination of germ line sequences in the ciliate Paramecium. Res Microbiol 155: 399–408.

5. Gratias A, Lepere G, Garnier O, Rosa S, Duharcourt S, et al. (2008) Developmentally programmed DNA splicing in Paramecium reveals short-distance crosstalk between DNA cleavage sites. Nucleic Acids Res 36: 3244–3251.

6. Mayer KM, Mikami K, Forney JD (1998) A mutation in Paramecium tetraurelia reveals functional and structural features of developmentally excised DNA elements. Genetics 148: 139–149.

7. Duharcourt S, Butler A, Meyer E (1995) Epigenetic self-regulation of developmental excision of an internal eliminated sequence on Paramecium tetraurelia. Genes Dev 9: 2065–2077.

8. Duharcourt S, Keller AM, Meyer E (1998) Homology-dependent maternal inhibition of developmental excision of internal eliminated sequences in Paramecium tetraurelia. Mol Cell Biol 18: 7075–7085.

9. Galvani A, Sperling L (2001) Transgene-mediated post-transcriptional gene silencing is inhibited by 3' non-coding sequences in Paramecium. Nucleic Acids Res 29: 4387–4394.

10. Ruiz F, Vayssie L, Klotz C, Sperling L, Madeddu L (1998) Homology-dependent gene silencing in Paramecium. Mol Biol Cell 9: 931–943.

11. Galvani A, Sperling L (2002) RNA interference by feeding in Paramecium. Trends Genet 18: 11–12.

12. Garnier O, Serrano V, Duharcourt S, Meyer E (2004) RNA-Mediated Programming of Developmental Genome Rearrangements in Paramecium tetraurelia. Mol Cell Biol 24: 7370–7379.

13. Lee SR, Collins K (2007) Physical and functional coupling of RNA-dependent RNA polymerase and Dicer in the biogenesis of endogenous siRNAs. Nat Struct Mol Biol 14: 604–610.

14. Lepere G, Nowacki M, Serrano V, Gout JF, Guglielmi G, et al. (2009) Silencing-associated and meiosis-specific small RNA pathways in Paramecium tetraurelia. Nucleic Acids Res 37: 903–915.

15. Nowacki M, Zagorski-Ostoja W, Meyer E (2005) Nowa1p and Nowa2p: novel putative RNA binding proteins involved in trans-nuclear crosstalk in Paramecium tetraurelia. Curr Biol 15: 1616–1628.

16. Bouhouche K, Gout JF, Kapusta A, Betermier M, Meyer E (2011) Functional specialization of Piwi proteins in Paramecium tetraurelia from post-transcriptional gene silencing to genome remodelling. Nucleic Acids Res 39: 4249–4264.

17. Malone CD, Anderson AM, Motl JA, Rexer CH, Chalker DL (2005) Germ line transcripts are processed by a Dicer-like protein that is essential for developmentally programmed genome rearrangements of Tetrahymena thermophila. Mol Cell Biol 25: 9151–9164.

18. Mochizuki K, Fine NA, Fujisawa T, Gorovsky MA (2002) Analysis of a piwi-related gene implicates small RNAs in genome rearrangement in tetrahymena. Cell 110: 689–699.

19. Mochizuki K, Gorovsky MA (2005) A Dicer-like protein in Tetrahymena has distinct functions in genome rearrangement, chromosome segregation, and meiotic prophase. Genes Dev 19: 77–89.

20. Schoeberl UE, Kurth HM, Noto T, Mochizuki K (2012) Biased transcription and selective degradation of small RNAs shape the pattern of DNA elimination in Tetrahymena. Genes Dev 26: 1729–1742.

21. Sandoval PY, Swart EC, Arambasic M, Nowacki M (2014) Functional Diversification of Dicer-like Proteins and Small RNAs Required for Genome Sculpting. Dev Cell 28: 174–188.

22. Mochizuki K, Gorovsky MA (2004) Conjugation-specific small RNAs in Tetrahymena have predicted properties of scan (scn) RNAs involved in genome rearrangement. Genes Dev 18: 2068–2073.

23. Mochizuki K, Gorovsky MA (2004) Small RNAs in genome rearrangement in Tetrahymena. Curr Opin Genet Dev 14: 181–187.

24. Aronica L, Bednenko J, Noto T, DeSouza LV, Siu KM, et al. (2008) Study of an RNA helicase implicates small RNA-noncoding RNA interactions in programmed DNA elimination in Tetrahymena. Genes Dev 22: 2228–2241.

25. Lepere G, Betermier M, Meyer E, Duharcourt S (2008) Maternal noncoding transcripts antagonize the targeting of DNA elimination by scanRNAs in Paramecium tetraurelia. Genes Dev 22: 1501–1512.

26. Liu Y, Taverna SD, Muratore TL, Shabanowitz J, Hunt DF, et al. (2007) RNAi-dependent H3K27 methylation is required for heterochromatin formation and DNA elimination in Tetrahymena. Genes Dev 21: 1530–1545.

27. Liu Y, Mochizuki K, Gorovsky MA (2004) Histone H3 lysine 9 methylation is required for DNA elimination in developing macronuclei in Tetrahymena. Proc Natl Acad Sci U S A 101: 1679–1684.

28. Taverna SD, Coyne RS, Allis CD (2002) Methylation of histone h3 at lysine 9 targets programmed DNA elimination in tetrahymena. Cell 110: 701–711.

29. Beisson J, Betermier M, Bre MH, Cohen J, Duharcourt S, et al. (2010) Silencing specific Paramecium tetraurelia genes by feeding double-stranded RNA. Cold Spring Harb Protoc 2010: pdb.prot5363.

30. Brown T (2001) Dot and slot blotting of DNA. Curr Protoc Mol Biol Chapter 2: Unit2.9B.

31. Beisson J, Betermier M, Bre MH, Cohen J, Duharcourt S, et al. (2010) Immunocytochemistry of Paramecium cytoskeletal structures. Cold Spring Harb Protoc 2010: pdb.prot5365.

32. Li H, Durbin R (2009) Fast and accurate short read alignment with Burrows-Wheeler transform. Bioinformatics 25: 1754–1760.

33. Finn RD, Bateman A, Clements J, Coggill P, Eberhardt RY, et al. (2014) Pfam: the protein families database. Nucleic Acids Res 42: D222–230.

34. Hildebrand A, Remmert M, Biegert A, Soding J (2009) Fast and accurate automatic structure prediction with HHpred. Proteins 77 Suppl 9: 128–132.

35. Nowak JK, Gromadka R, Juszczuk M, Jerka-Dziadosz M, Maliszewska K, et al. (2011) Functional study of genes essential for autogamy and nuclear reorganization in Paramecium. Eukaryot Cell 10: 363–372.

36. Durinck S, Moreau Y, Kasprzyk A, Davis S, De Moor B, et al. (2005) BioMart and Bioconductor: a powerful link between biological databases and microarray data analysis. Bioinformatics 21: 3439–3440.

37. Arnaiz O, Sperling L (2011) ParameciumDB in 2011: new tools and new data for functional and comparative genomics of the model ciliate Paramecium tetraurelia. Nucleic Acids Res 39: D632–636.

38. Le Mouel A, Butler A, Caron F, Meyer E (2003) Developmentally regulated chromosome fragmentation linked to imprecise elimination of repeated sequences in paramecia. Eukaryot Cell 2: 1076–1090.

39. Matsuda A, Forney JD (2006) The SUMO pathway is developmentally regulated and required for programmed DNA elimination in Paramecium tetraurelia. Eukaryot Cell 5: 806–815.

40. Matsuda A, Shieh AW, Chalker DL, Forney JD (2010) The conjugation-specific Die5 protein is required for development of the somatic nucleus in both Paramecium and Tetrahymena. Eukaryot Cell 9: 1087–1099.

41. Coyne RS, Lhuillier-Akakpo M, Duharcourt S (2012) RNA-guided DNA rearrangements in ciliates: is the best genome defence a good offence? Biol Cell 104: 309–325.

42. Swart EC, Wilkes CD, Sandoval PY, Arambasic M, Sperling L, et al. (2014) Genome-wide analysis of genetic and epigenetic control of programmed DNA deletion. Nucleic Acids Res 42: 8970–8983.

# Growth Hormone-Regulated mRNAs and miRNAs in Chicken Hepatocytes

**Xingguo Wang**[1,2,9], **Lei Yang**[1,9], **Huijuan Wang**[1], **Fang Shao**[1], **JianFeng Yu**[1], **Honglin Jiang**[3], **Yaoping Han**[1], **Daoqing Gong**[2], **Zhiliang Gu**[1]*

**1** Department of Life Science and Technology, Changshu Institute of Technology, Changshu, P R China, **2** College of Animal Science and Technology, Yangzhou University, Yangzhou, P R China, **3** Department of Animal and Poultry Sciences, Virginia Polytechnic Institute and State University, Blacksburg, Virginia, United States of America

## Abstract

Growth hormone (GH) is a key regulatory factor in animal growth, development and metabolism. Based on the expression level of the GH receptor, the chicken liver is a major target organ of GH, but the biological effects of GH on the chicken liver are not fully understood. In this work we identified mRNAs and miRNAs that are regulated by GH in primary hepatocytes from female chickens through RNA-seq, and analyzed the functional relevance of these mRNAs and miRNAs through GO enrichment analysis and miRNA target prediction. A total of 164 mRNAs were found to be differentially expressed between GH-treated and control chicken hepatocytes, of which 112 were up-regulated and 52 were down-regulated by GH. A total of 225 chicken miRNAs were identified by the RNA-Seq analysis. Among these miRNAs 16 were up-regulated and 1 miRNA was down-regulated by GH. The GH-regulated mRNAs were mainly involved in growth and metabolism. Most of the GH-upregulated or GH-downregulated miRNAs were predicted to target the GH-downregulated or GH-upregulated mRNAs, respectively, involved in lipid metabolism. This study reveals that GH regulates the expression of many mRNAs involved in metabolism in female chicken hepatocytes, which suggests that GH plays an important role in regulating liver metabolism in female chickens. The results of this study also support the hypothesis that GH regulates lipid metabolism in chicken liver in part by regulating the expression of miRNAs that target the mRNAs involved in lipid metabolism.

**Editor:** Michael Schubert, Laboratoire de Biologie du Développement de Villefranche-sur-Mer, France

**Funding:** This research was supported by the National Natural Science Foundation of China (31072025, 31272438). The funders had no role in study design, data collection and analysis, decision to publish, or preparation of the manuscript.

**Competing Interests:** The authors have declared that no competing interests exist.

* Email: zhilianggu88@hotmail.com

**9** These authors contributed equally to this work.

## Introduction

Growth hormone (GH) is a peptide hormone from the anterior pituitary gland [1,2]. It has many biological effects at both the whole body and tissue levels [3]. GH regulates animal growth, development and metabolism [3,4,5,6]. GH regulates the metabolism of not only protein but also that of lipid and carbohydrates [7]. GH initiates its function by binding to the GH receptor (GHR) [8,9,10]. Binding of GH to the GHR activates the receptor-associated tyrosine kinase JAK2 [11], and JAK2 then activates multiple proteins, including STAT1, STAT3, STAT5, MAPK, and PI3K [12]. These proteins in turn mediate GH-caused changes in gene expression or protein modification.

Liver is a key metabolic organ, and in chickens this is where most of the *de novo* synthesis of fatty acids occurs [13,14]. microRNAs (miRNAs) are a class of small non-coding RNAs about 22 nucleotides in length, and regulate gene expression by interacting with the 3′ untranslated regions (UTRs) of target mRNAs [15]. miRNAs have been shown to play important roles in many biological processes including liver metabolism. For example, miR-122, abundantly expressed in liver, modulates protein metabolism in liver by targeting cationic amino acid transporter 1 [16]; it regulates the synthesis of fatty acids and cholesterol by repressing the expression of aldolase-A, 3-hydroxy-3-methylglutaryl-coenzyme A reductase, and AMP-activated protein kinase [17,18]. miR-33 is another miRNA involved in liver metabolism: it regulates cholesterol efflux and high-density lipoprotein metabolism by targeting ATP-binding cassette, sub-family A (ABC1), member 1 and ATP-binding cassette, sub-family G (WHITE), member 1 [19], and it reduces fatty acid degradation by targeting multiple genes involved in fatty acid β-oxidation [20].

RNA sequencing (RNA-seq) is a novel gene expression profiling technology based on high-throughput DNA sequencing. The benefit of RNA-Seq over other large-scale gene expression profiling methods is its ability to measure mRNA expression in a single assay and, at the same time, reveal new genes and transcripts [21,22,23]. Similarly, RNA-seq can be also used to identify novel miRNAs and detect differentially expressed miRNAs between samples [24].

As in mammals [3,7,25,26], GH has metabolic effects in chickens: GH regulates lipid metabolism in chicken adipose tissue [27] and chicken liver [28]. However, it is unclear whether GH has the same growth-stimulating effect in chickens as in mammals because exogenous GH treatment to chickens causes no growth

**Table 1.** An overview of mRNA sequencing results.

| Sample | Raw Reads | Clean Reads | Reads mapped to chicken genome | Reads mapped to chicken genes |
|--------|-----------|-------------|-------------------------------|-------------------------------|
| chGH | 16,474,842 | 13,638,494 | 10,669,529 | 8,812,163 |
| PBS | 10,662,250 | 7,473,018 | 5,652,712 | 4,656,574 |

responses and because plasma GH concentrations in chickens are in general not correlated with growth rates [29].

The mechanism by which GH regulates lipid metabolism in chicken liver is not clear. In this study, we determined the effects of GH on the expression levels of all mRNAs and miRNAs in primary hepatocytes from female chickens by RNA-seq. We analyzed the differentially expressed mRNAs or genes (DEG) and differentially expressed miRNAs (DEM) with multiple bioinformatics tools to correlate the DEG and the DEM to the physiological functions of GH. The main hypothesis to be tested in this study was that GH regulates liver metabolism in the chicken in part by regulating the expression of miRNAs that target mRNAs directly related to liver metabolism.

## Materials and Methods

### 1. Culture of primary chicken hepatocytes

All procedures involving animals were approved by Changshu Institute of Technology Institutional Animal Care and Use Committee and conformed to the Guide for the Care and Use of Laboratory Animals of Jiangsu Province. All efforts were made to minimize suffering. Primary chicken hepatocytes were isolated from 4 female, 4-week-old Arbor Acres commercial chickens which were fasted 12 hours (h) before being anaesthetized by intraperitoneal injection of sodium thiopenthal (50 mg/kg) and anticoagulated by intraperitoneal injection of heparin (1750 U/kg). Livers were isolated and the chickens were sacrificed by removal of the hearts. Hepatocytes were isolated from the livers as previously described [30]. Hepatocytes from individual chickens were cultured separately. Hepatocytes were plated in 24-well plates or 10-cm dishes at a density of $1.3 \times 10^6$ cells/ml in Willam's E medium (Gibco, Grand Island, NY) supplemented with 5% chicken serum, 100 U/ml penicillin-streptomycin, 10 µg/ml insulin and 30 mmol/L NaCl in a humidified incubator at 37°C with 5% $CO_2$. Twenty-eight h later, hepatocytes were serum starved for 8 h, followed by 12 h of treatment with 500 ng/ml chicken GH (chGH) (Prospec, Ness-Ziona, Israel) or an equal volume of PBS. Cells were lysed and total RNA was isolated using TRIzol reagent (Invitrogen, Carlsbad, CA) following the manufacturer's directions.

**Table 2.** Top 10 GH up-regulated and top 10 GH down-regulated mRNA-encoding genes.

| Name | PBS (RPKM) | chGH (RPKM) | fold change (RPKM (chGH/PBS)) | P-value |
|------|------------|-------------|-------------------------------|---------|
| **up-regulated genes** | | | | |
| RPS28 | 0.52 | 21.63 | 40.74 | 0.0005 |
| TTLL3 | 1.09 | 23.57 | 21.41 | 0.028 |
| CISH | 1.97 | 27.94 | 13.99 | 1.38E-10 |
| SCN4A | 1.87 | 20.79 | 10.95 | 5.29E-21 |
| BLB1 | 0.74 | 7.13 | 9.53 | 0.009 |
| TGM2 | 0.20 | 1.78 | 8.88 | 0.037 |
| SS1R | 0.68 | 5.33 | 8.75 | 0.015 |
| SERPINA4 | 8.86 | 78.26 | 8.70 | 1.06E-18 |
| GGA.45581 | 35.97 | 278.79 | 7.64 | 9.82E-31 |
| CCK | 3.23 | 24.44 | 7.46 | 5.05E-05 |
| **down-regulated genes** | | | | |
| AFP | 38.74 | 0.10 | 0.0025 | 4.76E-40 |
| RGS6 | 1.92 | 0.09 | 0.047 | 0.014 |
| STAR | 27.75 | 5.01 | 0.18 | 1.28E-08 |
| CCKAR | 5.57 | 1.01 | 0.18 | 0.013 |
| KIAA0408 | 3.38 | 0.63 | 0.19 | 0.020 |
| BCL6 | 19.86 | 4.26 | 0.21 | 8.27E-08 |
| PDE10A | 7.07 | 1.57 | 0.22 | 0.002 |
| ENSGALG00000024377 | 26.06 | 6.18 | 0.23 | 0.044 |
| NECAB1 | 7.12 | 1.74 | 0.24 | 0.010 |
| TAGLN3 | 34.96 | 8.62 | 0.24 | 5.64E-06 |

**Table 3.** Real time RT-PCR validation of gene expression levels revealed by RNA-seq.

| Gene | Real-time RT-PCR | | RNA-seq | |
|---|---|---|---|---|
| | Fold change[1] | P-value | Fold change[2] | P-value |
| **Group 1[3]** | | | | |
| FABP1 | 4.30 | 0.034 | 3.96 | 3.82E-07 |
| FGFR3 | 2.37 | 3.93E-04 | 2.11 | 0.009 |
| FURIN | 1.71 | 0.012 | 2.30 | 2.82E-05 |
| IRF8 | 2.68 | 0.005 | 2.50 | 1.69E-05 |
| LPIN1 | 1.81 | 0.049 | 1.78 | 1.45E-04 |
| MAPKAPK3 | 1.68 | 0.021 | 2.24 | 0.010 |
| PHGDH | 1.12 | 0.748 | 1.54 | 0.004 |
| PKIG | 1.47 | 0.094 | 1.67 | 0.046 |
| THRSP | 1.44 | 0.565 | 1.76 | 0.007 |
| **Group 2[4]** | | | | |
| ABCG8 | 0.28 | 0.048 | 0.33 | 2.14E-04 |
| ALDH1A3 | 0.70 | 0.087 | 0.66 | 0.009 |
| BCL6 | 0.18 | 0.003 | 0.21 | 8.27E-08 |
| LPIG | 0.33 | 5.69E-04 | 0.34 | 2.67E-07 |
| NECAB1 | 0.43 | 0.447 | 0.24 | 0.010 |
| PDE10A | 0.30 | 0.042 | 0.22 | 0.002 |
| PPAP2B | 0.61 | 0.075 | 0.52 | 0.010 |
| RGS6 | 0.49 | 0.218 | 0.05 | 0.014 |
| STAR | 0.17 | 0.024 | 0.18 | 1.28E-08 |
| **Group 3[5]** | | | | |
| FGF1 | 1.19 | 0.743 | 1.62 | 0.661 |
| GPX7 | 1.15 | 0.760 | 3.66 | 0.552 |
| MEF2A | 0.90 | 0.665 | 6.79 | 0.461 |
| NME1 | 0.90 | 0.800 | 7.83 | 0.384 |
| PDGFB | 0.78 | 0.350 | 1.72 | 0.694 |
| PNPLA4 | 0.90 | 0.656 | 1.99 | 0.449 |
| PRKAR2B | 1.47 | 0.339 | 4.44 | 0.438 |
| PTK2B | 0.97 | 0.925 | 2.27 | 0.220 |
| PTPRC | 1.21 | 0.540 | 1.72 | 0.247 |
| ROMO1 | 0.84 | 0.205 | 2.13 | 0.168 |
| **Group 4[6]** | | | | |
| ACER2 | 1.30 | 0.163 | 0.46 | 0.437 |
| ADPGK | 1.02 | 0.759 | 0.56 | 0.371 |
| ATF7 | 1.14 | 0.602 | 0.31 | 0.445 |
| BIRC5 | 0.93 | 0.753 | 0.26 | 0.685 |
| CDK2AP1 | 1.23 | 0.613 | 0.28 | 0.384 |
| DUSP28 | 0.90 | 0.628 | 0.17 | 0.471 |
| FTO | 1.33 | 0.159 | 0.39 | 0.472 |
| HGF | 1.87 | 0.463 | 0.49 | 0.470 |
| MAFK | 0.83 | 0.074 | 0.50 | 0.144 |
| NT5M | 1.30 | 0.497 | 0.45 | 0.488 |
| RUNX2 | 0.98 | 0.939 | 0.157 | 0.272 |

[1]fold change by real-time RT-PCR is relative gene expression level (chGH/PBS).
[2]fold change by RNA-seq is RPKM (chGH/PBS).
[3]Group 1: fold change (RPKM (chGH/PBS)) >1.5 and P<0.05.
[4]Group 2: fold change (RPKM (chGH/PBS)) <2/3 and P<0.05.
[5]Group 3: fold change (RPKM (chGH/PBS)) >1.5 and P>0.05.
[6]Group 4: fold change (RPKM (chGH/PBS)) <2/3 and P>0.05.

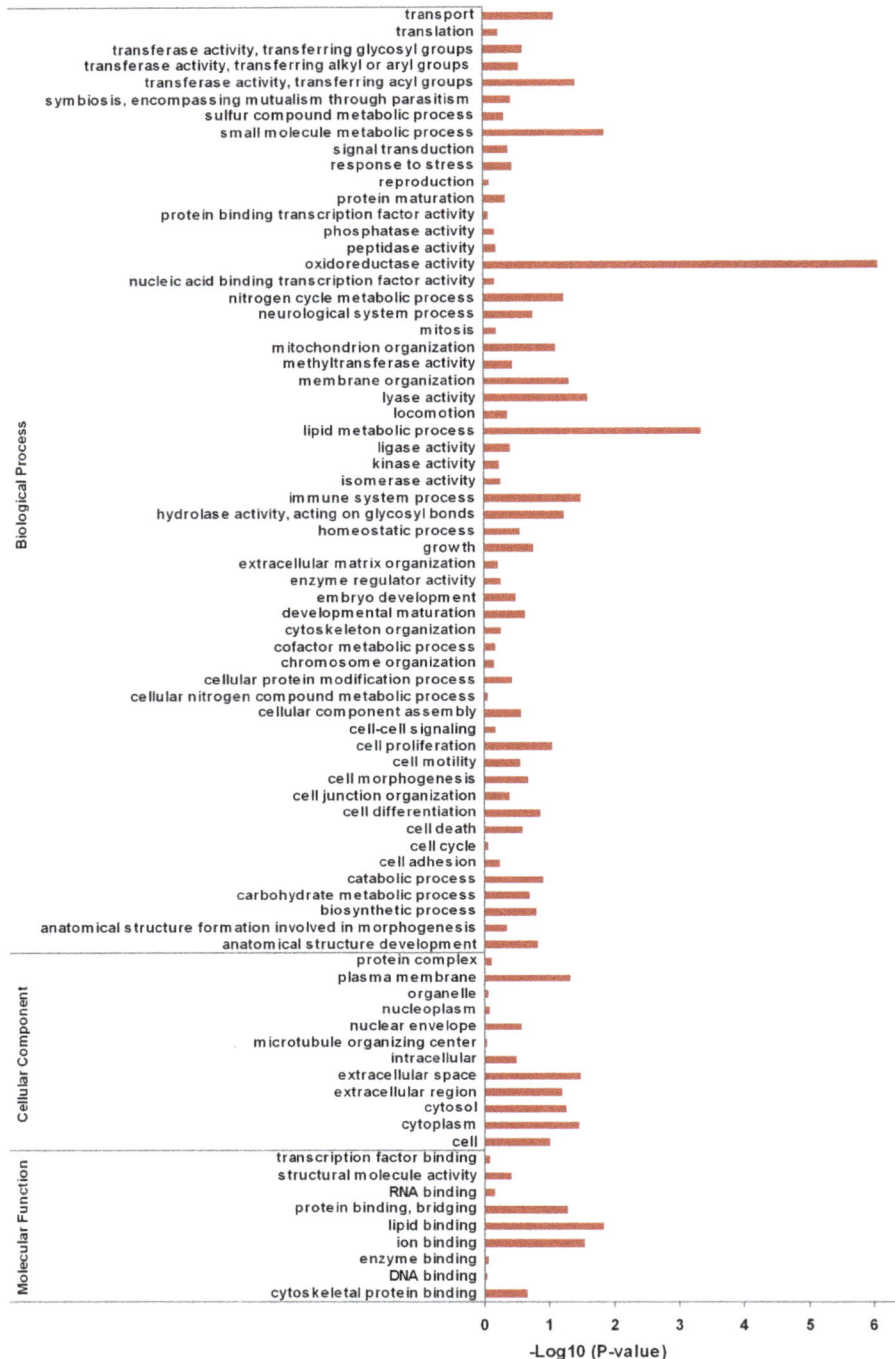

**Figure 1. GO enrichment analysis of genes regulated by chGH in chicken hepatocytes.** The GO terms are sorted by −Log10 of the enrichment P-value, which represents the enrichment significance of GO terms. The enrichment of GO terms is showed by comparing DEG with the whole genome from this figure.

## 2. Real time RT-PCR

Concentrations and quality of RNA samples were determined by NanoDrop ND2000 spectrophotometry (Thermo Scientific, Wilmington, DE) and formaldehyde-agarose gel electrophoresis. Five hundred ng of total RNA was reverse-transcribed to cDNA in a total volume of 10 μl using Takara PrimeScript RT reagent kit (Takara, Dalian, China). The relative expression levels of genes were quantified using SYBR Premix Ex Taq (Takara) on an

Applied Biosystems 7500 Thermocycler (Applied Biosystems) according to the manufacturer's directions. In this analysis, GAPDH was used as an internal control. All reactions were run in duplicate. Data are means ± SEM of 4 independent cell culture experiments (i.e., 4 chickens) and analyzed by student's t-test. The primers for real time RT-PCR were presented in Table S1.

**Table 4.** List of growth-related mRNAs whose expression in chicken hepatocytes was regulated by chGH.

| Name | PBS (RPKM) | chGH (RPKM) | fold change (RPKM (chGH/PBS)) | P-value |
|------|-----------|-------------|-------------------------------|---------|
| BCL6 | 19.86 | 4.26 | 0.21 | 8.27E-08 |
| CISH | 1.97 | 27.94 | 13.99 | 1.38E-10 |
| FOXP2 | 11.40 | 19.73 | 1.71 | 0.026 |
| IRF8 | 25.75 | 65.31 | 2.50 | 1.69E-05 |
| ROBO1 | 11.86 | 7.27 | 0.60 | 0.038 |
| ULK2 | 23.90 | 70.81 | 2.92 | 2.97E-09 |

RPKM was denoted as a formula: $RPKM = \dfrac{\text{Total exon reads}}{\text{Mapped Reads (millions} \times \text{exon lenth (KB))}}$.

## 3. mRNA sequencing

RNA samples for mRNA sequencing were prepared using Illumina TrueSeq RNA Sample Preparation Kit according to the manufacturer's directions. The liver RNA samples from 4 chGH-treated or PBS-treated chickens were pooled in equal volumes after their concentrations were adjusted to 10 nM. Sequencing of the cDNA was performed by Personalbio (Shanghai, China) using the Illumina Miseq system.

## 4. mRNA analysis

The reads obtained from sequencing were mapped to the chicken genome (WASHUC2.69) in Ensembl using Bowtie/Tophat (2.0.5) (http://tophat.cbcb.umd.edu), and the reads of each gene were normalized using reads per kilo bases per million reads (RPKM). The significance was determined by normalizing the raw reads and calculating the P-value using DESeq (http://bioconductor.org/packages/release/bioc/html/DESeq.html). Genes with fold change (RPKM (chGH/PBS)) >1.5 or <2/3 and P-value <0.05 were identified as DEG.

Gene Ontology (GO) enrichment analysis was performed using GOslim (http://www.geneontology.org/page/go-slim-and-subset-guide). The data were presented as −log (P-value). P-value = 1 − $\sum_{i=0}^{m-1} \dfrac{\dbinom{M}{i}\dbinom{N-M}{n-i}}{\dbinom{N}{n}}$, where N is the total gene number in genome, n is the number of all DEG, M is the number of genes in genome that are involved in certain GO, and m is the number of DEG that are involved in certain GO. The signaling pathways were analyzed using the KEGG database.

## 5. Small RNA sequencing

The RNA samples used for small RNA sequencing were the same as for mRNA sequencing. Small RNA libraries for small RNA sequencing were constructed according to the Miseq small RNA sequencing protocol. Briefly, total RNA was ligated to the 3′ adapter. Then it was ligated to the 5′ adapter. The ligated RNA was reverse-transcribed to cDNA. The cDNA which had a 3′ adapter and a 5′ adapter was amplified by PCR. The PCR products with appropriate length were extracted from polyacrylamide gels to construct the small RNA libraries. The libraries of four samples were pooled in equal volumes after first being normalized to 10 nM. Sequencing of the small RNA libraries was performed by Personalbio using the Illumina Miseq system.

## 6. miRNA analysis

The raw reads from small RNA sequencing were treated by trimming the adapters and removing the low-quality sequences to

obtain the clean reads. The clean reads with 15–30 nt were analyzed by counting and grouping identical sequences as unique reads. The unique reads were mapped to the chicken genome using Bowtie and BLAST searched against the ncRNA database Rfam (10.1) assess the quality of sequences and obtain ncRNA annotation. The sequence reads were first searched against the chicken miRNA database in miRBase (20.0) to identify known chicken miRNAs, then against the miRNA databases of other species in miRBase to identify chicken miRNAs homologous to known miRNAs in other species, and finally against the chicken genome to obtain their ~80 nt genomic sequences flanking the 5′ or 3′ end and analyzed using the program mireap (http://sourceforge.net/projects/mireap/) to predict potentially novel chicken miRNAs and their precursors according to the miRNA biogenesis principle [31].

The reads of miRNAs from control (PBS treated) samples and GH treated samples were normalized using reads per million (RPM). RPM = Actual miRNA count/Total count of clean reads*1000000. When the RPM of a certain miRNA in one type of samples was zero, it was revised to 0.01, and if the RPM of a certain miRNA in both types of samples was less than 1, the miRNA would not be used in further DEM analysis. Fold change and P-value were calculated from the RPM. Fold change = RPM (chGH)/RPM (PBS). P-value formula: $p(x|y) = \left(\dfrac{N_2}{N_1}\right)^y$

$\dfrac{(x+y)!}{x!\,y!\left(1+\dfrac{N_2}{N_1}\right)^{(x+y+1)}}$ $\begin{aligned} C(y \leq y_{min}|x) &= \sum_{y=0}^{y \leq y_{min}} p(y|x) \\ D(y \geq y_{max}|x) &= \sum_{y \geq y_{max}}^{\infty} p(y|x) \end{aligned}$. The x and

y represent normalized expression level, and the N1 and N2 represent total count of clean reads of a given miRNA in small RNA libraries of chGH and PBS groups, respectively. The miRNAs with fold change (RPM (chGH/PBS)) >1.5 or <2/3 and P-value <0.05 were identified as DEM [32].

Target genes of identified known chicken DEM were predicted in the DEG using miRanda with the TargetScan principle [30]. In brief, the target mRNAs of GH up-regulated miRNAs were predicted in GH down-regulated mRNAs; the target mRNAs of GH down-regulated miRNAs were predicted in the GH up-regulated mRNAs. Then GO enrichment analysis of GH-regulated mRNAs targeted by GH-regulated miRNAs was performed using the GOslim program.

## Results

### 1. mRNA expression profiling by RNA-seq

The RNA-seq generated 16,474,842 and 10,662,250 paired end reads from chGH-treated group and PBS-treated group, respec-

**Table 5.** List of metabolism-related mRNAs whose expression in chicken hepatocytes was regulated by chGH.

| Name | PBS (RPKM) | chGH (RPKM) | fold change (RPKM (chGH/PBS)) | P-value |
|---|---|---|---|---|
| Protein metabolism | | | | |
| AASS | 35.18 | 53.89 | 1.51 | 0.029 |
| ABCG5 | 15.51 | 7.35 | 0.47 | 0.018 |
| ABCG8 | 20.88 | 6.94 | 0.33 | 2.14E-04 |
| AKAP13 | 32.52 | 21.50 | 0.65 | 0.02311 |
| BCL6 | 19.86 | 4.26 | 0.21 | 8.27E-08 |
| CACNA1D | 9.77 | 4.62 | 0.47 | 6.47E-04 |
| CAMK1D | 16.62 | 30.05 | 1.78 | 0.049 |
| CEBPA | 2.90 | 13.19 | 4.48 | 0.018 |
| FGFR3 | 8.20 | 17.53 | 2.11 | 0.009 |
| FOXP2 | 11.40 | 19.73 | 1.71 | 0.026 |
| FZD5 | 22.61 | 46.43 | 2.02 | 0.021 |
| F1NQS2 | 76.69 | 122.69 | 1.58 | 0.013 |
| HAL | 165.14 | 260.84 | 1.56 | 0.005 |
| HistoneH4 | 12.98 | 36.83 | 2.79 | 0.0498 |
| ICER | 44.32 | 71.76 | 1.60 | 0.023 |
| IRF8 | 25.75 | 65.31 | 2.50 | 1.69E-05 |
| LPIN1 | 259.93 | 468.47 | 1.78 | 1.45E-04 |
| MC5R | 10.41 | 26.56 | 2.52 | 0.009 |
| OXA1L | 3.28 | 16.92 | 5.09 | 0.030 |
| PHGDH | 542.39 | 848.66 | 1.54 | 0.004 |
| PKIG | 163.59 | 277.14 | 1.67 | 0.046 |
| PPAP2B | 43.27 | 22.99 | 0.52 | 0.010 |
| RAPGEF2 | 30.14 | 55.26 | 1.81 | 7.43E-04 |
| RNL2 | 158.78 | 257.55 | 1.60 | 0.013 |
| SLC7A2 | 19.42 | 30.91 | 1.57 | 0.048 |
| UPP2 | 130.97 | 229.47 | 1.73 | 0.002 |
| Carbohydrate metabolism | | | | |
| CHIA | 59.33 | 141.04 | 2.35 | 3.04E-06 |
| E1BZP4 | 31.08 | 122.15 | 3.88 | 1.3E-10 |
| MAN2A2 | 2.87 | 6.39 | 2.20 | 0.041 |
| PCK2 | 51.79 | 102.40 | 1.95 | 0.003 |
| ST6GAL1 | 33.42 | 54.46 | 1.61 | 0.018 |
| Lipid metabolism | | | | |
| ABHD5 | 23.45 | 47.53 | 2.00 | 8.61E-04 |
| ACSL3 | 27.37 | 16.69 | 0.60 | 0.027 |
| AKR1D1 | 0.85 | 4.83 | 5.57 | 0.043 |
| ALDH1A3 | 148.38 | 99.67 | 0.66 | 0.009 |
| ALDH8A1 | 108.87 | 174.87 | 1.58 | 0.008 |
| CEBPA | 2.90 | 13.19 | 4.48 | 0.018 |
| ELOV6 | 196.41 | 117.72 | 0.59 | 0.005 |
| ICER | 44.32 | 71.76 | 1.60 | 0.023 |
| LIPG | 77.96 | 26.78 | 0.34 | 2.67E-07 |
| LPIN1 | 259.93 | 468.47 | 1.78 | 1.45E-04 |
| PLCD1 | 139.45 | 87.34 | 0.62 | 0.003 |
| PPAP2B | 43.27 | 22.99 | 0.52 | 0.010 |
| RETSAT | 18.49 | 37.63 | 2.01 | 0.024 |
| STAR | 27.75 | 5.01 | 0.18 | 1.28E-08 |
| THRSP | 74.53 | 133.21 | 1.76 | 0.007 |

**Table 6.** List of signal transduction-related mRNAs whose expression in chicken hepatocytes was regulated by chGH.

| Name | PBS (RPKM) | chGH (RPKM) | fold change (RPKM (chGH/PBS)) | P-value |
|---|---|---|---|---|
| Nucleic acid binding transcription factor | | | | |
| BCL6 | 19.86 | 4.26 | 0.21 | 8.27E-08 |
| CAMK1D | 16.62 | 30.05 | 1.78 | 0.049 |
| CEBPA | 2.90 | 13.19 | 4.48 | 0.018 |
| FOXP2 | 11.40 | 19.73 | 1.71 | 0.026017 |
| ICER | 44.32 | 71.76 | 1.60 | 0.023 |
| IRF8 | 25.75 | 65.31 | 2.50 | 1.69E-05 |
| PPAP2B | 43.27 | 22.99 | 0.52 | 0.010 |
| Signal transduction | | | | |
| AKAP13 | 32.52 | 21.50 | 0.65 | 0.023 |
| ALDH1A3 | 148.38 | 99.67 | 0.66 | 0.009 |
| ARL5A | 18.46 | 33.45 | 1.79 | 0.042 |
| BCL6 | 19.86 | 4.26 | 0.21 | 8.27E-08 |
| CACNA1D | 9.77 | 4.62 | 0.47 | 6.47E-04 |
| CBLB | 5.26 | 2.37 | 0.44 | 0.046 |
| CCKAR | 5.57 | 1.01 | 0.18 | 0.013 |
| CISH | 1.97 | 27.94 | 13.99 | 1.38E-10 |
| FGA | 140.08 | 247.03 | 1.74 | 5.29E-04 |
| FGFR3 | 8.20 | 17.53 | 2.11 | 0.009 |
| FGG | 111.03 | 176.06 | 1.56 | 0.009 |
| FURIN | 21.31 | 49.79 | 2.30 | 2.82E-05 |
| FZD5 | 22.61 | 46.43 | 2.02 | 0.021 |
| GPR1 | 49.72 | 24.24 | 0.48 | 0.005 |
| SS1R | 0.68 | 5.33 | 8.75 | 0.015 |
| MAPKAPK3 | 7.28 | 16.57 | 2.24 | 0.010 |
| MC5R | 10.41 | 26.56 | 2.52 | 0.009 |
| NCAM1 | 6.36 | 3.17 | 0.49 | 0.043 |
| PDE10A | 7.07 | 1.57 | 0.22 | 0.002 |
| PLCD1 | 139.45 | 87.34 | 0.62 | 0.003 |
| PPAP2B | 43.27 | 22.99 | 0.52 | 0.010 |
| RAPGEF2 | 30.14 | 55.26 | 1.81 | 7.43E-04 |
| RGS6 | 1.92 | 0.09 | 0.047 | 0.014 |
| SPTBN1 | 151.29 | 101.93 | 0.66 | 0.007 |

tively, with an average length of 151 bp. After filtering with Q30, 13,638,494 and 7,473,018 useful reads were obtained from the treated group and control group, respectively. We mapped the useful reads to the chicken genome using bowite/tophat software, and found that 75.64% and 78.23% of reads in the treated group and control group mapped to the chicken genome, respectively. Of them, 82.59% in the treated group and 82.38% in the control group can be mapped to genes (Table 1). A total of 16,736 genes were identified.

Among the identified 16,736 genes, 164 genes with fold change >1.5 and $p$-value <0.05 were identified as DEG, of which 112 were up-regulated and 52 were down-regulated by chGH treatment (Table S2). Table 2 shows the top 10 up-regulated DEG and the top 10 down-regulated DEG.

To confirm the result of RNA-seq, we classified the genes into four groups: 1) fold change (RPKM (chGH/PBS)) >1.5 and P< 0.05, 2) fold change (RPKM (chGH/PBS)) <2/3 and P<0.05, 3)

fold change (RPKM (chGH/PBS))>1.5 and P>0.05, 4) fold change (RPKM (chGH/PBS)) <2/3 and P>0.05. And real-time RT-PCR was performed on a total of 39 genes that were randomly selected from four groups. Among them, 21 genes that were not differentially expressed between chGH and PBS based on RNA-seq were also found to be not differentially expressed by real-time RT-PCR, and 11 of 18 genes that were found to be differentially expressed by RNA-seq were also confirmed to be differentially expressed by real time RT-PCR (Table 3). Thus, our RNA-seq results were in general confirmed by real-time RT-PCR.

## 2. Functional analysis of differentially expressed mRNAs

To correlate the differentially expressed mRNAs with biological functions, we analyzed the functional bias of the DEG according to Gene Ontology enrichment. The analysis showed that the DEG included many genes involved in lipid binding, ion binding, transferase activity, oxidoreductase activity, immune system

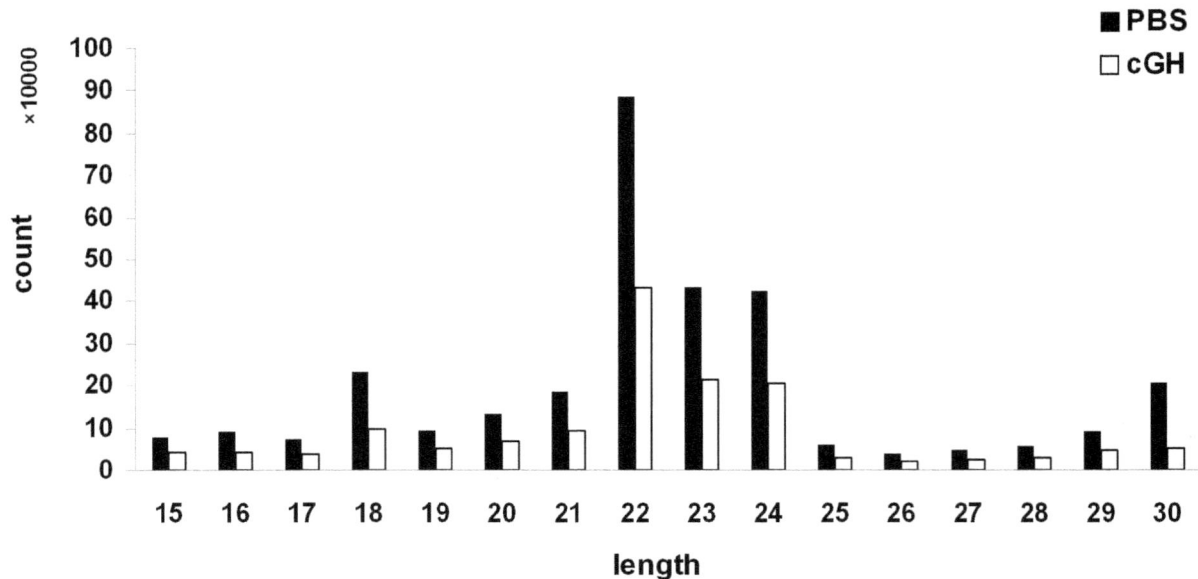

**Figure 2. Length distributions of small RNA sequences.**

process, lyase activity, and lipid metabolic process. Other enriched GO terms included growth, cellular nitrogen compound metabolic process, carbohydrate metabolic process, and biosynthetic process (Figure 1). Most of the KEGG Orthology (KO) terms were related to lipid metabolism (Table S3). Among the GH up-regulated genes, three were involved in the pathways controlling lipid metabolism: *AKR1D1* was involved in primary bile acid biosynthesis and steroid hormone biosynthesis pathway, *FABP1* was involved in PPAR and fat digestion and absorption pathway, and *LPIN1* was involved in glycerolipid metabolism pathway. Among the GH down-regulated genes, only one was involved in the pathway related to lipid metabolism: *PPAP2B* was involved in glycerolipid metabolism and fat digestion and absorption pathway.

**2.1. Differentially expressed mRNAs involved in growth.** We found that there were 6 GH-regulated genes involved in animal growth. Of them, *CISH*, *ULK2*, *IRF8*, and *FOXP2* were up-regulated by chGH, and *Bcl6* and *ROBO1* were down-regulated by chGH (Table 4). Both *Bcl6* and *ROBO1* are negative regulators of cell growth [33,34]. This indicates that chGH may regulate chicken growth by both increasing the expression of genes that positively regulate growth and reducing the expression of genes that negatively regulate growth.

**2.2. Differentially expressed mRNAs related to metabolism.** A total of 46 GH-regulated DEG were found to participate in protein, carbohydrate and lipid metabolism. Among them, 26 were related to protein metabolism as shown in Table 5, of which 20 were up-regulated and 6 were down-regulated by chGH, suggesting that GH plays an important role in protein metabolism in chicken hepatocytes. Further analysis showed that most of the 26 genes either promote protein synthesis or inhibit protein degradation, consistent with a previous study [7]. In addition to protein metabolism, 5 genes were related to carbohydrate metabolism, and all of them were up-regulated by chGH (Table 5). The specific functions of these genes include carbohydrate binding, transferase activity, and gluconeogenesis. As shown in Table 5, 15 DEG were related to lipid metabolism. Among these 15 genes, 8 showed increased expression after chGH treatment and all have oxidoreductase activity, suggesting that GH may promote lipid oxidation in chicken hepatocytes. Most of the 7

down-regulated genes were related to unsaturated fatty acid and long chain fatty acid biosynthesis, suggesting that GH may inhibit lipid synthesis in chicken hepatocytes.

**2.3. Differentially expressed mRNAs related to signaling pathways.** We also identified which GH-regulated genes might be involved in intracellular signaling. The analysis indicated that 7 GH-regulated genes have DNA binding activity, of which 5 were up-regulated and 2 were down-regulated by chGH, and 24 GH-regulated genes were components of signal transduction pathways, of which 11 were up-regulated and 13 were down-regulated by chGH (Table 6). These pathways include the STAT and MAPK pathways, which are known to be activated by GH [35,36] and pathways that are not known to be activated by GH.

### 3. miRNA profiling by RNA-seq

Small RNA libraries of chGH-treated and PBS-treated hepatocytes were deep sequenced, generating 8,248,771 raw reads from PBS-treated hepatocytes and 5,232,471 raw reads from chGH-treated cells. Of these sequences, 5,240,376 and 2,144,128 were clean reads. A total of 3,124,826 clean reads and a total of 1,487,885 clean reads from PBS-treated cells and chGH-treated cells, respectively, were found to be between 15 and 30 nt in length, and these reads were considered as potential miRNAs.

The majority of the 15–30 nt long reads ranged from 18 to 24 nt, and reads of 22 nt long were more abundant than reads of other lengths, indicating that the distribution of the small RNA sequences was consistent with the length range of miRNAs (Figure 2). In other words, most of the 15–30 nt long reads were likely miRNAs. A further analysis showed that most of the small RNAs could be mapped to the chicken genome. After the 15–30 nt reads were counted and grouped to generate unique reads, total reads or unique reads were compared to the sequences in Rfam database and classified. About 75% of total reads matched the sequences in Rfam. As shown in Figure 3, among the matched reads, ~80% of total reads represented miRNAs, and ~20% of unique reads represented miRNAs, indicating that the sequenced small RNA reads were enriched with miRNAs.

**A**

# Total reads

**B**

# Unique reads

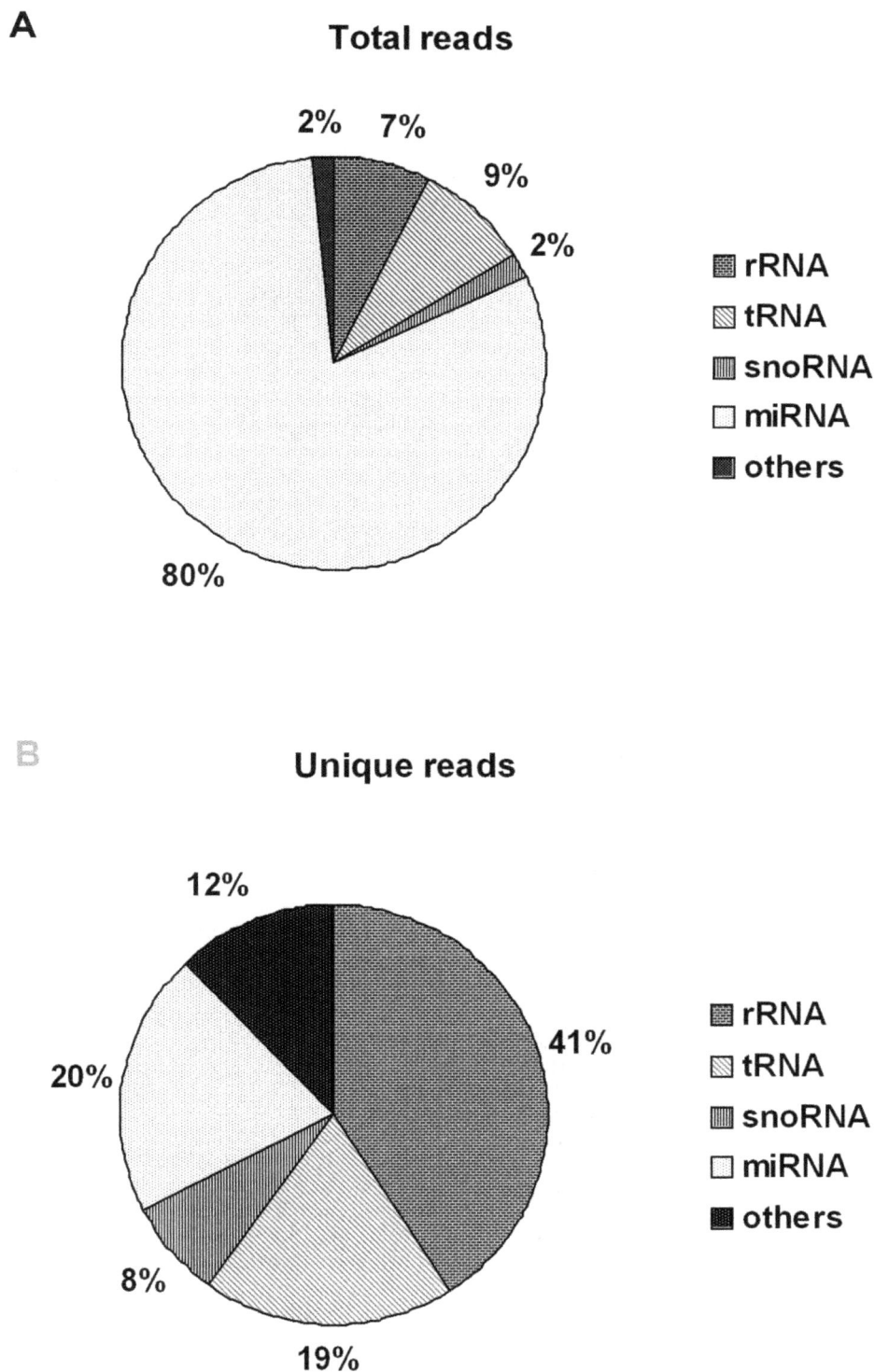

**Figure 3. Annotation of small RNA sequences.** (A) Annotation of total sequence reads. (B) Annotation of unique sequence reads.

The unique reads that matched miRNAs in Rfam were BLAST searched against chicken miRNAs in miRBase (20). Of 996 known chicken miRNAs in miRBase, 225 were identified in the present study. Among them, 219 known chicken miRNAs were identified from PBS-treated and 206 were identified from chGH-treated hepatocytes, which corresponded to 491 and 422 known chicken

miRNA precursors, respectively (Tables 7 and S4). The remaining reads that did not match known chicken miRNAs were BLAST searched against all the miRNAs in other species, and 264 homologous miRNAs were identified. Of them, 259 were from PBS-treated cells and 234 were from chGH-treated cells (Table S5). The small RNA reads that had no matches in Rfam were

**Table 7.** Identification of known chicken miRNAs from chicken hepatocytes.

|          | miRNA | precursor | miRNA (unique) | miRNA (total) | precursor (unique) | precursor (total) |
|----------|-------|-----------|----------------|---------------|--------------------|-------------------|
| miRBase  | 996   | 734       | -              | -             | -                  | -                 |
| PBS      | 219   | 491       | 5,010          | 1,552,789     | 8,447              | 1,958,885         |
| chGH     | 206   | 422       | 4,240          | 751,527       | 6,927              | 948,397           |

analyzed to predict novel miRNAs and their precursors. A total of 95 sequences were predicted to be novel miRNAs, of which 93 were sequenced from PBS-treated cells and 73 from chGH-treated cells (Table S6). We estimated the potential of the precursor sequences for these miRNAs to form stable stem-loop hairpin structures (Figure S1). Interestingly, some of these sequences are related to known chicken miRNAs. For example, the predicted novel chicken miRNA gga-m0015-3p is complementary to the known chicken miRNA gga-miR-126-5p, the 3′ portion of gga-m0016-5p overlaps with the 5′ portion of gga-miR-219b, and the 5′ portion of gga-m0028-3p overlaps with the 3′ portion of gga-miR-3525 (Figure S2).

## 4. GH-regulated miRNAs

We estimated the expression levels of miRNAs based on their read numbers. The 10 most abundantly expressed miRNAs in chicken hepatocytes were listed in Table S7. Among the identified known chicken miRNAs, 17 were identified as DEM between PBS-treated and chGH-treated hepatocytes, of which 16 were up-regulated and 1 was down-regulated by chGH. Among the identified chicken miRNAs homologous to miRNAs in other species, 21 were DEM between PBS-treated and chGH-treated hepatocytes, of which 15 were up-regulated and 6 were down-regulated in chGH-treated cells. Among the predicted novel chicken miRNAs, 7 were DEM between PBS-treated and chGH-treated cells, of which 5 were up-regulated and 2 were down-regulated by chGH (Table 8).

## 5. Functional analysis of GH-regulated miRNAs

The major function of miRNAs is to down-regulate the expression of target mRNAs. To investigate the function of GH-regulated miRNAs, we determined which of the GH-regulated mRNAs could be targeted by the GH-regulated miRNAs. Among the GH down-regulated mRNAs, 32 were predicted as targets of GH up-regulated miRNAs. Among the GH up-regulated mRNAs, 12 were predicted as target genes of GH down-regulated miRNAs (Table 9). A GO enrichment analysis showed that these miRNA target genes were enriched in lipid metabolism, lipid binding, cell motility, and small molecule metabolic process (Figure 4). The most significant GO term was the lipid metabolic process. As shown in Table 10, the GH up-regulated miRNAs targeted 7 genes related to lipid metabolism and the GH down-regulated miRNAs were predicted to target one gene related to lipid metabolism. Among these GH-regulated miRNAs, miR-15b had more predicted target genes related to lipid metabolism than had any other miRs. Several GH-regulated miRNAs were predicted to target the same genes (Table 10).

## Discussion

In this research through RNA-seq, we found a total of 164 DEG between chGH-treated and untreated chicken hepatocytes. Functional analyses showed most of the GH-regulated genes are involved in liver metabolism, indicating that GH regulates liver metabolism. Lipid metabolism is different between birds and mammals. In mammals lipogenesis occurs in liver, adipose tissue, and mammary gland, whereas in birds this occurs mostly in liver [37]. Lipid metabolic process was identified as a significant GO term among chGH-regulated genes, indicating that GH might play an important regulatory role in lipid metabolism in chicken hepatocytes. In the study of human adipose tissue, Zhao et al. (2011) found the GH-regulated DEG include those that stimulate triglyceride (TG)/ free fatty acid (FFA) cycle, and they also found a new TG hydrolase gene called patatin-like phospholipase domain containing 3 (*PNPLA3*), which could promote TG hydrolysis [38]. Comparing our DEG with those in the above report, we found there was little overlap, indicating the mechanism of GH regulation of lipid metabolism is different between birds and mammals, or between adipose tissue and liver. The chicken *PNPLA4* identified in our study is homologous to *PNPLA3* in mammals, suggesting this gene may play a role in regulating TG hydrolysis in the chicken liver.

In addition to genes that are involved in animal growth, many of the 164 DEG are involved in biosynthetic process (Table S8). This suggests that GH stimulates body growth in chickens not only by stimulating cell proliferation and hypertrophy but also by stimulating nutrient accumulation.

Recent research indicated that multiple signaling pathways are activated by GH, including the JAK2-STAT pathway, the MAPK-ERK1/2 pathway and the PI3K-AKT pathway [12]. In this study, 31 DEG are transcription factors or signaling molecules. Some of them belong to the pathways known to be activated by GH [35,36] while others belong to pathways that are not known to be activated by GH. Further study of the DEG showed that some of them, for example CAMK1D, had both DNA binding activity and tyrosine kinase activity. Another interesting GH-regulated transcription factor is BCL6. Studies showed that the expression of *BCL6* was controlled by GH through STAT5, and as a transcription factor BCL6 could mediate downstream gene expression [39,40]. We found that chicken *BCL6* gene has 11 potential STAT5 binding sites (data not shown), implying it could be regulated by chGH through STAT5.

The KEGG pathway analysis of GH-regulated DEG showed the KO terms were mainly related to metabolism, in particular, lipid metabolism. Two DEG were involved in fat digestion and absorption. One of them is *FABP1*, which is an important lipid-related gene [41] and has been shown to be up-regulated by GH in young chickens [42]. The other gene *PPAP2B* is involved in the glycerolipid metabolism pathway.

Among the identified miRNAs in this study, gga-miR-148a is the most abundant miRNA in chicken hepatocytes. This is consistent with the previous report, in which miR-148a was identified as the most abundant miRNA in porcine livers of different breeds [43]. This result indicated miR-148a might play important roles in liver. From many previous reports, miR-122

**Table 8.** chGH-regulated miRNAs among the identified known and predicted chicken miRNAs.

| miRNA | RPM (PBS) | RPM (chGH) | fold change (RPM (chGH/PBS)) | p-value |
|---|---|---|---|---|
| **DEM among the identified known chicken miRNAs** | | | | |
| up-regulated miRNA | | | | |
| gga-miR-15b | 35.3 | 63.9 | 1.8 | 2.27E-07 |
| gga-miR-19b | 263.9 | 451.0 | 1.7 | 5.49E-36 |
| gga-miR-29b | 31.5 | 49.0 | 1.6 | 4.71E-04 |
| gga-miR-99a-5p | 9.2 | 15.4 | 1.7 | 0.022 |
| gga-miR-146b-3p | 3.8 | 7.5 | 2.0 | 0.043 |
| gga-miR-181a-3p | 45.0 | 68.6 | 1.5 | 7.91E-05 |
| gga-miR-190 | 96.0 | 162.8 | 1.7 | 1.02E-13 |
| gga-miR-193a | 17.9 | 30.8 | 1.7 | 8.78E-04 |
| gga-miR-194 | 15.6 | 23.8 | 1.5 | 0.019 |
| gga-miR-223 | 55.7 | 94.2 | 1.7 | 1.71E-08 |
| gga-miR-455-3p | 15.5 | 24.7 | 1.6 | 0.008 |
| gga-miR-1306 | 0.6 | 3.7 | 6.5 | 0.003 |
| gga-miR-1618-5p | 1.1 | 4.2 | 3.7 | 0.011 |
| gga-miR-1628 | 0.4 | 2.3 | 6.1 | 0.019 |
| gga-miR-1699 | 1.5 | 5.6 | 3.7 | 0.003 |
| gga-miR-1731 | 35.9 | 56.0 | 1.6 | 1.72E-04 |
| down-regulated miRNA | | | | |
| gga-miR-1724 | 2.3 | 0.01 | 0.004 | 0.023 |
| **DEM among the miRNAs homologous to miRNAs in other species** | | | | |
| up-regulated miRNA | | | | |
| aca-miR-16b-5p | 3.2 | 7.9 | 2.4 | 0.009 |
| aca-miR-30c-3p | 11.1 | 19.6 | 1.8 | 0.005 |
| ahy-miR-3512 | 21.4 | 52.7 | 2.5 | 2.30E-11 |
| bta-miR-139 | 11.8 | 22.4 | 1.9 | 0.001 |
| ccr-miR-92a | 21.2 | 43.8 | 2.1 | 3.20E-07 |
| ccr-miR-99 | 1.1 | 3.7 | 3.3 | 0.024 |
| hsa-miR-34a-3p | 1.7 | 4.7 | 2.7 | 0.027 |
| hsa-miR-4792 | 852.8 | 1356.3 | 1.6 | 3.19E-81 |
| mmu-miR-6238 | 0.8 | 2.8 | 3.7 | 0.036 |
| oan-miR-1335 | 41.0 | 71.8 | 1.8 | 1.84E-07 |
| ola-miR-122 | 19.3 | 30.3 | 1.6 | 0.005 |
| ppt-miR-894 | 9.5 | 15.9 | 1.7 | 0.022 |
| tca-miR-3885-5p | 14.3 | 27.1 | 1.9 | 3.19E-04 |
| xtr-miR-210 | 11.8 | 24.3 | 2.0 | 1.60E-04 |
| xtr-miR-212 | 1.7 | 5.1 | 3.0 | 0.013 |
| down-regulated miRNA | | | | |
| ccr-miR-22b | 12.2 | 5.1 | 0.4 | 0.005 |
| ccr-miR-26a | 75.8 | 29.4 | 0.4 | 9.45E-15 |
| ccr-miR-130c | 10.9 | 4.7 | 0.4 | 0.009 |
| cgr-miR-425-5p | 13.2 | 4.7 | 0.4 | 7.79E-04 |
| ggo-let-7f | 71.2 | 37.8 | 0.5 | 4.46E-08 |
| ggo-miR-146a | 34.34 | 11.2 | 0.3 . | 4.96E-09 |
| **DEM among the predicted novel miRNAs** | | | | |
| up-regulated miRNA | | | | |
| gga-m0060-3p | 0.01 | 1.9 | 186.6 | 0.004 |
| gga-m0006-5p | 0.01 | 1.4 | 139.9 | 0.014 |
| gga-m0072-3p | 0.8 | 2.8 | 3.7 | 0.036 |
| gga-m0085-3p | 5.0 | 12.6 | 2.5 | 7.13E-04 |

**Table 8.** Cont.

| miRNA | RPM (PBS) | RPM (chGH) | fold change (RPM (chGH/PBS)) | p-value |
|---|---|---|---|---|
| gga-m0073-3p | 6.9 | 14.0 | 2.0 | 0.004 |
| down-regulated miRNA | | | | |
| gga-m0018-5p | 1.9 | 0.01 | 0.005 | 0.046 |
| gga-m0011-5p | 2.9 | 0.01 | 0.003 | 0.008 |

was known as the most abundant miRNA in liver. But in our study gga-miR-122-5p (previously named gga-miR-122) was identified as the fifth abundant miRNA in primary chicken hepatocytes, similar to the report by Li et al. which found miR-122 was the fourth abundant miRNA in porcine liver [43].

The differentially expressed miRNAs identified in this study might play various important roles in chGH regulation of growth, development and metabolism. miR-223 is expressed higher in the liver of Large White pig (lean type) than in the liver of Tongcheng pig (fatty type) [44]. In this study, we found gga-miR-223 was up-regulated by chGH, indicating it might play a role in lipid metabolism in chicken liver. miR-193a and -190 are expressed at higher levels in the liver of Erhualian pig than in the liver of Large White pig (The two breeds have different rate of lipid metabolism) [43]. In this study gga-miR-193a and 190 were also up-regulated by chGH, indicating they may play roles in lipid metabolism in chicken liver too. miR-15b was another up-regulated miRNA by the chGH treatment. This miRNA in pigs is expressed at a lower level in the liver of Erhualian pig than in the liver of Large White pig, opposite to miR-193a and -190 [43], implying it may play different roles between chickens and pigs. In male rat livers, miR-451 and miR-29b are down-regulated by GH [45]. gga-miR-451 was too down-regulated by GH in chicken hepatocytes in our study, indicating gga-miR-451 might mediate GH regulation of chicken liver metabolism but gga-miR-29b played different roles in GH regulation of liver metabolism comparing to miR-29b in rat.

The expression of miR-122 is not changed by GH in male rat liver [45]. We found the same for gga-miR-122 in this study, indicating this miRNA might not be involved in GH regulation of liver metabolism. In this study we also found that some of GH-regulated DEG were potential targets of GH-regulated DEM. GO enrichment analysis showed that these DEG were enriched with genes involved in lipid metabolism, indicating that GH may regulate lipid metabolism through miRNAs. Among the GH-regulated DEM, miR-15b was predicted to target more GH-regulated DEG related to lipid metabolism than any other DEM, including ACSL3, LIPG, PLCD1, PPAP2B and STAR. miR-15b expression was increased in NAFLD models and it lead to inducing the storage of intracellular triglyceride [46]. When ACSL3 increases, triglycerides and free fatty acids reduce [47]. Knock down of ACSL3 decrease hepatic lipogenesis [48]. LIPG has substantial phospholipase activity, and it can reduce plasma concentrations of HDL cholesterol [49]. PLCD1 can catalyze the hydrolysis of membrane lipid phosphatidylinositol 4,5-bisphosphate into second messengers inositol 1,4,5-trisphosphate and diacylglycerol [50]. PPAP2B bioactive lysophospholipids, including lysophosphatidic acid and sphingosine-1-phosphate, and thereby terminates their signaling effects [51]. STAR plays a critical role in the rapid translocation of cholesterol across the outer and inner mitochondrial membranes [52]. It also plays a key role in steroidogenesis by enhancing the metabolism of cholesterol into pregnenolone [53]. This finding indicates miR-15b might be a

**Table 9.** GH-regulated chicken genes predicted to be targeted by GH-regulated chicken miRNAs.

| miRNA | Target gene |
|---|---|
| gga-miR-15b | ABCG5, ABHD2, ACSL3, CBLB, ETNPPL, FAM3C, HEPHL1, HIVEP2, KIAA0408, KIAA1107, LIPG, MAST4, MYRIP, NCAM1, PANK1, PLCD1, PPAP2B, RGS6, STAR |
| gga-miR-19b | ABHD2, AKAP13, ALDH1A3, CBLB, LIPG, MYRIP, PPAP2B, SERPINB6 |
| gga-miR-29b | ABCG5, ABCG8, AKAP13, LIPG, MAST4, PPAP2B, PRRG1, ROBO1, TSKU |
| gga-miR-146b-3p | ABHD2, ANXA13, PLCD1 |
| gga-miR-190 | CBLB, ELOVL6, MYRIP |
| gga-miR-193a | ABHD2, KIAA0408, MYRIP, PLCD1, PRRG1 |
| gga-miR-194 | ABHD2, AKAP13, CBLB, MAST4, MYRIP, PLCD1 |
| gga-miR-223 | ALDH1A3, CBLB, MAST4, PLCD1 |
| gga-miR-455-3p | ABHD2, ELOVL6 |
| gga-miR-1306 | ABCG8, ANXA13, AKAP13, CCK1R |
| gga-miR-1618-5p | HEPHL1, MYRIP, SERPINB1 |
| gga-miR-1628 | ABHD2, CBLB, LIPG, PANK1, PLCD1, RGS6, SERPINB1, TPPP |
| gga-miR-1699 | ABHD2, AKAP13, HEPHL1, MAST4, PLCD1, PPAP2B, PRRG1, TPPP |
| gga-miR-1724 | AASS, CAMK1D, COQ4, DNAAF2, ECI1, EGLN3, FURIN, FZD5, PXDC1, RETSAT, TMEM229B, TOB2 |
| gga-miR-1731 | ABCG5, CCK1R, KIAA1107, PPAP2B, PRRG1, SPTBN1 |

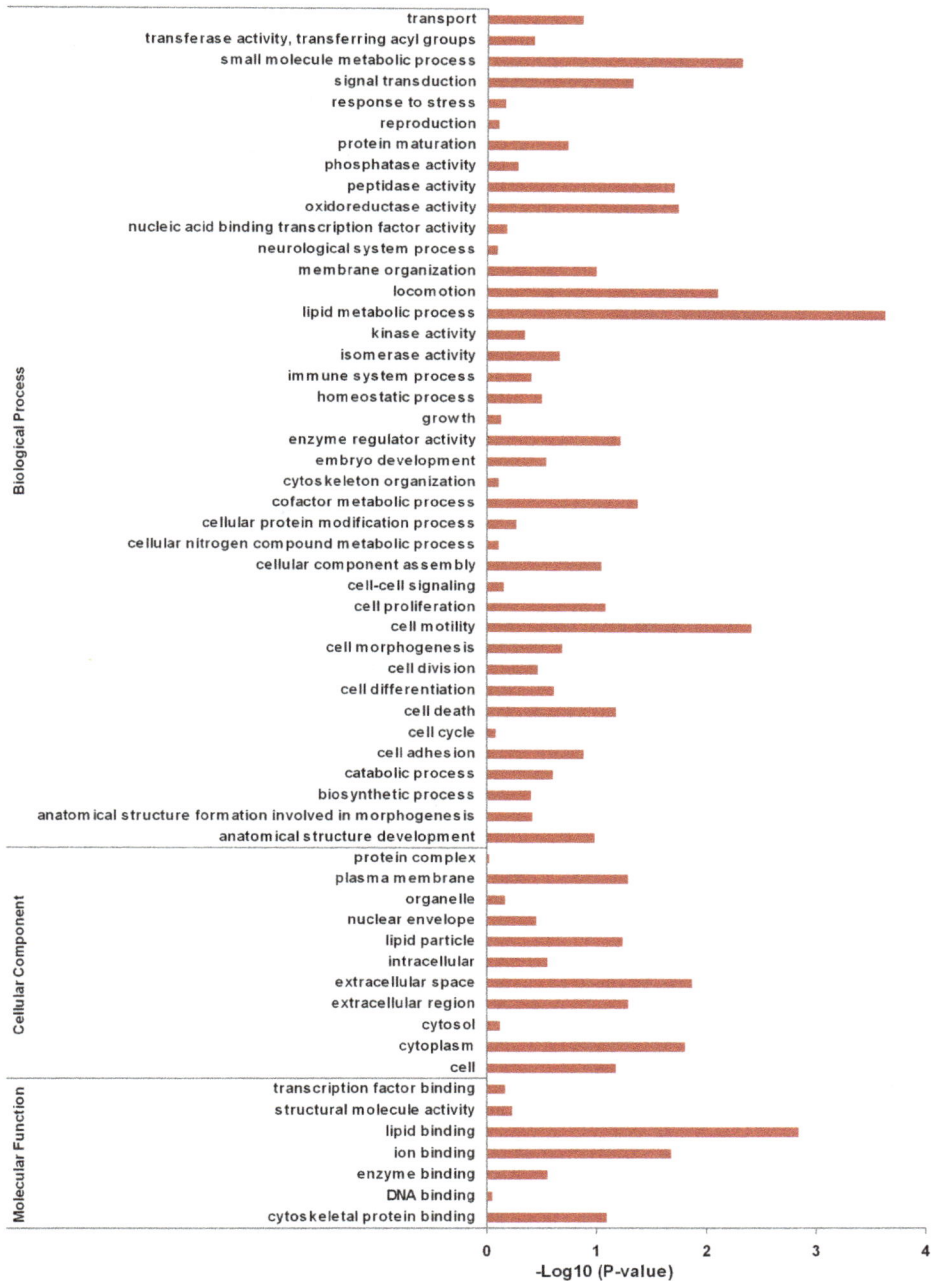

**Figure 4. GO enrichment analysis of chGH-regulated mRNAs predicted to be targeted by GH-regulated miRNAs.** The target mRNAs for GH up-regulated miRNAs were predicted in the GH down-regulated mRNAs; the target mRNAs for GH down-regulated miRNAs were predicted in the GH up-regulated mRNAs.

very important miRNA that mediates the effect of GH on lipid metabolism in chicken liver.

In mammals, the secretion of GH is characterized by an ultradian rhythm and sexual dimorphism. In the male rat, the major bursts of GH secretion occur at regular 3.3 h intervals separated by intervening trough periods with very low or undetectable basal GH levels [54,55]. GH is also secreted in a pulsatile manner in chickens [56,57]. In this study, hepatocytes from chickens were treated with one concentration of GH in culture. Therefore, a limitation of the study is that the observed

GH-regulated mRNA and miRNA expression profiles may not represent those that occur in the chicken liver *in vivo*.

## Conclusions

This study reveals that GH regulates the expression of many mRNAs involved in metabolism in female chicken hepatocytes, which suggests that GH plays an important role in regulating hepatic metabolism besides somatic growth. This study suggests that GH regulates lipid metabolism in female chicken liver in part

**Table 10.** Lipid metabolism-related genes predicted to be targeted by chGH-regulated miRNAs.

| miRNA | Predicted target gene |
|---|---|
| up-regulated miRNA | |
| gga-miR-15b | ACSL3, LIPG, PLCD1, PPAP2B, STAR |
| gga-miR-19b | ALDH1A3, LIPG, PPAP2B |
| gga-miR-29b | LIPG, PPAP2B |
| gga-miR-146b-3p | PLCD1 |
| gga-miR-190 | ELOV6 |
| gga-miR-193a | PLCD1 |
| gga-miR-194 | PLCD1 |
| gga-miR-223 | ALDH1A3, PLCD1 |
| gga-miR-455-3p | ELOV6 |
| gga-miR-1628 | LIPG, PLCD1 |
| gga-miR-1699 | PLCD1, PPAP2B |
| gga-miR-1731 | PPAP2B |
| down-regulated miRNA | |
| gga-miR-1724 | RETSAT |

by regulating the expression of miRNAs that target the mRNAs involved in lipid metabolism.

## Supporting Information

**Figure S1 Structures of predicted novel miRNAs.** For each novel miRNA, the precursor name, position, strand, length and energy are shown in the first line. The sequence of precursor and total reads are described in the second line. The brackets in the third line denote the secondary structure. The sequence of major unique reads, novel miRNA name and the number of reads are shown below with '*'. Then the sequence of each unique reads and the number of reads are shown below with '−'.

**Figure S2 Relationship between predicted novel miR-NAs and known chicken miRNAs.** For each novel miRNA, the name is listed in the first line. The sequence of known precursor and its name are described in the second line. The sequence of known miRNA and its name are shown below with '*'. Then the sequence of novel miRNA and its name are shown below with '−'. Note: The sequence of gga-m0015-3p is complementary to gga-miR-126-5p.

**Table S1 Primers for real time RT-PCR.**

**Table S2 Differentially expressed genes by chGH treatment.**

**Table S3 KEGG of differentially expressed genes.**

**Table S4 Identified known chicken miRNAs.**

**Table S5 Identified miRNAs homologous to other species.**

**Table S6 Predicted novel miRNAs.**

**Table S7 Ten most abundant miRNAs.**

**Table S8 Differentially expressed genes related with biosynthetic process.**

## Author Contributions

Conceived and designed the experiments: HJ ZG. Performed the experiments: XW LY HW FS JY. Analyzed the data: XW LY ZG. Contributed to the writing of the manuscript: XW LY HJ YH DG ZG. Wrote the paper: XW LY HJ YH DG ZG.

## References

1. Davidson MB (1987) Effect of growth hormone on carbohydrate and lipid metabolism. Endocr Rev 8: 115–131.
2. Press M (1988) Growth hormone and metabolism. Diabetes Metab Rev 4: 391–414.
3. Vijayakumar A, Novosyadlyy R, Wu Y, Yakar S, LeRoith D (2010) Biological effects of growth hormone on carbohydrate and lipid metabolism. Growth Horm IGF Res 20: 1–7.
4. Waters MJ, Brooks AJ (2012) Growth hormone and cell growth. Endocr Dev 23: 86–95.
5. Bartke A (2000) Effects of growth hormone on male reproductive functions. J Androl 21: 181–188.
6. Hull KL, Harvey S (2001) Growth hormone: roles in female reproduction. J Endocrinol 168: 1–23.
7. Moller N, Jorgensen JO (2009) Effects of growth hormone on glucose, lipid, and protein metabolism in human subjects. Endocr Rev 30: 152–177.
8. Wells JA (1996) Binding in the growth hormone receptor complex. Proc Natl Acad Sci U S A 93: 1–6.
9. Gent J, van Kerkhof P, Roza M, Bu G, Strous GJ (2002) Ligand-independent growth hormone receptor dimerization occurs in the endoplasmic reticulum and is required for ubiquitin system-dependent endocytosis. Proc Natl Acad Sci U S A 99: 9858–9863.
10. Brown RJ, Adams JJ, Pelekanos RA, Wan Y, McKinstry WJ, et al. (2005) Model for growth hormone receptor activation based on subunit rotation within a receptor dimer. Nat Struct Mol Biol 12: 814–821.
11. Argetsinger LS, Campbell GS, Yang X, Witthuhn BA, Silvennoinen O, et al. (1993) Identification of JAK2 as a growth hormone receptor-associated tyrosine kinase. Cell 74: 237–244.
12. Lanning NJ, Carter-Su C (2006) Recent advances in growth hormone signaling. Rev Endocr Metab Disord 7: 225–235.
13. O'Hea EK, Leveille GA (1968) Lipogenesis in isolated adipose tissue of the domestic chick (Gallus domesticus). Comp Biochem Physiol 26: 111–120.
14. Leveille GA, O'Hea EK, Chakbabarty K (1968) In vivo lipogenesis in the domestic chicken. Proc Soc Exp Biol Med 128: 398–401.
15. Bartel DP (2004) MicroRNAs: genomics, biogenesis, mechanism, and function. Cell 116: 281–297.
16. Chang J, Nicolas E, Marks D, Sander C, Lerro A, et al. (2004) miR-122, a mammalian liver-specific microRNA, is processed from hcr mRNA and may downregulate the high affinity cationic amino acid transporter CAT-1. RNA Biol 1: 106–113.
17. Krutzfeldt J, Rajewsky N, Braich R, Rajeev KG, Tuschl T, et al. (2005) Silencing of microRNAs in vivo with 'antagomirs'. Nature 438: 685–689.
18. Esau C, Davis S, Murray SF, Yu XX, Pandey SK, et al. (2006) miR-122 regulation of lipid metabolism revealed by in vivo antisense targeting. Cell Metab 3: 87–98.
19. Rayner KJ, Suarez Y, Davalos A, Parathath S, Fitzgerald ML, et al. (2010) MiR-33 contributes to the regulation of cholesterol homeostasis. Science 328: 1570–1573.
20. Gerin I, Clerbaux LA, Haumont O, Lanthier N, Das AK, et al. (2010) Expression of miR-33 from an SREBP2 intron inhibits cholesterol export and fatty acid oxidation. J Biol Chem 285: 33652–33661.
21. Mortazavi A, Williams BA, McCue K, Schaeffer L, Wold B (2008) Mapping and quantifying mammalian transcriptomes by RNA-Seq. Nat Methods 5: 621–628.
22. Cloonan N, Forrest AR, Kolle G, Gardiner BB, Faulkner GJ, et al. (2008) Stem cell transcriptome profiling via massive-scale mRNA sequencing. Nat Methods 5: 613–619.

23. Trapnell C, Roberts A, Goff L, Pertea G, Kim D, et al. (2012) Differential gene and transcript expression analysis of RNA-seq experiments with TopHat and Cufflinks. Nat Protoc 7: 562–578.

24. Creighton CJ, Reid JG, Gunaratne PH (2009) Expression profiling of microRNAs by deep sequencing. Brief Bioinform 10: 490–497.

25. Brooks AJ, Waters MJ (2010) The growth hormone receptor: mechanism of activation and clinical implications. Nat Rev Endocrinol 6: 515–525.

26. Ono M, Chia DJ, Merino-Martinez R, Flores-Morales A, Unterman TG, et al. (2007) Signal transducer and activator of transcription (Stat) 5b-mediated inhibition of insulin-like growth factor binding protein-1 gene transcription: a mechanism for repression of gene expression by growth hormone. Mol Endocrinol 21: 1443–1457.

27. Campbell RM, Scanes CG (1988) Pharmacological investigations on the lipolytic and antilipolytic effects of growth hormone (GH) in chicken adipose tissue in vitro: evidence for involvement of calcium and polyamines. Proc Soc Exp Biol Med 188: 177–184.

28. Cupo MA, Cartwright AL (1989) Lipid synthesis and lipoprotein secretion by chick liver cells in culture: influence of growth hormone and insulin-like growth factor-I. Comp Biochem Physiol B 94: 355–360.

29. Harvey S (2013) Growth hormone and growth? Gen Comp Endocrinol 190: 3–9.

30. Wang XG, Shao F, Wang HJ, Yang L, Yu JF, et al. (2013) MicroRNA-126 expression is decreased in cultured primary chicken hepatocytes and targets the sprouty-related EVH1 domain containing 1 mRNA. Poult Sci 92: 1888–1896.

31. Ambros V, Bartel B, Bartel DP, Burge CB, Carrington JC, et al. (2003) A uniform system for microRNA annotation. RNA 9: 277–279.

32. Chen C, Deng B, Qiao M, Zheng R, Chai J, et al. (2012) Solexa sequencing identification of conserved and novel microRNAs in backfat of Large White and Chinese Meishan pigs. PLoS One 7: e31426.

33. Albagli O, Lantoine D, Quief S, Quignon F, Englert C, et al. (1999) Overexpressed BCL6 (LAZ3) oncoprotein triggers apoptosis, delays S phase progression and associates with replication foci. Oncogene 18: 5063–5075.

34. Marlow R, Strickland P, Lee JS, Wu X, Pebenito M, et al. (2008) SLITs suppress tumor growth in vivo by silencing Sdf1/Cxcr4 within breast epithelium. Cancer Res 68: 7819–7827.

35. Herrington J, Smit LS, Schwartz J, Carter-Su C (2000) The role of STAT proteins in growth hormone signaling. Oncogene 19: 2585–2597.

36. Winston LA, Hunter T (1995) JAK2, Ras, and Raf are required for activation of extracellular signal-regulated kinase/mitogen-activated protein kinase by growth hormone. J Biol Chem 270: 30837–30840.

37. Bergen WG, Mersmann HJ (2005) Comparative aspects of lipid metabolism: impact on contemporary research and use of animal models. J Nutr 135: 2499–2502.

38. Zhao JT, Cowley MJ, Lee P, Birzniece V, Kaplan W, et al. (2011) Identification of novel GH-regulated pathway of lipid metabolism in adipose tissue: a gene expression study in hypopituitary men. J Clin Endocrinol Metab 96: E1188–1196.

39. Chen Y, Lin G, Huo JS, Barney D, Wang Z, et al. (2009) Computational and functional analysis of growth hormone (GH)-regulated genes identifies the transcriptional repressor B-cell lymphoma 6 (Bcl6) as a participant in GH-regulated transcription. Endocrinology 150: 3645–3654.

40. Zhang Y, Laz EV, Waxman DJ (2012) Dynamic, sex-differential STAT5 and BCL6 binding to sex-biased, growth hormone-regulated genes in adult mouse liver. Mol Cell Biol 32: 880–896.

41. Musso G, Gambino R, Cassader M (2009) Recent insights into hepatic lipid metabolism in non-alcoholic fatty liver disease (NAFLD). Prog Lipid Res 48: 1–26.

42. Wang X, Carre W, Saxton AM, Cogburn LA (2007) Manipulation of thyroid status and/or GH injection alters hepatic gene expression in the juvenile chicken. Cytogenet Genome Res 117: 174–188.

43. Li R, Sun Q, Jia Y, Cong R, Ni Y, et al. (2012) Coordinated miRNA/mRNA expression profiles for understanding breed-specific metabolic characters of liver between Erhualian and large white pigs. PLoS One 7: e38716.

44. Xie SS, Li XY, Liu T, Cao JH, Zhong Q, et al. (2011) Discovery of porcine microRNAs in multiple tissues by a Solexa deep sequencing approach. PLoS One 6: e16235.

45. Cheung L, Gustavsson C, Norstedt G, Tollet-Egnell P (2009) Sex-different and growth hormone-regulated expression of microRNA in rat liver. BMC Mol Biol 10: 13.

46. Zhang Y, Cheng X, Lu Z, Wang J, Chen H, et al. (2013) Upregulation of miR-15b in NAFLD models and in the serum of patients with fatty liver disease. Diabetes Res Clin Pract 99: 327–334.

47. Wu M, Cao A, Dong B, Liu J (2011) Reduction of serum free fatty acids and triglycerides by liver-targeted expression of long chain acyl-CoA synthetase 3. Int J Mol Med 27: 655–662.

48. Bu SY, Mashek MT, Mashek DG (2009) Suppression of long chain acyl-CoA synthetase 3 decreases hepatic de novo fatty acid synthesis through decreased transcriptional activity. J Biol Chem 284: 30474–30483.

49. Jaye M, Lynch KJ, Krawiec J, Marchadier D, Maugeais C, et al. (1999) A novel endothelial-derived lipase that modulates HDL metabolism. Nat Genet 21: 424–428.

50. Sidhu RS, Clough RR, Bhullar RP (2005) Regulation of phospholipase C-delta1 through direct interactions with the small GTPase Ral and calmodulin. J Biol Chem 280: 21933–21941.

51. Ren H, Panchatcharam M, Mueller P, Escalante-Alcalde D, Morris AJ, et al. (2013) Lipid phosphate phosphatase (LPP3) and vascular development. Biochim Biophys Acta 1831: 126–132.

52. Estabrook RW, Rainey WE (1996) Twinkle, twinkle little StAR, how we wonder what you are. Proc Natl Acad Sci U S A 93: 13552–13554.

53. Sugawara T, Lin D, Holt JA, Martin KO, Javitt NB, et al. (1995) Structure of the human steroidogenic acute regulatory protein (StAR) gene: StAR stimulates mitochondrial cholesterol 27-hydroxylase activity. Biochemistry 34: 12506–12512.

54. Tannenbaum GS, Martin JB (1976) Evidence for an endogenous ultradian rhythm governing growth hormone secretion in the rat. Endocrinology 98: 562–570.

55. Stroh T, van Schouwenburg MR, Beaudet A, Tannenbaum GS (2009) Subcellular dynamics of somatostatin receptor subtype 1 in the rat arcuate nucleus: receptor localization and synaptic connectivity vary in parallel with the ultradian rhythm of growth hormone secretion. J Neurosci 29: 8198–8205.

56. Johnson RJ (1988) Diminution of pulsatile growth hormone secretion in the domestic fowl (Gallus domesticus): evidence of sexual dimorphism. J Endocrinol 119: 101–109.

57. Buonomo FC, Lauterio TJ, Scanes CG (1984) Episodic growth hormone secretion in the domestic fowl (Gallus domesticus): alpha adrenergic regulation. Comp Biochem Physiol C 78: 409–413.

# PERMISSIONS

# LIST OF CONTRIBUTORS

**Philip Lössl, Knut Kölbel, Dirk Tönzler, Christian H. Ihling and Andrea Sinz**
Department of Pharmaceutical Chemistry and Bioanalytics, Institute of Pharmacy, Martin Luther University Halle-Wittenberg, Halle (Saale), Germany

**Manuel V. Keller and Frank Zaucke**
Center for Biochemistry, Medical Faculty, University of Cologne, Cologne, Germany

**Marian Schneider**
Research Group Artificial Binding Proteins, Institute of Biochemistry and Biotechnology, Martin Luther University Halle-Wittenberg, Halle (Saale), Germany

**Jens Meiler and David Nannemann**
Department of Chemistry and Center for Structural Biology, Vanderbilt University, Nashville, TN, United States of America

**Vendula Tvrdonova**
Department of Cellular and Molecular Neuroendocrinology, Institute of Physiology Academy of Sciences of the Czech Republic, Prague, Czech Republic
Department of Physiology of Animals, Faculty of Science, Charles University, Prague, Czech Republic

**Milos B. Rokic**
Department of Cellular and Molecular Neuroendocrinology, Institute of Physiology Academy of Sciences of the Czech Republic, Prague, Czech Republic
Section on Cellular Signaling, Program in Developmental Neuroscience, National Institute of Child Health and Human Development, National Institutes of Health, Bethesda, Maryland, United States of America

**Stanko S. Stojilkovic**
Section on Cellular Signaling, Program in Developmental Neuroscience, National Institute of Child Health and Human Development, National Institutes of Health, Bethesda, Maryland, United States of America

**Hana Zemkova**
Department of Cellular and Molecular Neuroendocrinology, Institute of Physiology Academy of Sciences of the Czech Republic, Prague, Czech Republic

**J. Joe Hull, Kendrick Chaney, Jeffrey A. Fabrick and Colin S. Brent**
USDA-ARS, Arid Land Agricultural Research Center, Maricopa, Arizona, United States of America

**Scott M. Geib**
USDA-ARS, Daniel K. Inouye Pacific Basin Agricultural Research Center, Hilo, Hawaii, United States of America

**Douglas Walsh and Laura Corley Lavine**
Dept. of Entomology, Washington State University, Pullman, Washington, United States of America

**Bruce J. Walker, Terrance Shea, Margaret Priest, Amr Abouelliel, Sharadha Sakthikumar, Christina A. Cuomo, Qiandong Zeng, Jennifer Wortman, Sarah K. Young and Ashlee M. Earl**
Broad Institute of MIT and Harvard, Cambridge, Massachusetts, United States of America

**Thomas Abeel**
Broad Institute of MIT and Harvard, Cambridge, Massachusetts, United States of America
VIB Department of Plant Systems Biology, Ghent University, Ghent, Belgium

**Congting Ye and Lei Li**
Department of Automation, Xiamen University, Xiamen, Fujian 361005, China
Department of Biology, Miami University, Oxford, Ohio 45056, United States of America

**Guoli Ji**
Department of Automation, Xiamen University, Xiamen, Fujian 361005, China
Innovation Center for Cell Biology, Xiamen University, Xiamen, Fujian 361005, China

**Chun Liang**
Department of Biology, Miami University, Oxford, Ohio 45056, United States of America

State Key Laboratory for Biology of Plant Diseases and Insect Pests, Institute of Plant Protection, Chinese Academy of Agricultural Sciences, Beijing 100193, China

**Akira Ishikawa and Sin-ichiro Okuno**
Laboratory of Animal Genetics, Graduate School of Bioagricultural Sciences, Nagoya University, Nagoya, Aichi, Japan

**Yiqing Liu**
Department of Ornamental Horticulture, China Agricultural University, Beijing 100193, China
College of Life Science & Forestry, Chongqing University of Art & Science,
Yongchuan 402160, China

**Yusong Jiang, Jianbin Lan and Yong Zou**
College of Life Science & Forestry, Chongqing University of Art & Science, Yongchuan 402160, China

**Junping Gao**
Department of Ornamental Horticulture, China Agricultural University, Beijing 100193, China

**Huei-Jiun Su**
Institute of Ecology and Evolutionary Biology, National Taiwan University, Taipei, Taiwan

**Saskia A. Hogenhout**
Department of Cell and Developmental Biology, John Innes Centre, Norwich, United Kingdom

**Abdullah M. Al-Sadi**
Department of Crop Sciences, Sultan Qaboos University, Al Khoud, Oman

**Chih-Horng Kuo**
Institute of Plant and Microbial Biology, Academia Sinica, Taipei, Taiwan
Molecular and Biological Agricultural Sciences Program, Taiwan International Graduate Program, National Chung Hsing University and Academia Sinica, Taipei, Taiwan
Biotechnology Center, National Chung Hsing University, Taichung, Taiwan

**Motomi Osato**
Institute of Molecular and Cell Biology, Agency for Science, Technology and Research (A*STAR), Singapore, Singapore
Cancer Science Institute of Singapore, National University of Singapore, Singapore, Singapore

Institute of Bioengineering and Nanotechnology, Agency for Science, Technology and Research, Singapore, Singapore

**Sydney Brenner**
Institute of Molecular and Cell Biology, Agency for Science, Technology and Research (A*STAR), Singapore, Singapore
Okinawa Institute of Science and Technology Graduate University, Onna-son, Okinawa 904-0495, Japan

**Byrappa Venkatesh**
Institute of Molecular and Cell Biology, Agency for Science, Technology and Research (A*STAR), Singapore, Singapore
Department of Paediatrics, Yong Loo Lin School of Medicine, National University of Singapore, Singapore, Singapore

**Giselle Sek Suan Nah and Boon-Hui Tay**
Institute of Molecular and Cell Biology, Agency for Science, Technology and Research (A*STAR), Singapore, Singapore

**Jian Ma**
CSIRO Agriculture Flagship, St Lucia, QLD, Australia
Triticeae Research Institute, Sichuan Agricultural University, Wenjiang, Chengdu, China

**Jiri Stiller, Donald Gardiner and John M. Manners**
CSIRO Agriculture Flagship, St Lucia, QLD, Australia

**Qiang Zhao, Qi Feng and Bin Han**
National Center for Gene Research, Shanghai Institutes for Biological Sciences, Chinese Academy of Sciences, Shanghai, China

**Colin Cavanagh and Penghao Wang**
CSIRO Agriculture Flagship, Black Mountain, ACT, Australia

**Frédéric Choulet and Catherine Feuillet**
INRAUBP Joint Research Unit 1095, Genetics, Diversity and Ecophysiology of Cereals, Clermont-Ferrand, France

**You-Liang Zheng and Yuming Wei**
Triticeae Research Institute, Sichuan Agricultural University, Wenjiang, Chengdu, China

**Guijun Yan**
School of Plant Biology, Faculty of Science and The UWA Institute of Agriculture, The University of Western Australia, Perth, WA, Australia

**Chunji Liu**
CSIRO Agriculture Flagship, St Lucia, QLD, Australia
School of Plant Biology, Faculty of Science and The UWA Institute of Agriculture, The University of Western Australia, Perth, WA, Australia

**Nadja Korotkova and Thomas Borsch**
Institut für Biologie/Botanik, Systematische Botanik und Pflanzengeographie, Freie Universität Berlin, Berlin, Germany
Dahlem Centre of Plant Sciences (DCPS), Berlin, Germany
Botanischer Garten und Botanisches Museum Berlin-Dahlem, Berlin, Germany

**Lars Nauheimer**
Institut für Biologie/Botanik, Systematische Botanik und Pflanzengeographie, Freie Universität Berlin, Berlin, Germany
Dahlem Centre of Plant Sciences (DCPS), Berlin, Germany

**Hasmik Ter-Voskanyan**
Botanischer Garten und Botanisches Museum Berlin-Dahlem, Berlin, Germany
Institute of Botany, National Academy of Sciences of Republic Armenia, Yerevan, Armenia

**Martin Allgaier**
The Berlin Center for Genomics in Biodiversity Research (BeGenDiv), Berlin, Germany

**Yoshiyuki Henning, Christiane Vole, Sabine Begall and Hynek Burda**
Department of General Zoology, Faculty of Biology, University of Duisburg-Essen, Essen, Germany

**Martin Bens, Arne Sahm and Karol Szafranski**
Genome Analysis, Leibniz Institute for Age Research - Fritz Lipmann Institute, Jena, Germany

**Martina Broecker-Preuss**
Department of Endocrinology and Metabolism and Division of Laboratory Research, University Hospital, University of Duisburg-Essen, Essen, Germany

**Philip Dammann**
Department of General Zoology, Faculty of Biology, University of Duisburg-Essen, Essen, Germany
Central Animal Laboratory, University Hospital, University of Duisburg-Essen, Essen, Germany

**Christian J. Slubowski, Scott M. Paulissen and Linda S. Huang**
Department of Biology, University of Massachusetts Boston, Boston, Massachusetts, United States of America

**Teruyo Ojima-Kato**
Hub of Knowledge Aichi, Aichi Science and Technology Foundation, Yakusa, Toyota, Aichi, Japan

**Naomi Yamamoto, Mayumi Suzuki and Hiroto Tamura**
School of Agriculture, Meijo University, Shiogamaguchi, Tenpaku-ku, Nagoya, Aichi, Japan

**Tomohiro Fukunaga**
Japan Food Research Laboratories, Osu, Naka-ku, Nagoya, Aichi, Japan

**Blanca Taboada, Marco A. Espinoza, Pavel Isa, Fernando E. Aponte, Susana López and Carlos F. Arias**
Instituto de Biotecnología, Universidad Nacional Autónoma de México, Cuernavaca, Morelos, Mexico

**María A. Arias-Ortiz, Jesús Monge-Martínez, Rubén Rodríguez-Vázquez, Fidel Díaz Hernández and Fernando Zárate-Vidal**
Colegio de Pediatría del Estado de Veracruz, Veracruz, Mexico

**Rosa María Wong-Chew and JoséI. Santos-Preciado**
Facultad de Medicina, Universidad Nacional Autónoma de México, México D.F., Mexico

**Verónica Firo-Reyes**
Hospital General de México, México D.F., Mexico

**Carlos N. del Río-Almendárez**
Hospital Pediátirco de Coyoacán, México D.F., Mexico

**Jesús Gaitán-Meza**
Nuevo Hospital Civil de Guadalajara "Dr. Juan I. Menchaca", Guadalajara, Jalisco, Mexico

**Alberto Villaseñor-Sierra**
Centro de Investigación Biomédica de Occidente, IMSS, Guadalajara, Jalisco, Mexico

**Gerardo Martínez-Aguilar and Ma. del Carmen Salas-Mier**
Unidad de Investigación Biomédica IMSS, Durango, Durango, Mexico

**Daniel E. Noyola**
Universidad Autónoma de San Luis Potosí, San Luis Potosí, Mexico

**Luis F. Pérez-Gónzalez**
Hospital Central "Dr. Ignacio Morones Prieto", San Luis Potosí, Mexico

**Miroslav Arambasic, Pamela Y. Sandoval, Cristina Hoehener, Aditi Singh, Estienne C. Swart and Mariusz Nowacki**
Institute of Cell Biology, University of Bern, Bern, Switzerland

**Xingguo Wang**
Department of Life Science and Technology, Changshu Institute of Technology, Changshu, P R China
College of Animal Science and Technology, Yangzhou University, Yangzhou, P R China

**Lei Yang, Huijuan Wang, Fang Shao, JianFeng Yu, Yaoping Han and Zhiliang Gu**
Department of Life Science and Technology, Changshu Institute of Technology, Changshu, P R China

**Honglin Jiang**
Department of Animal and Poultry Sciences, Virginia Polytechnic Institute and State University, Blacksburg, Virginia, United States of America

**Daoqing Gong**
College of Animal Science and Technology, Yangzhou University, Yangzhou, P R China

# Index

**A**

Alanine Scanning Mutagenesis, 16-17

Algorithm, 13-14, 53, 56, 59-60, 62-64, 66-67, 69, 72, 85, 89, 94, 107, 109, 125, 134, 146, 179

Anellovirus, 184

Arabidopsis Thaliana, 62, 64, 68-71, 86, 95, 131, 162-163

Astrovirus, 184, 188, 192, 194, 196

Atp-binding Cassette, 28, 44-46, 209

**B**

Binding Modes, 1, 11, 25

Bioinformatics Tools, 62-63

Body Weight, 73-85, 149

Bread Wheat, 121, 131

**C**

C. Sinensis, 93, 97-102, 105-106

Candidate Gene Prioritization, 73

Cartilaginous Fishes, 108, 117

Chicken Liver, 209-210, 218, 220-221

Chloroplast Genome, 97-103, 106-107, 146-147

Chromosome, 63-64, 68-71, 73-74, 77, 81-82, 84-85, 121-123, 126-128, 131, 208

Citrus, 97-99, 102, 105-107

Citrus Aurantiifolia, 97-98, 106

Complex Numbers, 62-64

Coronavirus-oc43, 184

Cross-linking, 1-7, 9-15

Cruciform, 62-63, 72

Cyclostomes, 108-109, 113, 115, 118

**D**

Deep Sequencing, 88, 95, 184-185, 191, 223

Detectir, 62-64, 68-71

Diploid, 48, 51-52, 56, 60, 124, 159, 198

Dna Methylation, 62

Dna Replication, 62, 71, 160, 171

Dna Transition, 62

Dorsal Fin, 16, 20, 23, 27

**E**

Ectodomain, 16, 26-27

Eucalyptus Dunnii, 86, 93

Eukaryotic Genomes, 62, 70, 107

**Evolution**, 44-45, 47-48, 60, 63, 97-98, 105-109, 112-113, 116-117, 119-120, 131-132, 134, 141, 143, 145-147, 150, 158, 182, 197

Exome Sequencing, 73-75, 77, 81-84

**F**

Fine Mapping, 73, 85, 121, 128

Fruits, 97, 106

Fusarium Crown Rot Resistance, 121, 131

**G**

Gene Search, 73, 77, 81

Genome Assembly, 48-49, 52, 60, 96, 99, 109-111, 113, 133, 145, 189

Genome Assembly Improvement, 48-49

Genus, 41, 56, 97, 105, 132, 134, 141, 143, 146, 149, 182-183, 191

Gnathostome, 108-113, 115-116, 118-119

Growth Hormone, 82, 209, 222-223

**H**

Haematopoiesis, 108-109, 119

Haploid, 48, 56, 131, 159, 166, 198

Hepatocytes, 209-210, 212-218, 220-221, 223

Homo Sapiens, 5, 33, 62-63, 68-71, 111, 115, 117-119, 156-157, 162

**I**

Immunocytochemistry, 200, 208

Inverted Repeats, 62-64, 66-72, 99, 133, 135-136, 138-139

Ivermectin, 16, 22, 26-27

**J**

Japanese Lamprey, 108-113, 115-119

Jawed Vertebrates, 108-109

Jawless Vertebrate, 108

**L**

Laminins, 1, 6, 12, 14-15

Lethenteron Japonicum, 108-109, 111, 115, 117-119

Lipid Metabolism, 209-210, 214, 216, 218, 220-223

Lygus Hesperus, 28-29, 33, 42, 45-47

**M**

Macronucleus, 198, 200, 203

Mass Spectrometry, 1-2, 10, 14-15, 162, 164, 172-173, 182-183
Microbial Variant Detection, 48
Mouse Genome Database, 73, 85
Mrna Sequencing, 210, 213, 222
Mus Musculus Castaneus, 73-74, 85

**N**
Neurogenesis, 108-109
Nidogen-1, 1-15
Nuclei, 159, 198, 202, 206
Nucleic Acid Extraction, 185

**O**
Obesity, 46, 73-74, 77, 81-85
Omani Lime, 97-98, 100
Open Source Software, 48, 59

**P**
Paramecium Tetraurelia, 198-200, 207-208
Pathogen, 49, 54, 95, 126, 128, 130-131, 173, 184-186, 188, 190-193, 197
Pilon, 48-60, 197
Plant Bug, 28-29, 45-46
Polypeptide Chains, 1
Prokaryotic Genomes, 182

**Q**
Quantitative Trait Locus, 73, 85, 131

**R**
Rat P2x4 Receptor, 16, 27
Respiratory Syncytial Virus, 184, 187, 190, 192, 194, 196
Rhinovirus, 184, 186-187, 190, 192, 194, 196
Rna Sequencing, 88-89, 93, 122, 209, 213
Rotavirus, 184, 188, 190, 192, 194, 196
Rutaceae, 97, 106

**S**
Saffold Virus, 184, 188-189, 192, 196-197
Sexual Progeny, 198-200
Skeletogenesis, 108-109, 118-119
Software, 3, 14, 17, 19, 42, 47-49, 59, 72, 75, 77, 83, 91, 94, 99, 124, 128, 146, 150, 161-163, 179, 182, 186, 200, 215
Subcongenic Strains, 73-75, 77, 82
Sulfosuccinimidyl, 1, 3

**T**
Transcriptome, 28-29, 41-42, 45-47, 86-88, 90-92, 95-97, 121-122, 127, 131, 149-150, 158, 222

**V**
Vector, 17, 42, 45, 62-64, 66-67, 106, 109, 168, 200-203, 205
Vector Calculation, 62-63

**Z**
Zea Mays, 62-63, 68-71, 95
Zebrafish P2x4 Receptor, 16
Zygotes, 198